ELECTRIC TRANSMISSION and DISTRIBUTION

ELECTRIC UTILITY LIBRARY

ELECTRIC GENERATION—STEAM STATIONS

ELECTRIC GENERATION—HYDRO, DIESEL, AND GAS-TURBINE STATIONS

ELECTRIC TRANSMISSION AND DISTRIBUTION

ELECTRIC SYSTEM OPERATION

BOOKS BY B. G. A. SKROTZKI AND W. A. VOPAT

APPLIED ENERGY CONVERSION

STEAM AND GAS TURBINES

ELECTRIC TRANSMISSION AND DISTRIBUTION

BERNHARDT G. A. SKROTZKI, Editor
ASSOCIATE EDITOR, POWER

CONTRIBUTORS

E. M. ADKINS — Jersey Central Power and Light Company

F. S. BENSON — Pacific Gas and Electric Company

E. L. MICHELSON — Commonwealth Edison Company

McGRAW-HILL BOOK COMPANY
New York Toronto London
1954

ELECTRIC TRANSMISSION AND DISTRIBUTION

Library of Congress Catalog Card Number: 54-6220

13 - MP - 9

57925

PREFACE

The electric-utility industry uses many crafts and engineering techniques in its operations. This volume covers the transmission and distribution of electric energy from the generating station to the ultimate consumer. While this book has been written *by* professional engineers expert in this field, it has been written *for* the reader who wishes to gain a broad understanding of the design and operation of the utility system as a whole.

Basic electricity is discussed in a simple manner to outline the theory on which the practical equipment has been designed. The mathematics has been limited to simple arithmetic and elmentary trigonometry needed to make the simpler calculations. A high-school graduate should have no trouble understanding the computations; a grammar-school graduate should be able to grasp it with a little study.

With the theory as a foundation, the major portion of the book describes the equipment used in transmission and distribution systems. All the description is given in such a way that the reader gets a general idea of the actual layout of the equipment and what it does in the system.

Electrical mechanics and technicians should find this book valuable in gaining a better fundamental understanding of the equipment they work with. Each chapter ends with a set of review questions which highlight the main points. The answer to each question can be found in the chapter. Those chapters which include mathematics contain some simple problems together with their solution.

While the books in the Electric Utility Library are not engineering books, they should prove helpful to the engineering student entering the utility industry. The library will give him a general picture of the utility system, helping to get him squared away in the new career he is starting.

BERNHARDT G. A. SKROTZKI

CONTENTS

CHAPTER 1

ELECTRICAL THEORY

By E. M. Adkins

Jersey Central Power and Light Company

1-1. Fluid and Electron Theories of Electricity. We do not try to define electricity but rather explain it by describing its effects and results. Our knowledge of electricity has been accumulating for centuries as the result of experiments and observations by many different people. We have no reason to believe that our knowledge is yet complete.

It has been known since about 600 B.C. that, when a rod of amber is rubbed with a piece of flannel cloth, it attracts and holds light objects, such as bits of paper. The reader has probably observed when combing his hair with a hard rubber comb that sometimes the comb attracts and holds small bits of paper. Such effects are among the earliest record of electric phenomena. In time it was noted that many substances had this property of becoming electrified. These electrified objects attracted certain other electrified objects while forcefully repelling some others. For instance, a glass rod which has been rubbed with a piece of silk repels a similarly electrified glass rod but will be mutually attracted by a rod of hard rubber electrified by rubbing with a piece of flannel. From these observations, it was thought there are two different kinds of electricity and that like charges exert a force of repulsion on each other while unlike charges exert a similar force of attraction. Our own Benjamin Franklin noted these facts and arbitrarily called one positive electricity and the other negative electricity. For example, he gave the name *positive electricity* to the charge carried by a glass rod rubbed with silk and *negative electricity* to the charge carried by a piece of hard rubber rubbed with flannel. Such electric charges are electricity at rest, or static electricity.

If a copper wire is connected to the terminals of an ordinary battery, together with a push-button switch, as shown in Fig. 1-1, electricity flows through the wire when the switch is closed and ceases flowing when the switch is opened. Early scientific observers thought that electricity was a fluid and that an electrical pressure caused it to flow. In the battery of Fig. 1-1 the electrical pressure is supplied by the difference in charge between the two terminals of the battery. This driving force, causing the flow of electricity, is called *electromotive force* (emf), and a unit of electromotive force is called a *volt.* The fluid theory assumed that the flow of electricity was from the point of high pressure, or positive, to the point of low pressure, or negative. The designation positive and negative was completely arbitrary, and consequently the direction of electricity flow was arbitrarily assumed.

Modern scientific research has increased our knowledge of the structure of matter enormously. Such research has led science to advance a new theory concerning the common structure of all matter. This new theory, called the electron theory, assumes that all matter is composed of combinations of but two things: positive charges of electricity called *protons* and negative charges of electricity called *electrons.* An atom of any substance is composed of a central core, or nucleus, carrying a positive electrical charge and a group of one or more electrons, or negative charges, rotating about the positively charged nucleus, somewhat like the solar system in which the earth and the other planets rotate about the sun. The number of planetary electrons rotating about the nucleus is controlled by the positive charge, or the number of protons, in the nucleus. The negative charge of an electron is equal to the positive charge of a proton, and there will be as many electrons rotating about the nucleus as there are protons in the nucleus. The complete atom is electrically neutral since the positive and negative charges are equal and opposite and cancel one another electrically.

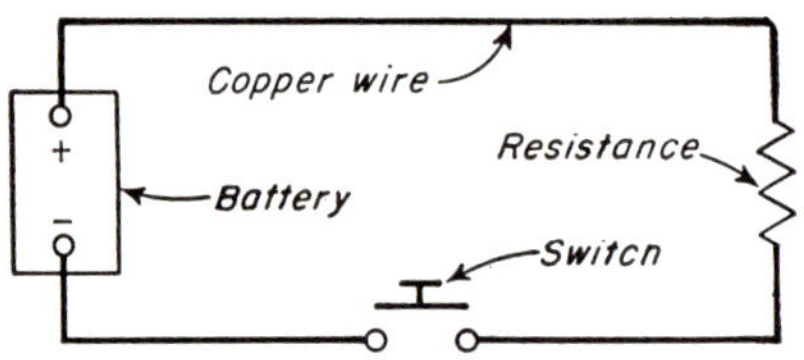

FIG. 1-1. The simplest electrical circuit has a battery as a source of electrical energy, a resistance as a user of electrical energy, wires to connect the resistance to the battery, and a switch to make and break the circuit controlling the flow of electrical current.

The nucleus of the atom is thought to be composed of positively charged particles, or protons, and other neutral particles carrying no charge, called *neutrons.* The neutron is sometimes considered to be a proton to which is attached an electron, or neutralizing negative charge. The protons and neutrons, forming the nucleus of the atom, are bound together in a tightly held cluster. This binding force is not yet fully explained but must overcome the force of repulsion between the like charges of the protons in the nucleus and provide a binding force of attraction between the positively charged protons and the neutral neutrons.

The difference between an atom of copper and an atom of gold, for example, depends upon the difference in the number of protons and neutrons in the nucleus and the number of electrons rotating about it. There are 92 chemical elements found in nature, from which all matter is formed. These 92 elements range in structure from hydrogen, the simplest, to uranium, the most complex. The hydrogen atom has a nucleus composed of a single proton, with no neutrons, about which rotates a single electron. The uranium atom, on the other hand, has a nucleus containing 92 protons and 146 neutrons with 92 electrons rotating about it.

All the planetary electrons of the atom do not spin about the nucleus at the same radius, except for the two simplest elements having one and two planetary electrons, respectively. In the more complicated atoms the rotating electrons are arranged in groups as though their orbits were in concentric shells about the nucleus, there being a separate shell for each group. These oribital shells vary in number from one for hydrogen, the first element, to seven for uranium, the ninety-second element. The proton, the neutron, and the electron, as well as the atom itself, are infinitesimal in size. The orbit in which the electrons nearest the nucleus rotate, however, is large when compared with the size of the nucleus. Any substance is composed of specks of solid matter held in a predetermined pattern with nothing between but space. The proton is about 1,840 times as heavy as the electron, and the neutron is slightly heavier than the proton. The nucleus of the atom provides most of the mass, or weight, of matter. The electrons in the outer orbital shell of the atom are bound less tightly to the positively charged nucleus, because of their greater distance from it. In most metals some of these outer electrons become separated from the atoms of which they are a part and become free electrons, drifting aimlessly through the interatomic spaces until attracted to another atom which has lost one or more electrons and is positively charged.

In those metals which are considered good conductors of electricity, one or more of the electrons of each atom is bound very loosely to its orbit and is easily freed by a slight attractive force external to the atom. In the battery of Fig. 1-1 there is an excess of electrons at the negative terminal compared to the positive terminal. This difference in charge, due to unbalance in electron supply between the positive and negative terminals, creates a potential difference, or emf, between them. The surplus of electrons at the negative terminal repels or repulses any other electron approaching the negative terminal. On the other hand, the deficit of electrons at the positive terminal attracts any free electrons within the field of attraction of its positive charge. When the switch of Fig. 1-1 is closed, the free electrons in the copper wire cease to drift and will be given direction in their movement while other electrons are separated from their atoms by the force of repulsion of the negative charge and the force of attraction of the positive charge at the negative and positive terminals.

This shift or exchange of electrons forms the flow of electricity from the negative to the positive terminal through the copper wire connecting them in a continuous circuit. This movement of electrons through the circuit is from the atoms in the wire to the positive terminal, from atom to atom in the wire, and from the negative terminal of the battery to the atoms of the wire. While this movement of electrons through the wire is slow, the quantity in motion is enormous. If the switch in the circuit of Fig. 1-1 remains closed, the flow of electricity continues, at a diminishing rate, until there is no longer an excess of electrons at one terminal and a deficit at the other. When this point of balance is reached, there is no difference of potential between positive and negative, the emf is gone, and the flow of electricity stops.

The flow of water through a pipe is measured by the number of gallons flowing past a given point per second. In a similar manner, the flow of electricity through a conductor is measured by the number of electrons passing a given point per second. The total flow of electricity is measured in *coulombs*. A coulomb equals about 6.28×10^{18}, or 6,280,000,000,000,000,000, electrons per second and represents a transfer of electric charge. The rate at which electrons pass the reference point measures the size of the electron stream. The practical unit for measuring the rate of flow of electricity is the *ampere*. When the rate of flow of electricity in a circuit equals one coulomb per second, or 6.28×10^{18} electrons per second, an electric current of one ampere is said to flow in the circuit.

It should be noted that the direction of electricity flow, according to the electron theory, is from negative to positive, while the flow of electricity according to the older theory of fluids is from positive to negative. The concept of current flow from positive to negative is well established in technical literature and among engineers and electrical workers, particularly in the power branch of the electrical industry. This discrepancy need cause no difficulty if it is kept in mind that when the flow of electric current is mentioned the assumed direction of flow is from positive to negative while the term *electron flow* designates transfer of electrons from negative to positive.

1-2. Ohm's Law. If water is forced through a pipe by a pump, the rate at which the water flows will be directly proportional to the driving force exerted by the pump and inversely proportional to the frictional resistance offered by the pipe to the flow of water. In a similar manner, the rate at which electric current flows through a circuit will be directly proportional to the emf and inversely proportional to the opposition or *resistance* which the circuit offers to the flow of current. This relation between electric current, emf, and circuit resistance is known as *Ohm's law:*

$$E = IR \qquad I = \frac{E}{R} \qquad R = \frac{E}{I}$$

where I = electric current, *amperes*
E = emf, *volts*
R = circuit resistance to current flow, *ohms*

The resistance offered by a conductor to current flow equals the ratio of emf to current, or the ratio of volts to amperes, as noted in the equations above. The unit of resistance is the *ohm.* Thus a conductor is said to have one ohm of resistance if one ampere of current flows when an emf of one volt is applied across its ends.

Example: The battery of Fig. 1-1 has an emf of 6 volts, and a current of 3 amp flows when the switch is closed. What is the resistance of the wire? What emf would be required to cause a current of 50 amp to flow?

Answer:

$$R = \frac{6 \text{ volts}}{3 \text{ amp}} = 2 \text{ ohms}$$
$$E = IR = \text{current in amperes} \times \text{resistance in ohms}$$
$$= 50 \text{ amp} \times 2 \text{ ohms} = 100 \text{ volts}$$

The relation of volts, amperes, and resistance expressed by Ohm's law refers specifically to direct current in which the direction of current flow stays constant.

1-3. Resistance. The electrical resistance of a conductor can be determined by Ohm's law, as was noted in Art. 1-2. It is also possible to figure the resistance of a conductor from its physical characteristics. The resistance of a wire 2 ft long is double the resistance of an identical wire 1 ft long because the total length of wire has doubled. The resistance of two wires, identical in cross-sectional area and length and laid side by side with their ends connected, to form a single conducting path, is equal to one-half the resistance of one of the wires, because the cross-sectional area of the conducting path is double that of a single wire. The resistance of an electrical conductor varies directly with its length and inversely with its cross-sectional area:

$$R = \rho \frac{l}{A}$$

where R = resistance of the conductor, *ohms*
l = length of the conductor, *feet*
A = cross-sectional area of the conductor, *circular mils*
ρ = resistivity of the conductor material, *ohms per circular mil-foot*

The resistivity of a conductor material is the resistance of a conductor of unit area and of unit length. In this country, length is measured in feet and conductor cross-sectional area in *circular mils*. A circular mil is the area of a circle one mil (0.001 in.) in diameter. The cross-sectional area of a conductor, in circular mils, is equal to the diameter in thousandths of an inch multiplied by itself. For example, the cross-sectional area of a conductor 0.100 in. in diameter (100 mils) is

$$100 \times 100 = 10{,}000 \text{ circular mils}$$

The resistivity is the resistance in ohms of a wire one foot long having a cross-sectional area of one mil, or the resistance of a mil-foot of wire.

The ohmic resistivity of wires can be found in handbook tables for given conductor materials. The resistance depends on the nature of the material. Temperature of a conductor affects its resistance. In most materials rising temperature slightly increases the resistance.

In alternating-current circuits the density of current flow tends to be greater near the surface of the conductor than at the center. This unequal distribution of current density causes a slight increase in apparent resistance called *skin effect*. Handbooks also give data on skin effect.

Example: A solid copper wire, 0.1620 in. in diameter, has a resistivity of 10.37 ohms per mil-ft. What is the resistance per 1,000 ft of length? What is its cross-sectional area in circular mils?

Solution:

$$R = \rho \frac{l}{A} = 10.37 \times \frac{1{,}000}{162.0 \times 162.0} = 0.3951 \text{ ohms}$$
$$A = 0.1620 \times 0.1620 \times 1{,}000 \times 1{,}000 = 26{,}250 \text{ cir mils}$$

1-4. Voltage Drop—Back EMF. Two resistors having a resistance of 0.25 ohm each and a third resistor having a resistance of 5.5 ohms are connected to a voltage source, a generator, supplying an emf of 120 volts, as shown in Fig. 1-2. All the resistance of the circuit is concentrated in the resistors. The total circuit resistance equals 0.25 ohm plus 5.5 ohms plus 0.25 ohm, or a total of 6 ohms. With an emf of 120 volts impressed, a current of 20 amp flows.

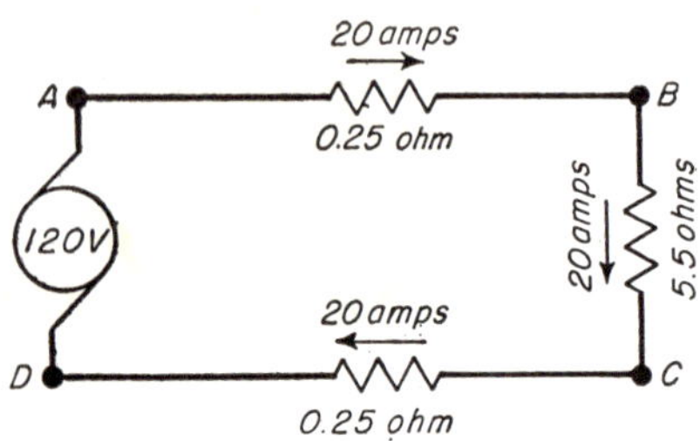

FIG. 1-2. A simple series circuit with a total of 6 ohms resistance connected to an emf of 120 volts has a current of 20 amp flowing.

All the 120 volts is used up in overcoming the opposition of the 6 ohms of resistance to the flow of 20 amp of current. By Ohm's law, it takes 20 volts to force 20 amp through each ohm of resistance. Then 5 volts are used in forcing 20 amp through each of the 0.25-ohm resistors (20 × 0.25) and the remaining 110 volts are used up in forcing 20 amp through the 5.5-ohm resistor (20 × 5.5). With an impressed emf of 120 volts across points A–D, there will be a loss of 5 volts (or a drop of 5 volts) from point A to B. In a similar manner, the voltage drop from D to C is 5 volts and from B to C, 110 volts. This reduction in voltage, or voltage drop, is the same as though an opposing voltage, tending to drive a current in the opposite direction, were introduced in the circuit by the current flowing. For this reason, voltage drop is often spoken of as a *back emf*, or as an emf opposing the flow of current.

1-5. Electric Power—Heating Effect of Electric Current. The primary function of an electric circuit is to do *work*. In a mechanical sense, work means the overcoming of opposition or resistance when moving an object. In lifting a weight of 100 lb, work is performed. The higher the weight is lifted, the more work is done. The work performed in lifting a weight is

$$\text{Force in pounds} \times \text{distance in feet} = \text{work in foot-pounds}$$

If a weight of 100 lb is lifted vertically a distance of 1 ft, 100 ft-lb of work is done. If a weight of 1 lb is lifted vertically a distance of 1 ft every second, 100 ft-lb of work will be done in 100 sec. If a weight of 100 lb is lifted vertically a distance of 1 ft in 1 sec, the same amount of work will be done but the *rate* at which work is done will be 100 times as great. *Power is the rate at which work is done.* The product of power and time is equal to work:

$$\text{Power in ft-lb per sec} \times \text{time in sec} = \text{work in ft-lb}$$

In electrical units, the force causing electricity to flow in a closed circuit is the applied emf measured in volts. The rate of flow of electricity is measured in amperes, and the quantity of electricity is equal to amperes multiplied by time. Electric power, then, is equal to the quantity of electricity flowing per second multiplied by the voltage causing the electricity to flow. The unit of electric power is the *watt.* Electric power measures the rate at which an electric circuit can perform work:

$$\text{Power in watts} = \text{volts} \times \text{amperes}$$
$$P = EI$$

The commonly used unit of power is a block of 1,000 watts called the *kilowatt.*

When current flows through a circuit, power is used in overcoming the resistance. The power so used converts to heat and is equal to

$$P = EI = (IR) \times I = I^2R \qquad \text{watts}$$

or

$$P = \frac{I^2R}{1{,}000} \qquad \text{kw}$$

The unit of electrical work is the *kilowatthour*, which is equal to the power in kilowatts multiplied by the hours during which work is performed. The above equations refer specifically to direct-current (dc) circuits. Similar equations for power in alternating-current (ac) circuits are shown in Arts. 1-23 and 1-24. The power consumed in forcing an electric current through a resistance is converted to heat energy in either a dc or an ac circuit.

Example: In Fig. 1-2, (1) what is the total power delivered to the circuit by the power source? (2) How much power is lost in the line if the two 0.25-ohm resistors represent line resistance? (3) If the resistor AB is a useful load, how much energy will be consumed in it in 12 hr? (4) What will be the line energy loss?

Solution:

1. Power $= P = EI = 120$ volts $\times$ 20 amp $=$ 2,400 watts $=$ 2.4 kw
2. Line power loss $= I^2R = 20 \times 20 \times (0.25 + 0.25) = 200$ watts
3. Load energy $= \dfrac{(120 \text{ volts} - 10 \text{ volts}) \times 20 \text{ amp} \times 12 \text{ hr}}{1{,}000} = 26.4$ kwhr
4. Line energy loss $=$ 0.200 kw $\times$ 12 hr $=$ 2.4 kwhr

The unit of mechanical power is the *horsepower*. One horsepower represents the ability to perform work at the rate of 33,000 ft-lb per minute. The relation between the horsepower and the kilowatt is given as follows:

$$1 \text{ hp} = 746 \text{ watts} = 0.746 \text{ kw}$$
$$1 \text{ kw} = 1.34 \text{ hp}$$

1-6. Magnetic Effect of Electric Current. A small bar magnet suspended by a string tied about its center of gravity will swing about so that it points north and south in line with the earth's magnetic field. The end pointing to the magnetic north of the earth is called the north pole, and the other end, pointing south, is called the south pole of the magnet. The region surrounding the poles of a magnet is called a magnetic field. If an ordinary small magnetic compass is placed at various positions in the magnetic field of a bar magnet, Fig. 1-3, the direction in which the north-seeking end of the compass needle points will vary with the location of the compass in the field. The complete field around a magnet can be outlined by the different positions of the compass needle, since the needle, being a magnet itself, will align itself with the direction of the field. The direction in which the north pole of the needle points is the direction of the field.

In Fig. 1-4 a wire connects the positive and negative terminals of a battery to form a complete circuit when the switch is closed. The wire runs north and south underneath a small magnetic compass. The compass needle will point north and south, parallel to the direction of the wire, until the switch is closed, permitting current to flow. If a sufficiently heavy current flows, the north pole of the needle will be deflected practically 90° to the right, as shown in *a*. If the compass is moved so that the wire passes over the compass, the needle will reverse and the

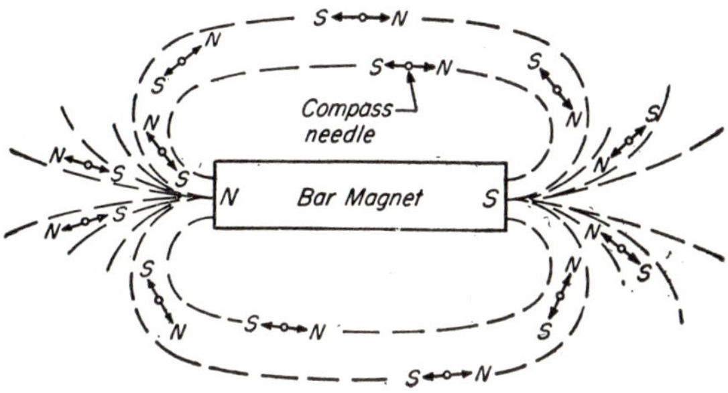

FIG. 1-3. A small compass needle can be used to map out the magnetic field of a permanent bar magnet.

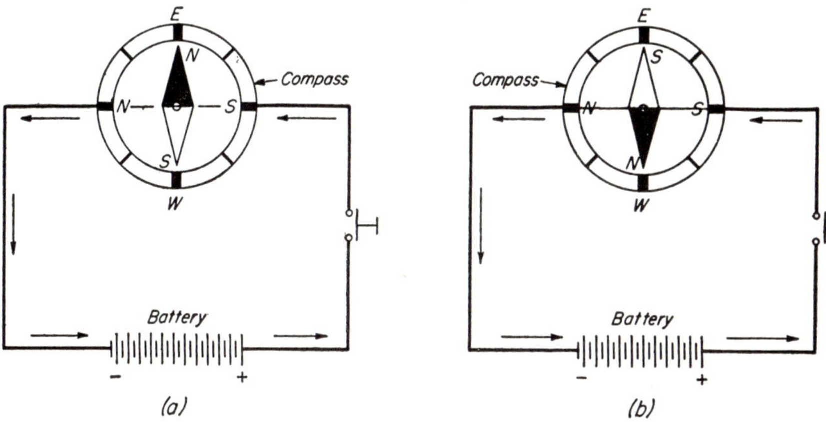

FIG. 1-4. A compass needle placed above and below a current-carrying wire shows magnetic forces set up by the electrical current. The compass points in opposite directions for two positions *a* and *b*.

north-seeking end will be deflected in the reverse direction, as shown in *b* of Fig. 1-4. If the compass is placed at different positions, completely circling the current-carrying wire, the compass needle will assume the positions shown in Fig. 1-5 when the current in the wire is flowing away from the reader or into the paper. The changing direction in which the needle points as the compass moves around the wire indicates the presence of a magnetic field surrounding the wire when current flows in it.

If the direction of current flow in the wire reverses, the direction in which the north pole of the compass needle points, in Figs. 1-4 and 1-5, will reverse. This indicates that the direction of a magnetic field about a conductor, carrying current, is determined by the direction of current flow. The direction of the magnetic field about a current-carrying conductor can be remembered by considering the direction of the field to be the same as the direction a nut on a right-hand-threaded bolt would have to be turned to drive the nut along the bolt in the same direction the current flows through the conductor.

1-7. Strength of a Magnetic Field. The strength or intensity of a magnetic field is measured by the force the field exerts on a magnet pole. Imagine that two isolated magnet poles of identical strength are concentrated in a point and placed 1 cm apart. If the two poles attract or

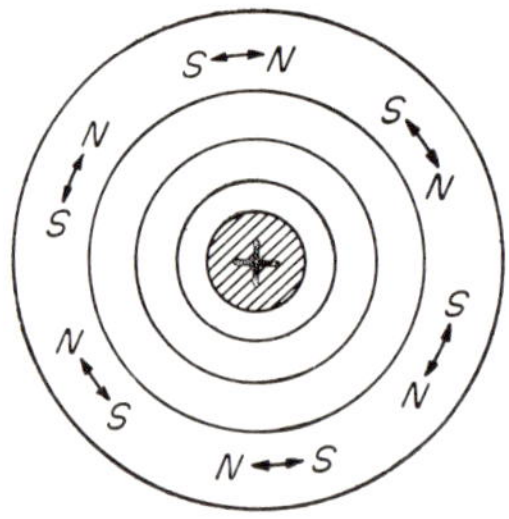

FIG. 1-5. A cross section of wire carrying current away from the observer; the compass needle maps out a circular electromagnetic field surrounding the wire while current flows. The + sign represents the tail of the arrow showing the direction of current flow.

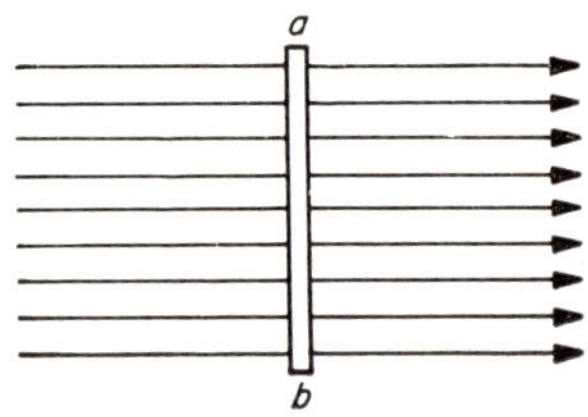

FIG. 1-6. A uniform magnetic field is described by the number of lines of force or flux passing through an area. Here we see the edge of a plane *ab* through which a uniform flux passes.

repel one another with a force of 1 dyne,* the poles are said to be of unit strength. The strength of the field surrounding a magnet pole varies inversely as the square of the distance from the pole. If a magnet pole exerts a force of 100 dynes on a unit pole at a distance of 1 cm, the magnet is said to have a strength of 100 units. The same magnet will exert a force of only 25 dynes on the unit pole at double the distance, or 2 cm. In other words, the strength of the field about the magnet pole is only one-fourth as great when the distance from the pole is doubled. The unit of magnetic field intensity is the *gauss*, which equals a force of one dyne per unit pole. A field has an intensity of one gauss if it exerts a force of one dyne upon a magnet pole of unit strength.

A magnetic field is defined in terms of its intensity and area. The product of field intensity in gauss and field area in square centimeters,

* 444,800 dynes = 1 pound of force.

perpendicular to the direction of the field, is called *magnetic flux*. The unit of magnetic flux is the *maxwell*, or the line of magnetic force as it is frequently called. Let Fig. 1-6 represent a uniform magnetic field, whose intensity is 100 gauss, in which *ab* represents a plane area of 10 sq cm perpendicular to the direction of the field. The total flux in the 10 sq cm area *ab* is equal to

$$100 \text{ gauss} \times 10 \text{ sq cm} = 1{,}000 \text{ maxwells, or lines of force}$$

1-8. Magnetizing Force. A wire through which a current is flowing will have a magnetic field set up about it as shown in Fig. 1-7. The

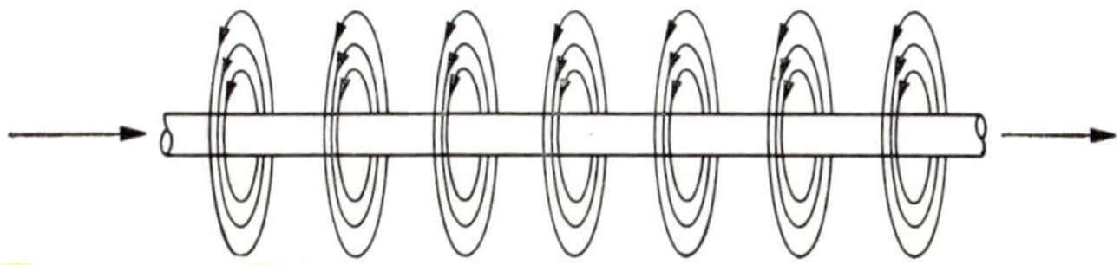

FIG. 1-7. The direction of electromagnetic field surrounding the current in a wire that creates it depends on the direction of current flow. If a right-hand screw thread is driven in the direction of current flow, the direction of rotation will be the direction of the magnetic field. Or if the wire is grasped in the right hand with the thumb pointing in the direction of current flow, other fingers point in the direction of magnetic flux.

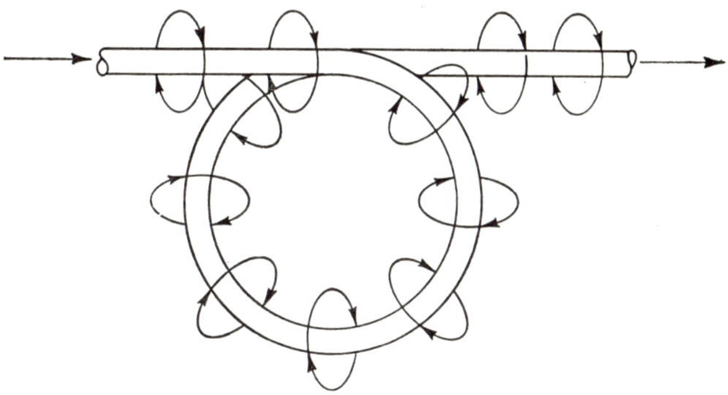

FIG. 1-8. Coiling a wire carrying dc concentrates the magnetic field at the inner part of the loop.

field will be at right angles to the direction the current flows, and the direction of the field around the wire will be as shown. The intensity of the field will vary inversely as the square of the distance from the wire. Now suppose the wire of Fig. 1-7 is coiled to form a single loop, as shown in Fig. 1-8. Inside the loop the flux lines are crowded together, and while the amount of magnetic flux has not changed it is concentrated in a smaller area. The intensity of the field, inside the loop, has increased while the area has decreased, keeping the total flux the same.

If the wire is now wound into several turns or loops, as shown in Fig. 1-9, it will be noted that the direction of the lines of magnetic flux, originating in each loop of the coil, is the same on the inside and outside of the coil and is in opposition between individual turns. Since the same current flows through each turn of the coil, the lines of flux cancel out between adjacent turns. The flux lines, originating in each turn, being in the same direction inside the coil, appear to add together and flow outward from one end of the coil and into the other end in exactly the same way as for the bar magnet in Fig. 1-3. Current flowing through the turns of the coil, in the direction shown in Fig. 1-9, forms an electromagnet with a north pole at the left end and a south pole at the right end as determined by the right-hand screw rule.

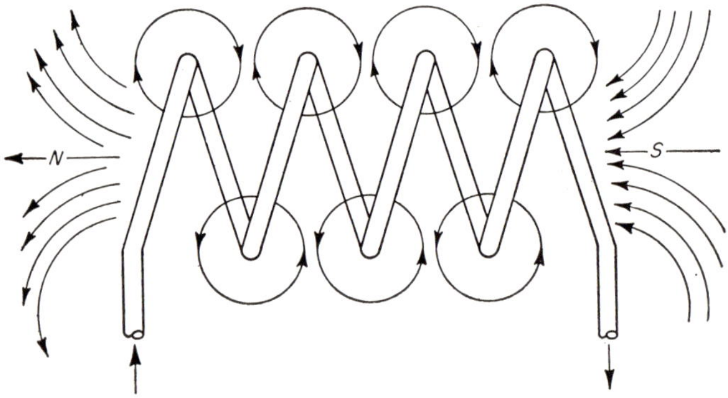

FIG. 1-9. A strong magnetic field is set up by current flowing through a coil of many turns.

The magnetic flux in a coil, such as that of Fig. 1-9, is proportional to the number of turns of wire and the current flowing in the wire. The intensity of the field is

$$H = \frac{0.4\pi NI}{l}$$

where H = average intensity of the field inside the coil, gauss per centimeter of coil length
N = number of turns in the coil
I = current flowing in the coil, amp
l = length of the coil, cm
π = 3.142

Since the total number of lines of flux set up in the coil equals the product of the field intensity by field area, perpendicular to the direction of the field, the flux emanating from the end of the coil is

$$\phi = AH = \frac{0.4\pi^2 r^2 NI}{l} \text{ lines}$$

where r = inside radius of the coil
A = cross-sectional area inside the coil

The intensity of the field and the number of lines of flux are proportional to the product of amperes by turns in the coil, or ampere-turns.

The intensity H of the magnetic field equals the magnetizing force per centimeter of coil length. The total magnetizing force producing lines of magnetic flux is called the *magnetomotive force*. The total magnetomotive force is

$$M = 0.4\pi NI = 1.26\ NI = lH = \text{magnetomotive force}$$

and

$$H = \frac{M}{l}$$

If the coil of Fig. 1-9 is wound about a highly magnetic material, such as iron, the intensity of the field and the number of lines of flux will be greatly increased. The ratio of the flux density of a coil wound about a core of magnetic material to the flux density of an identical coil wound about a core of nonmagnetic material is called the permeability of the magnetic material:

$$\mu = \frac{B}{H} = \text{permeability of magnetic material}$$

$$B = \mu H = \mu \frac{1.26NI}{l} = \text{flux density in magnetic material}$$

The permeability of a magnetic material varies with the intensity of magnetization. In the case of a coil wound about an air core, the intensity of the field in the coil increases in direct proportion to any increase in ampere-turns because air is nonmagnetic and its permeability is unity. For instance, if an air-core coil is wound with 100 amp-turns, a certain field intensity will result. If the ampere-turns about the air-core coil are now increased to 200, the field intensity will be doubled. If the coil is wound about a magnetic material, the increase in field intensity will be something less than double when the ampere-turns are doubled. The increase in field intensity with increased magnetizing force will become less and less until a point is reached where any further increase in magnetizing force will produce no further increase in magnetization. This point is called the saturation point of the particular magnetic material. The permeability is highest with a low value of magnetomotive force and decreases as the magnetomotive force increases.

Example: A coil of 500 turns has a length of 20 cm. The cross-sectional area inside the coil is 16 sq cm, and a current of 0.13 amp flows through the coil. What is the average flux density inside the coil? What is the total flux? What is the magnetomotive force? What would these values be if the coil were wound around a steel core, as shown in Fig. 1-10, whose permeability is 3,000?

Solution:

With an air core,

$$H = \frac{1.257NI}{l} = \frac{1.257 \times 500 \times 0.13}{20} = 4.084 \text{ gauss}$$

$$\phi = \frac{1.257NI}{l} \times A = AH = 16 \times 4.084 = 65.34 \text{ maxwells, or lines}$$

$$M = 1.257NI = lH = 20 \times 4.084 = 81.7 \text{ gilberts}$$

With a steel core,

$$B = \frac{1.257NI\mu}{l} = \mu H = 3{,}000 \times 4.084 = 12{,}250 \text{ gauss}$$

$$\phi = AB = 16 \times 12{,}250 = 196{,}000 \text{ maxwells, or lines of flux}$$

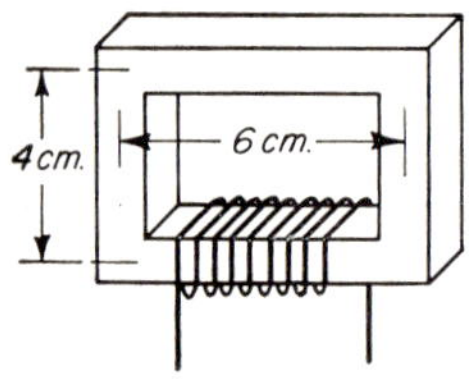

FIG. 1-10. Winding a current-carrying wire about a closed steel circuit multiplies the electromagnetic field thousands of times.

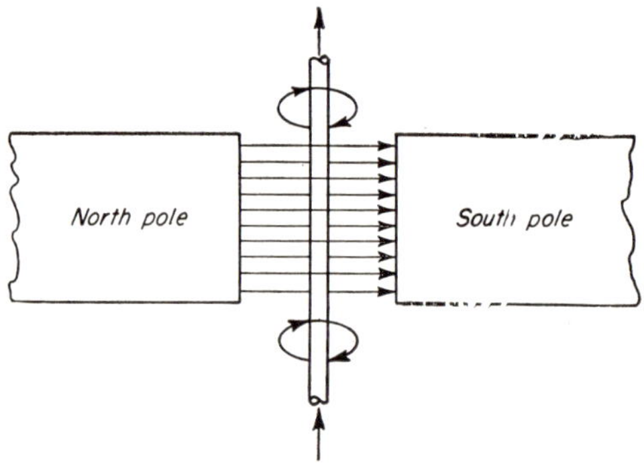

FIG. 1-11. When a wire carrying current passes through a magnetic field, the field reacts with the electromagnetic field set up around the wire by the current, to produce a side push on the wire. With the magnetic fields acting in the direction shown, the wire will be pushed away from the reader.

In the above example the total flux equals

$$\phi = \frac{\text{magnetomotive force}}{1/\mu A}$$

The term $1/\mu A$ is called the *reluctance* of the magnetic circuit. The reluctance of the magnetic circuit is similar to the resistance of the electrical circuit, where magnetomotive force is analogous to the emf and flux to the current of the electrical circuit. The unit of reluctance is the *oersted.* The reluctance of a cube which is 1 cm long and which has a cross-sectional area of 1 sq cm is equal to the reciprocal of the permeability of the material.

1-9. Effect of a Magnetic Field on a Conductor Carrying Current. If a wire passes through a magnetic field, at right angles to the direction of the field, Fig. 1-11, a force will be exerted on the wire when current flows through it. Current flowing through the wire sets up a cylindrical magnetic field around it. The strength of this field is proportional to the amount of current flowing. The direction of the magnetic field circling the wire will be the same as that of the field between the north and south magnet poles on one side and opposite on the other side of the wire. The resultant field, from the flux set up by the current flowing in the wire combining with the flux between the magnet poles, is congested and of a higher intensity on one side and of a lower intensity on the other side of the wire. The relative direction of the magnetic field between the magnet poles and around the wire, when current flows through the wire in a direction away from the reader, is shown in Fig. 1-12*a*. The strengthening of the field above the wire and the weakening of the field below the wire will be similar to that shown in Fig. 1-12*b*.

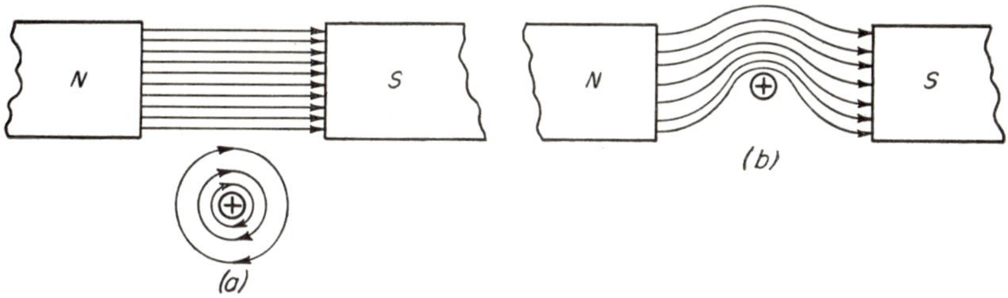

FIG. 1-12. When a current-carrying wire passes through a permanent magnetic field, the field is distorted by strengthening at the top and weakening at the bottom. The mechanical force between fields pushes the wire toward the bottom of the picture. This is the basic principle of the electric motor.

The lines of flux in a magnetic field act as though they were elastic fibers which seek to combine with the lines of flux about a current-carrying conductor all in one direction so that the flux lines in the established field will not be distorted. The resultant field pushes the wire in the direction of the weakened combined field. This is downward in Fig. 1-12 and away from the reader in Fig. 1-11. If the direction of current flow in the wire is reversed, or if the north and south poles are reversed, the direction of the push on the wire will be reversed. If a magnetic field were set up by winding wire about an iron core, forming an electromagnet, and a wire was then passed through the magnetic field set up by the electromagnet, a side push would be exerted on the wire in the same way.

This is the principle of the electric motor. The force exerted on the wire, because of the reaction between the current-carrying wire and the magnetic field, is proportional to the length of wire in the field, the strength or intensity of the magnetic field, and the amount of current flowing in the wire. The powerful forces in the electric motor, which are used to perform useful work, are obtained by using many wires carrying current at right angles to a very strong electromagnetic field.

1-10. Induced EMF. We have seen that an electric current produces magnetism. Under proper conditions, magnetism can produce electricity. In Fig. 1-13 is shown the wire and magnetic field of Fig. 1-11, but with no current flowing in the wire and with a galvanometer or sensitive type of millivoltmeter which will indicate the presence of small emf's, connected across the ends *AB* of the wire. If the wire is quickly pushed away from the reader, across the field between the north and south poles, the needle of the galvanometer will deflect, indicating a difference of potential or minute current flow between points *A* and *B*. If the wire is pulled across the field, toward the reader, the needle will deflect in the opposite direction.

If the zero point is in the center of the scale and the positive and negative terminals of the indicating instrument are as indicated in Fig. 1-13,

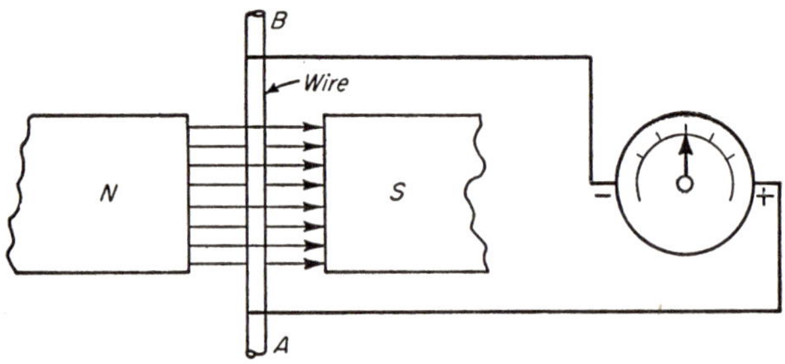

Fig. 1-13. A circuit for investigating directions of current flow, magnetic field, and wire motion through a field with the aid of a sensitive galvanometer.

the pointer will deflect to the right when the wire moves across the field in a direction away from the reader, and will deflect to the left when the wire is pulled across the field toward the reader. This indicates that emf is induced in the wire in such a direction that it will force a current from *A* through the instrument to *B* when the wire is moved across the field away from the reader, and from *B* to *A* when the movement of the wire is reversed. When current flows through the wire owing to the emf induced in it, there will be a deflecting force because of the reaction between the current in the wire and the magnetic field (see Art. 1-9). This deflecting force opposes the mechanical movement of the wire through the field. In moving the wire across the magnetic field, the mechanical force used must be great enough to overcome the opposing motor force developed by current flowing in the wire reacting on the magnetic field. In moving the wire across the magnetic field, mechanical power changes to electrical power. This is the principle of the electric generator, in which many wires are forced across a strong magnetic field to generate large values of emf. This emf can deliver large amounts of electrical power.

1-11. The DC Generator. If the wire of Fig. 1-13 moves in a direction from A toward B, or the reverse, no current will flow because no lines of flux are cut and no voltage is induced in the wire. If the wire moves *across* the field very quickly, the current flow will be greater than if the wire moves slowly. Increasing the number of lines of flux cut in a given time increases the emf. This can be demonstrated by the setup shown in Fig. 1-14, a coil of many turns of wire connected to a galvanometer. When a bar magnet in the coil center is at rest, all the lines of flux interlink all the coil turns, but there is no deflection of the indicating

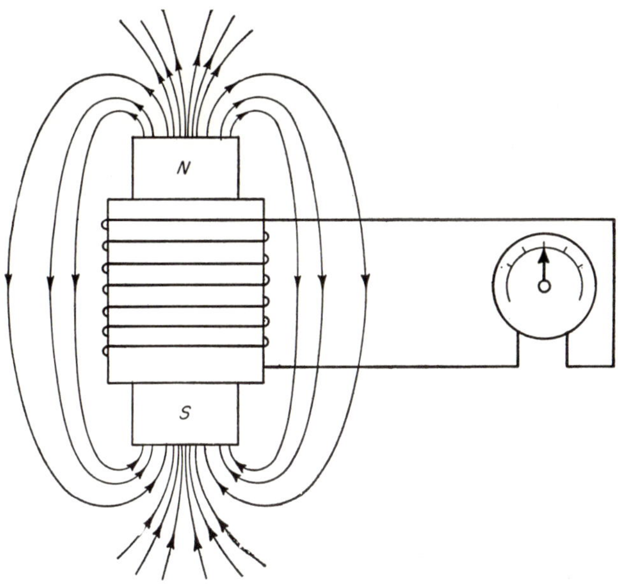

FIG. 1-14. Electric currents can also be generated by holding a wire coil stationary and moving a permanent bar magnet back and forth through the coil center.

needle of the instrument. Withdrawing the magnet deflects the needle, indicating an induced emf in the coil. If the number of turns of wire in the coil, the strength of the bar magnet, or the speed of motion of the magnet increases, the deflection of the needle will be proportionately greater. This indicates that the voltage induced is proportional to the strength of the magnetic field, the number of wires cutting across the field, or the total length of wire and the speed or rate at which the lines of flux are cut.

A direct current does not vary in magnitude with time and flows continuously in the same direction through the circuit. The current which flows from the positive terminal of a battery, through a load circuit, and back to the battery through the negative terminal is a direct current. Direct current (dc) is produced by the dc generator, also. The principle of the dc generator is shown in Fig. 1-15. N and S are poles of electromagnets magnetized by the field windings F_1, F_2. A is the armature, formed of magnetic material carrying the armature winding of many turns of wire. C is a cylinder called the commutator, formed of segments insulated from each other and insulated from the armature shaft on which they are mounted. Each commutator segment connects to a part of the armature winding.

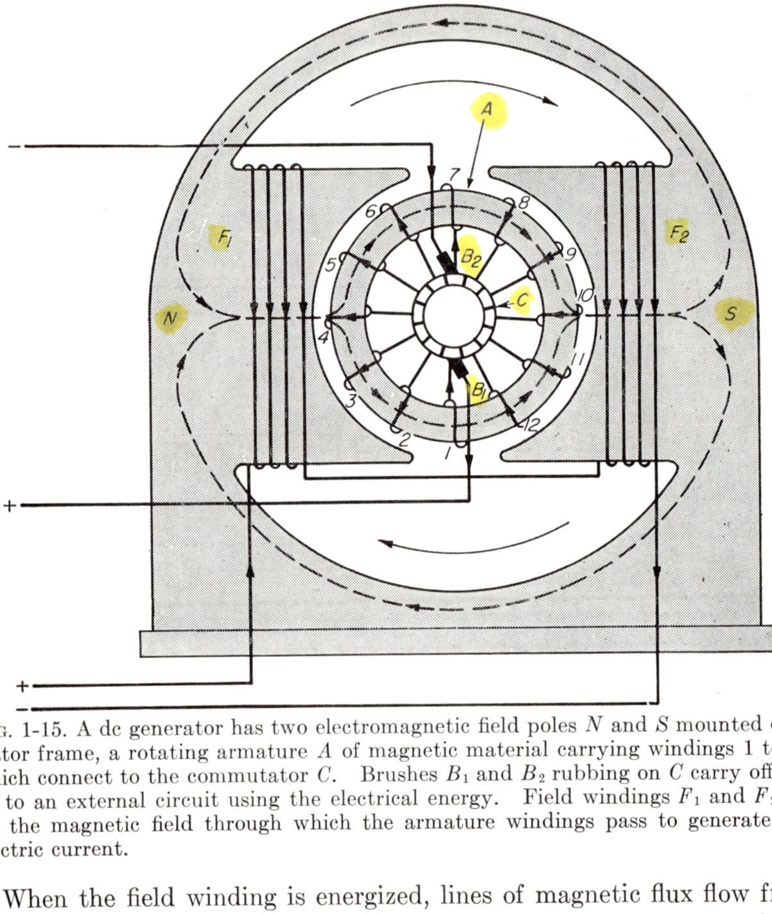

FIG. 1-15. A dc generator has two electromagnetic field poles N and S mounted on a stator frame, a rotating armature A of magnetic material carrying windings 1 to 12 which connect to the commutator C. Brushes B_1 and B_2 rubbing on C carry off the dc to an external circuit using the electrical energy. Field windings F_1 and F_2 set up the magnetic field through which the armature windings pass to generate the electric current.

When the field winding is energized, lines of magnetic flux flow from the north pole, across the small air gap between the pole face and the armature, through the armature core, across the air gap between the armature and the south pole, and through the frame or yoke carrying the poles back to the north pole, as shown by the dotted lines, Fig. 1-15. If the armature rotates in the direction indicated by the arrows by mechanical force applied to the shaft, the conductors on the face of the armature will cut across the magnetic field between the poles and an emf will be generated in them. At the instant that the armature is in the position shown in Fig. 1-15, all the conductors on the face of the armature are cutting across flux lines except those in the position of 1 and 7. Since the armature is circular in shape, the conductors on its outer surface are traveling in a direction tangent to its surface at any instant. Conductors 1 and 7 are traveling in a direction parallel to the direction of the field, at the instant shown, and therefore are cutting no lines of flux; consequently, no voltage or emf is induced in them. Conductors in position 2 are cutting lines of flux at a low rate because they are traveling diagonally across the field. Conductors in position 3 are cutting across flux lines at a faster rate, and conductors in position 4 are cutting flux lines at the maximum rate because they are moving at right angles to the direction of the field and travel the shortest possible distance between flux lines. Conductors in position 5 are cutting across flux lines at the same rate as conductors in position 3, and conductors in position 6 at the same rate as those in position 2.

Each conductor passing under the north pole is cutting lines of flux at a different rate, but each conductor is connected in series to the succeeding conductor so that the total difference in induced emf between points 1 and 7 is the sum of the emfs induced in conductors in positions 2 to 6 on the outer surface of the armature. Similar voltages are induced in conductors in positions 8 to 12 passing under the south pole so that the same voltage difference appears between points 1 and 7. The conductors shown on the inside of the armature in Fig. 1-15 have no voltage induced in them because practically all the flux passes through the armature frame and is not cut by the inner conductors.

B_1 and B_2 are conducting brushes bearing on the commutator segments. If wires lead off from B_1B_2 to an electrical load, such as a bank of incandescent lamps, the voltage difference between 1 and 7 will drive a current from B_1 through the load and back to B_2, since the magnetic field set up by the generated current flowing through the armature conductors must be in such a direction as to oppose the rotation of the armature by the resulting side push on the conductors. (If the generator of Fig. 1-15 were a motor and the current were flowing through the armature and field winding in the direction indicated, the motor would rotate in a direction opposite to that shown.) At an instant later than that shown in Fig. 1-15, brush B_1 will have passed to commutator segment 12, brush B_2 will have passed to commutator segment 6, and turns 6 and 12 of the armature winding will have no emf induced in them, but turns 1 and 7 will now have an induced emf. Voltage difference between B_1 and B_2 will be unchanged as long as the strength of the field and the armature speed stay constant. The terminal voltage of the generator will be a continuous dc voltage equal to the sum of the induced voltages between *positions* 1 and 7 less the voltage drop in the armature caused by the armature current and resistance.

1-12. The AC Generator. In Fig. 1-16 let us study a single conductor on the surface of an armature rotating in a clockwise direction at a constant speed. There is a uniform magnetic field from the north pole, through the armature, to the south pole. At the instant the armature conductor is in the position shown in Fig. 1-16*a*, no emf is induced in the conductor because it is traveling in a direction opposite but parallel to the direction of the magnetic field and not cutting any flux lines. In Fig. 1-16*b*, one-eighth revolution later, the conductor cuts *diagonally* across the field and an emf will be induced in proportion to the rate at which the con-

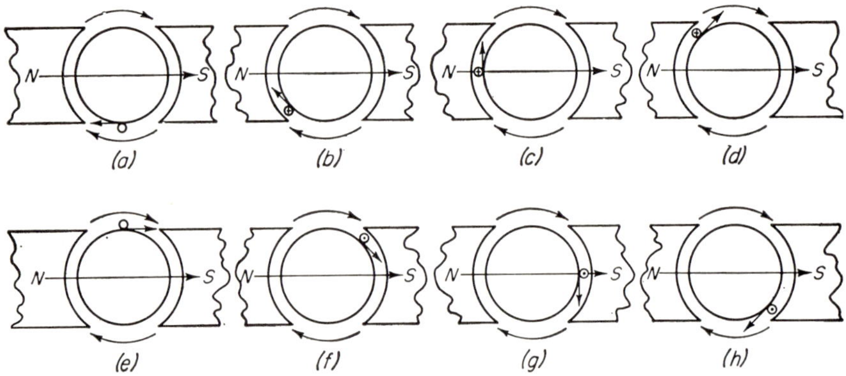

FIG. 1-16. As an armature conductor in a generator passes through one revolution, the direction of current flow alternates, depending on the direction in which the conductor passes through the magnetic field. The ⊕ sign (tail of arrow) shows the direction of current away from reader; the ⊙ sign (head of arrow) shows the direction of current toward the reader.

ductor cuts across lines of flux and in a direction tending to drive a current away from the reader. At the instant shown, the conductor travels at a 45° angle to the direction of the field. The rate at which lines of flux are cut, however, is equal to the product of field intensity and area *perpendicular* to the direction of the field across which the conductor would travel in unit time if velocity, direction of travel, or field intensity does not change (see Art. 1-7). It would have to travel a distance AB, of Fig. 1-17*a*, in order to move a distance equivalent to the line CB at right angles to the field. The emf induced, then, would be equal to the emf which would be induced in moving a distance CB, in unit time, at right angles to the field, or AB times the sine of the angle between the direction of conductor travel and the direction of field, or 45° in the case of Fig. 1-16*b*.

The values and relations of ac currents and voltages are frequently best expressed by a trigonometric formula, as was done in the preceding paragraph. It will be helpful to review the trigonometric relations of the right triangle at this point. The triangle of Fig. 1-17*a* is a right triangle. (A right triangle is one in which one of the angles is equal to 90°.) If the hypotenuse and sides of the right triangle are called *a*, *b*, and *c*, as shown in Fig. 1-17*b*, the ratios between *a*, *b*, and *c* are related to the angles between the hypotenuse and sides as shown by the following equations:

$$\frac{a}{c} = \text{sine } \theta \qquad \frac{c}{a} = \text{cosecant } \theta$$

$$\frac{b}{c} = \text{cosine } \theta \qquad \frac{c}{b} = \text{secant } \theta$$

$$\frac{a}{b} = \text{tangent } \theta \qquad \frac{b}{a} = \text{cotangent } \theta$$

Knowing one of the other angles of a right triangle and one of the sides, the third angle and the other two sides can be determined, since

$$a = c \text{ sine } \theta \qquad c = a \text{ cosecant } \theta$$

$$b = c \text{ cosine } \theta \qquad c = b \text{ secant } \theta, \text{ etc.}$$

Angles are sometimes measured in radians instead of degrees. An angle subtended by an arc equal in length to the radius of the circle is called a *radian*. In Fig. 1-17*c*,

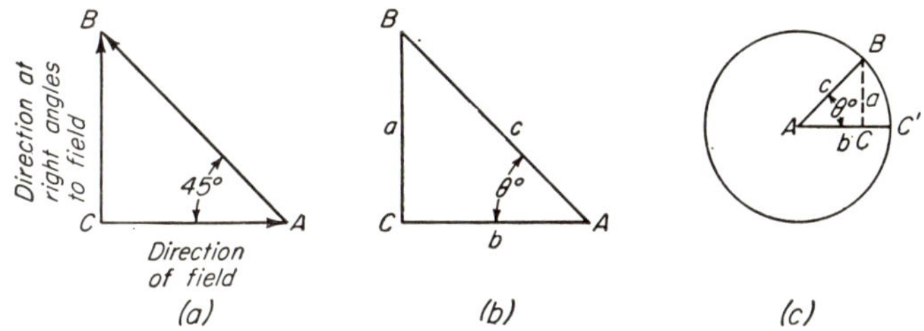

FIG. 1-17. The relations of a right triangle help to figure the rate of conductor travel through a magnetic field and the magnitude of the emf generated.

if a circle is inscribed with a radius equal to the hypotenuse *AB* and the length of the arc *C′B* is equal to the radius *AB*, the angle θ equals 1 radian. The size of an angle, measured in radians, equals the ratio of the arc subtending the angle to the radius of the circle of which the arc is a part measured along the circumference. There are 2π radians in any circle since the ratio of the circumference to the radius equals 6.2832, or 2π. The relation between the degree and radian is

$$1 \text{ radian} = 57.29578°$$

When the armature, Fig. 1-16*c*, has moved through one-quarter revolution, the conductor travels perpendicularly to the field and flux lines are being cut at the maximum rate. Maximum emf is induced in the conductor in a direction away from the reader. At succeeding intervals, of one-eighth revolution, the conductor will be at positions *d*, *e*, *f*, *g*, and *h*. When the conductor is at *d*, it again moves diagonally across the field and an emf is induced in it equal to and in the same direction as when at *b*. When the conductor is at *e*, it again moves parallel to the field and no emf is induced.

At *f* it will again be moving across the field at a 45° angle. The same emf will be induced as at *b* and *d*, but the direction will be reversed. This reversal is caused by the change in relative direction of conductor travel. The same emf will be induced in the conductor when in positions *b*, *d*, *f*, and *h*, but the relative direction of induced emf in *f* and *h* will be opposite to that in *b* and *d*. The same maximum emf will be induced in the conductor when in position *g* as at *c*, but the direction will be reversed.

Let's assign positive values to the emf in the conductor when traveling across the field under the north pole and negative values when cutting across under the south pole. The value of emf induced in the conductor at the different positions *a* to *h* can be plotted against time as shown in *a*

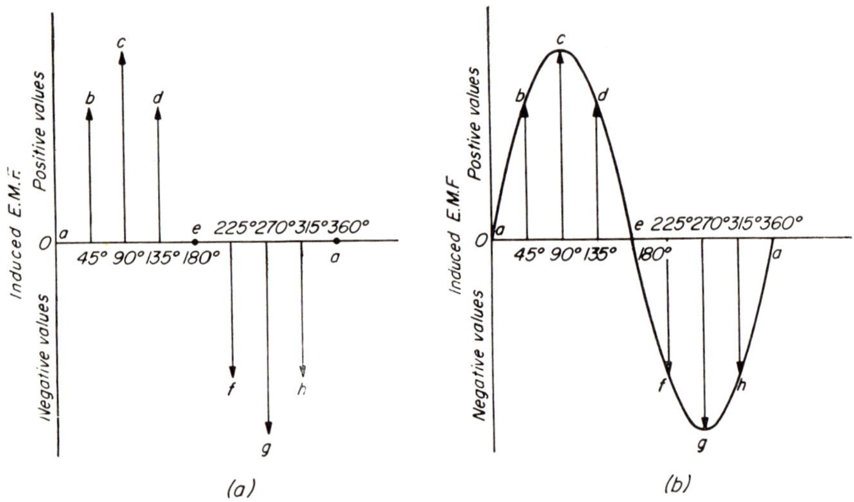

FIG. 1-18. An ac grows from zero to a maximum in one direction, then decreases to zero and grows to a maximum rate of flow in the opposite direction, after which it again decreases to zero and starts the cycle all over again. When plotting the rate of flow or magnitude of the induced emf against the angle of rotation (see Fig. 1-16) of the conductor, we get a sine wave.

of Fig. 1-18, where each division of time along the horizontal axis is equal to one-eighth revolution and the vertical arrows represent the induced emf measured along the ordinate or *y* axis. Drawing a smooth curve through these values as shown in *b* of Fig. 1-18, we get a sine wave. Vertical heights represent the magnitude of induced emf at any instant of time measured along the horizontal axis during a revolution of the armature.

As the armature rotates, the sine wave will be repeated for each revolution. The current flows in one direction through the circuit, increases to a maximum and then falls to zero, reverses and rises to a maximum value in the opposite direction and then falls to zero again. The direction of induced emf and current flow alternates cyclically, and an alternating current is said to flow.

The commutator on the shaft of the dc generator, Fig. 1-15, can be removed and two insulated rings R_1 and R_2 mounted in its place. Connecting them to the armature winding, as shown in Fig. 1-19, the generator will deliver ac. Armature conductors in positions 2 to 6 are cutting flux under the north pole, and conductors 8 to 12 are cutting flux at the same rate under the south pole, while conductors in positions 1 and

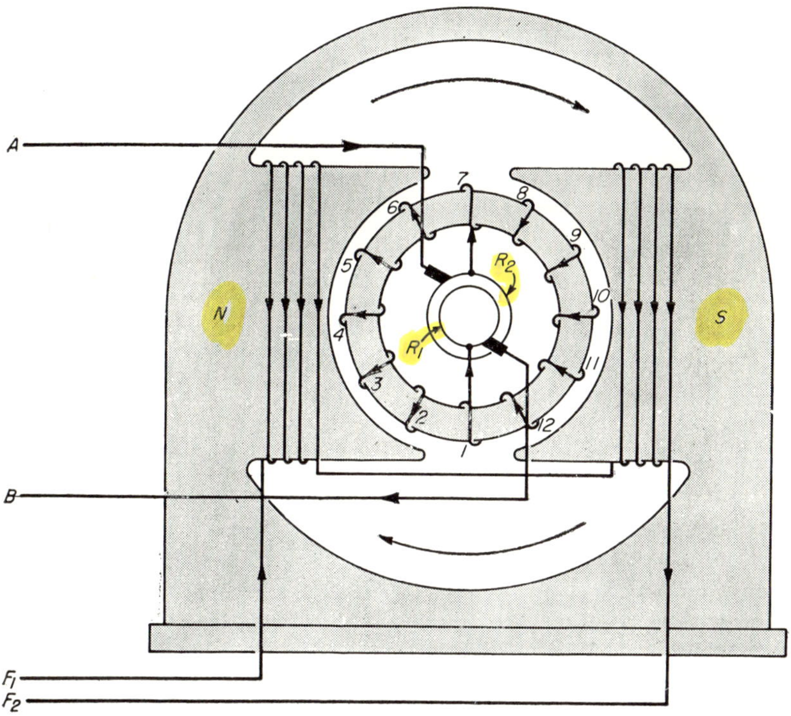

FIG. 1-19. An ac generator is constructed like a dc generator, except that the armature windings connect to solid slip rings (collector rings) instead of commutator segments. Field-pole windings must be connected to a dc source.

7 are cutting no lines of flux. The conductors under both the north and the south poles will have a current flow toward collector ring R_1 and out through terminal B, while current flows in through terminal A to ring R_2 and back into the armature at conductor position 7. When in the position shown in Fig. 1-19, maximum voltage difference is induced between collector rings R_1R_2 and between terminals A and B.

When the armature has turned through 90°, or one-quarter revolution, the conductors in position 7, the tap point to R_1, will be under the mid-point of the south pole while the conductors in position 1, the tap point to R_2, will be under the mid-point of the north pole. The induced-voltage difference between R_1 and R_2 is equal to zero at this instant because the voltage induced in the conductors under the north pole is equal to and in opposition to the voltage induced in the conductors under the south pole. When the armature has rotated through 180°, or one-half revolution, conductors 2 to 6 will be under the south pole while conductors 8 to 12 will be under the north pole. The voltage difference between R_1 and R_2 will again be a maximum, but the direction of current flow through the external circuit will be out from A and in at B.

The alternator in Fig. 1-19 has two magnetic field poles. An alternator may have more than two poles. The magnet poles, however, must be arranged in pairs so that there is an equal number of north and south poles. In the case of the two-pole machine, the armature winding is tapped at two points and connected to the collector rings so that there will be an equal number of turns under the north and south poles. If the machine were a four-pole machine, there would have to be two taps to ring R_1 and two taps to ring R_2 so that with the armature in the position shown in Fig. 1-19 (maximum emf being generated) current would flow from two taps into ring R_1 and from ring R_2 into two taps. If there were six poles, the armature winding would be divided, by taps, into six equal parts and into eight equal parts for eight poles, etc.

If the armature of Fig. 1-19 were held stationary and the field rotated, the same ac would be generated. Actually, the armature winding is usually placed on the stationary part, or stator, while the field winding is placed on the rotating part, or rotor, of ac generators. This permits delivering large amounts of electric power from an alternator without passing large currents through moving or sliding contacts. In dc machines, on the other hand, the field is stationary while the armature rotates; so the commutator rectifies the induced ac into a steady dc. The voltage wave induced by actual alternators is not a perfect sine wave but comes close to it.

1-13. Frequency. The sine wave of ac voltage and current from zero to a maximum, to zero, to a maximum in the opposite direction, and back to zero again is called a *cycle*. If a two-pole alternator makes 3,600 complete revolutions per minute, it will generate 3,600 complete cycles of ac voltage per minute, or 60 cycles per sec. The voltage appearing across the terminals of the machine would be said to have a frequency of 60 cycles per sec. The most common frequency of electrical supply in the United States is 60 cycles per sec, although some electrical energy is still supplied at other frequencies, particularly 50 and 25 cycles. The frequency of an alternator may be expressed as

$$\text{Frequency in cycles per second} = \frac{\text{number of poles} \times \text{rpm}}{2 \times 60}$$

Example: A six-pole alternator is driven at 1,200 rpm. What will be the frequency of the emf generated?

Solution:

$$\text{Frequency} = \frac{6 \times 1{,}200}{2 \times 60} = 60 \text{ cycles per sec}$$

1-14. The Electrical Degree. When the armature of Fig. 1-16 turns through one complete revolution, the conductor on its surface will have turned through 360° of circular measure. Each time the armature conductor of Fig. 1-16, or the armature winding of Fig. 1-19, turns through 360° of circular measure, the voltage induced will vary from zero, to maximum, to zero, to maximum in the opposite direction, and back to zero, completing one cycle. Values measured along the horizontal axis of Fig. 1-18, representing time for one complete cycle, can be represented in circular measure. Thus, in Fig. 1-18*b* the abscissa value of point *c* may be represented as 90° instead of one-quarter revolution.

The period of a sine wave is the time required for one complete cycle. If a voltage or current wave has a frequency of 60 cycles per sec, the period of the wave is $\frac{1}{60}$ sec. The period of a cycle, as shown in Fig. 1-18, can also be expressed in *electrical degrees.* For a two-pole generator, 360 circular degrees equals 360 electrical degrees. The period of one cycle, measured in electrical degrees, represents the time required for any point on the armature of Fig. 1-19 to move from the center of the north pole past the south pole and back to the center of the north pole. If the machine has more than one pair of poles, 360 electrical degrees would be measured from any point under a north pole to an identical point under the next north pole. In electrical measure, the angle between the center of a north pole and the center of the next adjacent south pole is equal to 180 electrical degrees, or π radians. Thus, when the rotor of a four-pole alternator turns through one complete revolution, it will have turned through 360° of circular measure and 720 electrical degrees, since two complete cycles will have been generated.

1-15. Single Phase—Polyphase. The alternator of Fig. 1-19 has two collector rings and two line leads, *A* and *B*. Such a machine is single phase. Suppose the alternator in Fig. 1-19 is fitted with three collector rings R_1, R_2, and R_3 and the armature winding tapped at three equidistant points *a*, *b*, *c*, as shown in Fig. 1-20. The separation between points *a*, *b*, and *c* equals 120 electrical degrees. The distance from *a* to *b*, *b* to *c*, and *c* to *a* is also equal to 120 circular degrees, since there is only one pair of poles. Point m_1 represents the mid-point of that portion of the armature winding from *a* to *b*, and m_2 and m_3 similarly represent mid-points of windings *bc* and *ca*.

In Fig. 1-20, the mid-point of coil *ab* (m_1) is under the center of the north pole. At this instant the maximum emf is induced in coil *ab*. In the previous instant, m_1 had not quite reached a position under the center of the north pole and the emf induced in coil *ab* had not quite reached its maximum value but was still increasing. An instant later the mid-point of coil *ab* will have passed the center of the north pole, and the emf induced in coil *ab* will have passed its maximum value and will have started to decrease to zero. Let's choose the positive half of the sine wave as being induced when the coil passes under a north pole. During the instant shown in Fig. 1-20 the maximum positive value of emf is being induced in coil *ab* and the direction of induced emf and current

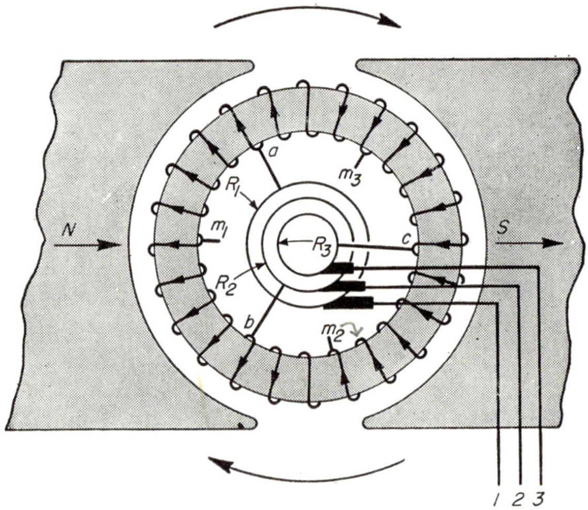

FIG. 1-20. An elementary ac-generator armature connected to three slip rings (collector rings) to generate three-phase current.

flow are from *a* to *b*, as indicated by the arrowheads. One-third of a period later, coil *bc* will be in the position of coil *ab* and the maximum positive value of emf will be induced. At this instant the direction of induced voltage and current flow will be from *b* to *c*. Two-thirds of a period later coil *ca* will be in the position of coil *ab*, and the maximum positive value of emf will be induced with the direction of current flow from *c* to *a*.

The voltages measured from *a* to *b*, from *b* to *c*, and from *c* to *a* are single-phase voltages which differ in phase time from one another by 120 electrical degrees. Separate single-phase leads, representing three separate single-phase circuits, could be taken from the alternator. This, however, would require six alternator terminals and six distributing wires. Since tap points *a*, *b*, *c* are each common to two of the three armature coils, giving rise to the three single-phase emfs, only three leads from the alternator need be provided. Thus, in Fig. 1-20, the single-phase emf induced in coil *ab* appears across wire leads 1 and 2. Similarly, the single-phase emf in coil *bc* appears across leads 2 and 3, while the emf in coil *ca* appears across leads 3 and 1. Such a machine which generates more than a single emf is a polyphase alternator. The alternator in Fig. 1-20 is a three-phase machine.

1-16. Sine-wave Representation of Alternating Currents. The value of emf induced in the phase windings of the alternator, Fig. 1-20, may be represented as in Fig. 1-21. The length of the arrow (vector) represents the maximum value of emf, and the phase relation between emfs is represented by the angular separation between the arrows. If the vectors rotate in a counterclockwise direction, their projection upon the y axis will represent the instantaneous value of induced emf at any instant

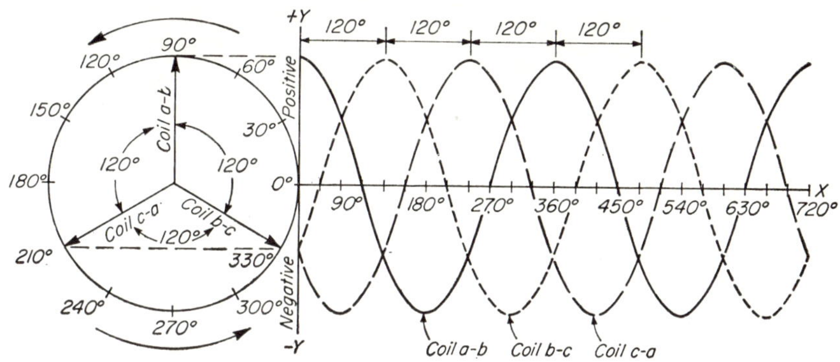

FIG. 1-21. Representing induced emfs of a three-phase circuit by individual rotating vectors, we can generate three sine waves plotted against the angle of rotation. The vertical height of the waves at a given angle shows the magnitude of the individual single-phase emfs at that instant.

where the angular rotation is in electrical degrees. If time is measured in electrical degrees, or radians, along the x axis, these projected values on the y axis will show the instantaneous values of the sine wave when plotted against time on the x axis. The three rotating vectors generate three sine waves representing the instantaneous values of emf generated in coils *ab*, *bc*, and *ca*. Values above the x axis are positive, and those below are negative.

The instantaneous value of induced emf at any instant is

$$e = E_M \sin \theta$$

where e = instantaneous value of emf
E_M = maximum value of induced emf
θ = angular advance from zero, electrical degrees

1-17. Effective Value of AC. In a dc circuit the rate of flow of electricity is constant as long as the emf and the resistance are unchanged. The ampere measures the rate of flow in a dc circuit. In an ac circuit on the other hand, the rate of flow keeps changing cyclically. The unit of measurement of an ac is expressed in terms of an *equivalent* dc.

To express an alternating current in terms of a direct current, they must be compared in terms of some property common to both. One such property is their heating effect when passing through a resistance. It was shown in Art. 1-5 that heat develops when an electric current flows through a resistance. The rate at which heat develops is proportional to the square of the current flowing. In an ac circuit, however, the rate at which heat develops is constantly changing because of the current variation. In a circuit in which current of sine-wave form is flowing, heat will be developed at a varying rate, but during one cycle the total heat developed will be equal to that developed by some *equivalent value* of dc.

Equivalent value of dc is

$$I = \frac{I_m}{\sqrt{2}} = 0.7071 I_m$$

where I_m = maximum instantaneous value of ac flow, amp

This equivalent value of dc is called the *effective value* or *root mean square* (rms) value of ac. It is the value read from ac ammeters and used in ac circuit calculations. When the maximum value of sine-wave current flowing is 100 amp, the effective or rms value equals 70.7 amp. This simply means that the sine-wave current of 100-amp maximum value is equivalent to a dc flow of 70.7 amp in heating effect. Since a sine-wave current flows through a circuit because of sine-wave voltage, values of emf are expressed in terms of effective or rms values also.

1-18. Inductance. It was shown in Arts. 1-10 to 1-12 that, when a conductor moves across a magnetic field cutting across lines of flux, an emf is induced in the conductor in proportion to the rate of movement. The same emf results if the conductor remains stationary and the field moves or changes in relation to the conductor. In Fig. 1-22, if A and B are two parallel conductors and an ac flows in A, an alternating emf of the same frequency will be induced in B. Assume that an ac of sine-wave form flows in conductor A in a direction away from the reader. As the current grows, the field due to the current moves outward from A and sweeps across B. The field continues to increase, and more lines of flux move out across B, until the current in A reaches a maximum at d and begins to decrease. The field from the current in A then begins to collapse and sweeps back across B in the reverse direction. If the current flowing in A were dc and the conductor B moved back and forth across the field, with the same relative time period, the induced emf would be the same. When the current in A increases and the field about A sweeps outward, Fig. 1-22a to d, an emf is induced in B in a direction to drive a current through *B toward* the reader, or in the *opposite direction to the flow of current through A.*

Since the field about *A* continually changes, more lines of flux sweep across *A* than across *B*. An emf, then, is also induced in *A* in proportion to the rate at which lines of flux cut across or link with conductor *A*.

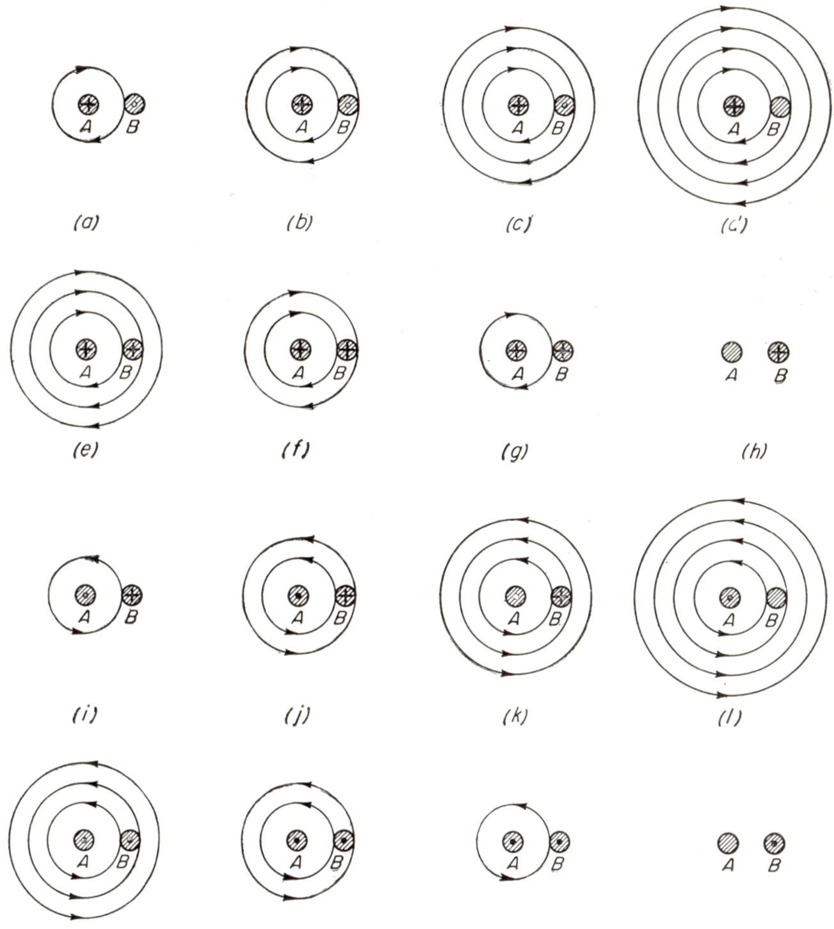

FIG. 1-22. With ac flowing in conductor *A*, a corresponding alternating magnetic flux is set up. In cutting across adjacent conductor *B*, the alternating flux generates an ac and emf in it. The induced emf in *B* lags behind the flow in *A*, reaching a maximum when flow in *A* is zero, and vice versa.

The emf induced in *A* when the current increases weakens the field about *A* and opposes the increase of current flowing, since the induced emf tends to drive a current in the opposite direction. When the current flowing in *A* reaches its maximum value, Fig. 1-22*d* and *l*, there is an instant when the field is unchanging and at rest. At this instant no emf is induced in A. An instant later the current through A begins to decrease from its maximum value toward zero. The field about A begins to collapse and sweeps back across A at a faster and faster rate. The emf now induced in A is in the opposite direction to the emf induced when the current was rising and tends to drive a current to increase the field strength and prevent the decrease of current flow. These effects, which are produced either by a conductor sweeping across a magnetic field or by a magnetic field sweeping across a conductor, are known by Lenz's law: "While the magnetic field about an electric circuit is changing, an emf is induced which opposes the change." When such a counter or back emf is set up by the current flowing in the circuit, we say that the circuit is *inductive*, or that it contains *inductance*.

When the relative motion between a magnetic field and a conductor is such that 100,000,000, or 10^8, lines of flux per second are cut, an emf of 1 volt will be induced in the conductor. If the ac in a circuit rises to a maximum in t sec, the average rate of change from zero to maximum value is I_M/t amp per sec. Since the magnetic flux ϕ is directly proportional to the current flowing, the average rate of change in flux is equal to ϕ_M/t lines of flux per second. In other words, when ac flows in a circuit and increases from zero to a maximum value in t sec or drops from a maximum value to zero in t sec, ϕ_M/t lines of flux per second cut across the conductor. The average emf induced in the conductor during this period of time equals

$$E_{av} = \frac{\phi_M}{t} \times \frac{1}{10^8} \qquad \text{volts}$$

Since the flux ϕ is proportional to the current flowing, it may be expressed in terms of the current as

$$\phi = kI \text{ lines of flux}$$

where k = a constant number

The average emf induced by an ac flow may be expressed in terms of the maximum value of current as

$$E_{av} = \frac{k}{10^8} \times \frac{I_M}{t}$$

or

$$E_{av} = L \times \frac{I_M}{t} \qquad \text{volts}$$

The value $k/10^8$ or L is the inductance of the circuit, and the relation between flux, inductance, and current is

$$\frac{\phi}{10^8} = LI \qquad \text{lines of magnetic flux}$$

Consider the coil shown in Fig. 1-9. The maximum flux produced equals

$$\phi_M = \frac{0.4\pi N \mu A I_M}{l}$$

If all the flux developed in the coil cuts each turn, the voltage induced in each turn equals the total number of lines of flux cutting across per second divided by 100,000,000. The average emf induced in the whole coil is

$$E_{av} = -\frac{\phi_M}{t} \times N \times \frac{1}{10^8} = \frac{-0.4\pi N^2 \mu A}{10^8 \times l} \times \frac{I_M}{t} \quad \text{volts}$$

Also

$$E_{av} = -L\frac{I_M}{t}$$

The minus sign is used because the voltage induced opposes the applied voltage and any change in the current flow. Now,

$$E_{av} = -L\frac{I_M}{t} = \frac{-0.4\pi N^2 \mu A}{10^8 l} \times \frac{I_M}{t}$$

from which

$$L = \frac{0.4\pi N^2 \mu A}{10^8 l} = \text{inductance of the coil}$$

The unit of inductance is the *henry*. When the current flowing through a circuit changes at the rate of one ampere per second and a counter emf of one volt is induced in the circuit, the circuit is said to have an inductance of one henry. The henry is a comparatively large unit of measure. A current which changes at the rate of 1 amp per sec is changing very slowly, and consequently a large amount of inductance must be present in the circuit for 1 volt to be induced. For convenience, a smaller unit, such as the millihenry, which is equal to one-thousandth henry, or the microhenry, which is equal to one-millionth henry, is sometimes used.

Inductance is not a material thing but a purely physical characteristic of the circuit. It depends upon the size and the shape of the circuit and upon the magnetic character, or permeability, of the surrounding medium. The calculation of inductance is complicated and can seldom be found by a simple formula. If a counter emf is induced in a circuit by a current flowing in the same circuit, the term *self-inductance* designates the presence of inductance. When an emf is induced in a circuit because of current flowing in another circuit, the term *mutual inductance* designates the relation between the rate of change of current and the voltage induced. The self- and mutual inductance of a circuit may be additive or subtractive according to their mutual effects.

1-19. Inductive Reactance. Inductance is the ratio between the average value of induced emf and the average *rate of change* of current flowing in a circuit containing inductance only, or

$$L = -\frac{E_{av}}{I_M/t}$$

The relation is true only as long as the value of current flowing in the circuit is changing. The effect of inductance is apparent in a dc circuit only while the current rises from zero to its steady value or falls from its steady value to zero. In an ac circuit the current never reaches a steady, constant value but is continually changing. If an ac circuit contained inductance only, however, the current would vary from zero to a maximum, to zero, and to a maximum in the opposite direction and back to zero again in exactly the same way it would if resistance only were present. (It should be noted that a resistance cannot be wound which is entirely free of inductance, nor can an inductance be formed which is entirely free of resistance. These limits can only be approached.) The ratio between the applied emf and the ac flowing in a circuit containing inductance only is called *inductive reactance.* The value of inductive reactance X_L in a circuit is given in ohms and found as

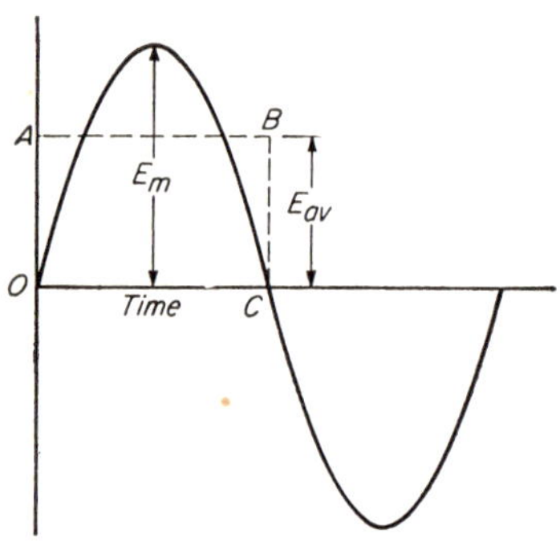

FIG. 1-23. The average value of an ac emf is the height E_{av} which equals 0.636 of the maximum sine-wave emf E_m.

$$X_L = \frac{E}{I}$$

The emf induced in an ac circuit with inductance is expressed in average values. For a sine-wave voltage the average value will be equal to instantaneous values of emf during one-half of the sine wave added together and averaged. This average value equals the area under one-half of the sine wave divided by the base of the half wave. Figure 1-23 shows this where the area of the rectangle $OABC$ equals the area under one-half the sine wave and in which the altitude BC of the rectangle equals the average of the instantaneous values of emf between O and C. Since the time period OC of the half wave, Fig. 1-23, equals π radians, the average emf may be expressed in terms of the maximum emf as

$$E_{av} = \frac{2}{\pi} E_M = 0.636 E_M$$

Since a sine-wave voltage will be induced by inductance in a circuit when a sine-wave current flows, the average induced emf may be expressed as

$$E_{av} = -L\frac{I_M}{t} = -\frac{2}{\pi} E_M$$

The time t equals one-fourth of a period, or $1/4f$ sec, where f is the frequency. The above equation may be rearranged as follows:

$$E_M = \frac{\pi L}{2t} I_M = \frac{\pi L}{2 \times \frac{1}{4f}} I_M = 2\pi f L I_M$$

But

$$E_M = X_L I_M$$

therefore,

$$X_L I_M = 2\pi f L I_M$$

from which

$$X_L = 2\pi f L \text{ ohms} = \text{inductive reactance}$$

1-20. Phase Difference between Current and Voltage. It was shown in Art. 1-4 that resistance in a dc circuit determines the amount of current flowing and that the applied voltage is used up in overcoming the opposition to the current flow offered by the resistance. In an ac circuit which contains resistance only, this same relation holds true, as expressed by Ohm's law, the current, at any instant, equaling the voltage divided by the resistance. The current varies as the voltage, the resistance remains constant, and the current is said to be *in phase with the voltage*. In other words, there is no time difference, expressed as an angular difference in degrees or radians, between the instant when the maximum value of voltage occurs and the instant when the maximum value of current in the circuit occurs. This *in-phase* relation between voltage and current is shown in Fig. 1-24.

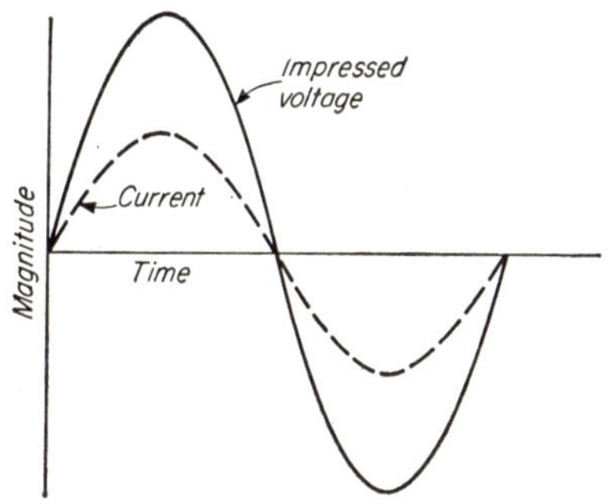

Fig. 1-24. When an ac emf acts across a circuit with resistance only, the current varies in exact step with the emf. The current is *in phase* with the emf.

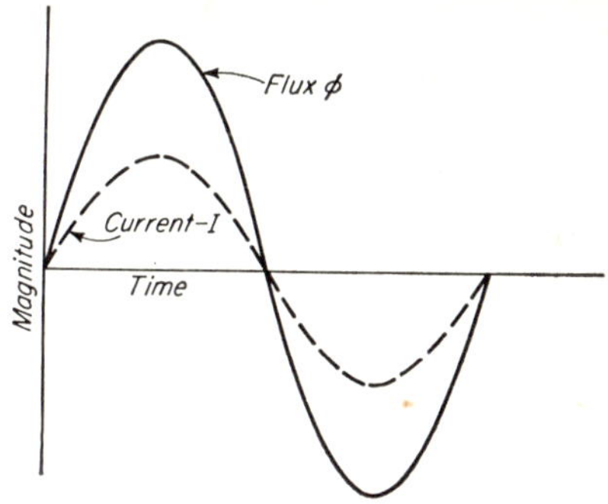

Fig. 1-25. The electromagnetic flux accompanying all currents is always in phase with the current.

In an ac circuit a sine-wave magnetic field is set up about any conductor through which a sine-wave current flows in proportion to the current and the inductive characteristics of the circuit. This is shown in Fig. 1-25. A sine-wave magnetic field will induce an emf in any conductor it sweeps across or links with, including the conductor through which the current flows.

Figure 1-26 represents a 60-cycle sine-wave current whose maximum value is 100 amp. The value of current at any instant is expressed by the following equation:

$$i = I_M \sin \theta$$

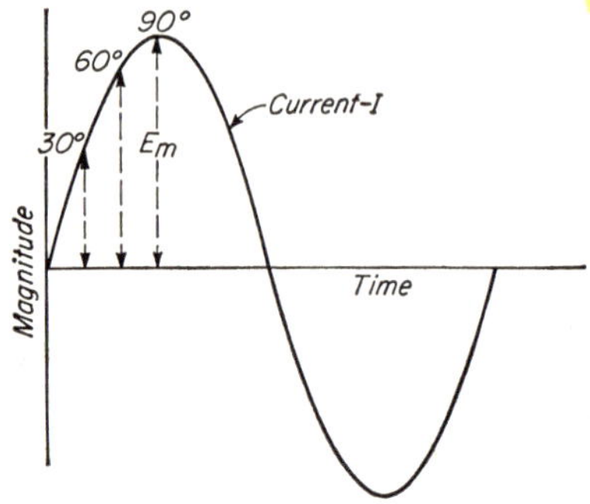

FIG. 1-26. The rate of change of ac flow for equal angular changes varies widely. The maximum rate of change occurs from 0 to 30°, a smaller rate from 30 to 60°, and the smallest rate from 60 to 90°. The *instantaneous* rate of change is maximum at 0° and zero at 90°.

The instantaneous value of current at times of 90, 60, and 30°, where time is measured in electrical degrees, is

$$i \text{ at } 90° = 100 \text{ amp} \times \sin 90° = 100 \text{ amp}$$
$$i \text{ at } 60° = 100 \text{ amp} \times \sin 60° = 86.6 \text{ amp}$$
$$i \text{ at } 30° = 100 \text{ amp} \times \sin 30° = 50 \text{ amp}$$

Thirty electrical degrees represents $30/360$, or $1/12$ period. In a sine wave of 60-cycle frequency, 30 electrical degrees represents $1/12$ of $1/60$, or $1/720$ sec. During the first 30° of a cycle the current increases from zero to 50 amp. Between 60 and 90° the current increases from 86.6 to 100 amp, or a total increase of 13.4 amp. During these times the average rate of current increase, in amperes per second, is

$$\text{From 0 to 30°} \qquad \frac{I}{t} = \frac{50}{1/720} = 36{,}000 \text{ amp per sec}$$

$$\text{From 60 to 90°} \qquad \frac{I}{t} = \frac{13.4}{1/720} = 9{,}648 \text{ amp per sec}$$

Between 0 and 1° the current grows from zero to 1.75 amp. Between 89 and 90° the current increases from 99.98 to 100 amp, or a net increase of 0.02 amp. The time represented by 1 electrical degree, with a frequency of 60 cycles per sec, equals 1/(60 × 360), or 1/21,600 sec. The average rate of current change during these time intervals is

$$\text{From 0 to 1°} \qquad \frac{I}{t} = 1.75 \text{ amp} \times 21{,}600 = 37{,}800 \text{ amp per sec}$$

$$\text{From 89 to 90°} \qquad \frac{I}{t} = 0.02 \text{ amp} \times 21{,}600 = 432 \text{ amp per sec}$$

At 90° the current reaches a maximum and begins to decrease toward zero. For an instant, at this point, the current is constant and unchanging, and while the magnetic flux is a maximum it, also, is constant.

During this instant no voltage is induced in the conductor because, since the magnetic field is constant and unchanging, no lines of magnetic flux are being cut by the conductor. At the other extreme, the current changes at the maximum rate when it passes through zero. This is indicated by the rate at which current changes from 0 to 1° as compared with the rate of change from 89 to 90°. The maximum voltage will be induced by the changing magnetic field when the current wave is passing through zero. This self-induced voltage, due to the inductance in the circuit, will lag 90° behind the current and flux waves because it must be negative in value and oppose the change of current when the current is increasing in value, as noted in Art. 1-18. This relation is shown in Fig. 1-27*a*.

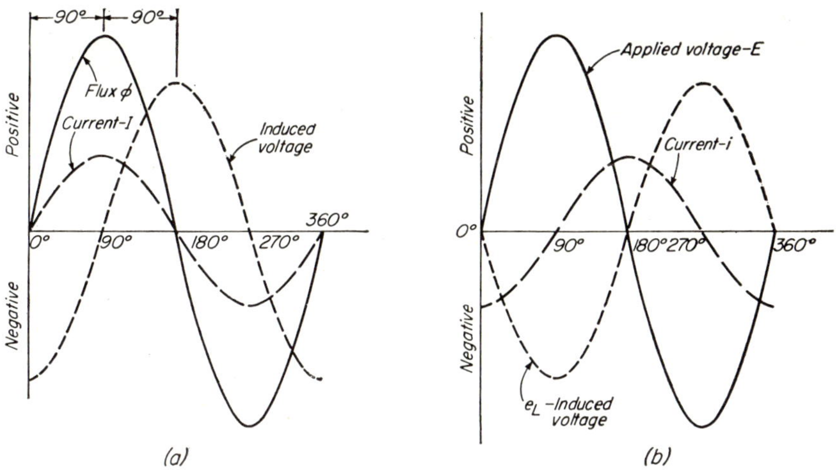

FIG. 1-27. In a circuit containing inductance only, the current lags behind the impressed emf by 90°. Current flow is zero when the emf is maximum, and vice versa.

The induced voltage delays the change in current by acting in opposition to the impressed voltage which drives the current through the circuit. Since the voltage induced in the circuit must lag 90° behind the current and at the same time act in direct opposition to the impressed voltage, the current flowing in a circuit which contains inductance only will lag 90° behind the impressed voltage, as shown in Fig. 1-27*b*.

1-21. Capacitance. In addition to resistance and inductance, a third factor, *capacitance*, is present in all circuits which affects the amount of current a given ac voltage will cause to flow. The capacitance is not large enough to be considered in the comparatively low voltages and short lines used for general distribution. At the higher voltages and with the longer transmission lines, the effect of capacitance in the circuit can be considerable.

Consider two metal plates separated by a good insulating material, in the form of a thin sheet. These plates connect to a 120-volt dc source, as shown in Fig. 1-28. When the switch is closed, a sensitive ammeter A indicates that current flows in the circuit. A maximum current flows when the switch is first closed. The current then gradually decreases until it completely ceases.

If ac voltage instead of dc is used, a continuous flow of ac becomes established in the circuit. The current flowing will be minute unless the plates are very large. A high-voltage transmission line, open at the distant end from the generator, is similar to the circuit of Fig. 1-28. The conductors correspond to the metal plates, and the air separating the conductors corresponds to the insulator, or *dielectric*, separating the plates. In a transmission line 100 miles long with 110,000 volts between conductors, about 36 amp of ac will flow in the circuit, even when the line is open on the load end. Such current flow caused by capacitance of the line is

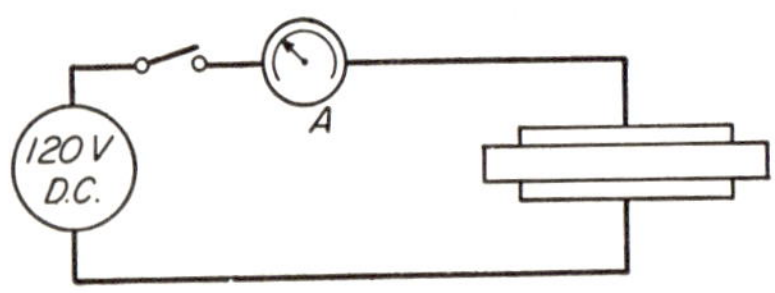

FIG. 1-28. A circuit with a condenser connected across an impressed emf has capacitance which acts differently from resistance or inductance.

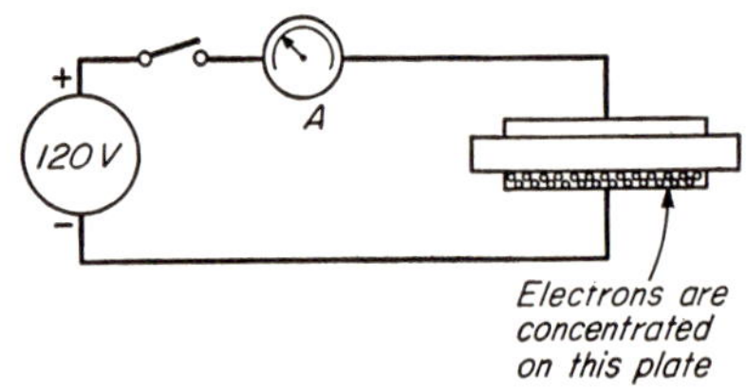

FIG. 1-29. On its plates a condenser holds static charges which attract each other across the insulating dielectric which prevents electron flow.

called the *charging current*. The device shown in Fig. 1-28 is a *capacitor* or *condenser*.

Capacitance is best explained by the electron theory of current flow. Before closing the switch of Fig. 1-28, the circuit is in equilibrium and the plates, separated by the dielectric, are electrically neutral. When the switch is closed, the electrical equilibrium of the circuit is upset by the potential difference of the battery or generator connected to the plates. The potential difference of the power source is brought about by a surplus of electrons at the negative terminal and an equal deficit of electrons at the positive terminal. This potential difference transfers to the plates of the capacitor when the circuit is closed. Since the plates are close together, the positive charge on one plate exerts a powerful attraction on the electrons in the other plate. The dielectric, being an insulator, has very few free electrons; hence, very few electrons will pass through it from one plate to the other. The action of the dielectric in blocking the passage of electrons between the plates will cause a movement of free electrons from the upper plate toward the positive terminal of the power source and from the negative terminal of the power source toward the lower plate of the capacitor, as shown in Fig. 1-29.

When sufficient electrons concentrate on the lower plate so that the potential difference between the plates equals the potential difference between the terminals of the power source, the flow of electrons, and consequently the flow of current, stops and the condenser is said to be charged. If the voltage source of Fig. 1-29 is ac, the plates will be alternately positive and negative and there will be an alternating shift of electrons, or current flow, around the circuit from one plate to the other. If the condenser of Fig. 1-29 is fully charged by the dc supply and then disconnected from the power supply, it will remain charged. If a wire is then connected between the plates, equilibrium will be reestablished by the flow of electrons from the negatively charged to the positively charged plates. The flow of electrons will cease when the excess of electrons on the negative plate transfers to the positive plate.

Since the charge on the condenser of Fig. 1-29, at any instant, is proportional to the voltage impressed, the charge will increase and decrease as the voltage increases and decreases. Let's assume that Fig. 1-29 shows the instant when the ac voltage wave has reached a maximum positive value so that the upper plate is positive and the lower plate negative. At this instant the charge on the condenser is a maximum and the maximum excess of electrons is accumulated on the lower plate. The flow of electricity through the circuit, charging the condenser, will have decreased to zero, and consequently the flow of current in the circuit will have ceased. An instant later the impressed ac sine-wave voltage will start to decrease from its positive maximum value toward zero. The charge, likewise, will start to decrease and the flow of electrons will reverse and start to flow from the bottom plate to the upper plate of Fig. 1-29, even though the upper plate is still positively charged. If it is remembered that current flow is considered to be opposite in direction to electron flow, the flow of current up to the instant when the condenser is fully charged will be into the upper plate and out of the lower plate while electron flow will be the opposite.

The circuit of Fig. 1-29 is considered to have zero resistance and inductance, or capacitance only. With capacitance only in the circuit the charge will be a maximum when the voltage is a maximum, and the current flow will be zero at the same instant. When the voltage has decreased from a positive maximum to zero, the charge will likewise have decreased from a maximum value to zero. When the voltage wave passes through zero, it changes at the maximum rate, as shown in Art. 1-20. The charge, or rate of flow of electricity through the circuit and into the condenser, is also changing at the maximum rate, and consequently the flow of current will be a maximum when the voltage wave passes through zero since current measures the *rate* of flow of electricity. To satisfy these conditions the current flowing in a circuit containing capacitance only must reach a maximum value 90° ahead of the sine-wave voltage. In other words, the current leads the voltage by 90 electrical degrees in a circuit containing capacitance only. The effect of capacitance in a circuit, then, is to oppose any change in the voltage, which results in the voltage lagging behind the current. The phase relation of a sine-wave voltage, current, and charge, in a circuit containing capacitance only, is shown in Fig. 1-30.

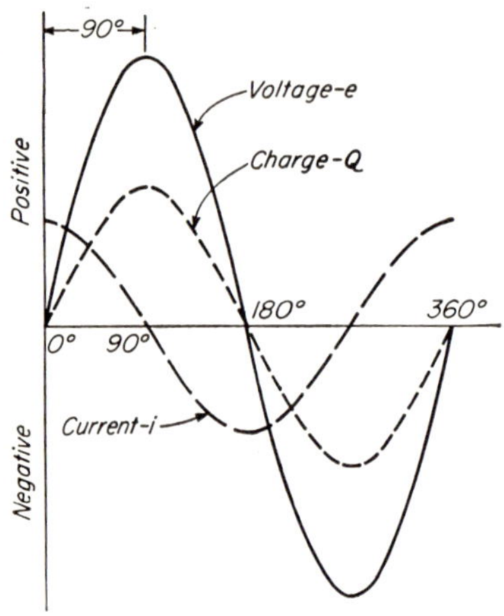

FIG. 1-30. In an ac circuit containing capacitance only, the current leads the impressed emf by 90°. Current flow is zero when the emf is maximum, and vice versa.

The charge of a condenser is the total quantity of electricity which has flowed through the circuit and into the plates of the condenser. The current being the rate at which electricity is flowing, then

$$Q = It$$

where Q = charge, coulombs
I = charging current, amp
t = charging time, sec

The charge is proportional to the charging voltage (see Fig. 1-30), and the capacitance is the proportionality factor between the charge and the voltage. This relation is expressed by the following equations:

$$Q = CE$$

and

$$It = CE$$

from which

$$I = \frac{CE}{t}$$

where E = impressed or charging voltage
C = capacitance, farads

The unit of measure of capacitance is the *farad*. A circuit is said to have one farad of capacitance if one coulomb per second of electricity, or one ampere, is caused to flow when an emf of one volt is applied. The farad is a very large unit of measurement. The more commonly used unit is the *microfarad*, which is equal to one-millionth farad. The effect of inductance in a circuit is dependent upon the rate of change of current. Similarly, the effect of capacitance in a circuit is dependent upon the rate of change of voltage. It should be noted that the relative effects of inductance and capacitance in a circuit are directly opposite.

1-22. Capacitative Reactance. The condenser of Fig. 1-29 charges when the voltage rises from zero toward a maximum, in either direction, and discharges when the voltage wave decreases from a maximum toward zero. Thus, the condenser charges twice and discharges twice each cycle. The total charge, or the maximum quantity of electricity forced into the condenser, is equal to the following:

$$Q = \text{average value of charging current} \times \text{time in seconds} = I_{av} \times t$$

$$t = \frac{1}{4}\text{ cycle} = \frac{1}{4f}\text{ sec} = \text{time to charge or discharge}$$

The total quantity of electricity forced into the condenser when it charges, or forced out of the condenser when it discharges, is also equal to

$$Q = CE_M$$

where E_M = maximum value of impressed emf
C = capacitance of the condenser, farads

These two expressions for the total charge being equal to the same thing,

$$CE_M = I_{av}t$$

and since

$$I_{av} = \frac{2}{\pi} I_M$$

$$CE_M = \frac{2}{\pi} \cdot \frac{1}{4f} I_M = \frac{2I_M}{4\pi f} = \frac{I_M}{2\pi f}$$

from which

$$E_M = \frac{1}{2\pi fC} I_M$$

The proportionality factor between the impressed ac voltage and the current in a circuit containing capacitance only is called the *capacitative reactance* of the circuit. Thus,

$$E = X_C I$$

from which

$$X_C I = \frac{1}{2\pi fC} I \qquad \text{and} \qquad X_C = \frac{1}{2\pi fC} \qquad \text{ohms}$$

It was pointed out in Arts. 1-18 and 1-20 that current due to inductance in the circuit lags behind the impressed voltage by 90 electrical degrees while current due to capacitance leads the voltage by 90 electrical degrees. The two, then, are in direct opposition. In order to take this into account it is customary to assign a negative value to capacitative reactance while a positive sign is given to inductive reactance.

1-23. Power in an AC Circuit. In Art. 1-5 it was shown that power, in watts, developed in a dc circuit equals the product of volts times amperes. In an ac circuit the current and voltage are continually changing. The power delivered from the ac source also continually changes. In an ac circuit the same relation holds as in a dc circuit. That is, the product of volts times amperes, at any instant, is equal to the output of power into the circuit *at that instant.* In an ac circuit containing resistance only, the current varies exactly as the voltage. It reaches its maximum value at the same instant and reaches its zero value at the same instant the voltage does. In other words, the current and voltage are in phase.

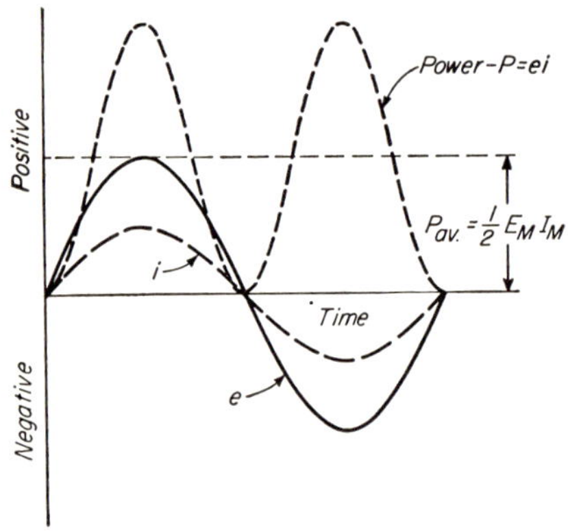

FIG. 1-31. Power flow, in an ac circuit with resistance only, absorbed by the resistance reaches a maximum twice during each cycle and drops to zero twice during each cycle.

The relation between current, voltage, and power in an ac circuit containing resistance only is shown in Fig. 1-31. The power at any instant in such a circuit is

$$p = ei$$

Since the current is positive when the voltage is positive and negative when the voltage is negative, the product of voltage times current, or power, will be positive at each instant during each cycle, as indicated by the curve marked P in Fig. 1-31. The average power delivered during a cycle is equal to the average of all the instantaneous values of power during the cycle. The average power is the constant and unchanging value of power which would be capable of doing the same amount of work, during one cycle or any given period of time, as was done by the constantly varying ac power shown in Fig. 1-31. The average power, then, is equal to the rate at which an unchanging dc voltage and current, equivalent to the ac voltage and current, could perform work. The dc equivalent value of ac current and voltage is the effective, or rms, value. The average value is equal to the following:

$$P = EI = \text{effective value of voltage} \times \text{effective value of current}$$

The average value of power delivered is equal to one-half the maximum value also, since

$$E = \text{effective value} = \frac{E_M}{\sqrt{2}} \qquad \text{and} \qquad I = \text{effective value} = \frac{I_M}{\sqrt{2}}$$

Then,

$$P = EI = \frac{E_M}{\sqrt{2}} \times \frac{I_M}{\sqrt{2}} = \frac{E_M I_M}{2}$$

The *phase relation* between voltage, current, and power in a circuit containing inductance only is shown in Fig. 1-32. In such a circuit the current lags 90° behind the voltage. During the first 90°, or quarter period, the voltage is shown rising from zero to a positive maximum while the current is changing from a negative maximum value to zero. The power, during this quarter period, rises from zero to a negative maximum and then falls from the negative maximum value back to zero, since power is equal to ei and i is negative while e is positive during this period. During the next quarter cycle the voltage changes from a positive maximum to zero while the current changes from zero to a positive maximum. The power during this quarter cycle is positive, since both e and i are positive. The instantaneous value of power rises from zero to a positive maximum and then decreases to zero again. This sequence of power variation repeats itself each half cycle. In a circuit containing inductance only, the instantaneous power varies as a sine wave but with twice the frequency of the current and voltage waves, as shown in Fig. 1-32. In such a circuit the positive half waves of power are exactly equal to the negative half waves. The average power in a circuit *containing inductance only*, then, is equal to zero. As shown in Fig. 1-31, the positive value of power represents the output of the power source into the circuit. The negative loops of power in Fig. 1-32, then, must represent power which is returned by the circuit to the power source. During one quarter cycle, power is forced out and stored in the magnetic field. During the next quarter cycle, the power stored in the magnetic field of the circuit returns to the power source. Since power is exchanged back and forth between the magnetic field and the power source in equal amounts, the net output into a circuit containing inductance only equals zero.

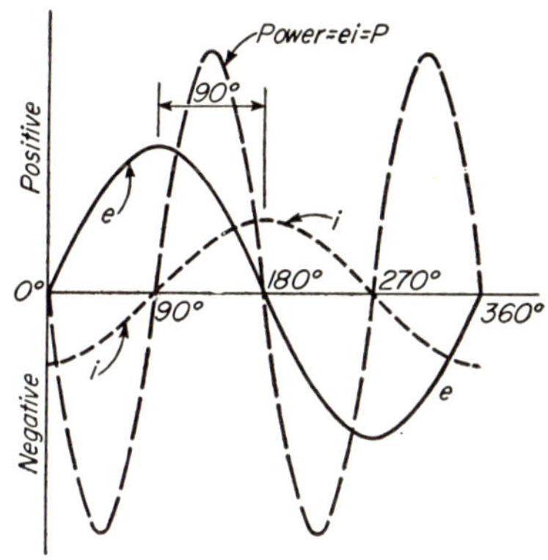

FIG. 1-32. In an ac circuit with inductance only, the amount of power flow out of the source equals the amount returned to the source without any net delivery.

The phase relation between voltage, current, and power in a circuit containing capacitance only is shown in Fig. 1-33. In such a circuit the current leads the voltage by 90 electrical degrees. The power varies as a sine wave of double frequency, its average value is zero and acts exactly opposite to power in a circuit containing inductance only, as can be seen by comparing Figs. 1-32 and 1-33. Current flowing in a capacitative circuit is dependent upon capacitance and the rate of change of voltage. Power is alternately pushed out and stored in the dielectric and then returned to the power source. This alternation of the power wave occurs twice each cycle of the voltage wave.

1-24. Power Factor. Suppose that a circuit has flowing in it a resistive current of 12.25 amp maximum value in phase with the voltage and an inductive current of 7.07 amp maximum value which lags the voltage by 90°. The phase relation between the resistive current, the voltage, and the power wave is similar to that shown in Fig. 1-31. The phase relation between the inductive current, the voltage, and the resulting power wave is similar to that shown in Fig. 1-32. However, since these two currents are in the same circuit, they will add together to form a single resultant current, Fig. 1-34, just as two streams of water can flow together to form a single stream. Since the resistive current is in phase with the voltage while the inductive current is 90° out of phase with the voltage, the two currents will combine to form a resultant current whose *maximum value* is less than the algebraic sum of the two.

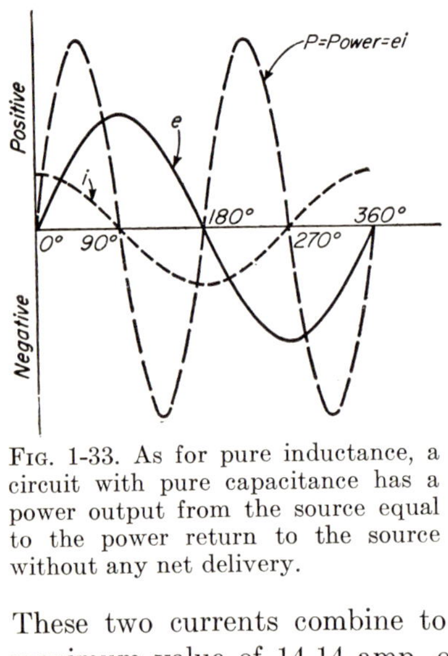

FIG. 1-33. As for pure inductance, a circuit with pure capacitance has a power output from the source equal to the power return to the source without any net delivery.

These two currents combine to form a resultant current which has a maximum value of 14.14 amp, or 10 amp rms value, and which lags the resistive current by 30° while leading the inductive current by 60°.

The phase relation between the resultant current, the circuit voltage, and the power is shown in Fig. 1-35. Here the positive power loops are larger than the negative power loops since they consist of the power delivered to the resistance of the circuit plus the power stored in the magnetic field, while the negative loops represent the power returned to the source by the magnetic field. The average, or equivalent dc, power is less than that represented by the two positive loops each cycle, since the values in the two negative loops must be subtracted. The average power is equal to the rms value of resistive current times the rms value of applied voltage, since the inductance consumes no power. The average power, then, is something less than the product of rms volts times the rms total amperes.

The out-of-phase voltage and current waves of Fig. 1-35 may be thought of as being generated by rotating vectors, as was shown in Art. 1-16. If the vectors of Fig. 1-36 represent maximum values of voltage and current, they will generate the solid-line portions of the sine waves shown, in rotating through 90° to the position shown. Here, the instantaneous value is equal to the projection of the vector upon the y axis—or the maximum value times the sine of the angle rotated through from 0°. If the vectors continue to rotate through the remainder of the cycle, the dotted portion, completing the cycle, will be generated.

Figure 1-34 indicates that any current which lags behind the voltage by less than 90° is the resultant of a current in phase with the voltage

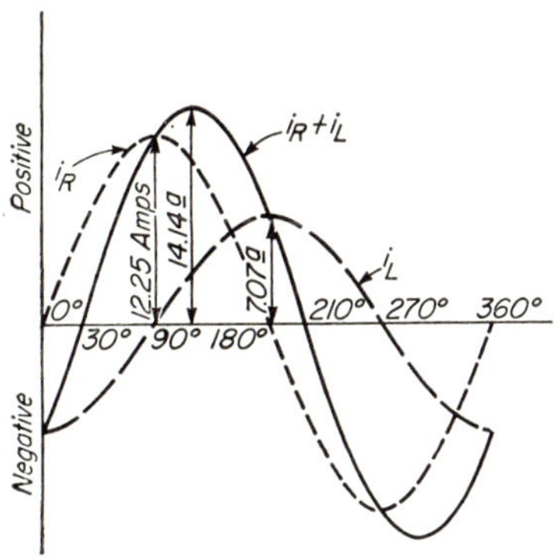

FIG. 1-34. An emf across an ac circuit containing resistance and inductance causes certain currents to flow through each that are 90° out of phase and which combine to form a single resultant current $i_R + i_L$ having a different phase relation from the two component currents.

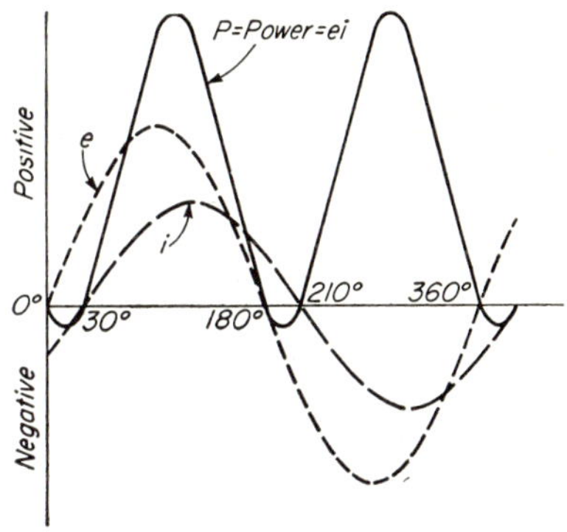

FIG. 1-35. Net power delivery to a circuit with both resistance and inductance depends on the relative magnitudes of each and equals the difference of positive and negative values of power flows.

and a current lagging the voltage by 90°. Similarly, any current leading the voltage by less than 90° is the resultant of a current in phase with the voltage and a current leading the voltage by 90°. The real power in a circuit, then, is equal to the product of the voltage and the portion of the resultant current which is in phase with the voltage. Inspection of Fig. 1-36 will show that the resultant current lags 30° behind the voltage for the case illustrated. The maximum value of the resultant current being 14.14 amp, the portion in phase with the voltage is the projected value of the current vector upon the voltage vector. In trigonometric terms the maximum value of the current in phase with the voltage is equal to the resultant current times the cosine of the time-lag angle between the voltage and current. In the case illustrated,

$$I_{MP} = I_R \cos \theta = 14.14 \cos 30° = 14.14 \times 0.866 = 12.25 \text{ amp}$$

where I_{MP} = maximum value of current in phase with the voltage
I_R = maximum value of resultant current
θ = angular difference between voltage and current

Since the average power in watts equals the rms value of emf in volts times the rms current in phase with the voltage, the average power in an ac circuit is

$$P = EI \cos \theta$$

where P = average power, watts
E = effective, or rms, value of emf, volts
I = effective, or rms, value of current, amp
θ = angular difference between current and voltage

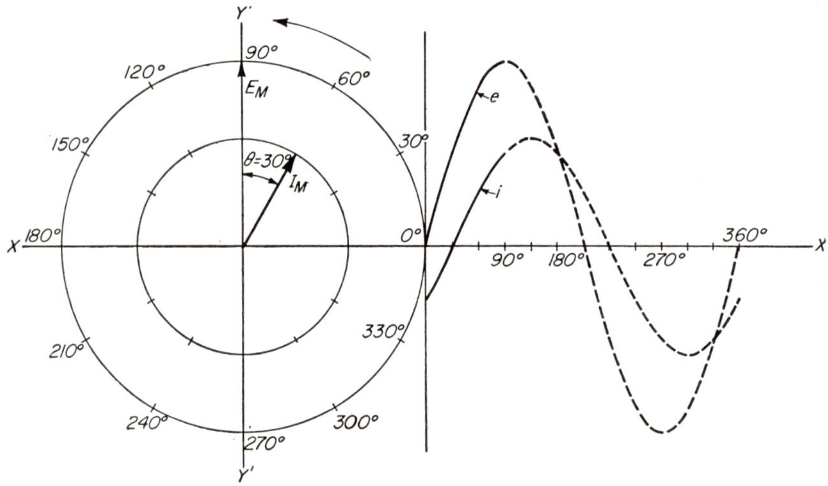

FIG. 1-36. Rotating vectors are convenient forms for calculating ac circuit behavior. Projection on the vertical axis plotted against degrees rotation (or electrical degrees) produces the familiar sine wave.

The above equation represents the mathematical expression for real power in a single-phase ac circuit, since a single voltage and current is involved. In a polyphase system, the above equation would have to be modified to account for the power delivered by the additional phases. The equation for power in a symmetrical three-phase system is given as follows:

$$P = \sqrt{3}\, EI \cos \theta$$

where E = effective, or rms, value of voltage between phase wires
I = effective, or rms, value of current in the phase wires

In the case illustrated, if the emf equals 240 volts, rms value, and the circuit is single phase, the average power equals

$$P = EI \cos \theta = 240 \text{ volts} \times 10 \text{ amp} \times \cos 30^\circ = 2{,}078.4 \text{ watts} = 2.0784 \text{ kw}$$

Rounding off the figures, let us call it 2,080 watts or 2.08 kw. If the circuit were three phase, with the same voltage between phase wires and the same rms value of current, the output of power would be equal to

$$P = \sqrt{3}\, EI \cos\theta = \sqrt{3} \times 240 \text{ volts} \times 10 \text{ amp} \times 0.866 = 3{,}599.8 \text{ watts}$$

or, in round figures, 3,600 watts.

In a circuit in which the current and voltage waves reach a maximum at different time intervals, the product of current and voltage is called the *apparent power*. The ratio of the *real power* to the *apparent power* is called the *power factor* (pf) of the circuit. The *power factor* is also *equal to* the *cosine of the angular difference* between the circuit voltage and the resultant current in the circuit. Since the current in phase with the voltage equals the resultant current times the cosine of the angular difference, the current 90° out of phase with the voltage will equal the product of the resultant current times the sine of the angular difference. The power which performs no useful work, but which periodically shifts back and forth between the generator and the magnetic or dielectric field, is called the *reactive power*. Reactive power is

$$P_X = EI \sin\theta \qquad \text{for single-phase circuits}$$

$$P_X = \sqrt{3}\, EI \sin\theta \qquad \text{for three-phase circuits}$$

1-25. Vector Representation. Quantities which are adequately defined and described by magnitude alone are called *scalar quantities*. The weight of an object, its temperature, its physical size, etc., are scalar quantities. Quantities which vary not only in magnitude but in direction as well are called *vector quantities*. A vector quantity is defined by magnitude and direction.

Alternating currents and voltages are vector quantities because they vary both in magnitude and time phase or direction. The relation between magnitude and time in determining quantities in an ac circuit is clearly shown by the time-phase relation of sine waves. It is difficult and laborious, however, accurately to construct sine waves of voltage and current magnitude and time-phase relation in an ac circuit. Thus, Fig. 1-34 shows that a resistance current of 12.25 amp maximum value combines with an inductive current of 7.07 amp maximum value to form a resultant current of 14.14 amp maximum value, which lags behind the resistance current and voltage by 30°. These are vector quantities since they are defined not only by their magnitude but by their time-phase relation as well. The relation between the resistive current, the inductive current, and the resultant current, showing the relative magnitude and time-phase relation, may be pictorially represented as shown in Fig. 1-37. Here, the length of vector OA represents the *maximum* value of the resistive current, the length of vector OC represents the *maximum* value of the inductive current, and the length of vector OB represents the *maximum* value of the resultant current which will flow in the circuit. The time-phase relation is represented by the direction of the vectors and the angular difference between them. The inductive current OC and the resistive current OA are drawn at an angle of 90° with each other and form the altitude and base of a right triangle where the hypotenuse equals the resultant current. The resultant current is found by solving for the hypotenuse of the right triangle.

$$OB = \sqrt{(OC)^2 + (OA)^2}$$

or,

$$I_{R+L} = \sqrt{I_R^2 + I_L^2} = \sqrt{(12.25)^2 + (7.07)^2} = 14.14 \text{ amp}$$

These are vector quantities since I_R, I_L, and I_{R+L} the resultant not only have magnitude but bear a time-phase relation. Since the rms values

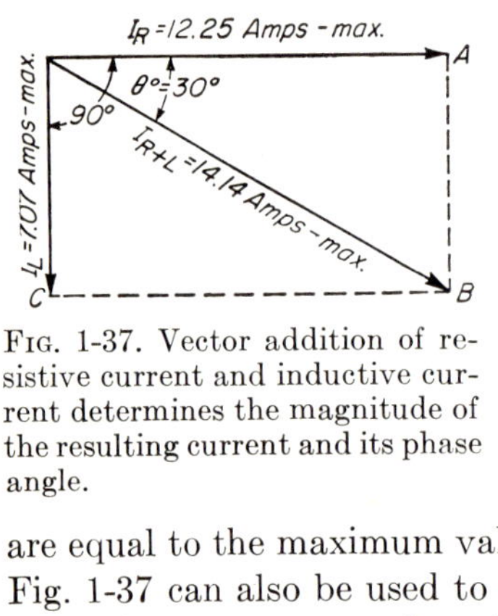

FIG. 1-37. Vector addition of resistive current and inductive current determines the magnitude of the resulting current and its phase angle.

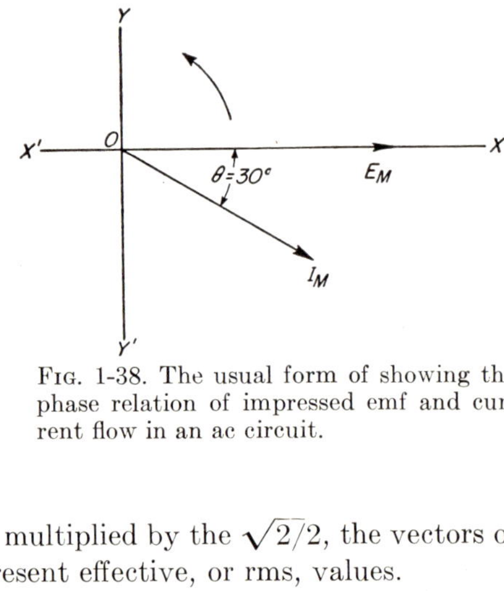

FIG. 1-38. The usual form of showing the phase relation of impressed emf and current flow in an ac circuit.

are equal to the maximum values multiplied by the $\sqrt{2}/2$, the vectors of Fig. 1-37 can also be used to represent effective, or rms, values.

The instantaneous value of current and voltage can be read from sine waves, as shown in Fig. 1-36. The same values can be more easily determined by the use of the rotating-vector method. In Fig. 1-38, E_M represents the maximum value of voltage shown in Fig. 1-36, while I_M represents the maximum value of current. The two vectors are considered to rotate at an angular velocity, in a counterclockwise direction, such that one complete revolution of the vectors will be completed for each sine wave cycle per second. Each revolution of the vectors is equal to 2π radians in angular measure. The total number of radians turned through in 1 sec is equal to 2π times the cycles per second. The precise point in any part of a cycle at which the instantaneous value is required is measured in terms of angular measure. The total angle rotated through is equal to

$$2\pi ft = \omega t \text{ radians}$$

where

$$2\pi f = \omega \text{ radians per sec}$$

The angular rotation measured in radians corresponds to time measured in electrical degrees. Thus, if the instantaneous values of current and voltage shown in Fig. 1-36 are desired after a rotation of 60°, the vectors would be in the position shown in Fig. 1-39. The instantaneous values

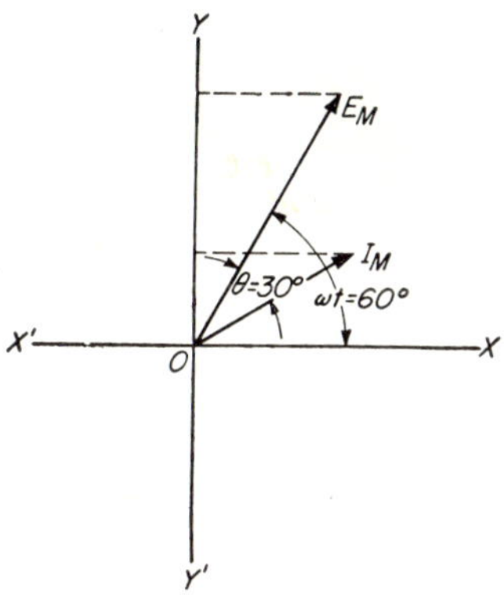

FIG. 1-39. To study relationships at a given instant of one cycle, vectors might take a position such as this.

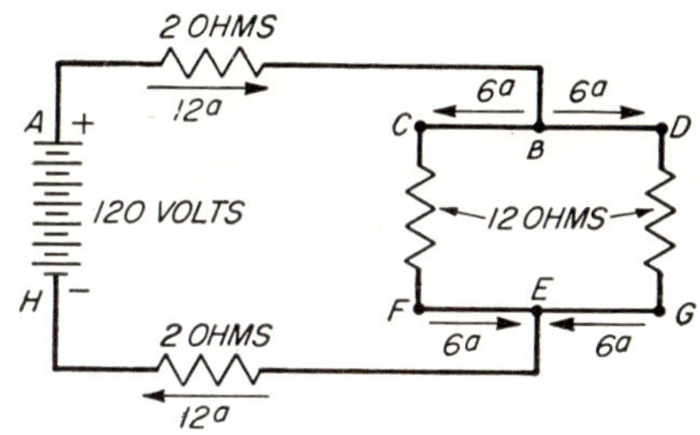

FIG. 1-40. Kirchhoff's laws help to figure the current flowing through different parts of a circuit.

of voltage and current would be equal to the following:

$$e = E_M \sin \omega t = E_M \sin \frac{\pi}{3} = E_M \sin 60°$$

$$i = I_M \sin (\omega t - 30°) = I_M \sin \left(\frac{\pi}{3} - \frac{\pi}{6}\right) = I_M \sin 30°$$

Where the rotating-vector method is used, counterclockwise rotation is considered to be positive. A positive angle is measured from the reference axis, which is considered to be the x axis, in a counterclockwise direction.

1-26. Kirchhoff's Laws. The relation between voltage, current, and resistance is given by Ohm's law. The distribution of currents and voltages in a circuit is expressed by Kirchhoff's laws. Kirchhoff's first law states the fundamental natural relation governing the distribution of current in a complex network: "At any point in a circuit the sum of the currents flowing toward the point is equal to the sum of the currents flowing away from the point." Kirchhoff's second law concerns the distribution of emfs around any closed path in a circuit: "The sum of the IR voltage drops around any one path in a closed circuit is equal to the sum of the emfs impressed on that path."

The meaning of these fundamental laws is illustrated by Fig. 1-40. Kirchhoff's first law is illustrated by considering the currents flowing into and away from points B and E. Solution of the circuit shows that 12 amp flows from the battery toward point B, divides, and flows away from point B so that 6 amp flows from B toward point C and 6 amp flows from B toward point D. The sum of the currents flowing toward point B equals the sum of the currents flowing away from point B, and Kirchhoff's first law is satisfied. There is a drop in voltage of 24 volts across each of the 2-ohm resistors and 72 volts drop from point B to point E.

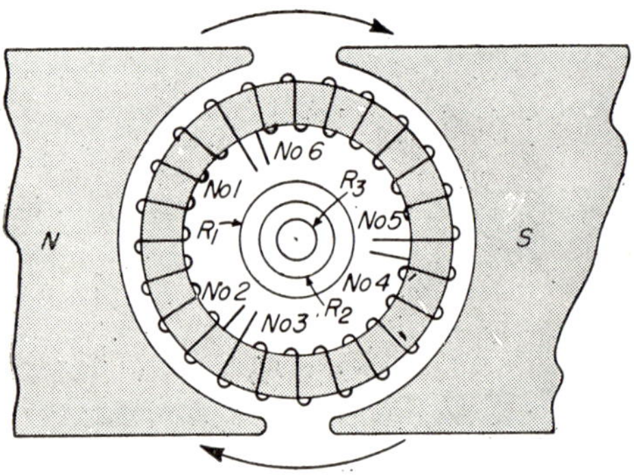

FIG. 1-41. Elementary ac-generator armature with windings to produce three-phase currents and emfs 120° out of phase with each other.

These voltage drops are back emfs, and their sum is equal to 120 volts. The impressed voltage is 120 volts and satisfies Kirchhoff's second law. Since the back emfs are in opposition to the impressed voltage, the sum of the voltages around the circuit in either direction is equal to zero.

Kirchhoff's two laws refer to dc. The sum of currents and voltages refers to the algebraic sum, or the sum with reference to direction. It is usual to refer to currents flowing toward a point as positive and currents flowing away from a point as negative. In ac circuits the phase relations as well as the magnitude of the various currents must be considered. This means that the instantaneous values of current and voltage in an ac circuit, or the vector sum of the vectors representing currents and voltages, satisfy the requirements of Kirchhoff's laws when stated for ac.

1-27. Delta and Wye Circuits. It was shown in Art. 1-15 that if three equal armature windings are symmetrically spaced, as shown in Fig. 1-41, emfs will be induced in each winding which will be out of phase with each other by 120 electrical degrees. The emf induced in the three symmetrically spaced windings may be represented by vectors, as shown in Fig. 1-42. The length of the vectors represents either the maximum or the rms value of induced voltage. The phase difference of 120 electrical degrees is represented by the angular separation of 120° between the vectors. The arbitrarily chosen positive direction of induced voltage in the winding is indicated by the direction of the vector arrowheads. The voltage induced in three such symmetrical windings will always differ in phase by 120 electrical degrees.

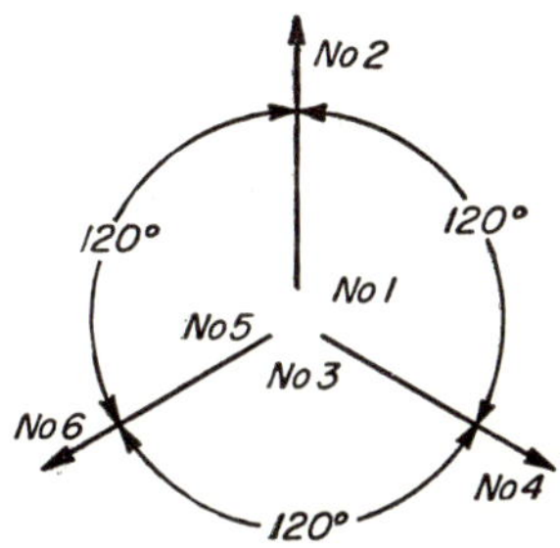

FIG. 1-42. Vectors used to study three-phase currents and emfs of the armature in Fig. 1-41.

If terminals 1 and 6, of Fig. 1-41, are connected to collector ring R_1 and terminals 2 and 5 are connected to collector rings R_2 and R_3, respectively, the vectors representing the voltages induced in these two windings will be added as shown in Fig. 1-43. The voltage between rings R_2 and R_3 is the vector sum of the voltages in windings 1-2 and 5-6, and this vector sum is equal and opposite to the voltage in winding 3-4 as can be seen by examining Figs. 1-42 and 1-43. If terminal 3 of the remaining winding is now connected to collector ring R_2, the voltage measured between terminals 4 and 5, or between terminal 4 and collector ring R_3, will be equal to zero since the voltage in coil 3-4 is equal and

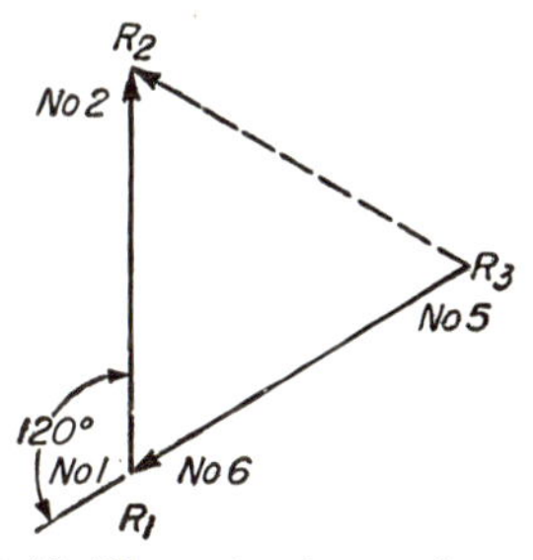

FIG. 1-43. The net voltage of two windings connected in series of a three-phase ac generator.

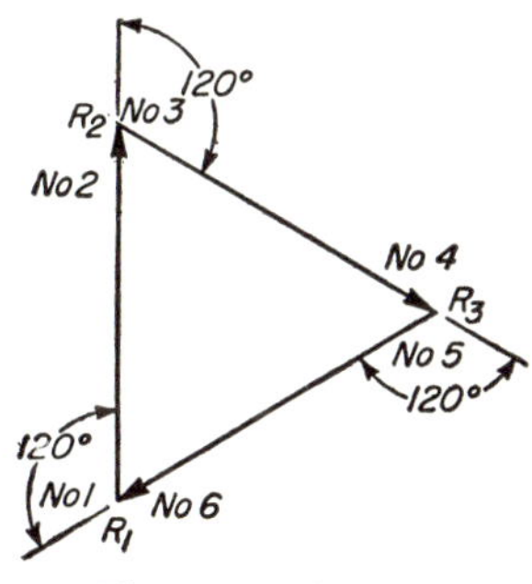

FIG. 1-44. The vector diagram of three-phase windings of an ac generator connected in series are said to be connected in delta.

opposite to the resultant of the voltages in the other two windings. If terminal 4 is now connected to collector ring R_3, all three windings will be connected in series and the vector sum of their induced voltages will be equal to zero, which is in accord with Kirchhoff's second law as applied to ac. The vector diagram of the voltages which would be impressed across a three-phase line tapped from collector rings R_1, R_2, and R_3 is shown in Fig. 1-44. When the three windings are connected in series to form such a closed loop, they are said to be connected in *delta,* or in *closed delta.* The name is derived from the resemblance of the connection diagram, or the vector diagram, to the Greek letter delta. The symbol used to indicate a delta connection is Δ.

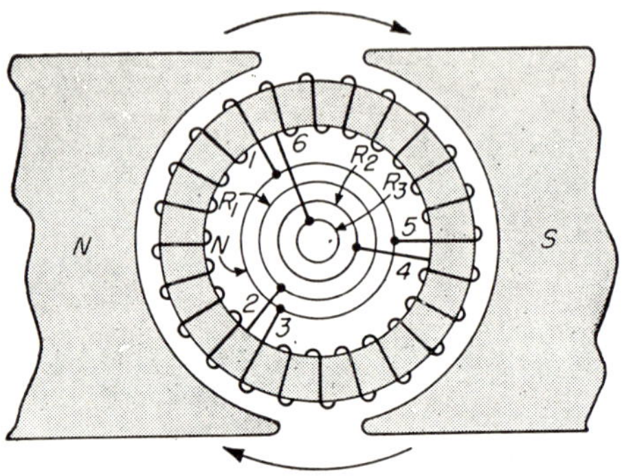

FIG. 1-45. Three-phase ac-generator windings connected to four slip rings (collector rings) as shown form a wye connection. The ring marked N is the neutral, a lead common to all three windings.

If terminals 2, 4, and 6 are connected to collector rings R_1, R_2, and R_3, as shown in Fig. 1-45, while terminals 1, 3, and 5 are connected to a common terminus, the windings are said to be connected in *wye.* The symbol for such a connection is the letter Y. The vector diagram for this connection is shown in Fig. 1-46. Upon examination of Figs. 1-44 and 1-46 it will be noted that the voltage difference between R_1 and R_2 of the delta connection shown in Fig. 1-44 is equal to the voltage induced in coil 1-2 while the voltage difference between R_1 and R_2 of the wye connection shown in Fig. 1-46 is composed of the voltages in coils 1-2 and 3-4 and is considerably greater than the voltage induced in either. When vectors are added, the tail of the vector added must join the head of the vector to which it is added. When the tails of two vectors, or the heads of two vectors, are joined, the opposite of addition is indicated. In other words, one of the vectors is subtracted from the other. The vector diagram of Fig. 1-44, then, shows the vector sum of three voltages, and the voltage difference between R_1 and R_2 is equal and opposite to the vector sum of the voltages in windings 3-4 and 5-6, as was noted above. On the other hand, the voltage difference between R_1 and R_2

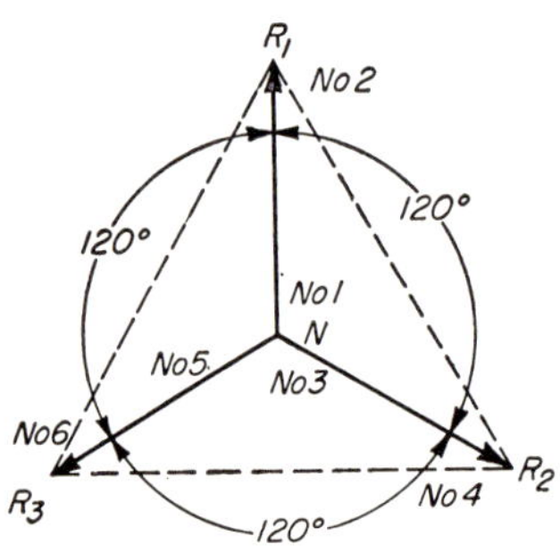

FIG. 1-46. Vector diagram of the wye-connected ac generator of Fig. 1-45.

of the wye-connected windings of Fig. 1-46 is equal to the vector difference of the voltages in windings 1-2 and 3-4.

When one terminal of each winding is connected to a common terminus, like terminals 1, 3, and 5 of Fig. 1-45, to form a wye connection, the point where all three windings join is called the *neutral point* of the wye. The vector relations between phase voltages, measured between a phase wire and neutral, and phase-to-phase voltages, measured between

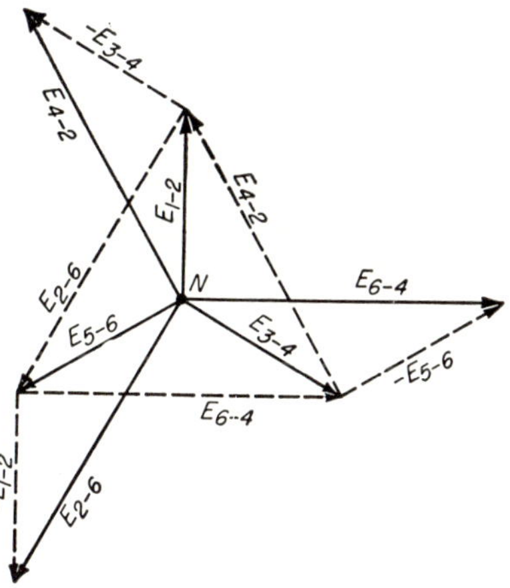

FIG. 1-47. Vector relations between phase voltages measured between phase wire and neutral, and phase-to-phase voltages measured between phase wires of a three-phase ac generator.

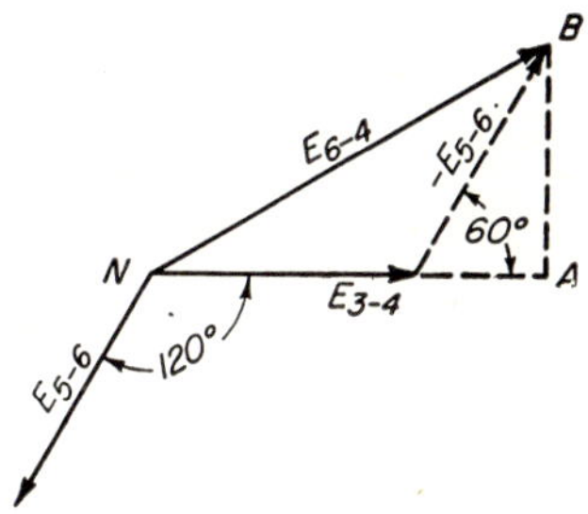

FIG. 1-48. How voltages E_{34} and E_{56} are combined to find E_{64}.

phase wires, is shown in Fig. 1-47. For the assumed positive direction of voltages shown, the line-to-line voltages are equal to

$$E_{64} = E_{34} - E_{56}$$
$$E_{42} = E_{12} - E_{34}$$
$$E_{26} = E_{56} - E_{12}$$

The quantities shown are vector quantities and the numbered subscripts refer to the terminal designations of Figs. 1-45 and 1-46. In Fig. 1-47 the chosen positive direction of phase-to-phase emfs is counterclockwise around the triangle representing the resultant phase-to-phase voltages. If this chosen direction of voltages had been reversed, the vector differences in the equations above would also have been reversed:

$$E_{46} = E_{56} - E_{34}$$

When two vectors are of equal value, such as E_{34} and E_{56} of Fig. 1-47, and one vector is subtracted from the other, the ratio of the resultant vector value to either of the equal vectors can be determined by simple trigonometry. Thus, in Fig. 1-48, the resultant value of E_{64} can be determined by the solution of a right triangle. The base of the equivalent triangle is equal to

$$NA = E_{34} + E_{56} \cos 60°$$

while the altitude of the equivalent triangle is equal to

$$AB = E_{56} \sin 60°$$

Since E_{64} is equal to the hypotenuse of the equivalent triangle of Fig. 1-48, and since E_{34} is equal to E_{56}, the value of E_{64} can be determined by the following equation:

$$\begin{aligned}(E_{64})^2 &= (E_{34} + E_{34} \cos 60°)^2 + (E_{34} \sin 60°)^2 \\ &= \left(E_{34} + \frac{1}{2} E_{34}\right)^2 + \left(\frac{\sqrt{3}}{2} E_{34}\right)^2 = 3(E_{34})^2 \\ E_{64} &= \sqrt{3}\, E_{34}\end{aligned}$$

From this it is seen that in a symmetrical three-phase wye system the phase-to-phase voltage equals $\sqrt{3}$ times the phase-to-neutral voltage.

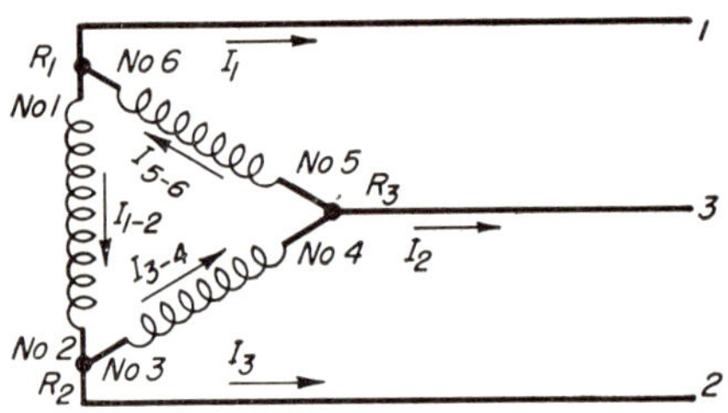

FIG. 1-49. Current flows for three-phase windings connected in delta.

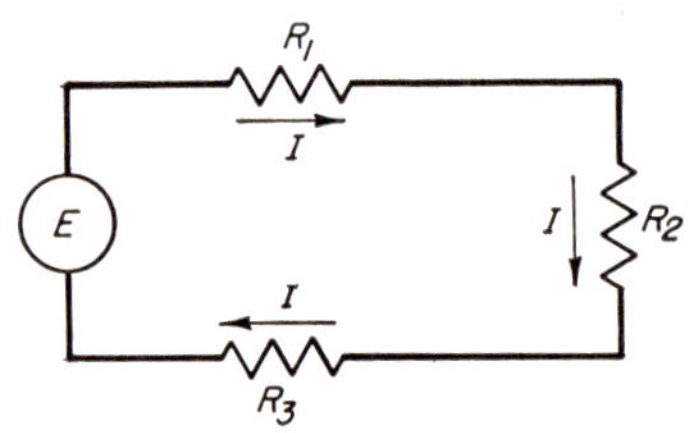

FIG. 1-50. The same current flows through all components of a simple series circuit.

The manner of connecting ac generator windings has been used to illustrate the difference between delta and wye connections, but other things, such as electrical loads, transformers, etc., may also be connected in delta or wye if all three phases of a three-phase electric system are involved. The relation between the current flowing in a line and the current flowing between phases of a delta-connected generator, or the phase winding of a three-phase delta-connected transformer bank or a delta-connected load is similar to the relation between the phase-to-phase and phase-to-neutral voltage of a wye-connected system. In Fig. 1-49, suppose that the diagram shown represents the delta-connected windings of Fig. 1-41 and that the arrows represent the assumed direction of current flow. According to Kirchhoff's first law, the line currents are equal to the following:

$$\begin{aligned}I_1 &= I_{56} - I_{12} \\ I_2 &= I_{12} - I_{34} \\ I_3 &= I_{34} - I_{56}\end{aligned}$$

The phase-to-phase voltage in a wye system is equal to the $\sqrt{3}$ times the phase-to-neutral voltage, while in a delta system the line current is equal to the $\sqrt{3}$ times the current in a phase-to-phase winding.

1-28. Series and Parallel Circuits. Figure 1-50 shows three resistances R_1, R_2, and R_3 connected so that the same current flows through all three. Such a circuit is called a *series circuit.* In a series circuit the current is the common factor and the voltage, forcing the current through the circuit, is equal to the sum of the voltage drops in the different parts of the circuit. Thus, in Fig. 1-50:

$$E = IR_1 + IR_2 + IR_3 = I(R_1 + R_2 + R_3) = IR$$

where $R = R_1 + R_2 + R_3 =$ equivalent resistance

In drawing the vector diagram of a series circuit, the current is used as the reference vector because it is the only element common to the whole circuit.

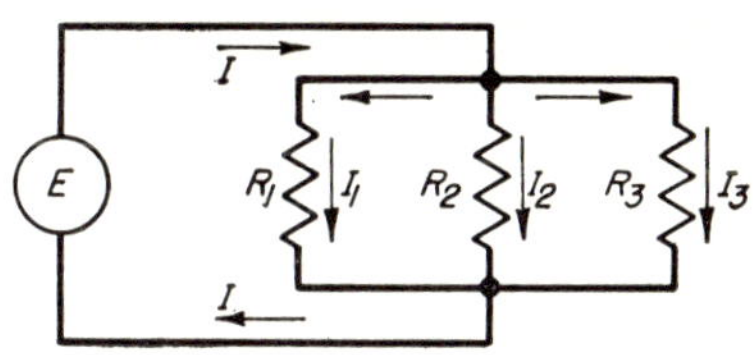

FIG. 1-51. Resistances connected in parallel (or multiple) all have the same impressed voltage.

If the three resistances of Fig. 1-50 are connected as shown in Fig. 1-51, they are in *parallel.* In such a circuit the voltage is the common factor. In the series circuit it was noted that the total voltage equals the sum of the component voltage drops. In a parallel circuit the voltage is common to all the parallel branches, but the total current equals the sum of the currents in the parallel branches. Thus, in Fig. 1-51,

$$I = I_1 + I_2 + I_3$$

and

$$\frac{I}{E} = \frac{I_1}{E} + \frac{I_2}{E} + \frac{I_3}{E}$$

from which

$$\frac{I}{E} = \frac{1}{R} = \frac{1}{R_1} + \frac{1}{R_2} + \frac{1}{R_3}$$

$$R = \frac{1}{\dfrac{1}{R_1} + \dfrac{1}{R_2} + \dfrac{1}{R_3}}$$

equivalent value of resistance.

For any number of resistances connected in parallel, the equivalent single resistance is less than the smallest parallel resistance. The equivalent resistance, for any group of n resistors, is equal to

$$R = \frac{1}{\dfrac{1}{R_1} + \dfrac{1}{R_2} + \dfrac{1}{R_3} + \cdots + \dfrac{1}{R_n}}$$

If the circuit is ac and resistance, inductance, and capacitance are present in the circuit, the solution of the circuit as to current flow and voltage drops is considerably more difficult. The solution of series and parallel circuits, involving ac circuit constants, is discussed in Art. 1-29. In drawing the vector diagram of voltages and currents in a parallel circuit, the voltage is usually taken as the reference vector because it is the common factor.

1-29. AC Circuit Constants—Impedance and Admittance. Resistance, inductive reactance, and capacitative reactance all affect the magnitude of current flowing in an ac circuit. Where all three are present in a circuit, their cumulative effect will be similar to a single factor which bears the same relation to voltage and current in an ac circuit as resistance alone in a dc circuit. It was shown in Art. 1-2 that resistance, as stated by Ohm's law, is equal to voltage divided by current in a dc circuit. The ratio of voltage to current in an ac circuit is called the *impedance* of the circuit and expresses a relation similar to Ohm's law for ac:

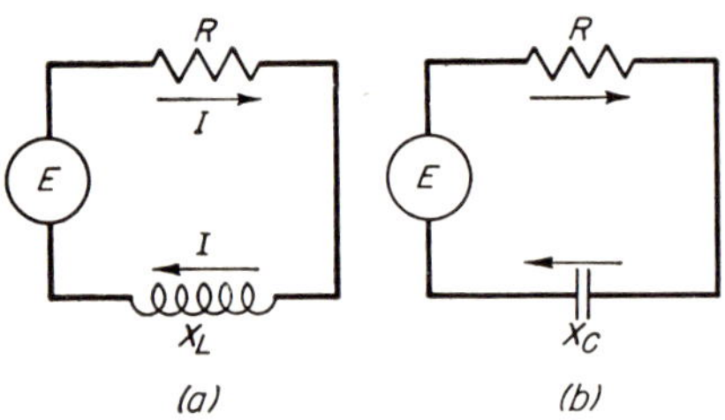

FIG. 1-52*a*. Series circuit with resistance and inductance. *b*. Series circuit with resistance and capacitance.

$$E = IZ \qquad \text{or} \qquad Z = \frac{E}{I}$$

where Z = impedance, ohms

Voltage across an inductive reactance is 180° out of phase with a voltage drop across a capacitative reactance, while both are 90° out of phase with the voltage drop across a resistance. Since impedance equals the joint effect of resistance, inductive reactance, and capacitative reactance present in an ac circuit, it may be expressed by

$$Z = \sqrt{R^2 + (X_L - X_C)^2} = \sqrt{R^2 + X^2} \text{ ohms}$$

where Z = impedance, ohms
X_L = inductive reactance in ohms, rotated 90° ahead of R
X_C = capacitative reactance in ohms, rotated 90° behind R

The impedance of a circuit multiplied by the current flowing in the circuit equals the voltage which causes the current to flow. If the voltage is multiplied by the reciprocal of the impedance, the product is equal to the current. The reciprocal of the impedance is called the *admittance* of the circuit. The relation between the current, voltage, and admittance of the circuit is

$$I = YE$$

where $Y = \dfrac{1}{Z}$ = reciprocal of the impedance = admittance, mhos

Admittance is composed of one component in phase with the voltage and a second 90° out of phase. The in-phase factor R/Z^2 is called the *conductance* of the circuit and is usually designated by the letter G, while the out-of-phase factor X/Z^2 is called the *susceptance* of the circuit and is usually designated by the letter B, just as resistance is the in-phase component and reactance the quadrature component of impedance. The product IR represents the component of circuit voltage, in phase with the current, necessary to overcome the opposition of resistance, while the product IX represents the component of circuit voltage, in quadrature with the current, necessary to overcome the opposition of inductance or capacitance in the circuit. Similarly the product GE represents the component of current, in phase with the voltage, due to resistance in the circuit, while BE represents the component of current, in quadrature with the voltage, due to inductance or capacitance in the circuit.

The equations for impedance specify a positive value for reactance when inductance is predominant in the circuit and a negative value when the reactance of the circuit is capacitative. The reasons for this sign convention, for reactance, is obvious when it is remembered that inductive current lags the voltage, capacitative current leads the voltage, and the current vector is the reference vector with respect to the component voltage vectors. The sign of X must be positive for inductive reactance, in order for the current to lag the voltage, and negative in the case of capacitative reactance, in order for the current to lead the voltage. Since the susceptance of a circuit is equal to $X/(R^2 + X^2)$, the sign of B is positive in the case of an inductive circuit and negative for a capacitative circuit.

The voltage vector is the reference vector with reference to the in-phase component of current GE and the quadrature component of current BE. The unit of measurement of admittance, conductance, and susceptance is the *mho*. Thus, a circuit which has an impedance of 5 ohms has an admittance of $\frac{1}{5} = 0.20$ mho. The related expressions and equations for the various circuit constants are given as follows:

$$R = \frac{G}{G^2 + (B_L - B_C)^2} = \frac{G}{Y^2} = GZ^2 = \sqrt{Z^2 - (X_L - X_C)^2}$$

= ohms resistance

$$X = \frac{(B_L - B_C)}{G^2 + (B_L - B_C)^2} = \frac{B_L - B_C}{Y^2} = (B_L - B_C)Z^2 = \sqrt{Z^2 - R^2}$$

= ohms reactance

$$G = \frac{R}{R^2 + (X_L - X_C)^2} = \frac{R}{Z^2} = RY^2 = \sqrt{Y^2 - (B_L - B_C)^2}$$

= mhos conductance

$$B = \frac{X_L - X_C}{R^2 + (X_L - X_C)^2} = \frac{X_L - X_C}{Z^2} = Y^2(X_L - X_C) = \sqrt{Y^2 - G^2}$$

= mhos susceptance

$$Z = \sqrt{R^2 + (X_L - X_C)^2} = \frac{1}{Y} = \frac{1}{\sqrt{G^2 + (B_L - B_C)^2}} = \sqrt{\frac{R}{G}}$$
$$= \text{ohms impedance}$$

$$Y = \sqrt{G^2 + (B_L - B_C)^2} = \frac{1}{Z} = \frac{1}{\sqrt{R^2 + (X_L - X_C)^2}} = \sqrt{\frac{G}{R}}$$
$$= \text{mhos admittance}$$

These expressions are useful in solving complex electrical circuits. A complex circuit may consist of both series and parallel branches. In a series circuit ohmic values of resistances and of reactances may be each added numerically to form single values of resistance and reactance which are equivalent to actual total circuit values. In a parallel circuit neither the resistances nor reactances of the paralleled circuits can be added numerically, for the reason that the product of impedance by current, in any parallel branch, is equal to the total circuit voltage. The value of conductance, in mhos, in the paralleled branches, however, may be added numerically, which together with the sum of the branch susceptances in quadrature forms the equivalent value of admittance. The product of the admittance of any parallel branch and the circuit voltage is equal to the current through the branch, while the product of the equivalent circuit admittance and the circuit voltage is equal to the total current in the circuit. The use of the circuit constants are illustrated in the following example.

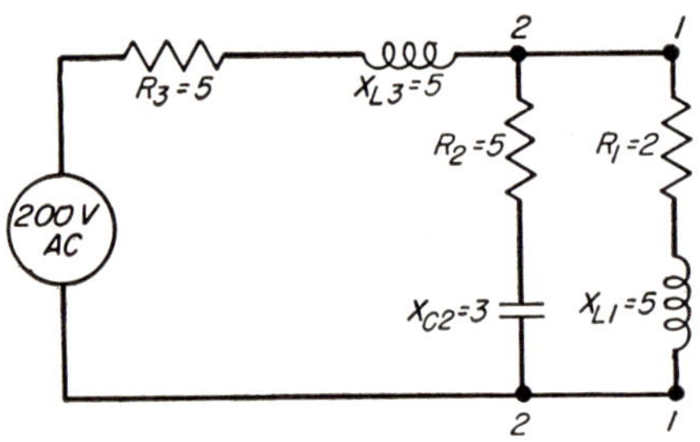

FIG. 1-53. Series-parallel circuit with resistance, inductance, and capacitance.

Example: A series-parallel circuit, shown in Fig. 1-53, has an ac voltage of 200 volts impressed across it. What is the total current flowing in the circuit? What is the power factor? What is the voltage drop across impedance Z_3 and across the parallel branches 1-1 and 2-2? What is the current flowing in 1-1 and 2-2?

Solution:

$$G_1 = \frac{2}{4 + 25} = 0.069 \text{ mho} \qquad B_{1L} = \frac{5}{4 + 25} = 0.172 \text{ mho}$$

$$Y_1 = \sqrt{\frac{G_1}{R_1}} = \sqrt{\frac{0.069}{2}} = 0.1857 \text{ mho}$$

$$G_2 = \frac{5}{25 + 9} = 0.147 \text{ mho} \qquad B_{2C} = \frac{-3}{25 + 9} = -0.088 \text{ mho}$$

$$Y_2 = \sqrt{\frac{G_2}{R_2}} = \sqrt{\frac{0.147}{5}} = 0.1714 \text{ mho}$$

The equivalent values which are equal to the combined constants of circuit branches 1-1 and 2-2 are equal to the following:

$$G_{12} = G_1 + G_2 = 0.216 \text{ mho} \qquad B_{12} = B_{1L} + (-B_{2C}) = 0.084 \text{ mho}$$
$$Y_{12} = \sqrt{G_{12}^2 + (B_{1L} - B_{2C})^2} = \sqrt{0.216^2 + 0.084^2} = 0.232 \text{ mho}$$
$$R_{12} = \frac{G_{12}}{Y_{12}^2} = \frac{0.216}{(0.232)^2} = 4.02 \text{ ohms} \qquad X_{12} = \frac{B_{12}}{Y_{12}^2} = \frac{0.084}{(0.232)^2} = 1.56 \text{ ohms}$$

The equivalent value of resistance, reactance, and impedance in the whole circuit is

$$R = R_3 + R_{12} = 5.0 + 4.02 = 9.02 \text{ ohms}$$
$$X = X_3 + X_{12} = 5.0 + 1.56 = 6.56 \text{ ohms}$$
$$Z = \sqrt{R^2 + X^2} = \sqrt{(9.02)^2 + (6.56)^2} = 11.1 \text{ ohms}$$

The total current flowing in the circuit is

$$I = \frac{E}{Z} = \frac{200 \text{ volts}}{11.1 \text{ ohms}} = 18 \text{ amp}$$

The power factor of the circuit is

$$\text{pf} = \frac{R}{Z} = \frac{9.02}{11.1} = 0.811$$

The voltage drop across the series impedance and across the parallel impedances and the current flowing through the parallel impedances are

$$E_3 = I\sqrt{R_3^2 + X_3^2} = 18\sqrt{5^2 + 5^2} = 127.28 \text{ volts}$$
$$E_{12} = I_{12}\sqrt{R_{12}^2 + X_{12}^2} = 18\sqrt{4.02^2 + 1.56^2} = 77.62 \text{ volts}$$
$$I_1 = Y_1E_1 = 0.1857 \times 77.62 = 14.4 \text{ amp}$$
$$I_2 = Y_2E_2 = 0.1714 \times 77.62 = 13.3 \text{ amp}$$

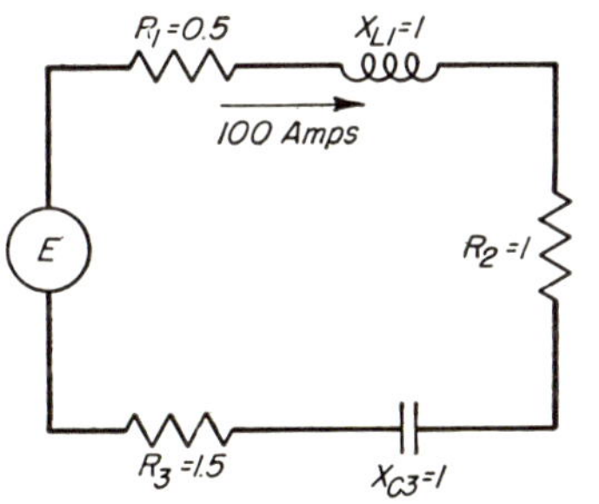

FIG. 1-54. Series circuit with resistance, capacitance, and inductance (Problem 1-23).

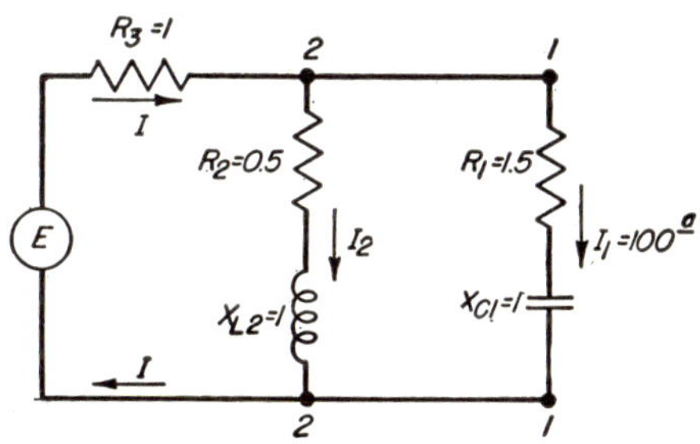

FIG. 1-55. Series-parallel circuit with resistance, inductance, and capacitance (Problem 1-24).

1-30. The Transformer. In Art. 1-18 it was shown that, when an ac flows in a circuit, its alternating flux will induce an emf in an adjacent conductor across which the flux sweeps. If two coils of wire are wound about a core of magnetic material, Fig. 1-56, and an alternating emf E is applied across the terminals PP' of one coil, causing ac to flow, an alternating flux passes through the magnetic circuit, indicated by the dotted line, linking with both coils PP' and SS'. These coils must be insulated electrically from each other and the steel core. The cyclically varying flux, originating in coil PP' and cutting across the turns of coil SS' each time the flux density changes, induces an emf in coil SS' of the same frequency as the impressed emf E, and proportional to the rate at which the flux linkages with coil SS' change. If a resistance, or other electrical load, such as R of Fig. 1-56, is connected across coil SS', forming a closed circuit, current will flow through the circuit. Coils wound, in such a manner, about a common core of highly magnetic material are the principal elements of a *transformer.*

1-31. Primary and Secondary Coils. The coil of the transformer connected to the source of supply is the *primary,* and the coil supplying power to the load circuit is the *secondary.* Thus, coil PP', Fig. 1-56, is the primary winding and coil SS' the secondary winding. If E'_S, the voltage induced in the secondary winding, is greater than E_P, the voltage impressed across the primary winding, the transformer is a *step-up transformer.* If E'_S is less than E_P, the transformer is a *step-down transformer.* If E'_S equals E_P, the transformer is an *insulating transformer* since it does not change the voltage but only insulates the secondary from the primary.

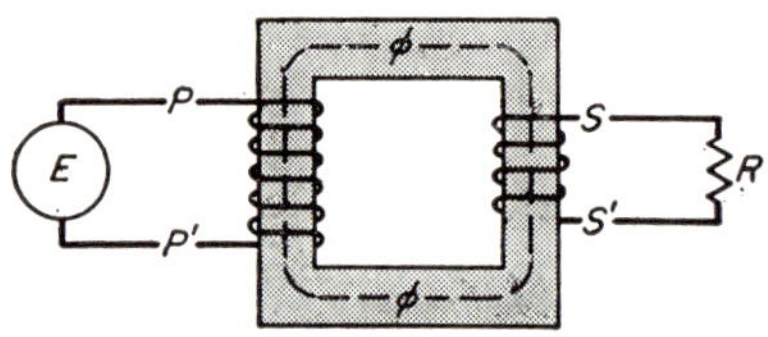

FIG. 1-56. Elementary transformer with electrical and magnetic circuits.

In a large electric system many different voltages may be used, ranging from, perhaps, 230,000 volts or more on the high-voltage transmission system to 120 volts for the ordinary residential consumer's service. The current that a wire can carry depends upon the wire size. The power transmitted through a circuit is proportional to the product of volts times amperes. To transmit thousands of kilowatts over a line, without conductors of excessive size, a high transmitting voltage must be used. It would be unwise to construct distributing lines along congested city streets and provide electric service to homes, stores, factories, etc., at the high voltages used for transmission. The transformer provides the link between all these voltages. The only link between the two windings is the mutual magnetic field through which electric power transfers from the primary to the secondary circuit.

In Arts. 1-18 and 1-19 it was shown that, when a sine-wave ac flux links with a single coil of wire, the *average* value of emf induced in rising from zero to a maximum or in falling from a maximum to zero is

$$e_{av} = \frac{4f\phi_m}{10^8} \qquad \text{volts}$$

If there are N turns of wire in the coil and the same flux links with each turn, the average value of the total emf induced in the coil is

$$E_{\text{av}} = Ne_{\text{av}} = \frac{4fN\phi_m}{10^8} \qquad \text{volts}$$

The ratio of the rms or effective value of a sine-wave voltage or current to the average value is called the *form factor* and equals 1.11. The total rms value of voltage induced in the primary and secondary coils of a transformer is

$$E'_P = \frac{4.44fN_P\phi_m}{10^8} = \frac{4.44fN_PB_mA}{10^8} \qquad \text{volts}$$

$$E'_S = \frac{4.44fN_S\phi_m}{10^8} = \frac{4.44fN_SB_mA}{10^8} \qquad \text{volts}$$

where E'_P = rms value of back emf induced in the primary coil
E'_S = rms value of emf induced in the secondary coil
f = sine-wave frequency, cycles per sec
ϕ_m = maximum value of flux lines linking with primary and secondary
= B_mA
B_m = maximum flux density in core, lines per sq cm
A = cross-sectional area of core, in square centimeters, at right angles to the direction of the flux
N_P = number of turns in the primary winding
N_S = number of turns in the secondary winding

The ratio of the induced voltages is equal to the ratio of the number of turns:

$$\frac{E'_P}{E'_S} = \frac{N_P}{N_S} \qquad E'_P = E'_S\frac{N_P}{N_S} \qquad E'_S = E'_P\frac{N_S}{N_P}$$

The induced voltage E'_P is so nearly equal to the impressed voltage E_P that, for practical use, the ratio of the impressed primary voltage to the open-circuit secondary voltage may be considered equal to the ratio of N_P to N_S.

1-32. Exciting Current. If the secondary coil is left open, so that no current flows in coil SS' of Fig. 1-56, the primary coil PP' will act as an inductance coil similar to that of Fig. 1-9. The coil will contain a high inductance and a low resistance. In this case, the voltage E_P, impressed across the primary winding, will cause a small current to flow through coil PP' of such a magnitude that the *vector sum* of the back emf induced in coil PP' plus the resistance drop in the coil just equals the impressed voltage. This no-load or *exciting current* flows in the primary circuit at a low power factor and is quite small because of the large inductance owing to the iron in the magnetic circuit. If inductance only were present in the circuit and no power were consumed in the primary when the secondary is open-circuited, the exciting current in the primary winding would lag 90° behind the impressed voltage and the induced voltages E'_P and E'_S would lag 180° behind the impressed voltage. This, however, is not quite the case. The primary, in addition to having a small resistance, must deliver a small quantity of power which is consumed and released as heat energy in the iron core of the magnetic circuit. This no-load energy loss, or iron loss, of the transformer is made up of two parts: the *hysteresis loss* and the *eddy-current loss*.

1-33. Hysteresis Loss. When an ac exciting voltage is impressed across the primary winding, a flux establishes in the core, alternating at the same frequency as the exciting voltage and lagging 90° behind it. The core alternately magnetizes in one direction and then the other.

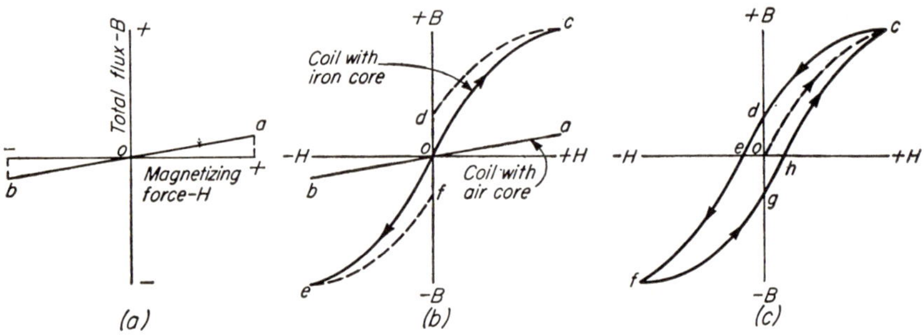

FIG. 1-57. The hysteresis effect in the magnetic circuit of a transformer arises from the residual magnetism left in the magnetic core after the electromagnetic field reduces to zero.

Figure 1-57*a* shows how a coil with an air core establishes a flux. As the magnetizing force *H* increases from 0 to *a*, the flux increases proportionately. As the current in the coil drops back to zero, the magnetizing force does likewise and so does the magnetic flux. At zero current flow there is zero flux. The current then reverses, the flux grows proportionately in an opposite direction as shown by the straight line 0 to *b*, then diminishes again to zero with current decrease. The flux is directly proportional to the magnetizing current and the number of turns.

When the coil has an iron core, the flux multiplies many times because of the iron permeability, Fig. 1-57*b*. For the same current flow the flux rises to *c* while it would only be at *a* with an air core. The curve 0*c* shows the flux rise when the iron *initially* has no magnetism. When the magnetizing force reduces to zero by current flow dropping to zero, the magnetic flux reduces according to the dotted curve *cd*. At zero magnetizing force the iron retains residual magnetism of magnitude *d*. The curves 0*e* and *ef* show what happens when this series of events takes place in the opposite direction.

This residual magnetism must be removed on each current reversal in a transformer, causing what's known as hysteresis loss. Figure 1-57*c* shows how the flux varies with one cycle of ac flow. At *c*, primary magnetizing current has reached its maximum and then reduces to zero, the flux falling from *c* to *d*. The magnetizing force *H* reverses to *e* before the flux falls to zero. The flux then rises in an opposite direction to *f* when the current flow and magnetizing force is a maximum in the opposite direction. As current flow falls to zero, flux magnitude drops to *g*, representing the residual magnetism in the opposite direction. Then as current rises again in the original direction, the flux first reduces to zero at *h* and reaches its peak at *c*. This cycle of events keeps repeating as long as current flows.

The loop of Fig. 1-57*c* is known as the hysteresis loop. The greater the area in the loop, the greater the hysteresis loss. If the loop should reduce to the curved line *c*0*e* in Fig. 1-57*b*, the hysteresis loss would be zero. The supply circuit must supply the hysteresis loss.

The hysteresis loss is proportional to the quantity of iron in the magnetic circuit, the frequency of the magnetizing supply current, and the flux density. Hysteresis causes a distortion of the magnetizing current from a sine-wave form, even though the exciting voltage, the induced voltage, and the flux are sine wave in form. This distortion is such that, if an *equivalent sine wave* is substituted for the actual distorted wave of magnetizing current, it will be shifted ahead so that the equivalent sine wave of magnetizing current leads the sine wave of flux. In this case, the equivalent sine wave of magnetizing current is thought of as being composed of two components: I_M in phase with and causing the flux ϕ and I_h supplying the hysteresis loss and leading the flux ϕ by 90°. Under this assumption the component I_M causes all of flux ϕ and supplies none of the hysteresis loss while I_h supplies all the hysteresis loss and causes no flux.

1-34. Eddy Currents. When a harmonically varying (sine-wave form) magnetic field of high intensity is set up in a steel core, as by coil *PP'* of Fig. 1-56, a harmonically varying emf will be induced in any conductor across which the field sweeps. Thus, an emf will be induced in the core itself since iron or steel is a conducting material. These induced emfs cause small currents to flow along conducting paths within the core. Such currents are *eddy currents*. The conducting paths these eddy currents seek out through the steel core have electrical resistance and consequently cause heat energy to be released within the steel core in proportion to the product of core resistance and the square of the eddy current. Since the resistance of any conducting path is inversely proportional to its cross-sectional area, the resistance to the eddy current is much higher in a thin sheet than in a solid block of material. So that eddy currents may be held to a minimum, the core is made up of thin sheets, or laminations. These thin sheets of core steel are placed parallel to the direction of the flux and insulated from one another so that eddy currents cannot easily pass from one sheet to another. This reduces the cross-sectional area and increases the length of the conducting paths over which eddy currents flow.

Opposition to eddy currents, in the conducting paths through the steel core, may be thought of as being resistance only. The induced voltage within the core and, consequently, the eddy currents lag 90° behind the flux. The eddy currents, themselves, produce a flux within the core which modifies and reduces the flux ϕ. This effective reduction of flux ϕ causes a reduction of E'_P, the back emf induced in the primary coil. The resultant excess of impressed voltage E_P over the balancing back emfs causes an additional current to flow in the primary winding. This current will be of such a value and at such a phase angle that it will produce a flux exactly equal and opposite to the flux produced in the core by the eddy currents. This current flowing in the primary circuit will lead the flux ϕ by 90° and will be the component of no-load current supplying the eddy-current energy loss. Eddy-current energy loss is proportional to the quantity of iron, to the square of the thickness of the laminations, frequency, and flux density.

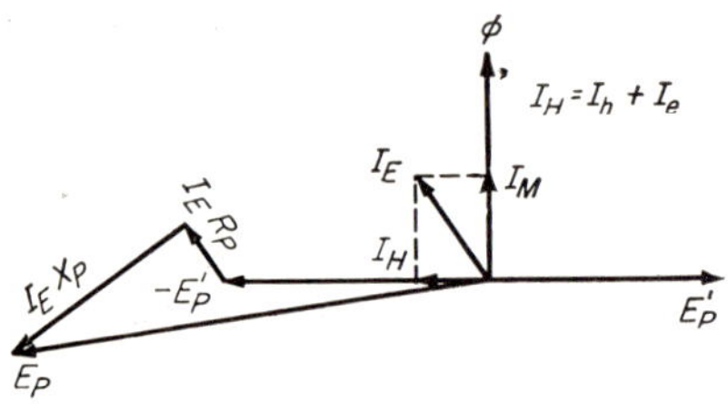

FIG. 1-58. The magnetizing current I_M in a transformer lags behind the impressed primary voltage E_P by more than 90°. The exciting current I_E is the resultant of the magnetizing current I_M, the eddy current I_e, and the hysteresis effect I_h.

1-35. Current Components. Figure 1-58 shows the total current flowing in the primary circuit of the transformer at no load is composed of I_e the eddy-current component, I_h the hysteresis component, and I_M the magnetizing component. I_M is in phase with the flux and lags 90° behind the component of impressed voltage necessary to overcome E'_P the induced voltage in the primary. I_h and I_e are in phase with the component of impressed voltage used in overcoming the primary induced voltage E'_P and, consequently, lead the flux ϕ by 90°. Since I_h and I_e are in phase with each other, they are generally combined into a single component I_H sometimes called the *magnetic power component* or *hysteresis component*, even though it represents eddy current as well as hysteresis current. The two components I_H and I_M, which are at right angles to each other, combine to form a resultant current I_E, which is called the *exciting current* and equals the total no-load current of the transformer.

The exciting current is quite small, generally not exceeding 4 or 5 percent of the rated full-load current of the transformer.

When the secondary circuit, Fig. 1-56, is closed, the induced emf, E'_S in the secondary forces a current through load R and coil SS'. The secondary load current I_S causes a modifying magnetic flux, in phase with I_S, to be established in the core. This has an effect upon the total flux similar to the effect of eddy currents. The flux caused by I_S tends to reduce flux ϕ so that the induced primary voltage E'_P is lowered. The impressed voltage E_P is now greater than the opposing induced primary voltage E'_P plus the voltage drop $I_E R'_P$, and this resultant voltage difference forces additional current through the primary winding PP'. This additional current is the primary load current I'_P. Since E'_S is in direct opposition to the component of E_P necessary to overcome E'_P, the phase relation of I'_P to $-E'_P$, the induced primary voltage, will be the same as the phase relation of I_S to E'_S. The magnetizing effect of I'_P, then, will be in direct opposition to the magnetizing effect of I_S. Thus, the primary load current will increase until its magnetizing force exactly equals and cancels out the magnetizing force of I_S, in other words, until $N_P I'_P$ is equal and opposite to $N_S I_S$. The total current in the primary I_P is equal to the vector sum of the load current I'_P and the exciting current I_E. The exciting current I_E is so small in comparison to the load current that it may, in general, be neglected and I_P may be thought of as being equal to $\frac{N_S}{N_P} I_S$.

In addition to the no-load energy loss, there is an additional energy loss when the transformer is loaded. This loss is called the *copper* or *load loss* and depends upon the load and the resistance of the primary and secondary windings:

$$\text{Copper loss} = I'^2_P R'_P + I_S^2 R'_S \qquad \text{in watts}$$

The no-load energy loss, or transformer iron loss,

$$P_H = E_P I_H \qquad \text{in watts}$$

1-36. Leakage Reactance. In Fig. 1-56 it was assumed that all the flux stays within the steel core and is mutual to coils PP' and SS'. Actually this is not the case. A small part of the flux caused by any current flowing in either coil is forced out of the steel core and finds a return path through the air without linking with the other coil. This flux is called *leakage flux*. In Fig. 1-59, ϕ_{LP} and ϕ_{LS} represent leakage flux from the primary and secondary, respectively, while ϕ represents the mutual flux. The leakage flux ϕ_{LP} induces an emf in the primary but not in the secondary, while flux ϕ_{LS} induces an emf in the secondary but not in the primary.

The effect of the leakage flux may be expressed in terms of the *leakage inductance* (see Art. 1-18) or the *inductive leakage reactance.* The primary winding and the secondary winding, both, have an impedance, composed of resistance and inductive reactance (see Art. 1-29), which consumes a part of the impressed voltage equal to $I_P Z'_P$ in the primary and a part of the induced voltage equal to $I_S Z'_S$ in the secondary when the transformer is loaded. The vector diagram illustrating the phase

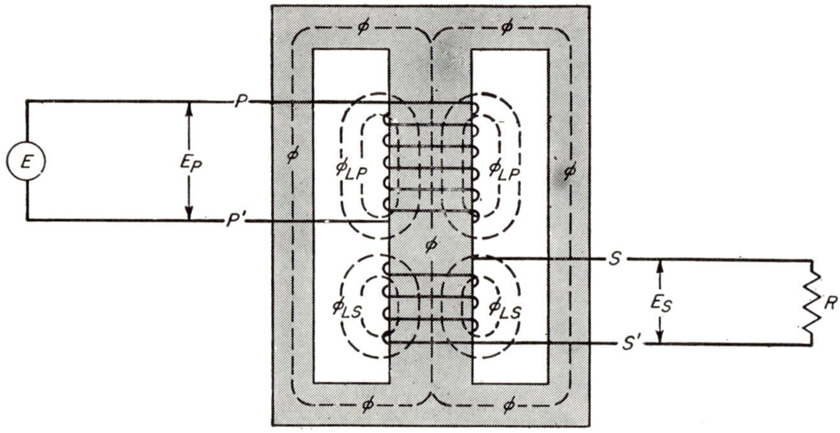

FIG. 1-59. Transformer circuit showing the nature of leakage flux for primary and secondary windings.

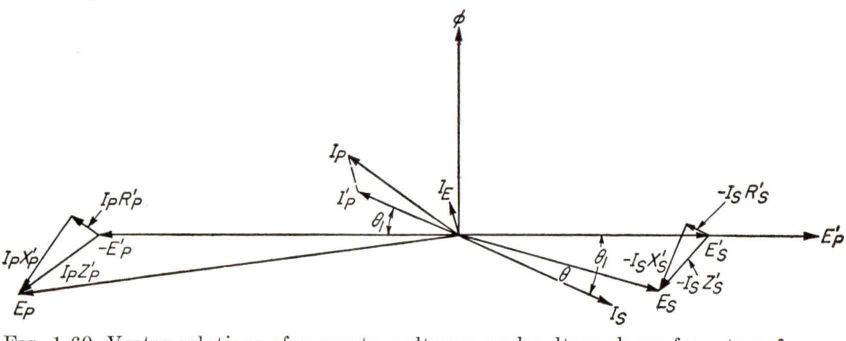

FIG. 1-60. Vector relations of currents, voltages, and voltage drops for a transformer carrying rated load.

relations between impressed voltage, induced voltage, and current, when the transformer is loaded, is shown in Fig. 1-60 where θ_1 is the power-factor angle, including the impedance of the secondary, and θ is the power-factor angle of the load circuit, excluding the impedance of the secondary winding.

1-37. Transformer Impedance. The easiest and most practical method of determining the impedance of a transformer is by measuring the vector portion of the impressed primary voltage used in overcoming the opposition of resistance and leakage reactance in the primary and the secondary. In Fig. 1-59, load R is disconnected, coil SS' is short-circuited across its terminals, and a voltage is impressed across the terminals of coil PP' just large enough to cause rated full load current I_S to flow in coil SS'. Then the voltage impressed across coil PP' is the vector voltage necessary to overcome opposition of primary and secondary impedance to the flow of rated-load current. The effect is the same as though the resistance and leakage reactance of coil SS' were removed and a separate resistance and leakage reactance were added to coil PP' of such a value that the secondary terminal voltage E_S will be the same as in the actual transformer at rated full load. If the voltage thus measured is divided by the rated primary full-load current, the result equals the *equivalent impedance* of the transformer in ohms. If coil PP' is short-circuited and sufficient voltage is impressed across coil SS' to cause rated full-load current I'_P (equal to $(N_S/N_P)\ I_S$) to flow in coil PP', the result will be the same, except that the impressed voltage will be changed in proportion to the ratio of N_S to N_P and the equivalent impedance will appear as though it were in coil SS'.

When referring to the impedance of a transformer, its equivalent impedance is usually meant. Transformer-impedance values are usually given in percent, which simply means that transformer-impedance voltage drop is expressed as a percent of rated voltage. Suppose that a 25-kva transformer has a measured impedance voltage of 72 volts on the primary and is rated 2,400-volts primary to 240-volts secondary. Its impedance, expressed in percent, is equal to

$$\%\,Z = \frac{72 \text{ volts}}{2{,}400 \text{ volts}} \times 100 = 3\%$$

The corresponding impedance voltage on the 240-volt side is 7.2 volts. If the percent impedance is known, the ohmic value on either the primary or secondary side can be determined by converting percent Z to impedance volts and dividing by rated full-load current. In the case of the 25-kva transformer cited above, the ohmic value of impedance is found as follows:

$$Z = \frac{\%\,Z \times E}{100} \div \frac{\text{kva} \times 1{,}000}{E} = \frac{\%\,Z \times E^2}{\text{kva} \times 10^5} \quad \text{ohms}$$

and

$$Z_P = \frac{3 \times 2{,}400 \times 2{,}400}{25 \times 10^5} = 6.91 \text{ ohms} \qquad Z_S = \frac{3 \times 240 \times 240}{25 \times 10^5} = 0.0691 \text{ ohms}$$

From the above it is seen that the equivalent impedance in ohms, on the secondary side, is equal to $(E_S/E_P)^2\ Z_P$. This relation may be expressed in terms of the turn ratio as follows:

$$Z_P = \frac{N_P}{N_S} Z_S \quad \text{ohms} \qquad \text{and} \qquad Z_S = \frac{N_S}{N_P} Z_P \quad \text{ohms}$$

The equivalent impedance is used because of the difficulty in separating primary and secondary impedances and the ease with which their combined effect is determined. The ohmic value of resistance in the primary and secondary can be measured, but there is no practical method of measuring primary and secondary leakage reactance separately. If the equivalent resistance and the equivalent impedance are known, however, the equivalent value of the combined primary and secondary leakage reactance can be figured as follows:

$$Z = \sqrt{R^2 + X^2} \quad \text{ohms}$$

on either the primary or secondary side, and

$$R = \sqrt{Z^2 - X^2} \quad \text{ohms} \qquad \text{and} \qquad X = \sqrt{Z^2 - R^2} \quad \text{ohms}$$

The transformer resistance determines the copper loss, since the copper loss equals I^2R. The equivalent resistance of the transformer is usually given in percent. The percent resistance may be determined from the full-load copper loss as follows:

$$\% R = \frac{\text{rated full-load copper loss in watts}}{\text{rated transformer kva} \times 10}$$

The power input to a transformer, when measuring the impedance voltage, is called the *rated copper loss*, or *load loss*, since rated full-load current flows in the primary and secondary and since the exciting impedance voltage is so small as to make the no-load loss, or iron loss, negligible. The value of equivalent resistance determined from the rated-watts copper loss, however, will be slightly higher than the equivalent value determined from the measured dc resistance of the primary and secondary windings. This difference is caused principally by eddy currents generated by the leakage flux of the windings.

1-38. Transformer Regulation and Efficiency. The effect of transformer impedance is measured by the regulation defined as the voltage drop at full load expressed as a percentage of rated full-load secondary terminal voltage:

$$\% \text{ regulation} = \frac{E'_S - E_S}{E_S} \times 100$$

The leakage reactance is usually several times as great as the resistance. At unity power factor the IR drop, being in phase with the current, subtracts directly from the no-load voltage and consequently is the predominant element of impedance causing voltage drop. With lagging power factors, the IR drop swings out of phase with the voltage while the IX drop swings more nearly in phase with the voltage, Fig. 1-60. With low lagging power factors the leakage reactance becomes predominant in lowering the terminal voltage, and since X is larger than R the regulation is increased. If the power factor is leading, the IX drop, which lags 90° behind the current, will swing toward alignment with the voltage. This will tend to cause a voltage increase so that a negative voltage drop may result, or an increase of terminal voltage at full load. Regulation is a variable which changes with power factor. The regulation can be determined, at any power factor, by the following approximate equation:

$$\% \text{ regulation} = \% R \cos\theta + \% X \sin\theta + \frac{(\% X \cos\theta - \% R \sin\theta)^2}{200}$$

where θ is the power-factor angle.

The efficiency of a transformer equals the ratio of output to input. Efficiency is based upon a load of 100 percent power factor, unless specifically stated otherwise. Since copper loss varies as the square of the load, the efficiency of the transformer, for any load either greater or less than the rated output, is

$$\% \text{ efficiency} = \frac{K \times P \times 100}{KP + W_i + K^2 W_C}$$

where K = ratio of load to rated transformer output
P = rated transformer output, volt-amp = kva × 1,000
W_i = rated transformer iron loss, watts
W_C = rated transformer copper loss in watts, at full load

Example: A manufacturer's test report shows the following average performance characteristics of a group of three transformers: (1) rated output = 667 kva, (2) high voltage = 33,000 volts, (3) low voltage = 4,800 volts, (4) rated full-load primary current = 20.21 amp, (5) rated full-load secondary current = 139 amp, (6) rated iron loss = 3,063 watts, (7) rated copper loss (75°) = 4,027 watts, (8) per cent impedance = 4.64, (9) primary coil resistance = 5.29 ohms, (10) secondary coil resistance = 0.0919 ohm, (11) exciting current = 3.14% of full-load current.

Using these test data, the following characteristics can be calculated:
Primary exciting current $I_E = 0.0314 \times 20.21 \text{ amp} = 0.634 \text{ amp}$
No-load voltage drop $I_E R'_P = 5.29 \text{ ohms} \times 0.634 \text{ amp} = 3.35 \text{ volts}$

Turn ratio $\dfrac{N_P}{N_S} = \dfrac{33{,}000}{4{,}800} = 6.875$

Equivalent primary resistance from measured coil values

$$R_P = R'_P + \left(\frac{N_P}{N_S}\right)^2 R'_S = 5.29 + (6.875)^2 0.0919 = 9.63 \text{ ohms}$$

Equivalent primary resistance, calculated from rated copper loss

$$R_P = \frac{4{,}027 \text{ watts}}{(20.21 \text{ amp})^2} = 9.86 \text{ ohm}$$

$$\% \text{ resistance} = \frac{\text{watts copper loss}}{\text{rated kva} \times 10} = \frac{4{,}027}{667 \times 10} = 0.60\%$$

$$\text{Impedance } Z_P = \frac{\% Z_P \times E^2}{\text{kva} \times 10^5} = \frac{4.64 \times (33{,}000)^2}{667 \times 10^5} = 75.76 \text{ ohms}$$

$$\text{Primary reactance } X_P = \sqrt{Z_P^2 - R_P^2} = \sqrt{(75.76)^2 - (9.86)^2} = 75.11 \text{ ohms} = 4.60\%$$

$$\text{Hysteresis component of exciting current } I_H = \frac{\text{iron loss watts}}{33{,}000 \text{ volts}} = \frac{3{,}063 \text{ watts}}{33{,}000 \text{ volts}} = 0.093 \text{ amp}$$

$$\text{Power factor of exciting current } \cos\theta_1 = \frac{I_H}{I_E} = \frac{0.093}{0.634} = 0.1467$$

Angle exciting current lags behind exciting voltage $\theta_1 = 81°34'$

Magnetizing current $I_M = I_E \sin\theta_1 = 0.634 \sin 81°34' = 0.627$ amp

% regulation at unity power factor

$$= 0.6 \cos 0° + 4.6 \sin 0° + \frac{(4.6 \cos 0° - 0.6 \sin 0°)^2}{200} = 0.71\%$$

% regulation – 80% pf

$$= (0.6 \times 0.8) + (4.6 \times 0.6) + \frac{[(4.6 \times 0.8) - (0.6 \times 0.6)]^2}{200} = 3.30\%$$

$$\% \text{ efficiency at full load} = \frac{1 \times 667 \times 1{,}000 \times 100}{(1 \times 667{,}000) + 3{,}063 + (1)^2 4{,}027} = 98.95\%$$

$$\% \text{ efficiency at half load} = \frac{0.5 \times 667 \times 1{,}000 \times 100}{(0.5 \times 667{,}000) + 3{,}063 + (0.5)^2 4{,}027} = 98.79\%$$

Impressed primary voltage necessary to deliver rated full load with 4,800 volts across the secondary terminals

$$E'_P = \left(\frac{\% \text{ regulation} \times E_S}{100} + E_S\right)\frac{N_P}{N_S} = \left(\frac{3.30 \times 4{,}800}{100} + 4{,}800\right)6.875 = 34{,}100 \text{ volts at } 80\% \text{ pf}$$

$$E'_P = \left(\frac{0.71 \times 4{,}800}{100} + 4{,}800\right)6.875 = 33{,}200 \text{ volts at } 100\% \text{ pf}$$

1-39. Autotransformer. If terminal P', of Fig. 1-59, is connected to terminal S and primary voltage is applied across terminals PS' while load is supplied from terminals SS', the arrangement will be similar to Fig. 1-61. This shows an *autotransformer* with a single winding performing the functions of both primary and secondary in a two-winding transformer. The autotransformer is a *step-down* transformer in which a higher primary voltage E_P is stepped down to a lower voltage E_S to supply load R. Autotransformers can also work as *step-up* transformers. As for the two-winding transformer, an alternating flux sets up in the core when an ac exciting voltage is applied across its terminals. This flux induces an opposing voltage in each turn, such that the total induced voltage E'_P is almost equal to E_P, the difference being just great enough to drive an exciting current through coil PP' large enough to produce the flux. If the transformer has N_P total turns, the effect will be the same as though the impressed voltage E_P were divided into E_P/N_P volts per turn. If there are N_S turns between terminals SS', the voltage across terminals SS', then, is equal to $(E_P/N_P)\,N_S$, or

$$E_S = \frac{N_S}{N_P} E_P \qquad \text{volts}$$

For example, the primary voltage E_P, in Fig. 1-61, equals 2,400 volts while E_S equals 240 volts and load R equals 100 amp. The load current of 100 amp will divide as shown since the ampere-turns from A to

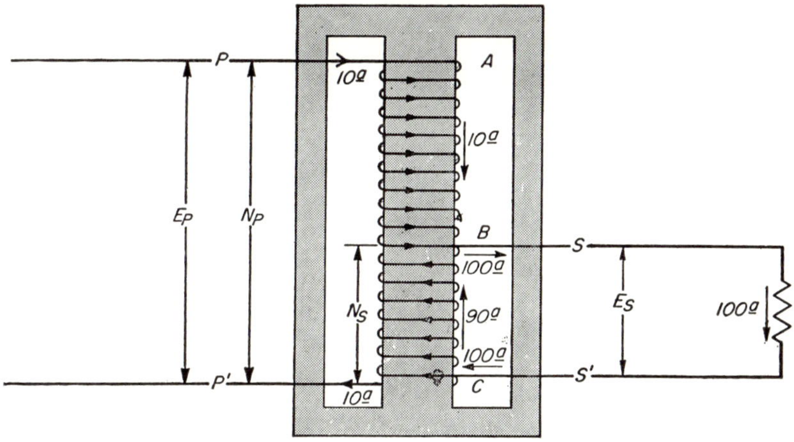

FIG. 1-61. Electric and magnetic circuit for an autotransformer.

B must equal but be in opposition to the ampere-turns from B to C. These same relations hold if the transformer steps the voltage up and the primary is the lower voltage. In the two-winding transformer, the product of volts times current in the primary is equal to the product of volts times current in the secondary, neglecting losses. In other words, the entire power output is transformed. In the autotransformer the power output may be greater than the power transformed. In Fig. 1-61, the turns from A to B are called the series winding and are a part of the primary supply only, while the turns from B to C are called the common winding since they are common to both primary and secondary. The power passing from the primary supply circuit to the secondary load circuit by transformer action is equal to the voltage from A to B times the current, or the voltage from B to C times the current. In the case illustrated in Fig. 1-61, the power transformed is equal to

$$P = (2{,}400 \text{ volts} - 240 \text{ volts})10 \text{ amp} = 21.6 \text{ kva}$$

or

$$P = (240 \text{ volts} \times 90 \text{ amp}) = 21.6 \text{ kva}$$

The remaining power passes directly from the primary to the secondary without transformation. Thus, for the case cited, a transformer capacity of 21.6 kva is required to supply a load of 24 kva. The rated autotransformer capacity required for any load is

$$\text{Kva autotransformer capacity} = \frac{E_P - E_S}{E_P} \times \text{kva load}$$

The autotransformer is lower in first cost and more economical to operate, but its use is objectionable in some cases because the primary and secondary are part of a common winding and cannot be insulated from each other.

1-40. Instrument Transformers. Instrument transformers supply current and voltage to meters for measuring electrical quantities, to control or operate regulating devices, and for the operation of various types of circuit protective devices. It would be impractical to build a voltmeter for connection directly across a 33,000-volt line, or to build an ammeter for operation directly in series with a line which may carry several hundred amperes at thousands of volts. An accurate indication of voltage and current in any line can be secured by a potential transformer and a current transformer which will step the high line values down to lower values, safe for meters or other devices. Potential transformers usually step down from rated primary voltage to 120 volts if the rated primary voltage is less than 34,500 volts. If the rated primary voltage is 34,500 volts or more, the secondary rated voltage is usually 115 volts. Current transformers usually step down from rated primary current to 5 amp secondary. Instrument transformers are designed for low power-output capacity—usually 500 volt-amp or less—but for high ratio accuracy.

The potential transformer is identical, in theory of operation, to the large power transformer and connects in shunt across the line. It is intended, however, for near-constant ratio of impressed to output voltage between no load and full load as can be attained and not for power output. This is of especial importance when used for metering. For high accuracy, the potential transformer is designed for a low value of regulation and for a small angular difference between the impressed primary voltage and the secondary terminal voltage reversed.

The current transformer consists of a primary winding, a secondary winding, and a magnetic circuit just as in the case of the power transformer or the instrument potential transformer. The primary of the current transformer, however, is connected in series with the line so that a complete electrical circuit is made only when the primary line, in which the current transformer is connected, makes a complete circuit. When the primary line circuit is complete, current flows in the line and through the primary of the current transformer. The line current flowing through the transformer primary establishes a magnetomotive force which produces magnetizing flux in the steel core. This magnetic flux causes a voltage to be induced in the secondary winding sufficient to cause a current to flow in the secondary circuit. The current produced by the voltage induced in the secondary is large enough so that the magnetomotive force in the secondary, which is in opposition to the magnetomotive force in the primary, is so nearly equal to the primary magnetomotive force that the vector difference is just sufficient to maintain the magnetizing flux in the magnetic circuit through the core. The principal result sought in the current transformer, especially when used for metering, is to produce a current in the secondary which is, as nearly as possible, an exact multiple of the primary current (in proportion to the ratio of turns) for all values of primary current. In other words, so that I_S is equal to $N_P I_P / N_S$ for all values of I_P within the current range of the transformer. The transformer is designed to operate at a lower value of flux density to reduce the total exciting component of primary current, so that the ratio of I_S to I_P is as nearly as possible equal to the ratio of N_P to N_S, and so that the angle between I_P and I_S reversed will be a minimum.

When the current through the primary increases, a higher effective flux results and a higher value of emf induced in the secondary drives the higher value of I_S through the secondary so that $N_S I_S$ approaches $N_P I_P$ in value. For this reason, the secondary winding should never be left open-circuited, since in this case there will be no opposing ampere-turns developed in the secondary and all the flux produced by $N_P I_P$ will induce a high emf in the secondary. This will be dangerous to the winding insulation and to any personnel which may come in contact with it. Current transformers are made in two types. The through-type current transformer corresponds to a transformer having a single turn in the primary winding realized by carrying the line conductor straight through an opening in the transformer. The through-type transformer is used for high line current. The wound type has a wound primary winding of more than one turn and is usually used for lower line current. The overall accuracy of the wound type is greater than that of the through type.

1-41. Transformer Polarity. Relative direction of the induced secondary voltage, with respect to the primary, at the external terminals of the windings must be known when transformers are to be operated in parallel or where single-phase transformers are to be interconnected to form a three-phase bank. The relative directional relation of primary to secondary is called the *polarity* of the transformer. Figure 1-62*a* shows a transformer with the primary and secondary coils wound in the same direction around the core, assuming that the primary coil starts from P and the secondary coil from S. Suppose voltage E_P is impressed across the primary terminals and current I_S flows through the secondary. During the instant when I_P flows in at terminal P and out at terminal P', the direction of emf induced in the secondary will make I_S flow out at terminal S, through load R, and back into the transformer

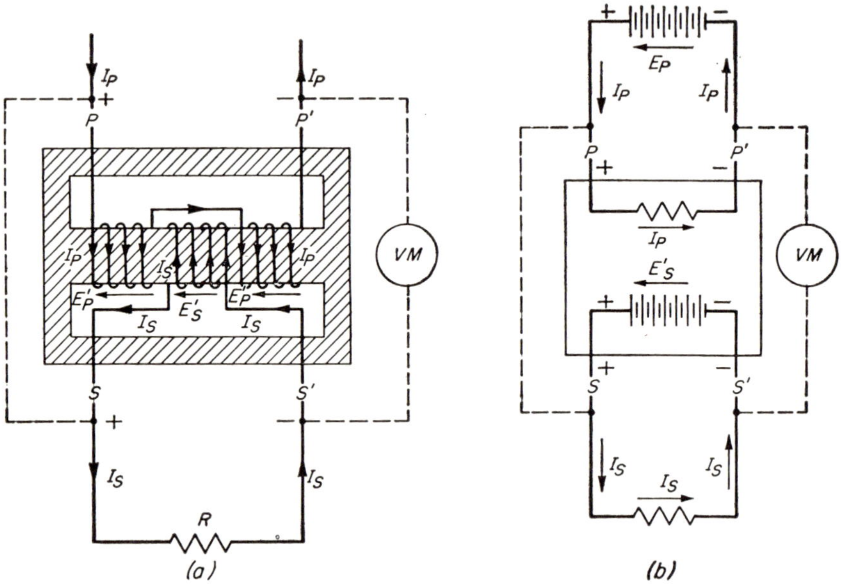

FIG. 1-62. Transformer polarities must be determined when they are to be connected in parallel or banked for three-phase operation. Here actual windings and the equivalent circuit for subtractive polarity are compared.

through terminal S'. This direction of current flow in the secondary is determined by the requirement that flux set up by I_S must be equal and opposite to the flux set up by the load component of I_P. The direction of current flow through the secondary must be such that the magnetizing force in ampere-turns set up by the flow of current in the secondary is equal and opposite to the magnetizing force in ampere-turns set up by I'_P, the load component of primary current, so that one cancels out the other.

During the instant when ac flows through the transformer in the direction shown in Fig. 1-62*a*, the transformer is similar to a dc circuit, Fig. 1-62*b*, in which the primary supply voltage E_P is replaced by a dc battery feeding a load current I_P to a load replacing the primary winding and the secondary winding, in which the voltage E'_S is induced, is replaced by a dc battery supplying current I_S to load R. If the adjacent primary and secondary terminals P and S are connected and a voltmeter is placed across terminals P' and S', as shown by the dotted lines, the voltmeter will read the difference between E_P the impressed primary voltage and E'_S the induced secondary voltage, or $E_P - E'_S$. Such a transformer has subtractive polarity.

If the secondary coil is wound in the reverse direction with respect to the primary winding, the direction of current flow with respect to the secondary terminals and the direction of induced emf will be reversed,

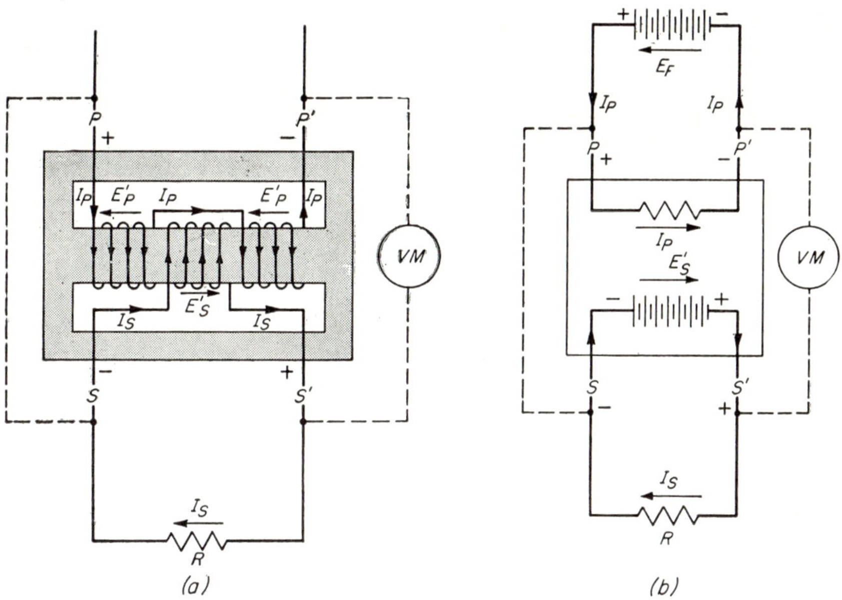

FIG. 1-63. Comparison of actual transformer circuit and equivalent circuit for additive polarity.

as shown in Fig. 1-63a. This reversal, with respect to the secondary terminals, is necessary in order that the magnetizing effect of the secondary ampere-turns may counteract and cancel out the magnetizing effect of the primary load ampere-turns. Comparing Fig. 1-63a and b with Fig. 1-62a and b shows that the secondary induced emf E'_S is in additive series with the impressed primary voltage E_P. If the adjacent primary and secondary terminals P and S are connected and an exciting voltage E_P is impressed across the terminals PP', a voltmeter placed across terminals $P'S'$ will read the sum of the impressed voltage plus the secondary induced voltage, or $E_P + E'_S$. Such a transformer has additive polarity.

When transformers are to be banked, or operated in parallel, the directional relation of the voltage vectors is of first importance and the direction of current flow need not be considered since this depends upon the voltage. When considering transformer voltages it must first be specified whether impressed or induced voltages are meant. It is generally confusing to speak of impressed voltage in the primary and induced voltage in the secondary since it is not always clear which is primary and which is secondary. The induced voltage, in the primary and in the secondary, is customarily meant when referring to transformer polarity. Studying Figs. 1-62 and 1-63, note that the plus and minus polarity designation of the transformer terminals is unchanged whether impressed voltage, external to the tank, or induced voltage, from within the tank, is intended.

The external leads of a transformer are identified by the letters H_1, H_2, etc., on the high-voltage side and X_1, X_2, etc., on the low-voltage

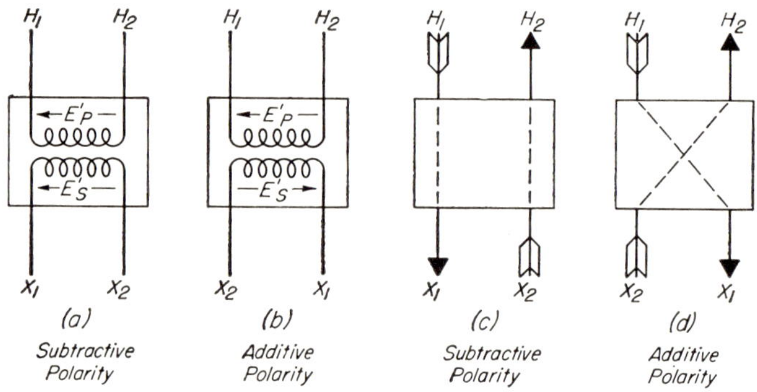

FIG. 1-64. Usual name-plate designations of transformers for additive and subtractive polarities.

side, on name-plate and connection diagrams, to indicate corresponding polarity of the primary and secondary leads. On power transformers the convention is such that the H_1 lead is on the right when standing facing the high-voltage side of the transformer. If the transformer is of subtractive polarity, the X_1 lead is on the same side of the tank and adjacent to the H_1 lead, as shown in Fig. 1-64*a*. If the transformer is of additive polarity, the X_1 lead is on the opposite side of the tank and adjacent to the H_2 lead, as is shown in Fig. 1-64*b*. If the relative direction of current flow is considered, current will be flowing out from terminal X_1 when current is flowing in at terminal H_1, with the result that the apparent path of current through the transformer is similar to that shown in *c* and *d* of Fig. 1-64. In Figs. 1-62 to 1-64 the direction of induced voltages, relative to the terminals, is the same through the primary and secondary when the polarity is subtractive and opposite when the polarity is additive.

If two subtractive polarity transformers are to be paralleled, they should be connected as shown in Fig. 1-65*a*. If two transformers are to be connected in parallel and one is subtractive and the other is additive polarity, they should be connected as shown in Fig. 1-65*b*. If the H_1 leads are connected to the same phase wire on the primary side, the X_1 leads must likewise be connected to the same wire on the secondary side. In the case of three-phase transformers the relative polarity is also represented by voltage diagrams since the lettering of the leads does not fully describe the voltage vector relations between the different phases of the bank. The standard practice for identification of transformer polarity specifies that

1. All instrument transformers, both current and potential, shall be of subtractive polarity.

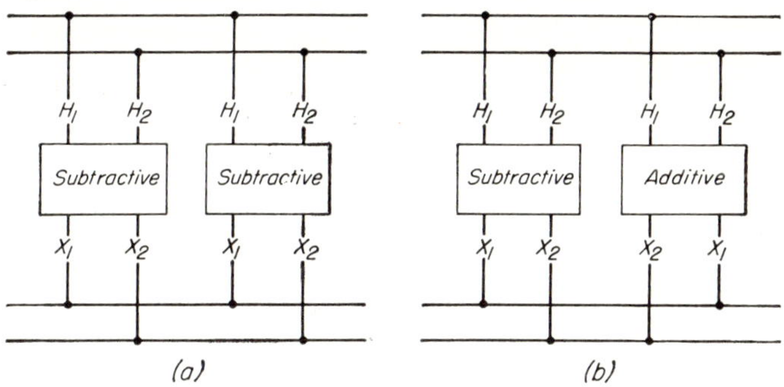

FIG. 1-65. How transformers of various polarities are paralleled for correct operation.

2. All single-phase power transformers that are 200 kva or less in size and whose high-voltage winding is 8,660 volts or less shall be of additive polarity.

3. All single-phase power transformers whose size exceeds 200 kva, regardless of voltage rating, or whose high-voltage winding exceeds 8,660 volts, regardless of kva size, shall be of subtractive polarity.

CHAPTER REVIEW TOPICS

1. What is the electron theory of matter?
2. What are electrons, protons, and neutrons?
3. How do electrons behave in a good conductor of electricity?
4. What is the difference between electron and electricity flow?
5. Write Ohm's law in its three equation forms.
6. How can wire resistance be calculated from resistivity, length, and cross-sectional area?
7. What is back emf?
8. How is power figured from volts and amperes to obtain kilowatts?
9. Write the formula for power loss from resistance in a conductor.
10. How many horsepower are there in 1 kw?
11. How is the direction of a magnetic field determined?
12. How does a compass needle act when brought near a wire carrying dc?

13. How are the directions of an electromagnetic field and its creating current related?
14. What happens to the strength of an electromagnetic field around a wire as its distance from the wire increases?
15. Describe how an electromagnet is made and the direction of the flux as related to the direction of current flow.
16. What happens to the flux of an electromagnet when an iron bar is used as the core?
17. What happens when a wire forming a closed loop is passed through a magnetic field?
18. How can an emf be induced in a conductor?
19. Sketch the principal parts and connections of a dc generator.
20. What is the purpose of the commutator on a dc generator?
21. Trace the variation in emf of a conductor on the surface of an armature rotating in a uniform magnetic field.
22. How many angular degrees are there in 1 radian?
23. Draw the principal features and connections for a simple ac generator.
24. Does the armature of an ac generator always rotate?
25. What is meant by a cycle of ac emf or current?
26. What is meant by the frequency of an ac flow?
27. Write the formula relating current frequency and generator rpm.
28. What is meant by an electrical degree?
29. What is a sine wave?
30. How is a rotating vector related to a sine wave?
31. What is the difference between a single-phase and a polyphase generator?
32. What is the effective value of an ac?
33. How is the effective value of an ac related to its maximum value?
34. What is meant by inductance?
35. Explain Lenz's laws.
36. What is the unit of inductance?
37. What is inductive reactance? How is it calculated?
38. What does inductive reactance do to relative phasing of impressed emf and current flow in an ac circuit?
39. What are the phase relations of ac flow and its electromagnetic field?
40. What is capacitance? In what units is it measured?
41. How does a condenser work when connected to a source of dc emf?
42. What does a condenser do to the phase relations between impressed emf and current for an ac circuit?
43. How are coulombs and amperes related?
44. What is capacitive reactance? How is it measured?
45. How are inductive and capacitive reactances related in an ac circuit?
46. How does electrical power flow vary in an ac circuit?
47. What happens to power flow in circuits containing either inductance or reactance only?
48. What is power factor? How is it used in computing kilowatts power flow?
49. Write the formula for figuring three-phase power output knowing phase voltage and current and power factor.
50. What is apparent power?
51. What is reactive power? How is it related to real and apparent power?
52. State Kirchhoff's laws.
53. Draw typical delta and wye three-phase circuits.
54. Draw the vector diagram for three-phase delta and wye connections.

55. Draw a sample series circuit, a sample parallel circuit, a series-parallel circuit.
56. What is an ac-circuit impedance? Admittance?
57. How are resistance, inductive reactance, capacitive reactance, and impedance related?
58. What is meant by conductance and susceptance?
59. Draw the electrical and magnetic circuits of a simple transformer.
60. What are the functions of a transformer?
61. Explain the features which cause the exciting current of a transformer.
62. Explain the mechanics of hysteresis loss in a transformer.
63. What are eddy currents? How are they minimized in a transformer?
64. Draw the vector relations for a transformer operating at no load.
65. What is the leakage flux in a transformer?
66. How is transformer impedance measured?
67. What is meant by transformer regulation? How is it measured?
68. How do you figure transformer efficiency?
69. What is an autotransformer?
70. What is the relation between the winding ratio and voltage ratio of a transformer?
71. What is a potential transformer used for? A current transformer?
72. Show the circuits for a potential and a current transformer.

PROBLEMS

Problem 1-1: A coil of wire, to be used as a resistor, is connected to a dc supply with an ammeter connected between one terminal of the coil and the source and with a voltmeter connected across the terminals of the coil. The ammeter reads 6 amp, and the voltmeter reads 117 volts. What is the resistance of the resistor?

Solution: By Ohm's law the resistance is equal to

$$R = \frac{E}{I} = \frac{\text{volts}}{\text{amp}} = \frac{117}{6} = 19.5 \text{ ohms}$$

Problem 1-2: A rectangular bar is 0.10 in. square in cross section. What is the equivalent area in circular mils?

Solution:

$$\text{Area of rectangular bar} = 0.1 \times 0.1 = 0.01 \text{ sq in.}$$

$$\text{Area of equivalent circle} = \frac{\pi d^2}{4} = 0.01 \text{ sq in.}$$

$$d^2 = \text{diameter squared} = \frac{4 \times 0.01}{\pi} = 0.012732 \text{ in.}^2$$

$$\text{Area in circular mils} = 0.012732 \times 1{,}000 \times 1{,}000 = 12{,}732$$

Problem 1-3: What is the resistance of the metal bar in Prob. 1-2 if the bar is 20 ft long and the resistivity per mil-foot is 500?

Solution:

$$R = \frac{\rho L}{A} = 500 \times \frac{20}{12{,}732} = 0.7854 \text{ ohm}$$

Problem 1-4: The emf across the terminals of a dc generator is 120 volts. The generator supplies 50 amp to a bank of lamps 500 ft away and 50 amp to a second lamp bank 500 ft beyond the first. The resistance of the wire, over which the current is transmitted, is equal to 0.05 ohm per 1,000 ft. (1) What is the total voltage drop from the generator to the first bank of lamps? (2) What is the line voltage drop from the first bank to the second bank? (3) What is the voltage supplied to each bank of lamps?

Solution:

1. The current for both banks flows between the generator and the first group of lamps, 500 ft away. The total voltage drop between the generator and the first lamp bank equals

$$\text{Voltage drop} = 2RI_1 = 2\left(\frac{500}{1{,}000} \times 0.05\right) 100 \text{ amp} = 5 \text{ volts}$$

2. The line voltage drop between the first and second lamp banks is equal to

$$2RI_2 = 2\left(\frac{500}{1{,}000} \times 0.05\right) 50 \text{ amp} = 2.5 \text{ volts}$$

3. The voltage across the terminals of the first lamp bank is equal to

$$E - 2RI_1 = 120 \text{ volts} - 5 \text{ volts} = 115 \text{ volts}$$

The voltage across the terminals of the second lamp bank is equal to

$$E - 2RI_1 - 2RI_2 = 120 \text{ volts} - 5 \text{ volts} - 2.5 \text{ volts} = 112.5 \text{ volts}$$

Problem 1-5: A toaster has a resistance of 20 ohms and draws a current of 6 amp. (1) How much power will be required? (2) What will be the impressed voltage? (3) What will be the energy consumption per hour of operation?

Solution:

1. Power $= RI^2 = 20 \text{ ohms} \times 6 \text{ amp} \times 6 \text{ amp} = 720 \text{ watts}$
2. Impressed voltage $= E = RI = 20 \text{ ohms} \times 6 \text{ amp} = 120 \text{ volts}$, or

$$E = \frac{P}{I} = \frac{720 \text{ watts}}{6 \text{ amp}} = 120 \text{ volts}$$

3. Energy consumption $= \text{power} \times \text{time} = 720 \text{ watts} \times 1 \text{ hr} = 720 \text{ watt-hr}$

$$= \frac{720}{1{,}000} = 0.720 \text{ kwhr}$$

Problem 1-6: A 5 hp-motor is supplied from a 240-volt dc power source. There are 100 watts heat loss in the motor at full load and 320 watts heat loss in the line supplying the motor. (1) What is the power output of the source in kilowatts? (2) What is the voltage across the motor terminals?

Solution:

1. Power from source $= \dfrac{(5 \text{ hp} \times 746 \text{ watts}) + 100 \text{ watts} + 320 \text{ watts}}{1{,}000} = 4.15 \text{ kw}$
2. Voltage at motor $= 240 \text{ volts} - \text{line drop}$

$$= 240 \text{ volts} - \left(320 \text{ watts} \div \frac{4{,}150 \text{ watts}}{240 \text{ volts}}\right)$$

$$= 240 \text{ volts} - 18.5 \text{ volts} = 221.5 \text{ volts}$$

Problem 1-7: A two-pole ac generator produces 2,400 volts, rms value. The machine is rotating at normal speed, and the dc field is at normal strength. (1) What is the maximum value of emf generated? (2) What will be the instantaneous value of emf generated after 0.1 revolution if zero time is taken when the instantaneous value of emf is zero and the generated voltage is rising in the positive direction?

Solution:

1. Effective value, or rms value $= 2{,}400$ volts $= 0.707E_M$

$$E_M = \frac{2{,}400}{0.707} \text{ volts} = 3{,}395 \text{ volts}$$

2. One revolution $= 360$ electrical degrees
0.1 revolution $= 0.1 \times 360 = 36$ electrical degrees
Instantaneous value of emf $= e = E_M \sin \theta$
$e = 3{,}395 \sin 36° = 1{,}996$ volts

Problem 1-8: (1) What would be the rpm of the generator of Prob. 1-7 if the emf generated is 60 cycle? (2) What would the frequency be if the machine were eight-pole?

Solution:

1. $$f = \frac{\text{poles} \times \text{rpm}}{2 \times 60}$$

$$\text{rpm} = \frac{2 \times 60 \times f}{\text{poles}} = \frac{2 \times 60 \times 60}{2} = 3{,}600$$

2. $$f = \frac{8 \times 3{,}600}{2 \times 60} = 240 \text{ cycles per sec}$$

Problem 1-9: A coil of 400 turns is wound on a cylindrical iron core. The coil is 24 in. long, and the diameter of the core is 10 in. (1) What is the inductance of the coil if the permeability of the iron is 1,000? (2) What would the inductance be if an air core is used?

Solution:

Length of coil $= 24$ in. $\times 2.54 = 61$ cm
Diameter of coil $= 10$ in. $\times 2.54 = 25.4$ cm

1. $$L = \frac{0.4\pi N^2 \mu A}{l \times 10^8} = \frac{0.4 \times 3.1416 \times 400 \times 400 \times 1{,}000 \times \pi \times 25.4^2/4}{61 \times 10^8}$$
$$= 16.7 \text{ henrys with an iron core}$$

2. $$L = \frac{0.4 \times 3.1416 \times 400 \times 400 \times 1 \times 506.7}{61 \times 10^8} = 0.0167 \text{ henry with an air core}$$

Problem 1-10: What will be the back emf induced in the coil of Prob. 1-9 if a 60-cycle current of 1 amp, maximum value, flows in the coil?

Solution: The time required for the current to grow from zero to 1 amp, maximum value, is equal to

$$t = \frac{1}{4f} = \frac{1}{4 \times 60} = \frac{1}{240} \text{ sec}$$

The average rate of change of current

$$\frac{1 \text{ amp}}{t} = \frac{1 \text{ amp}}{1/240} = 240 \text{ amp per sec}$$

The average value of induced emf:

$$E_{av} = L\frac{I}{t} = 16.7 \text{ henrys} \times 240 \text{ amp} = 4{,}010 \text{ volts with iron core}$$
$$= 0.0167 \times 240 = 4.01 \text{ volts with air core}$$

The rms value of induced emf:

$$E = E_{av}\frac{0.707}{0.636} = 4{,}010 \times \frac{0.707}{0.636} = 4{,}450 \text{ volts with iron core}$$
$$= 4.45 \text{ volts with air core}$$

Problem 1-11: What will be the inductive reactance of the coil in Prob. 1-9 in a 60-cycle circuit?

Solution:

$$X_L = 2\pi fL = 2 \times 3.1416 \times 60 \times 16.7 = 6{,}290 \text{ ohms with iron core}$$
$$= 6.29 \text{ ohms with air core}$$

Problem 1-12: A 60-cycle transmission line has an inductive reactance of 0.6722 ohm per mile of line. What is its inductance per mile, in henrys?

Solution:

$$L = \frac{X_L}{2\pi f} = \frac{0.6722}{2 \times 3.1416 \times 60} = 0.00178 \text{ henry}$$

Problem 1-13: The capacitance of a condenser is 30 μf (microfarads). How much electricity will be stored in the plates if 240 volts dc is applied?

Solution:

$$Q = CE = \frac{30 \times 240}{10^6} = 0.0072 \text{ coulomb or amp-sec}$$

Problem 1-14: A three-phase 60-cycle transmission line has 110,000 volts, rms value, between conductors. When the line is open on the load end, a current of 16 amp flows in the line. (1) What is the capacitance of a single conductor? (2) What is its capacitative reactance?

Solution:

1. $I = \dfrac{E}{X_C} = 2\pi fCE$

$$C = \frac{I}{2\pi fE} = \frac{\sqrt{3} \times 16}{2 \times \pi \times 60 \times 110{,}000} = 0.000000669 \text{ farad} = 0.669\ \mu\text{f}$$

2. $X_C = \dfrac{1}{2\pi fC} = \dfrac{1 \times 10^6}{2 \times \pi \times 60 \times 0.669} = 3{,}970 \text{ ohms}$

$$X_C = \frac{E}{I_C} = \frac{110{,}000}{\sqrt{3} \times 16} = 3{,}970 \text{ ohms}$$

Problem 1-15: Eight 150-watt lamps and twelve 100-watt lamps are connected to a 120/240-volt single-phase power source so that the load is balanced across the source. What current flows in each leg of the supply circuit?

Solution:

$$\text{Total power} = (8 \times 150 \text{ watts}) + (12 \times 100 \text{ watts}) = 2{,}400 \text{ watts}$$
$$\text{Power} = EI\cos\theta = 240 \text{ volts} \times I \times \cos 0° = 240 \text{ volts} \times I \times 1 = 2{,}400 \text{ watts}$$
$$I = \frac{2{,}400 \text{ watts}}{240 \text{ volts}} = 10 \text{ amp}$$

Problem 1-16: A single-phase induction motor takes a current of 24 amp at 220 volts, which lags 30° behind the voltage. (1) What is the power factor? (2) What is the apparent power input? (3) What is the real power input? (4) What is the reactive power input?

Solution:

1. Power factor $= \cos 30° = 0.8666 = 86.66\%$
2. Apparent power $= EI = 220 \text{ volts} \times 24 \text{ amp} = 5{,}280 \text{ va} = 5.28 \text{ kva}$
3. Real power $= EI \cos\theta = 220 \text{ volts} \times 24 \text{ amp} \times \cos 30° = 4{,}572 \text{ watts}$ $= 4.572 \text{ kw}$
4. Reactive power $= EI \sin\theta = 220 \text{ volts} \times 24 \text{ amp} \times \sin 30° = 2{,}640 \text{ var}$ $= 2.64 \text{ kvar}$

Problem 1-17: A balanced load is connected to a three-phase power source. An ammeter in the circuit shows 182 amp going to the load. A voltmeter reads 2,400 volts between phase wires, while a wattmeter shows 600 kw delivered to the load. (1) What is the power factor of the load? (2) What is the angle between current and voltage? (3) What is the reactive power being delivered?

Solution:

1. $$\text{Apparent power} = \sqrt{3}EI = \frac{\sqrt{3} \times 2{,}400 \text{ volts} \times 182 \text{ amp}}{1{,}000} = 756.6 \text{ kva}$$

$$\text{Power factor} = \frac{\text{real power}}{\text{apparent power}} = \frac{600 \text{ kw}}{756.6 \text{ kva}} = 0.793 = 79.3\%$$

2. $\cos\theta = 0.793$ therefore,
   $\theta = 37°32' =$ power-factor angle

3. $$\text{Reactive power} = \sqrt{3}\, EI \sin\theta = \frac{\sqrt{3} \times 2{,}400 \text{ volts} \times 182 \text{ amp} \times \sin 37°32'}{1{,}000}$$
$$= 461 \text{ kvar}$$

Problem 1-18: The current flowing to the load of Prob. 1-17 is the resultant of what currents (1) in phase with the voltage and (2) lagging the voltage by 90°?

Solution:

$I \cos\theta = 182 \text{ amp} \times \cos 37°32' = 144.3$ amp in phase with the voltage
$I \sin\theta = 182 \text{ amp} \times \sin 37°32' = 110.9$ amp lagging the voltage by 90°

Problem 1-19: A circuit has a resistance of 8 ohms and an inductive reactance of 6 ohms. Fifteen amp ac flows through the circuit. (1) What will the voltage drop

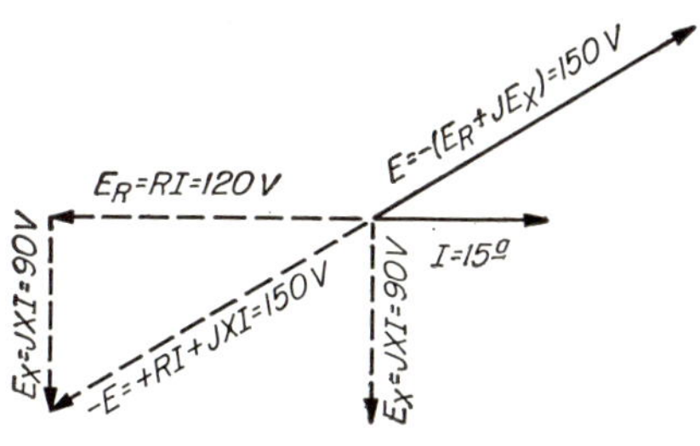

across the resistance and across the reactance be? (2) What voltage must be applied to cause 15 amp to flow? (3) Draw the vector diagram.

Solution:

1. Resistance drop $= RI = 8 \times 15 = 120$ volts
   Reactance drop $= XI = 6 \times 15 = 90$ volts
2. Applied voltage $= \sqrt{(RI)^2 + (XI)^2}$
   $= \sqrt{120^2 + 90^2} = 150$ volts

3. Vector diagram: E_R, the voltage drop across the resistance, must be in the opposite direction to the current since it opposes the flow of current. E_X, the voltage drop across the reactance, must lag the current by 90° since it opposes any change in current flow. The total voltage drop must be equal and opposite to the applied voltage.

Problem 1-20: If 6 ohms of capacitative reactance were added to the circuit of Prob. 1-19, what applied voltage would be required to maintain a flow of 15 amp? Show the vector diagram.

Solution: Since the effect of capacitance is exactly opposite to inductance in a circuit and since X_C is equal to X_L, in this example, the voltage drop across the capaci-

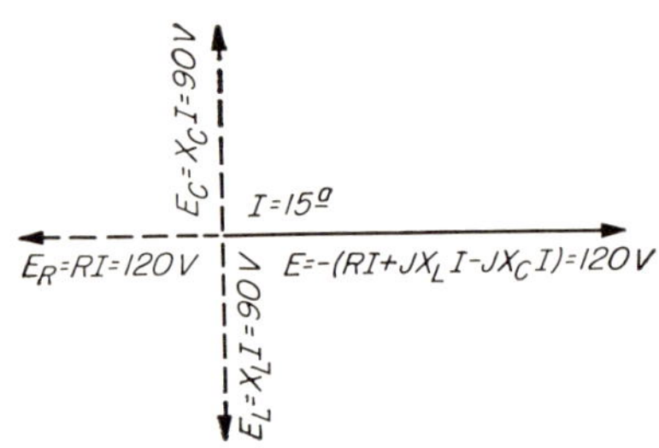

tance will be equal to and will cancel out the voltage drop across the inductance. The applied voltage, then, will only have to be equal to the voltage drop across the resistance, or 120 volts.

Problem 1-21: What is the impedance of the circuit of Probs. 1-19 and 1-20?

Solution:

1. $Z = \sqrt{8^2 + 6^2} = \sqrt{64 + 36} = 10$ ohms (Prob. 1-19)
2. $Z = \sqrt{8^2 + 6^2 - 6^2} = 8$ ohms (Prob. 1-20)

Problem 1-22: A three-phase 4,157-volt distribution circuit, with 2,400 volts from phase to neutral, serves a load of 918 kw at 85 percent power factor, lagging. (1) What is the kva? (2) What is the line current? (3) What is the resistance component of current? (4) What is the inductive component of current, lagging the voltage by 90°? (5) What is the total resistance, reactance, and impedance of the circuit, including the load?

Solution:

1. $\text{kva} = \dfrac{918 \text{ kw}}{0.85} = 1{,}080 \text{ kva}$
2. $I = \dfrac{\text{kva} \times 1{,}000}{3 \times E_N} = \dfrac{1{,}080 \times 1{,}000}{3 \times 2{,}400 \text{ volts}} = 150 \text{ amp}$
3. $I_R = I \cos\theta = 150 \text{ amp} \times 0.85 = 127.5 \text{ amp}$
4. $I_L = I \sin\theta = 150 \text{ amp} \times \sin(\cos^{-1} 0.85) = 76.1 \text{ amp}$
5. $E_N = ZI$

$$Z = \frac{E}{I} = \frac{2{,}400 \text{ volts}}{150 \text{ amp}} = 16 \text{ ohms}$$

$$RI = E \cos \theta$$

$$R = \frac{E \cos \theta}{I} = \frac{2{,}400 \times 0.85}{150} = 13.6 \text{ ohms, or}$$

$$R = Z \cos \theta = 16 \text{ ohms} \times 0.85 = 13.6 \text{ ohms}$$

$$X_L = Z \sin \theta = Z \sin (\cos^{-1} 0.85) = 8.1 \text{ ohms}$$

Problem 1-23: A series circuit is connected as shown in Fig. 1-54. (1) What is the total circuit impedance? (2) What is the impressed ac voltage necessary to drive 100 amp through the circuit?

Solution:

1. $Z = \sqrt{(R_1 + R_2 + R_3)^2 + (X_{L1} - X_{C3})^2} = \sqrt{3^2 + (1 - 1)^2} = 3 \text{ ohms}$
2. $E = ZI = 3 \times 100 = 300 \text{ volts}$

Problem 1-24: Suppose impedances are connected in series-parallel as shown in Fig. 1-55. (1) If 100 amp flow through branch 1-1, what current will flow through branch 2-2? (2) What will the total current in the circuit be? (3) What will the impressed voltage be? (4) What is the power factor of the entire circuit? (5) What is the total power?

Solution:

1. $Z_1 = \sqrt{R_1^2 + (-X_{C1})^2} = \sqrt{1.5^2 + (-1)^2} = 1.803 \text{ ohms}$

$$E_{11} = I_1 Z_1 = 100 \times 1.803 = 180.3 \text{ volts}$$

$$Z_2 = \sqrt{R_2^2 + X_{L2}^2} = \sqrt{0.5^2 + 1^2} = 1.12 \text{ ohms}$$

$$I_2 = \frac{E}{Z_2} = \frac{180.3}{1.12} = 161 \text{ amp}$$

2. $G_1 = \dfrac{R_1}{(R_1^2 + X^2)} = \dfrac{1.5}{(1.5^2 + 1^2)} = 0.461 \text{ mho}$

$$B_1 = \frac{-1}{(1.5^2 + 1^2)} = -0.308 \text{ mho}$$

$$G_2 = \frac{R_2}{(R_2^2 + X_2^2)} = \frac{0.5}{(0.5^2 + 1^2)} = 0.400 \text{ mho}$$

$$B_2 = \frac{1}{(0.5^2 + 1^2)} = 0.800 \text{ mho}$$

$$Y_{12} = \sqrt{(G_1 + G_2)^2 + (B_2 - B_1)^2} = \sqrt{(0.461 + 0.400)^2 + (0.800 - 0.308)^2} = 0.993 \text{ mho}$$

$$I = Y_{12}E_1 = 0.993 \times 180.3 = 179.1 \text{ amp}$$

3. $E_3 = R_3 I = 1 \times 179.1 = 179.1 \text{ volts}$

$$R_{12} = \frac{G_{12}}{Y_{12}^2} = \frac{0.861}{0.993^2} = 0.875 \text{ ohm}$$

$$X_{12} = \frac{B_{12}}{Y_{12}^2} = \frac{0.492}{0.993^2} = 0.50 \text{ ohm}$$

$$E = ZI = \sqrt{(0.875 + 1)^2 + 0.5^2} \times 179.1 = 1.94 \times 179.1 = 347 \text{ volts}$$

4. Power factor $= \dfrac{RI}{E} = \dfrac{(0.875 + 1)179.1}{347} = 0.968 = 96.8\%$
5. Total power $= EI \cos \theta = 347 \times 179 \times 0.968 = 60{,}000 \text{ watts}$

Problem 1-25: A transformer has 1,000 turns in the high-voltage winding and 100 turns in the low-voltage winding. The maximum flux, linking both windings, is equal to 916,000 lines at 60 cycles. (1) What is the rms value of emf induced per turn? (2) What is the total rms value of voltage induced in both coils? (3) What is the ratio of the transformer?

Solution:

1. Induced emf per turn $= \dfrac{4.44 \times f \times \phi_M}{10^8}$

$$= \frac{4.44 \times 60 \times 916{,}000}{10^8} = 2.44 \text{ volts rms}$$

2. Total emf $= 1{,}000 \times 2.44 = 2{,}440$ volts in the high-voltage winding
$= 100 \times 2.44 = 244$ volts in the low-voltage winding

3. Transformer ratio $= \dfrac{N_1}{N_2} = \dfrac{1{,}000}{100} = 10$

Problem 1-26: If the transformer of Prob. 1-25 were designed for an induced voltage of 1.5 volts per turn, what would the total flux have to be?

Solution:

$$\phi_M = \frac{e \times 10^8}{4.44 \times f} = \frac{1.5 \times 10^8}{4.44 \times 60} = 564{,}000 \text{ lines}$$

Problem 1-27: In Prob. 1-25, what would the induced voltage per turn be if the frequency were 25 cycles?

Solution: The induced voltage varies directly as the frequency. Then,

$$\frac{E_{25}}{N} = \frac{f_{25}}{f_{60}} \times \frac{E_{60}}{N} = \frac{25}{60} \times 2.44 = 1.017 \text{ volts per turn}$$

Problem 1-28: A 1,000-kva transformer has a rated no-load energy loss of 2,170 watts and a rated loss of 11,170 watts at full load, at 75 C. The impedance is equal to 5.8 percent. What is the reactance equal to?

Solution:

$$\% R = \frac{\text{copper loss in watts}}{\text{kva} \times 10} = \frac{(11{,}170 - 2{,}170)}{1{,}000 \times 10} = 0.90\%$$

$$\% X = \sqrt{\% Z^2 - \% R^2} = \sqrt{(5.8)^2 - (0.9)^2} = 5.73\%$$

Problem 1-29: A 25-kva 2,400-volt to 120/240-volt transformer has the following rated characteristics: iron loss, 115 watts; regulation at unity power factor, 1.59%; copper loss at rated load, 390 watts. (1) What is its reactance? (2) What is its impedance?

Solution:

1. $\%$ regulation $= \% R \cos\theta + \% X \sin\theta + \dfrac{(\% X \cos\theta - \% R \sin\theta)^2}{200}$

$\theta = 0°$ at unity power factor $\qquad \cos\theta = 1 \qquad \sin\theta = 0$

$$\% R = \frac{\text{watts copper loss}}{\text{kva} \times 10} = \frac{390}{25 \times 10} = 1.56\%$$

$$1.59\% = (1.56 \times 1) + (\% X \times 0) + \frac{[(\% X \times 1) - (1.56 \times 0)]^2}{200}$$

$$1.59\% = 1.56\% + \frac{\% X^2}{200}$$

$$\% X = \sqrt{(1.59\% - 1.56\%)200} = 2.45\%$$

2. $\% Z = \sqrt{\% R^2 + \% X^2} = \sqrt{(1.56)^2 + (2.45)^2} = 2.9\%$

Problem 1-30: What is the efficiency of the transformer of Prob. 1-29 at 100 percent load? At 150 percent load?

Solution:

$$\frac{KP \times 100}{KP + W_i + K^2W_C} = \frac{1 \times 25 \times 1{,}000 \times 100}{25{,}000 + 115 + 390} = 98.02\% \text{ at } 100\% \text{ load}$$

$$= \frac{1.5 \times 25{,}000 \times 100}{(1.5 \times 25{,}000) + 115 + (1.5 \times 1.5 \times 390)}$$

$$= 97.42\% \text{ at } 150\% \text{ load}$$

Problem 1-31: What is the impedance, in ohms, of the transformer in Prob. 1-29 (1) on the high-voltage side? (2) On the 240-volt side?

Solution:

1. $Z = \frac{\%\, Z \times E^2}{\text{kva} \times 10^5} = \frac{2.9 \times 2{,}400 \times 2{,}400}{25 \times 10^5} = 6.68 \text{ ohms at } 2{,}400 \text{ volts}$

2. $Z_s = \left(\frac{N_s}{N_P}\right)^2 Z_P = \left(\frac{240}{2{,}400}\right)^2 \times 6.68 = 0.0668 \text{ ohm at } 240 \text{ volts}$

CHAPTER 2

TRANSMISSION TERMINAL FACILITIES

By Francis S. Benson

Pacific Gas and Electric Company

2-1. The central-station system is a highly sensitive arrangement of equipment which enables power to be produced at one location for immediate use at another location many miles away. Central generation of power has many advantages from investment and operating-cost standpoints and because noise and dust are confined to areas where the community as a whole will not be disturbed.

In most regions a transmission system must be used to carry the energy from the generating station to a series of substations. The transmission system carries the large quantities of energy which can be most economically transported on relatively thin conductors at high voltage over long distances. Then at the areas of energy consumption the transmission lines deliver energy to the substations which reduce the voltage to safer values for circuits running through inhabited localities, Fig. 2-1.

The distribution circuits in turn feed transformers, carried on poles or located in underground manholes, that finally reduce the voltage to the magnitude at which it will be used by the consuming apparatus.

In this chapter we will talk about the equipment connected to each end of a transmission line (1) at the generating station and (2) at the substation. In Chap. 3 we'll talk about the transmission line itself. In "Steam-electric Central Stations,"* Chaps. 10 and 11, there will be more detail of bus and switch arrangements in the generating station.

2-2. Generating stations manufacture electrical energy from chemical energy stored in raw fuel or from large quantities of water falling through a vertical distance. Steam-electric plants use coal, oil, or gas as a source of energy. Internal-combustion engines use oil or natural gas. Hydraulic stations use falling water.

Energy transformations need (1) raw energy in the form of fuel or water, (2) equipment to convert this energy to a mechanical form (a

* B. G. A. Skrotzki, "Steam-electric Central Stations," McGraw-Hill Book Company, Inc.

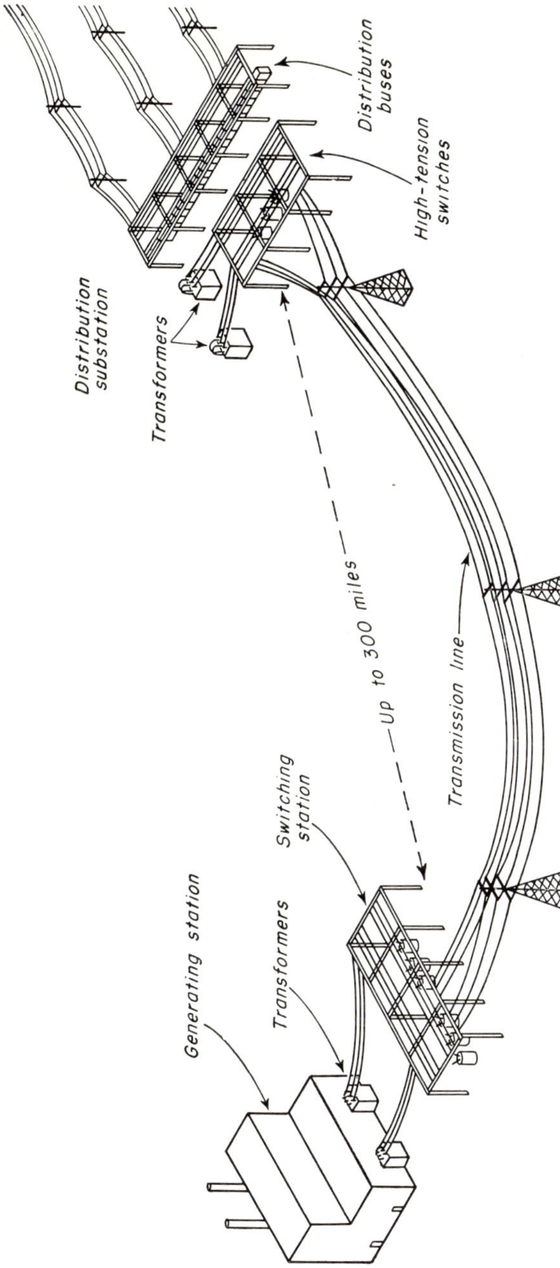

FIG. 2-1. Suggestive outline of elementary overhead high-tension transmission system. Actual systems have several generating stations and many distribution substations tied together through several transmission lines.

rotating shaft), (3) apparatus to convert mechanical to electrical energy, (4) a system to convey and control the electrical energy to the point of use. In addition, auxiliary devices use some of the electrical energy to keep the station running.

A simple example of the steps involved in energy generation is a single-unit diesel electric station, Fig. 2-2. A fuel-oil tank feeds raw energy to the reciprocating engine which converts the chemical energy to mechanical form. The ac generator coupled to the engine changes mechanical energy to electrical energy. The electrical energy flows to the bus from which it is distributed to the consuming areas through overhead or underground feeders.

Motor-driven auxiliaries use some of the electrical energy to keep the station running. These include pumps for engine cooling and for lubri-

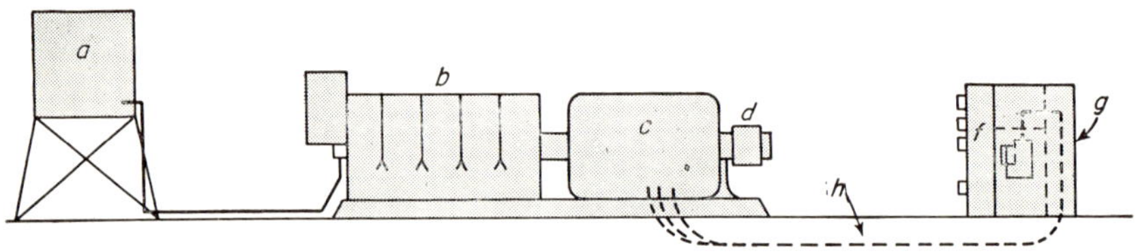

FIG. 2-2. Schematic arrangement of a simple diesel-electric generating station. Fuel oil in tank *a* feeds reciprocating diesel engine *b* which drives electric generator *c*. Cables *h* carry electric energy to bus switchgear *g* controlled by instruments and push buttons on switchboard *f*.

cating-oil supply, fuel-transfer pumps, lighting, ventilating fans, and so forth.

Details of steam-electric stations are given in "Steam-electric Central Stations," and information about hydro, reciprocating-engine, gas-turbine, and mercury-steam stations will be found in "Hydro, Diesel, and Other Stations."*

From an electrical standpoint, the transmission system might be considered to start with the electrical generator. In many cases, though, the transmission system is assumed to start at the terminals of the high-tension transformers. In any event, all the electrical equipment must be integrated properly to achieve a practical workable system.

The usual generating station has more than one generator driven by a prime mover (steam turbine, diesel engine, hydro turbine). There are several reasons for this. (1) An electrical system grows with advancing years—as demand for more energy increases, additional units are installed to raise station capacity. (2) There is a limit to the maximum capacity that can be built in a single prime mover or generator. When the total load is greater than the capacity of a single unit, several units must be

* B. G. A. Skrotzki, "Hydro, Diesel, and Other Electric Stations," McGraw-Hill Book Company, Inc.

installed. (3) Prime movers must be taken out of service periodically for maintenance and inspection; so it often pays to install several small units rather than one large unit. They can then be taken out in rotation without reducing the ability of the station to carry the load.

So we find that thousands of individual varying demands for electrical energy converge on the distribution substations through the distribution system. These then appear as single large demands that must be supplied by the transmission lines from the generating station. The generating station then simply knows it must supply a given total electrical demand which is the sum of the demands of the several transmission lines radiating from the station.

The generating-station electrical system (bus) must be arranged so that any of its generators can supply this energy. In one sense all generators dump their output into a common pool, the station bus system, and the transmission circuits or lines draw out the energy as needed. But, there is no "reserve" of energy in a bus system, in the sense that if all generators were cut off simultaneously, the transmission lines could still draw some energy. Electric generators normally never supply more energy than that needed by the energy-consuming apparatus.

The bus system must be built to meet several conditions. (1) It must be flexible enough so that any generator can feed any transmission line. (2) It must have adequate capacity to carry all loads placed upon it. (3) It must be reliable enough to give substantially continuous service when properly operated and maintained. (4) Construction must be adequate to withstand abnormal electrical and mechanical forces that can be foreseen and simple enough to avoid unnecessary cost. (5) The bus must be able to withstand the effects of short circuits or grounds by the aid of adequate relaying and circuit-breaker protection and without endangering personnel or equipment. (6) Space needs should be a minimum without sacrificing reliability, safety, and maintenance convenience.

2-3. Indoor AC Bus Structures. Indoor buses may be separated into two types, open and enclosed. The open type is used for low-voltage (up to 5,000 volts) for relatively low capacity, that is, short-circuit duties of 100,000 kva or less. Circuit breakers are rated by their ability to interrupt short circuits. Unless otherwise indicated, the usual interrupting capacity of a circuit breaker is based on the three-phase instantaneous short-circuit kva with a symmetrical waveform. The buses, insulators, and disconnecting switches must all withstand the mechanical forces produced by such short circuits.

Almost all systems in the United States are three-phase. There are a few two-phase systems in operation. The two-phase bus requires four assemblies of insulators and copper bars. The three-phase system has only three assemblies of insulators and copper bars. The buses are mounted on top of the frame which braces and supports the switchboard. The oil circuit breakers are mounted behind switchboard panels on the wall or on a separate pipe or angle-frame structure. Old switchboards had small circuit breakers mounted directly on the rear of the switchboard panel, Fig. 2-3.

Electrical equipment may be enclosed in structures built of concrete or some inorganic material such as transite. There is a variety of such

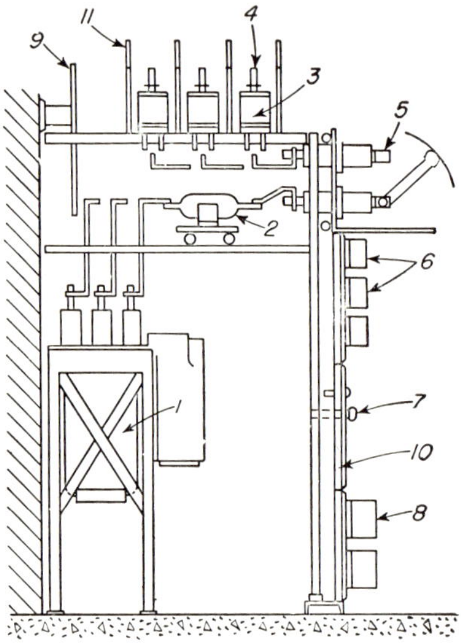

Fig. 2-3. Older type of indoor three-phase ac bus arrangement. 1, oil circuit breakers; 2, current transformers; 3, bus supports; 4, bus bars; 5, disconnecting switches; 6, ammeters; 7, control switch; 8, relays; 9, outgoing feeder bars; 10, ebony asbestos panels; 11, insulating barriers between bus bars.

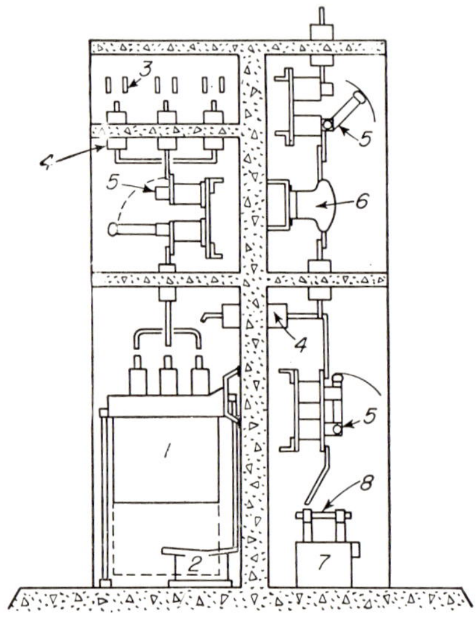

Fig. 2-4. Indoor concrete cell structure for three-phase ac bus. 1, indoor oil circuit breaker; 2, solenoid; 3, main ac bus; 4, porcelain bushing; 5, single-pole disconnecting switches; 6, current transformer; 7, potential transformer; 8, transformer fuse.

designs for single-bus arrangements, double-bus arrangements, etc. The buses may be mounted horizontally or vertically. Disconnecting switches isolate the circuit breaker so that maintenance work may be done on it without deenergizing the bus. The circuit breaker is installed in a separate cell, Fig. 2-4.

Modern switchgear is enclosed in steel cells called cubicles or it may be metal-clad and have a removable circuit breaker. The removable elements are of the draw-out or lift-up type. The metal-enclosed cells have the advantage of being grounded and provide an effective strong barrier between adjacent cells and buses. Circuit breakers may be of the magnetic air-break type, the compressed-air type, or oil circuit breakers. They are built with voltage ratings up to 15 kv and interrupting duties up to 1½ million kva. Bus assemblies may be a single bus, a double bus, or a main bus and a transfer bus. The buses may be built in one enclosure, or each bus may be in a separate metal enclosure. Cell structures may be on different floor levels, the main and transfer buses being on one floor and the steel cells containing the circuit breakers, current transformers, and potential transformers being on another floor. When disconnecting switches are used, they are interlocked with the circuit breaker, so that they cannot be opened if the circuit breaker is closed. The cell doors are usually interlocked so that the doors cannot be opened unless the circuit breaker is open.

The buses may be made of flat bars, rods, or tubes of copper or aluminum. Tubes may be round or square. The square copper tubes are

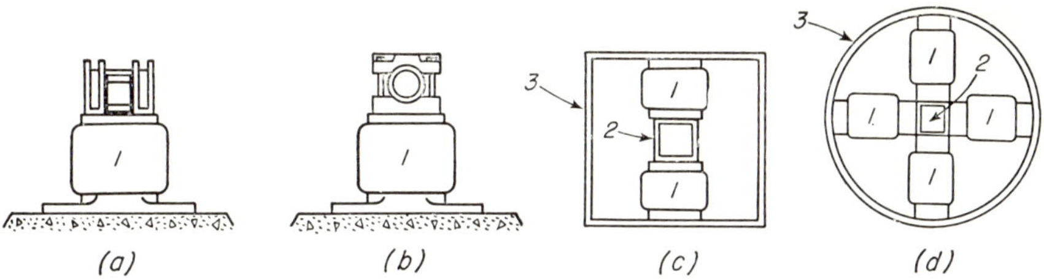

Fig. 2-5. Bus-bar supports must withstand mechanical forces created by short-circuit current flows. (*a*) Four copper bars make up this single-phase bus. (*b*) Support for tubular copper bus. (*c*) Two supports give needed strength to rectangular bus in square metal housing. (*d*) Four supports hold square bus in round housing: 1, porcelain bus supports; 2, hollow square bus; 3, steel enclosure.

either ventilated or nonventilated. The ventilated tubes have holes drilled through the surface which allow the passage of air, thereby cooling the bus and permitting a greater current to be carried.

Bus bars must be designed to withstand the enormous mechanical forces produced by large short-circuit currents, Fig. 2-5. Bus-bar assembly *a* has four copper bars separated by a 2- or 2½-in. spacer. The spacer allows more current to be carried. Insulator supports have various mechanical ratings and may be obtained with a strength to resist a cantilever force of 6,000 lb at a point 2 in. above the top of the insulator.

Bus-bar assembly *b* shows a copper tube. These are usually iron-pipe size and have a wall thickness equal to standard pipe. They may be obtained, however, in heavy and extra-heavy designs.

Assembly *c* shows two bus-bar insulators used to develop the necessary mechanical strength.

Assembly *d* is a heavy-duty bus type. Each phase stands within a circular nonmagnetic-metal housing. The four insulators develop the required mechanical strength. This is the isolated-phase type. The bus housing may be rectangular as well as round.

In a segregated bus assembly, a single metal panel or sheet separates one phase from another, Fig. 2-6. Segregated buses are popular because they provide excellent phase separation and occupy less space than the isolated-phase type.

A short circuit between the bus copper and the metal barrier separating the phases may burn a hole through the phase barrier causing a bus-to-bus (or line-to-line) short circuit instead of a line-to-ground short circuit.

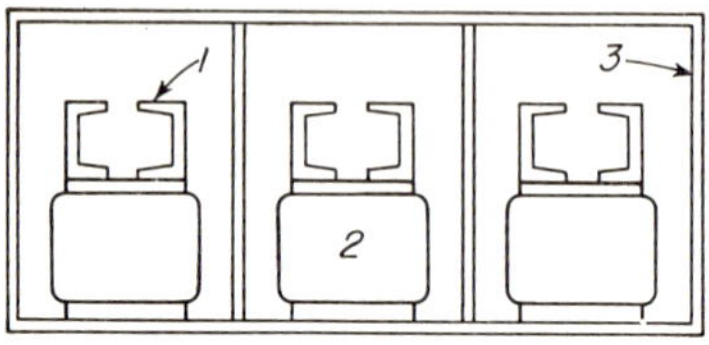

Fig. 2-6. In segregated bus assemblies metal panels separate phases. 1, channel bus bar; 2, porcelain bus supports; 3, steel enclosure.

Metal-enclosed bus structures have been built of aluminum, copper, or nonmagnetic steel. The latter has become more popular since tools were developed to cut and drill this extremely hard steel, Fig. 2-7.

Bus structures are sometimes separated from the circuit-breaker equipment by masonry or concrete walls, or they may be installed in a prefabricated bus structure containing the bus, circuit breakers, bushing current transformers, and disconnecting switches.

Bus structures may be arranged with one bus, Fig. 2-8, or with a double bus, Fig. 2-9, or with a main and transfer bus, Fig. 2-10. The single bus

Fig. 2-7. Ventilated hollow-square bus mounted in aluminum housing. Holes in the bus bar provide natural draft ventilation. A strip of bar copper serves as a ground bus for aluminum cells.

is becoming quite popular with modern switchgear as the insulation levels are so high that bus failure is infrequent. Also, buses are now built near the enclosure top instead of near the floor, as in older structures. Mice and rats caused many failures when the buses were near the floor. The double bus is used for high reliability or at generating stations where switching may be needed to supply power to a feeder circuit from different generators. The main and transfer bus is now used for most distribution work. One circuit breaker serves each circuit. The transfer bus is a standby for emergency use. For more bus arrangements, see "Steam-electric Central Stations," chapter on Central Station Switching.

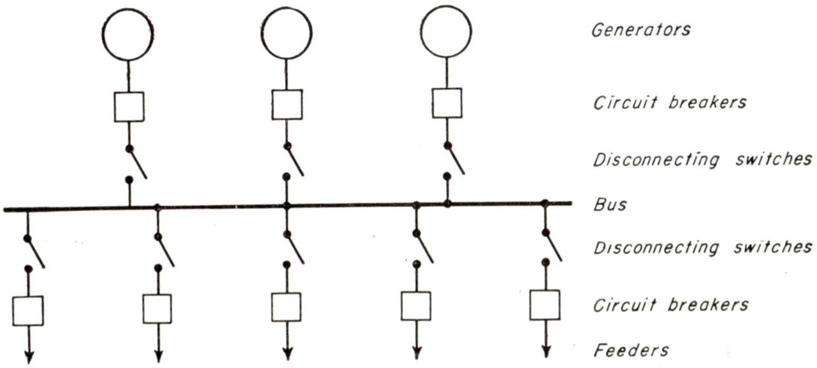

FIG. 2-8. Single-bus arrangement needs minimum equipment, can be used for small energy flows or where equipment has high reliability for larger energy amounts.

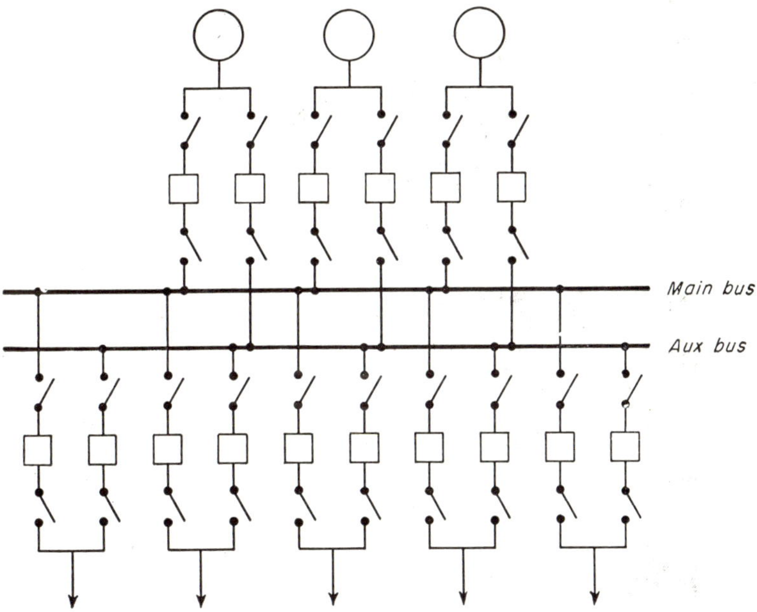

FIG. 2-9. Double-bus double-breaker arrangement has high degree of reliability and more flexibility in operation.

Prefabricated switchgear can be purchased with the three bus arrangements mentioned, making a flexible station design, Fig. 2-11. Many buses in metal-enclosed structures, particularly those having circuit breakers of comparatively low interrupting duty (500,000 kva), are insu-

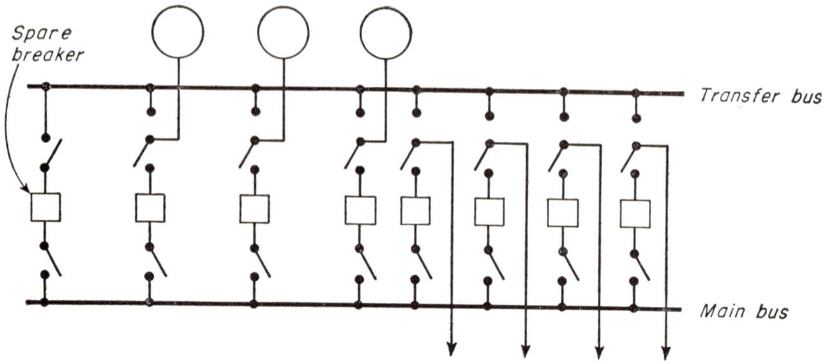

FIG. 2-10. Main and transfer bus achieves flexibility and service continuity with fewer circuit breakers.

FIG. 2-11. One room of double-bus structure using metal-enclosed bus and disconnecting switches. 4,000-amp bus stands behind locked doors, right, while cells on left enclose fused potential transformers. Bus is designed for 1,000,000-kva short-circuit capacity. (*Courtesy of Pacific Gas and Electric Company.*)

lated for full voltage, Fig. 2-12. The capacity of such bus structures is limited to 2,000 amp, which is the capacity of the largest circuit breaker that can be installed in such switchgear. The buses are insulated with material molded around the bar assembly or by a taping method, Figs. 2-13 and 2-14. The insulated buses can be placed very close together, separated and supported by Micarta or Herkolite insulation. The entire assembly of the three buses is enclosed in metal sheeting, Fig. 2-15.

2-4. Outdoor Structures and Buses. Electrical generating stations and substations use outdoor buses where local conditions, ordinances, or

FIG. 2-12. Metal-enclosed bus for 12 kv has magnetic air circuit breakers of 500,000-kva interrupting duty. The front of the structure used as a switchboard panel holds ammeters, voltmeters, relays, control switches, etc. Metal-enclosed bus bars rise from the top of the structure and pass through the wall to the equipment on the other side. (*Courtesy of General Electric Company.*)

property values prove them to be economical. An outdoor-bus structure requires much greater land area than an indoor bus. The outdoor bus has similar functions to the indoor bus, that is, it collects and distributes electrical energy via connecting feeders and station tie lines. Since outdoor stations require a much larger parcel of property, they are usually economical only in rural areas, suburban areas, or on the fringe of dense residential areas.

Electrical clearances, line-to-line and line-to-ground, are much greater for outdoor installations, requiring rather wide and high structures.

There are many types of outdoor stations, some are quite small, using a framework made of 1½- and 2-in. pipe. Such a structure supports buses, disconnecting switches, etc. Other outdoor stations may be very large, the structure using lattice-steel poles or structural-steel members for supporting buses, disconnecting switches, etc.

Outdoor buses use either copper or aluminum as conductors. Aluminum has the advantage of being much lighter in weight, while copper has greater electrical conductivity. Aluminum has about 67 percent of the conductivity of copper per unit area. That is, an aluminum bus is about 150 percent greater in area than a copper bus for the same current. As the price of copper advances, aluminum becomes more attractive. Care must be taken when joining copper to aluminum, to exclude air or moisture from the joint and prevent excessive corrosion and oxidation.

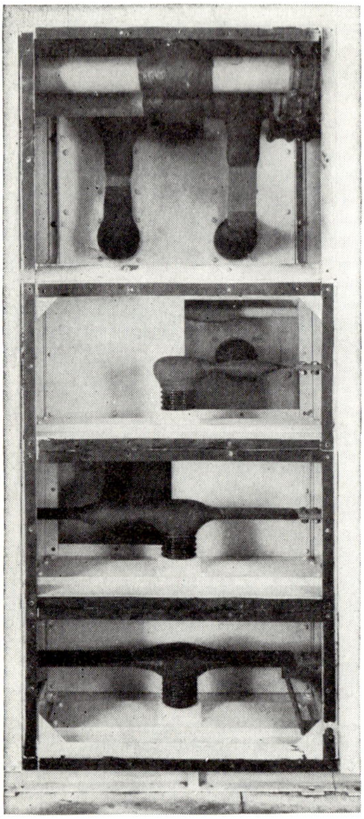

FIG. 2-13. Steel bus structure with insulated buses. Three lower buses are flat bar insulated for 15 kv. Bare splicing connection will be taped for full voltage. Upper round copper tube bus is insulated with molded material for 5 kv; all three phases are in one compartment. (*Courtesy of Electric Power Equipment Corporation.*)

FIG. 2-14. Section of copper tube insulated with molded tubing for 5 kv. Bare copper supporting member will be taped for full voltage when bus is assembled in the metal enclosure. (*Courtesy of Electric Power Equipment Corporation.*)

The term "bus" refers to the main collectors of electrical energy, which may be bars, rods, tubes, or channels. In older days, copper bar was used exclusively for buses. The derivation of bus is from the word omnibus, meaning a collector of things. Thus an electric bus collects or assembles electrical energy at one location, and the bus bar is the conductor through which the electric current flows.

Outdoor buses may be copper bar, rods, pipe, or thin-walled tubing. Generally, less electrical current flows through outdoor buses than indoor buses; therefore the cross-sectional area of the outdoor bus is smaller. Copper pipe or thin-walled tubing is favored because it has enough stiffness to span rather long distances, 15 to 20 ft, without too much sag.

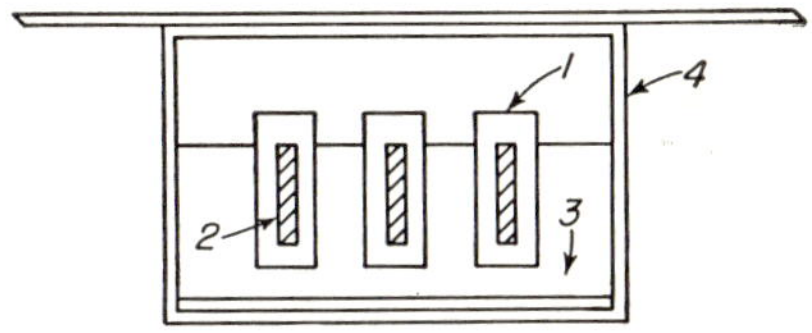

FIG. 2-15. Insulated three-phase bus arranged in one enclosure. 1, molded insulating material; 2, bus bar; 3, insulating support and spacer; 4, steel enclosure.

Outdoor buses are widely spaced to prevent birds with a large wingspread from causing short circuits. Wide spacing also reduces the mechanical strain on insulator supports when short-circuit currents flow in the bus. These produce an extremely dense magnetic field which creates mechanical forces of repulsion and attraction, see Art. 1-9. These forces become very large if the buses are close together and tend to pull them together and force them apart at a rate of 60 times per second in a 60-cycle system. The supporting insulators must have enough cantilever strength to prevent breakage of the porcelain insulators under such abnormal transient conditions.

FIG. 2-16. Low-voltage pin-type insulators for low-capacity substations in rural territory. Small copper-rod buses fasten to the biscuit at the top of the insulator with wires. (*Courtesy of Ohio Brass Company.*)

Since the mechanical forces produced by electrical currents vary inversely as the square of the distance, the mechanical forces decrease very rapidly as the distance between buses increases. If the spacing is doubled, the force is only one-fourth as great; spacing of three times reduces the force to one-ninth; etc.

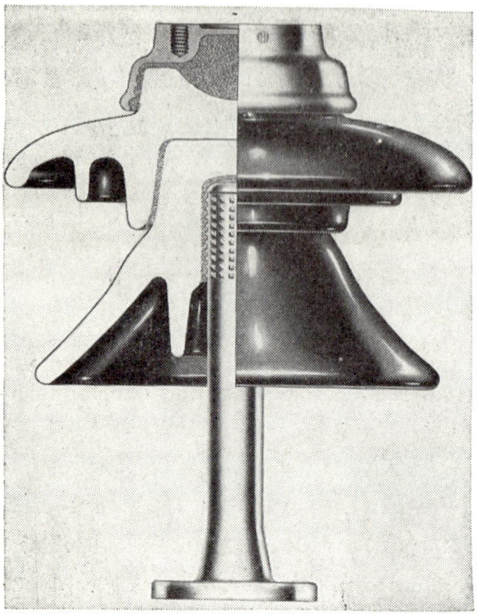

FIG. 2-17. Station pillar-type insulators may be stacked on top of each other. The Top biscuit of insulator cements into the metal cap with four tapped holes equally spaced. Cement fastens the pin into the bottom of the insulator. The base of the pin has four holes matching those in the cap above. (*Courtesy of Ohio Brass Company.*)

Porcelain insulators are used exclusively for outdoor bus supports. They may be the pin-insulator type, Fig. 2-16, the pillar type, Fig. 2-17, or the suspension type, Fig. 2-18. Low-voltage outdoor stations use the pin type or the post type up to 34.5 kv. Above that level, multiple pillar or suspension insulators are used. The insulation of outdoor buses must withstand the atmospheric conditions in the area.

At the seashore, where the prevailing winds blow from the ocean, salt fog, fine salt spray, or mist and vapor are carried many miles inshore, depositing a fine layer of salt on insulators. Under such atmospheric conditions, the station buses must be overinsulated; for example, 23-kv insulators are often installed for 12-kv operation. The voltages are line-to-line.

In industrial areas where soot, acid gas, or dust, such as magnesium or cement dust, contaminates the air, the buses must also be overinsulated.

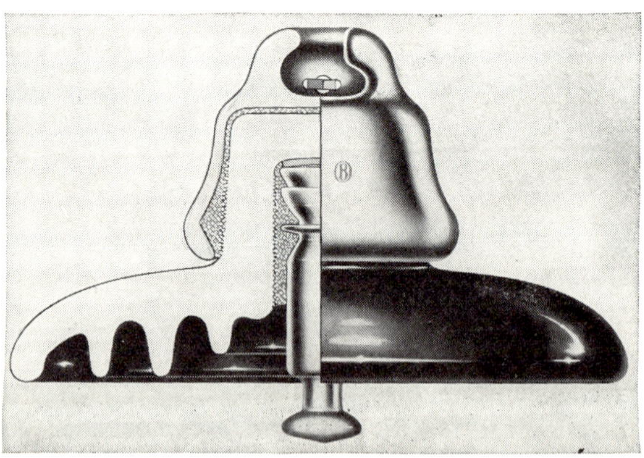

FIG. 2-18. Ball-and-socket type suspension insulator used for station voltages above 60 kv. (*Courtesy of Ohio Brass Company.*)

However, if the area has frequent rainstorms, especially driving rains, the insulators will be washed relatively clean and little or no overinsulation is needed.

In areas that have a cyclic dry period, for 3 or 4 months yearly, buses must be overinsulated as an accumulation of dust during dry months causes a decrease of effective leakage distance over the insulator from conductor to pin support. If the first precipitation of the rainy season is a fine mist or drizzle, the lowered effective insulated leakage distance may allow an excessive leakage current to flow which may result in an insulator flashover. These areas also require overinsulation of the buses to maintain trouble-free operation and low insulator maintenance.

Some operating companies establish a practice of washing or wiping station-bus insulators periodically. This may be done as often as three times a year, depending on the contamination of the air in the area. The bus is usually deenergized when insulators are wiped. This is laborious and costly, but may be necessary when the contaminated matter adheres tightly to the procelain surface. When insulators are washed, the bus does not need to be deenergized. A jet of water impinging on an insulator will not cause a flashover unless the insulator is cracked or otherwise defective. Care must be taken to ground the hose nozzle. Water having a high electrical conductivity, such as salt water, must not be used. A gasoline-driven high-pressure pump (600 to 850 psi) is generally used. A hose nozzle of special design produces a round jet of water with a high velocity, striking the insulator surface with impact. A good impact produces enough mechanical force to remove stubborn material adhering or sticking to the insulator surface.

The hose nozzle should be grounded for several reasons: (1) to drain off the static charge produced by the water flowing through the nozzle, (2) to shunt the current flowing to ground from the live conductor through the stream of water. The threshold of feeling of electric shock of the average individual is 1 milliampere. Electrical currents of more than 6 ma may be dangerous, depending upon the body resistance of the individual.

Many areas have frequent rainfall with little or no contamination in the air. Electric stations built in these areas do not overinsulate but use standard insulators for bus supports.

Supporting structures for buses may be wood or steel. Wood is often used for low-voltage or temporary stations. Permanent structures are generally of galvanized pipe or structural-steel members.

Stations built with wood supports are relatively simple. The transformers and regulators rest on wood cribbing, serving as a foundation. Wood poles support other wood members, which in turn act as a support for fuses, disconnecting switches, and buses. These substations are usually low voltage, although they have been used for stations with a high-voltage rating of 115,000 volts, line-to-line. They are relatively low capacity, 1,500 to 10,000 kva, and usually of the single-circuit type having only one bus. A high-voltage fuse, Fig. 2-19, protects the high-voltage winding of the transformer. The outgoing circuit may be equipped with a single disconnecting switch or an oil circuit breaker with its complement of disconnecting switches and by-pass switch.

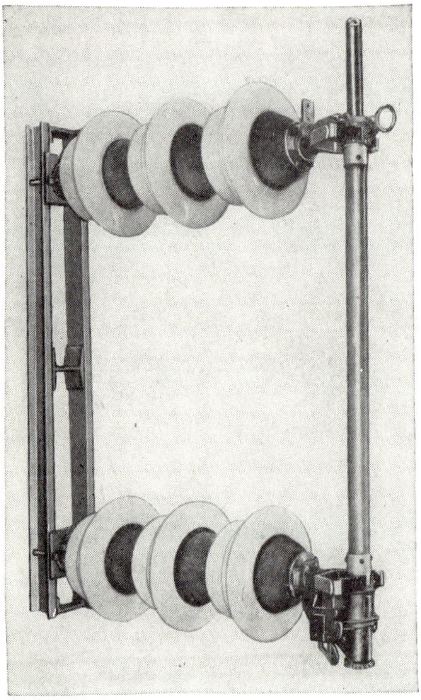

FIG. 2-19. A 115-kv fuse with drop-out feature. When the fuse blows, it disengages itself from the upper clip and hangs from the lower clip. (*Courtesy of Schweitzer and Conrad Company.*)

Structures use steel because it has strength with lightness, is rigid and permanent. Steel takes many forms, such as pipe, tubular poles, lattice poles, and structural members, such as columns, beams, and girders.

Iron-pipe structures are popular for low-voltage low-capacity stations. The height of such structures usually does not exceed 15 to 20 ft. Pipe-frame assemblies are generally braced with clamp-type fittings. If pipe structures become too high or long, it is difficult to brace them securely for stiffness as the pipe slips through the clamp-type fittings.

Structural shapes are also used for low substation structures, Fig. 2-20. They are more rigid than pipe frames. They have the disadvantage, however, that they must be riveted or welded while being erected. It is often difficult to obtain riveters in outlying areas where substations are built. The structural frame is a completely designed unit manufactured in a steel-fabricating shop. It is very difficult to make additions, alterations, or changes in the field. Sometimes, interferences develop in the field which require an immediate modification of the erection of the completed structure. In such cases, a structural-steel frame may cause considerable delay in the completion of the station. On the other hand, if a pipe-frame structure is used, alterations may be made quickly in the field. If additional steel members are required, the pipe can be purchased at the nearest rural hardware or general store.

Large outdoor stations use lattice-steel poles and members such as shown in Fig. 2-21. These are relatively light and strong. Such a structure is completely engineered; alterations cannot be made in the field.

Some high-voltage bus structures have the buses mounted high in the air well above the disconnecting switches. Others have the buses mounted on short steel pedestals on the ground. The low-bus structures usually require more ground area than do the high ones. They must be fenced in so that all personnel are kept out of the bus area.

Fig. 2-20. Small outdoor substation using structural-steel shapes to support insulators for incoming and outgoing lines. A three-phase outdoor transformer connects to the outdoor metal-enclosed switchgear. (*Courtesy of General Electric Company.*)

2-5. Bus Arrangements. Common outdoor bus arrangements are the single-bus, double-bus, main-and-transfer-bus, and ring-bus. Each bus arrangement has its own special application. The single bus is the one in common use and has the simplest design. It is generally used in small outdoor stations having relatively few outgoing or incoming feeders and lines. The single-bus design, Fig. 2-22, may be used for either distribution stations or switching stations. Distribution stations are step-down stations, reducing voltage to that suitable for distribution of electrical energy to customers, whether they be industrial, commercial, agricultural, or residential. Switching stations have one common high voltage, or they reduce voltage from a very high voltage to a transmission voltage, such as from 230 kv to 138 or 115 kv.

The double-bus design, Figs. 2-23 and 2-24, is used with large switching stations. Because of the higher cost of oil circuit breakers only one circuit breaker is installed for each double-bus position.

The double-bus design has the advantage that a given power system may be separated or divided into two isolated systems for operation, if

FIG. 2-21. Typical steel structure for a high-voltage transmission substation. Latticed steel members make up columns and beams. Oil circuit breakers, right, have potential devices mounted on sides of tanks, replacing potential transformers. The stack of insulators, left foreground, is coupling capacitor for carrier-current equipment. (*Courtesy of Pacific Electric Manufacturing Company.*)

necessary. One system is connected to one bus, the other part of the system to the remaining bus. This may be desirable to reduce short-circuit duty on the breakers. The double-bus design also allows a power system to be changed from a loop-system arrangement to a radial system or to two radial systems.

A switching station is also used to bring electrical energy to a common area by two or more lines and then retransmit this energy to other localities over other tie lines. It may also be used to open or break up a long high-voltage circuit (several hundred miles) into several shorter ones. When two long parallel lines transmit a large block of power (100,000 to 200,000 kva), for several hundred miles, switching stations are often provided every 75 to 100 miles. This allows a faulty section of line to be cleared, or sectionalized, without losing system stability. The remaining circuit in the faulted section can carry the load with only slight loss

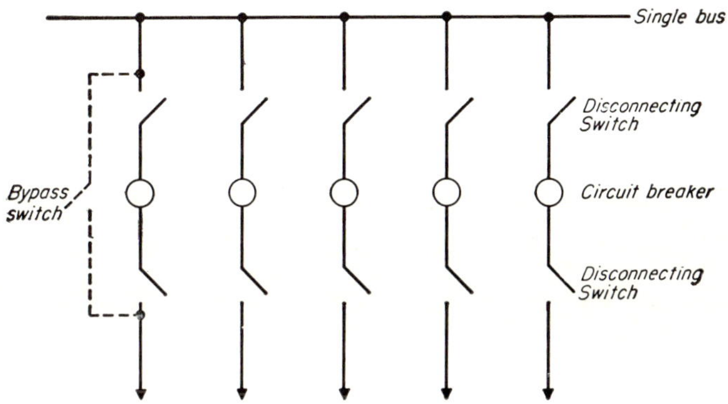

FIG. 2-22. Single bus has circuits connected through single oil circuit breakers. The breakers may be by-passed by connections shown dotted. When the bus fails, it goes out of service until faulty insulators are replaced.

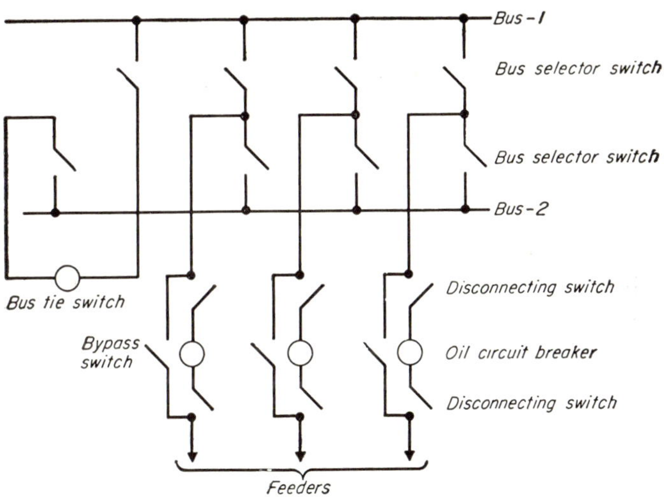

FIG. 2-23. Double bus design for large switching stations. Bus selector switches are gang operated and used to select the proper bus. Bus tie switch connects two buses together while changing buses so that no current is broken by selector switches, also used as spare feeder circuit breaker when regular feeder breaker is out of service.

of total capacity. Such intermediate line-switching stations are controlled automatically by carrier-current relays and by high-speed supervisory control. Two single-line diagrams of such intermediate switching stations are shown in Figs. 2-25 and 2-26.

Figure 2-25 shows a single-line diagram of a circuit-breaker arrangement for switching two parallel lines into two sections. This arrangement requires five circuit breakers while that of Fig. 2-26 needs only four to accomplish the same result.

The high-voltage ring bus, Fig. 2-27, is not so flexible an arrangement as the double bus. It is, however, a useful design for limited areas with both indoor and outdoor bus installations. It is economical with respect to the number of circuit breakers needed for satisfactory operation. Relaying, however, is more complicated and requires careful planning and designing for this type of bus installation. It has the disadvantage that, when a line relays out of service, two circuit breakers open,

FIG. 2-24. Double-bus 110-kv switching station with bus selector switches mounted on horizontal framing below buses. Oil-circuit-breaker structures stand on each side of the bus. Circuit-breaker disconnecting switches are mounted on each side of switch tower above breakers; bypass switch is mounted on top of tower. (*Courtesy of Pacific Gas and Electric Company.*)

destroying the ring effect. The bus arrangement is then changed into a sectionalized single bus.

The main-and-transfer bus, Figs. 2-28 and 2-29, used for stations supplying power to distribution circuits, is in effect a modified double-bus type. It is used more commonly for low-voltage design but has been applied to stations having voltages up to 70 kv.

The sectionalized single-bus arrangement is convenient for switching stations where two transmission lines supply energy into a sectionalized bus and two or more lines radiate out from the bus to other substations within the area.

The bus-section switch is valuable for splitting or sectionalizing a bus, thereby reducing the short-circuit duty on the circuit breakers. Sectionalizing a bus may be used to control the flow of reactive kilovolt-amperes (kvar) which is often important. The bus-section switch in some systems may close a long loop of transmission lines, while opening the loop will change the system to a conventional radial system.

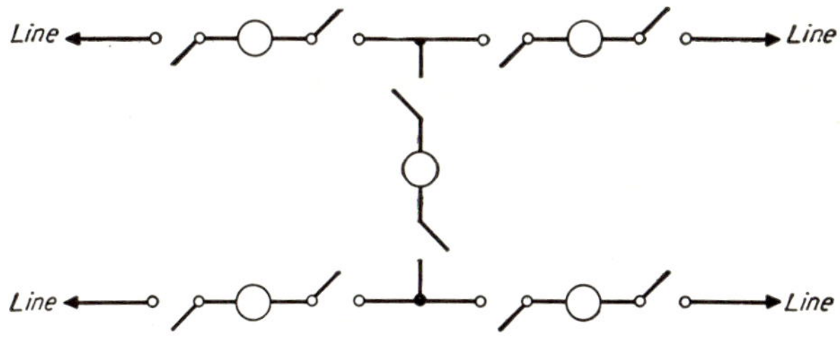

FIG. 2-25. Intermediate switching station for two long high-voltage transmission lines. Each oil circuit breaker can be isolated by opening disconnecting switches.

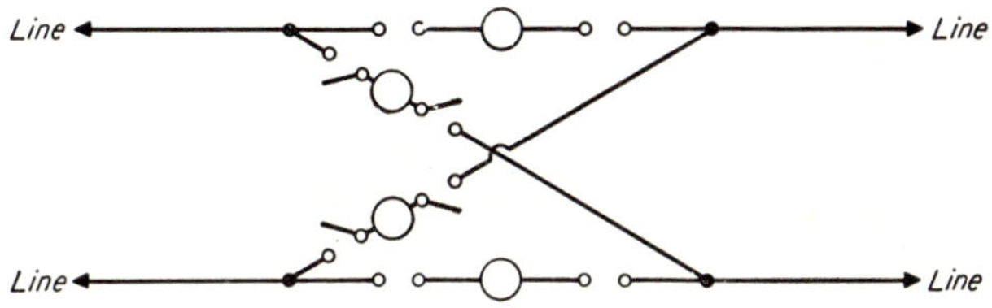

FIG. 2-26. Intermediate switching station for two long high-voltage transmission lines using only four oil circuit breakers.

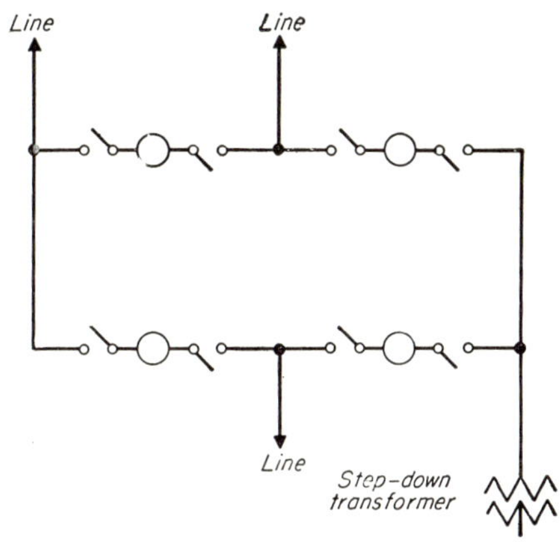

FIG. 2-27. Ring bus tying together three lines and one step-down transformer.

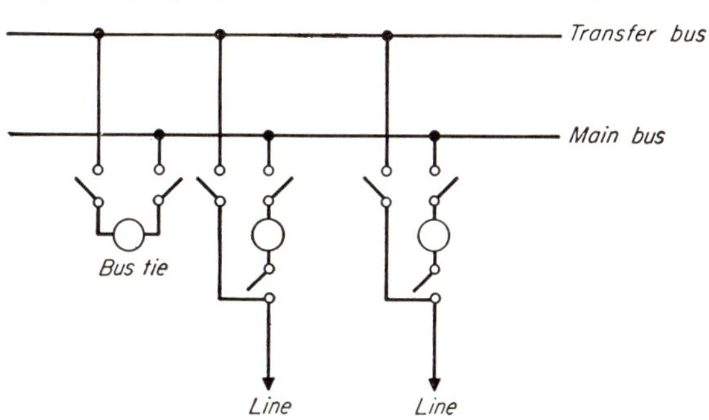

FIG. 2-28. Main and transfer bus is economical from circuit-breaker standpoint; the bus tie breaker acts as a spare to the line circuit breakers.

2-6. Substation Layouts. Transmission substations are those classed as transmitting electrical energy at 60 kv and above, including 287 kv. Transmission substations have a different design than distribution substations, which supply electrical energy at relatively low voltages (up to 22 kv) for distribution to customers or to their substations. These stations supply electrical energy in small blocks of power such as 1,500 kva for a 4-kv feeder, 6,000 kva for a 12-kv feeder, while transmission substations are designed to receive and transmit large blocks of electrical energy amounting to several hundred thousand kva.

Transmission substations may also control power-system voltage and power factor by synchronous condensers or static capacitor banks, Figs.

Fig. 2-29. A 60-kv substation using main and transfer bus scheme. Oil circuit breakers stand under bus structure supporting disconnecting switches. (*Courtesy of Pacific Electric Manufacturing Company.*)

2-30 and 2-31. Synchronous condensers are special synchronous motors with large field windings that can supply leading or lagging kvar by adjusting the field current, see Art. 1-29. A static capacitor bank consists of groups of outdoor condensers (similar to a large radio condenser) connected in series-parallel combinations to produce the desired capacitive kilovolt amperes (ckva). Capacitor banks are usually divided into small groups that may be switched in or out of service by a proper size oil circuit breaker. An oil circuit breaker of the proper size is one designed with arc-interrupting contacts that interrupt a current *leading* the voltage by 90°. Some circuit breakers are not capable of interrupting such a leading current without showing considerable distress, such as throwing oil.

Transmission substations, Fig. 2-32, are often the terminus of long high-voltage lines, such as 230-kv lines several hundred miles long. Such lines exhibit a capacitive effect that results in the receiving-station voltage being considerably greater than the sending-station voltage during light loads, see Art. 1-21. This excessive voltage is caused by the capacitance of the lines. As the load current increases, the capacitance voltage is counteracted by the inductive voltage so that at full load the receiving-station voltage is less than the sending-station voltage. Synchronous condensers regulate the receiving-station voltage by supplying

FIG. 2-30. A 230-kv bus and switch structure often located at generating stations. Capacitor-type potential devices are mounted on switch tanks at front. The operating mechanism for the breakers is mounted between two phase tanks. (*Courtesy of Pacific Electric Manufacturing Company.*)

lagging kvar to the high-voltage lines during light loads, thus reducing the receiving-station voltage. They also supply leading kvar to the lines during full-load periods, thus increasing the receiving-station voltage.

Regulating transformers control the direction of flow of electrical energy by varying the voltage. Energy flows from a higher to a lower potential. These transformers are particularly effective for regulating voltage when used with autotransformers, Fig. 2-33. An autotransformer may be used to supply power from a 220-kv to a 110-kv network. It may also have a synchronous condenser connected to and operating off a delta-connected tertiary (third) winding. The regulating transformer would be connected in the 110-kv lead which regulates the volt-

FIG. 2-31. A hydrogen-cooled synchronous condenser with shaft exciter and collector rings is sealed in a hydrogen-filled tank. Gas pressure may vary from ½ to 15 psig. Hydrogen causes less windage loss than air and has a higher heat-transfer rate. Pipes on the foundation side carry cooling water to heat exchangers in the casing. Electrical leads run from the top of the casing to the bus. (*Courtesy of Westinghouse Electric Corporation.*)

FIG. 2-32. A 230-kv switching station with two buses and selector air switches. One oil circuit breaker is used for each line or position. These breakers have an interrupting duty of 2,500,000 and 3,500,000 kva. (*Courtesy of Pacific Gas and Electric Company.*)

age to hold the 110-kv network at a constant voltage while the 220-kv network voltage varied up and down with load changes. The regulating transformer could raise or lower the 110-kv network voltage so that energy can flow from the 220-kv to the 110-kv system or from the 110-kv to the 220-kv system.

Phase-shifting transformers transmit power from one system to another system or regulate electrical energy flowing in a loop circuit. They may be thought of as a special design of regulating transformer built with two regulating windings. One regulating winding connects in phase with the main winding and acts to regulate voltage up or down as in a regulating transformer or voltage regulator. The other regulating winding connects 90° out-of-phase with the main winding. This winding acts to change the phase relation of the three-phase winding, increasing or decreasing the phase angle a given number of degrees. As an

FIG. 2-33. These form-fit transformers have very small tanks with little space between core and case. Large banks of radiators connect to the sides. Main high-voltage bushing stands on top of transformer at front; autotransformer high-voltage bushing is just in rear; low-voltage tertiary winding bushings stand at rear of tank cover. (*Courtesy of Pacific Gas and Electric Company.*)

example, one phase-shifting transformer may have a rating of 25,000 kva three-phase with an in-phase winding designed for 10 percent voltage, boost or buck, and a quadrature winding designed for a 15° angle range, advance or retard.

Synchronous condensers are designed for indoor or outdoor use. The present trend is to use outdoor condensers, the interior of the casing being filled with hydrogen gas. Hydrogen gas takes the place of air because (1) it removes heat faster than air, (2) it is less dense and reduces the loss caused by windage, (3) it is nonflammable, a fire cannot exist within its atmosphere. It is not explosive unless air is mixed with it. Since the gas is kept at ½ to 15 psi above atmospheric pressure, air cannot leak into the condenser casing. The latest design of condensers have the shaft exciter and its commutator operating within the hydrogen-gas atmosphere.

A typical transmission substation layout will include a high-voltage bus (perhaps 230 kv), a bank of autotransformers with a tap for 110 kv, Fig. 2-34. The tertiary winding of the transformer bank may be connected to a synchronous condenser. The transformer banks may consist of seven 33,333-kva single-phase transformers with a spare making up two 100,000-kva transformer banks. Each bank may be connected to one 50,000-kva synchronous condenser. The incoming 230-kv lines each

Fig. 2-34. A 230-kv switching station with bank of three autotransformers taking current at 230 kv, with autotransformer tap at 110 kv and low-voltage winding of 12 kv. The 230-kv oil circuit breakers stand in the center with isolating air switches. By-pass switch at the top is in open position. A 230-kv bus structure stands at right. Building in background is transformer untanking structure for removing core from cases. (*Courtesy of Pacific Gas and Electric Company.*)

would have a capacity of 200,000-kva (the line length being 100 miles), while four 110-kv lines would leave the station. Each 110-kv line would have a nominal capacity of 50,000 kva. A small building would be built with a steel frame, concrete walls, and roof or corrugated transite side walls and roof. The building encloses the switchboard-station battery, station-house bank, battery-charging sets, and various carrier-current relaying and telephone panels. Such a station may be controlled from a remote point by supervisory control, or it may be manually operated.

Another type of transmission substation is simply a switching station, Fig. 2-35. High-voltage lines entering and leaving the station high-voltage buses may be switched into radial groups or parallel-line groups supplying one load or another. High-speed carrier-current relays open a section switch of a double bus to improve the transient stability limit of the system. The term "stability" when applied to an electrical network refers to its ability to hold the system in synchronism during load changes and sudden disturbances, such as line short circuits. The word "transient" refers to sudden short-time electrical disturbances. The stability

FIG. 2-35. A 110-kv switching station with double buses, bus selector switches, and one circuit breaker per line. Power system may be split into two groups by station like this. Building housing switchboard is not shown. (*Courtesy of Pacific Gas and Electric Company.*)

FIG. 2-36. Current-limiting reactors at left reduce the flow under short-circuit conditions, ease burden on oil circuit breakers under maximum duty. (*Courtesy of Pacific Gas and Electric Company.*)

limit of a network of circuits is less when a sudden electrical disturbance occurs than when gradual load changes take place. When system load changes are slow, steady-state stability limits apply.

Transmission substations may receive electrical energy at a high voltage, Fig. 2-36, but also step down energy to a distribution voltage, 4, 6.9, 12, or 22 kv, to serve local customers within the area.

Station layouts generally include a tall building equipped with a crane for untanking a transformer that requires maintenance or may have failed in service. These high-voltage transmission stations cover an extensive area, and often many acres of land are required for the installation.

High-voltage buses for transmission stations may be of the double-bus type, with selector switches using one circuit breaker for each position:

FIG. 2-37. Two 110-kv oil circuit breakers with their disconnecting switches serve one half of a sectionalized single bus. Each breaker protects 25,000-kva three-phase outdoor forced-air-cooled 110/12-kv transformer, background. Insulating-oil storage tank for breakers and transformers stands at right. (*Courtesy of Pacific Gas and Electric Company.*)

a single bus or a sectionalized single bus, Fig. 2-37, or a ring bus. The double-bus design provides more flexibility for system operation. The ring bus occupies the least ground area; however, when a circuit is taken out of service, the ring bus is broken, forming a single bus.

2-7. Instrumentation. The instruments used for switchboards, Fig. 2-37*a*, are of several general types, such as indicating, recording, and integrating.

Indicating instruments include ammeters, measuring the magnitude of electrical current, and voltmeters measuring the value of electrical potential. Ammeters and voltmeters may be of the dc or ac type. If ammeters are dc, shunts are used to measure the current flowing in the conductor. A shunt has a number of thin copper fins placed parallel to each other; the voltage drop between the copper blocks supporting the thin fins is proportional to the current flowing through them. A dc ammeter is really a millivoltmeter calibrated in amperes. The commercial shunts generally have a standard voltage drop of 50 to 100 millivolts (mv). A 4,000-amp dc shunt would probably have a drop of 100 mv.

Ammeters for ac may be self-contained, up to 25 amp. Above that current magnitude, an instrument called a current transformer is used.

FIG. 2-37*a*. Benchboard controls 110-kv and 12-kv lines radiating from generating station. At left stands control board for boilers and turbines in station. (*Courtesy of Pacific Gas and Electric Company.*)

Current transformers built with one or two turns in the primary winding and many turns in the secondary winding usually have a secondary current rating of 5 amp. Current transformers may be built with one or two secondary windings. Always short-circuit the secondary winding of a current transformer if you find it necessary to open the circuit connected to it. A current-transformer circuit is always a series circuit and must not be opened, as a high voltage is induced, which may reach several thousand volts. The voltage waveform is sharply peaked; ordinary voltmeters will not read accurately when the voltage waveform is very distorted.

Another indicating instrument is the wattmeter, which may be of the center-scale type to measure the power in kilowatts flowing IN or OUT of a circuit. Kvar meters are also used, these measure reactive kilovolt-amperes; they may be of the center-scale type. Their scales are marked LEAD or LAG; however, IN and OUT is a more modern method of marking. As an example, a generator should have an indicating wattmeter with one scale, indicating the kilowatts flowing OUT, while the kvar meter will have a center scale marked both IN and OUT. If the generator field current is large and the generated ac is *lagging*, the kvar indicator would show kvar flowing OUT. As the field current is decreased, the generator ac passes through the unity power-factor position into the *leading* position, then the kvar indicator would show kvar flowing INTO the generator. It is possible for kilowatts to be flowing *out* of a generator while kvar are flowing *in*. Incidentally, a generator with a leading field (underexcited) is not too stable and may drop out of synchronism with the system with any slight load change.

Other indicating instruments are frequency indicators, power-factor meters, synchronism indicators, etc.

Recording instruments record kilowatts, kvar, volts, frequency, amperes, and power factor. In addition a totalizing wattmeter may be used for summing up the generated kilowatts. A totalizing instrument adds up the output of all the generators in a given plant. The totalized output is recorded on a strip chart or by a large indicating scale so that the plant attendants can see at a glance what is the total output of the station.

Recording voltmeters and wattmeters are most commonly used. They give a continuous record of the plant voltage and output in kilowatts. The recording devices may have charts for 24-hr periods, 7-day periods, or for 30 days. Also, some recording instruments which have chart speeds of ½ or 1 in. per hr will instantly speed up to 3 in. per min when a system frequency or voltage disturbance occurs. These recording devices allow ample space on the chart during system disturbances, so that personnel can study the recorded data carefully.

Automatic recording oscillographs are also used; they are known as the quick-start type. They begin recording on the chart from ½ to 1 cycle after system disturbances occur. A cycle is 1/60 sec, based on a frequency of 60 cycles per sec. Oscillographs have a variety of current and voltage elements. Some have four elements, others have six or seven galvanometer elements; three are used for recording ac phase currents, one is used for recording the neutral current, and the remaining two are used for recording two of the delta-phase ac voltages. Oscillographs may be of the portable type or the laboratory type. The film holder usually has a capacity of 200 ft. However, most system disturbances use only 2 to 3 ft of film. The oscillograph may be started by a neutral-current high-speed relay or an undervoltage high-speed relay. This equipment is invaluable for carefully analyzing electrical-system disturbances.

Temperature-recording devices are used to record transformer oil temperature, generator-stator-coil temperature, and generator-field-coil temperature. The generator stator usually has a number of thermocouples embedded between the coils; a selector switch completes an electrical circuit between the thermocouples and the temperature-recording instrument. The recorder is usually left connected to the particular thermocouple that consistently indicates the highest temperature.

Integrating devices include watthour meters and kvar meters. These integrating meters may be installed singly, or they may be multiplexed or electrically added together to give a total reading.

Gages and pressure-recording devices are also necessary equipment. Gages are used to indicate oil pressure, air pressure, etc. Recording devices make a record of pressure readings, vacuum readings, etc. Many of the automatic devices are operated by air pressure or by electrical circuits. Indicating and recording devices are installed so that the plant attendants can determine whether or not the plant operation is satisfactory.

2-8. Operation. The operation of most transmission substations and all generating stations requires the attendance of skilled personnel. The instruments installed in such stations require accurate interpretation; or the plants will not be operated at their best efficiency. Operating procedure requires that an accurate running record be kept of all plant or station operation; in addition, many instruments and gages must be read at a definite time interval, usually readings are taken each half hour.

Operation may be manual or by supervisory control. Manual control requries personnel at the station to do switching and fill out the log. Supervisory control means that the equipment is controlled from a remote point; instrument readings may be taken by telemetering equipment. This station control and telemetering may be done over two telephone lines or by carrier current, which is really wired-wireless stations controlled by supervisory personnel. Controls require maintenance inspection usually once every week or every two weeks.

Transmission substations may also be designed to be completely automatic. If the stations are fully automatic, and a circuit locks open after a given number of reclosures, an indication is sent to a dispatching office by a carrier-current signal so that the personnel may be advised that a circuit is inoperative. Men will then patrol the circuit to determine the cause of interruption.

Large transmission stations require several men on watch; one operator remains in the switchboard room, while another man, the second operator, does the necessary switching or grounding in the high-voltage switchyard. When he completes his assignment, he reports back to the first operator. Hydroelectric plants in general require less men than do steam-electric generating plants.

2-9. High-voltage Circuit Breakers. High-voltage circuit breakers in the United States are of the oil type, that is, the arc-interrupting devices

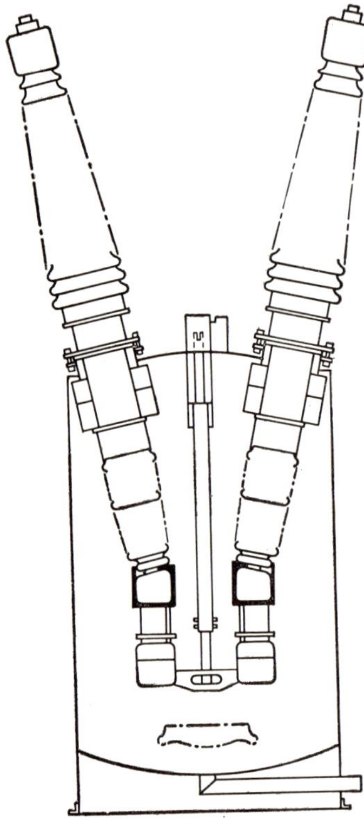

FIG. 2-37*b*. Important elements of a high-voltage circuit breaker. (*Courtesy of Westinghouse Electric Corporation.*)

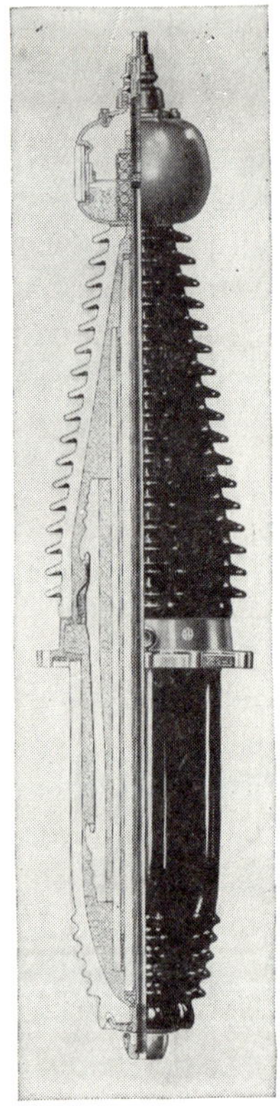

FIG. 2-38. Outdoor bushing similar to those used for oil circuit breakers. This has metal ground sleeve part way down from flange, far enough to go into oil. Oil reservoir is mounted on top of the bushing. This gage indicates oil level and allows space above oil for breathing. Concentric porcelain shells around conducting rod are submerged in oil. (*Courtesy of Ohio Brass Company.*)

are immersed in tanks filled with oil. In Europe, the air-blast high-voltage circuit breaker has been used for many years, in both indoor and outdoor installations. The important parts of an outdoor circuit breaker are the porcelain bushings, the tank, the contacts, the arc-interrupting devices, and the operating mechanism, Fig. 2-37*b*.

Bushings insulate the copper leads passing through the tank wall.

There are two types of bushings, the oil-filled and the condenser. The oil-filled bushing, Fig. 2-38, is a hollow formed by clamping two halves together, one half outside the tank, the other half inside. Several concentric porcelain cylinders are placed over the copper rod inside the main bushing. These are so placed in relation to the center clamping ring of the bushing that they add insulation to the bushing and substantially increase the leakage distance from the conductor to the steel switch tank. The interior of this bushing assembly is filled with oil. A round oil sight glass acting as a reservoir of oil is used to place a slight head of oil

FIG. 2-39. Oil circuit breakers with oil-filled bushings. Operating mechanism is enclosed in box at right, with gears and rods in pipe on top of breaker tanks. Steel box at front encloses terminal board for wires from bushing current transformers. (*Courtesy of Pacific Electric Manufacturing Company.*)

on the bushing to seal the oil thoroughly and prevent breathing during load cycles. Breathing draws in cool moist air, which condenses and forms droplets of water in the bushing. These could eventually cause the bushing to fail. The oil sight glass also gives a visual indication that the entire bushing is filled with oil.

The condenser bushing is also made up of hollow pieces of porcelain clamped or bolted together by metal rings cemented on the base of the bushings. A copper rod is placed in the center of the bushing. This rod is wrapped with concentric rings of paper, stepped at each end. Layers of foil are placed between groups of paper layers, the entire assembly is filled with a hot insulating compound or oil, sealing the entire interior of the bushing, and a watertight capnut is cemented to the top of the assembly. Sometimes grading rings are placed at the top of the bushings to equalize the voltage distribution over the end of the bushing.

Tanks for high-voltage circuit breakers are all made of steel formed into a cylinder, Fig. 2-39. The steel walls are comparatively thin, ¼ to ½ in. thick. A manhole on the side of the tank near the bottom allows inspection of the contacts and interrupting devices. It also provides access into the tank if it becomes necessary. The tanks are bolted or clamped to a concrete foundation. Steel rods and members brace the three tanks to each other, to keep the three phases in alignment. Otherwise, the operating shafts and gears may bend and prevent satisfactory opening and closing of the circuit breaker. Large quantities of a high flash-point (about 450 F) insulating oil are needed for each breaker. A high flash-point oil is necessary as the temperature of the electric arc drawn when the contacts part decomposes the oil into gases and vapors which might ignite if a low-flash oil were used.

The contacts conform to three general designs: the first type are butt contacts, in which the main masses of conducting metal butt together against a stiff spring backing, forming a positive contact under pressure. The second type uses a flat bar sliding between laminated fingers that are pushed together by springs. These contacts have a good wiping action. Their contact surfaces must be larger than the butt contact as the sliding contact does not make a pressure contact. Last is the rod type in which the contact assembly is made up of a cylindrical laminated assembly held together with springs and the conducting rod or bayonet is forced into the cylinder, giving a very good wiping action and a positive contact; the contact surfaces are long and adjust themselves by action of their springs to conform to the surface of the bayonet.

The main contacts described above *part* first, when the circuit breaker opens, and *make* last, when the circuit breaker closes. The main contacts carry the load current or short-circuit current, while the interrupting devices or arcing contacts *interrupt* or break the current or arc when the circuit breaker opens and *make* the initial contact when the circuit breaker closes.

Interrupting devices in early designs of oil circuit breakers consisted of arcing contacts immersed in oil. The greater the interrupting duty, the higher the head of oil above the contacts. Also multibreak contacts were designed which broke the arc into four or six parts, making its interruption much easier. Modern arc-interrupting devices are of two general designs. One is the *deion* type and the other the *oil-blast* type. There are variations of each one. The deion type consists of steel pieces separated by insulating pieces, such as micarta or herkolite; these pieces are shaped like a horseshoe, with one side open. The arcing contacts pass through the space in the center of the horseshoe assembly of alternate steel pieces separated by insulating pieces. When the arcing contacts part, an arc is drawn in the oil inside the horseshoe assembly, the field flux from the arc magnetizes the iron shapes, producing a strong repulsion effect which forces the arc through the horseshoe slot into the fresh oil. The arc is unable to ionize the oil (that is, knock one or more electrons off individual oil molecules, thereby leaving both positive and negative charges in the area) and is extinguished. Ionization of the oil molecules produces a conducting path which maintains the electric arc. As the arc is forced through fresh oil, the individual oil molecules do not break down so rapidly, and those which have recombined form neutral and stable molecules; in other words, the molecules are deionized.

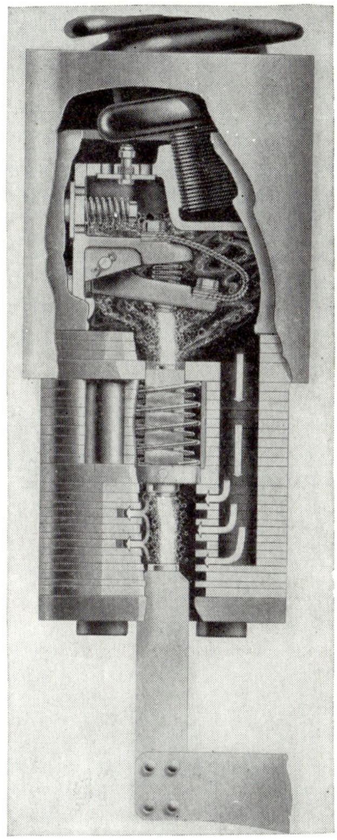

FIG. 2-39*a*. In oil-blast circuit breakers the upper arc occurs first, generating oil vapor to put oil under pressure and force it to extinguish lower interrupting break arc occuring an instant later. (*Courtesy of Westinghouse Electric Corporation.*)

The oil-blast principle does just the opposite, Fig. 2-39*a*. The oil is forced through the arc stream, thereby producing stable molecules, and the arc is extinguished. Early designs of the oil-blast type had a small piston operate the lever or rods that open the switch. When the breaker was opened, the mechanical action of the piston squirted a jet of oil through the arc. Modern designs do not use the piston as a separate device but do use the moving arcing contacts to act as a piston, the action of which is to attempt to form a partial vacuum. The fresh oil rushes in the evacuated space, the gas bubble formed by the arc is allowed to move upward, and the arc stream is extinguished by the fresh cool oil moving into that area.

Arcing contacts interrupt the circuit, drawing a powerful arc that tends to disintegrate the copper terminal. When the breaker is closed, an arc forms when the arcing tips come close enough together to break down the oil. This arc is severe, tending to burn the contacts; therefore, a dense and hard alloy called elkonite is placed on the surfaces of the arcing tips, prolonging their life many times and requiring much less inspection and maintenance.

The operating mechanisms built for opening and closing the breakers are of three general types: (1) solenoid-operated, (2) air-operated, and (3) spring-operated. The solenoid type uses a heavy solenoid for closing the circuit breaker and a smaller solenoid for tripping or opening it.

A solenoid is a large cylindrical coil having many turns of insulated wire. A movable iron rod is placed in the open center of the coil. When the coil is energized, the intense magnetic field produced pulls the iron rod into such a position that a maximum of flux lines pass through the rod, developing a powerful pull. This motion is translated in useful motion for opening and closing the breaker by lever systems. The disadvantage of a solenoid is the large current required momentarily for closing the breaker.

FIG. 2-40. Operating mechanism for an air-operated circuit breaker; air tank is at bottom of cabinet. (*Courtesy of General Electric Company.*)

The air-operated breaker, Fig. 2-40, consists of an air tank or receiver with a motor-driven compressor. The compressor maintains the required air pressure. The air tank usually has enough volume of air for several closures and openings of the circuit breaker. The air pressure operates pistons which in turn produce the required motion. The air-operating mechanism is fast, requires low electrical energy, and the breaker does not have the impact that a solenoid type gives.

The spring-actuated breaker has powerful springs compressed by a small motor. The springs open and close the breaker. These operating mechanisms are fast and require little electrical energy. The spring assembly allows the circuit to make an immediate reclosure (about 20 cycles).

A solenoid is a dc device. It is possible to operate a solenoid off an ac circuit if a rectifier is used. Usually the rectifier is the dry type, such as copper-oxide or selenium. AC can be used for the small trip coil, although many operating companies install small storage batteries (24 or 48 volts) for the tripping circuit. This is required, as the ac voltage at the substation bus drops sharply when a short circuit occurs on a line near the station. The ac voltage may be so low that the trip coil will not produce enough energy to trip or open the circuit breaker, thereby defeating the purpose of the circuit breaker to clear or open a short circuit.

A circuit breaker for three-phase work has six terminal bushings.

Each porcelain bushing has a bushing-type current transformer wound around it. A current transformer reduces the magnitude of the current. As an example, if a circuit breaker carries 600 amp, it is too much current to place on the back of a switchboard or to conduct through ammeters; therefore the current is reduced to 5 amp. This can safely be taken to the switchboard and conducted through various instruments, such as ammeters, wattmeters, varmeters, etc. The current transformer in the example would reduce the current from 600 to 5 amp, a ratio of 120 to 1 (divide 600 by 5). Current transformers are constructed by wrapping many turns of small insulated wire around a laminated steel core which loops around one large insulated conductor. The current-transformer secondary windings have taps, so that a number of ratios may be obtained and used, when necessary. Since the secondary circuit of a current transformer is a series circuit, it must *never* be disconnected from its load unless the winding is short-circuited. If the current-transformer secondary terminals are not short-circuited and the load circuit is opened, a dangerous rise of voltage occurs which is high enough to cause fatalities.

A modern circuit breaker usually has six bushing-type current transformers installed. All top connections are taken out to a terminal block mounted in a weatherproof box on the side of the circuit breaker. Modern circuit breakers are equipped with potential devices; each phase has a small capacitor producing a constant 240 or 120 volts with a specified maximum external load (voltmeters, wattmeters, varmeters, etc.). Capacitance potential devices take the place of high-voltage potential transformers. The potential devices take their energy from a small plate acting as a condenser plate placed around the bushing inside the tank, Fig. 2-41.

The modern oil breaker is of the conventional type, using three vertical mounted tanks, one for each phase, each tank containing large quantities of insulating oil. These breakers have an interrupting duty of 3,500,000 kva at 110 kv; other designs have a rating of 10,000,000 kva at 230 kv. The full opening time is 3 cycles for the fastest circuit breaker and 5 cycles for the slightly slower types. One electrical manufacturer has designed and built a new style oil circuit breaker. The tanks are eliminated and in their stead are three horizontal hollow porcelain cylinders mounted on porcelain pedestal insulators. Inside the large hollow horizontally mounted porcelain insulator are a series of arcing contacts, making a multiple-break switch using the oil-blast principle for extinguishing the arc. This special design is called an oil-poor circuit breaker, as the quantity of oil in each porcelain cylinder is very small. The main and arcing contacts are opened and closed by the action of powerful spring assemblies. These breakers are of the three-cycle type with an interrupting duty of 3½ and 10 million kva.

European circuit designers have built and installed, over a period of years, an air-blast high-voltage circuit breaker that does not use oil or large tanks. It is a unique design but has not been received with favor by operating companies in this country.

It has been customary to open and close all three phases of a circuit breaker simultaneously. But under certain stability conditions, single-phase switching has been used on high-capacity high-voltage transmission circuits. As over 90 percent of the transmission line short circuits

FIG. 2-41. A 230-kv oil circuit breaker with oil-filled bushings has operating mechanism mounted between two phase tanks. (*Courtesy of Pacific Electric Manufacturing Company.*)

are the line-to-ground type, it is much better to open the faulty phase, clear the fault, and immediately reclose the phase that had opened. The transient disturbance to the power-system network is greatly reduced as the two phases remaining in service act as a path to interchange synchronizing power and therefore keep the power-system network in step. This method of single-phase switching has one disadvantage, it causes electric customers' three-phase motors to operate single phase. If the single-phase breaker cannot be reclosed in a short time (30 to 40 cycles), the other two phases remaining closed must be immediately tripped out.

Automatic reclosing of a three-phase three-pole circuit breaker is common practice. It is necessary to delay the reclosing for 20 cycles or longer, as the arc of a line-to-ground fault on the line has ionized a large mass of air around the line conductor that must be deionized (become electrically neutral) before the circuit is reclosed. Otherwise the line voltage will immediately form another short-circuit current path through the mass of ionized air to ground.

The modern high-voltage circuit breaker is a very reliable device requiring infrequent inspection and little maintenance. Maintenance consists of checking the electrical circuits to the operating mechanism for

FIG. 2-42. Upper part of high-voltage line circuit-breaker tower with by-pass air switch at top of double-break type. Isolating or disconnecting switches are mounted on front and back of tower; all are shown in open position. (*Courtesy of Pacific Electric Manufacturing Company.*)

continuity and the compressed-air valves for tightness and for freedom from sticking; the bushings must be cleaned and inspected for cracks. If a severe short circuit has been interrupted, it may be necessary to drain the oil, inspect the main contacts and arcing contacts and clean them; it is also necessary to check the alignment of the circuit breaker and test the oil for dielectric strength before replacing it in the circuit-breaker tanks.

2-10. Oil Circuit-breaker Installation. Oil circuit breakers for high-voltage substation installations are mounted on a solid-concrete foundation. A tall steel tower supports disconnecting switches having two separate and distinct uses. The first group isolates the circuit breaker from any live or "hot" connections. This may be accomplished by installing one six-pole gang-operated disconnecting switch or two three-pole gang operated disconnecting switches. One switch opens the circuit-breaker connections to one side of the breaker. Gang-operated switches are disconnecting-switches operating mechanisms that are mechanically connected together. All single-pole elements open or close together by means of one operating lever. This lever may be interlocked with the oil circuit-breaker mechanism so that the disconnecting air switches

FIG. 2-43. 110-kv single-pole single-break disconnecting switches. Blades turn partly as switch opens or closes. This produces leverage at jaws to break away ice or sleet accumulations on switch contacts. (*Courtesy of Electric Power Equipment Company.*)

cannot be opened or closed if the oil circuit breaker is in the closed position.

The other air switch, Fig. 2-42, is a three-pole gang-operated disconnecting switch connected to act as a by-pass air switch, that is, to by-pass the oil circuit-breaker assembly electrically when the by-pass switch is in the closed position. This air switch may be interlocked with the oil circuit-breaker operating mechanism, if necessary.

A gang-operated three-pole switch is often installed as a grounding switch for the transmission line. This gang-operated air switch may be so interlocked that it cannot be closed unless the main oil circuit breaker and the by-pass air switches are in the open position.

A disconnecting air switch of the single-break type is shown in Fig. 2-43. This design is arranged for gang operation and has a voltage rating of 110 kv. These switches generally have an ampere rating of 600 or 1,200 amp. Low-voltage installations require a much greater current-carrying capacity, Fig. 2-44. There are many designs of disconnecting air switches.

2-11. DC Bus Layouts. DC bus layouts are usually quite simple. There are two classes: commercial dc buses and railway dc buses.

Commercial dc is supplied to lighting and power customers. It was very popular in years gone by, but ac is slowly taking over the load for-

FIG. 2-44. Single-break air switch for 3,000 amp at 34.5 kv. Blade makes a half turn to give wiping action when making contact with jaws under great pressure. (*Courtesy of Electric Power Equipment Company.*)

merly supplied by dc. Buses for dc are energized at two voltage levels, 250 and 500 volts. The 250 volt system is a three-wire system with 250 volts between lines and 125 volts to neutral. The neutral wire is grounded, Fig. 2-45. The 500-volt system is usually a five-wire system. The voltages are 500, 250, and 125 volts, Fig. 2-46. There are two neutral wires as the 500-volt system is really two three-wire dc systems.

The bus structure for the three-wire 250-volt dc system is built of pipe framework or structural-iron framework. The buses are flat copper bars. They may be $\frac{1}{4}$, $\frac{3}{8}$, or $\frac{1}{2}$ in. thick and 3, 4, or 6 in. wide. The copper bars are mounted vertically so that copper spacers between the bars allow natural ventilation. This allows the bars to carry more current. Disconnect switches separate the feeder cable from the bus. Two switches are required for each feeder, one for the positive connection and the other for the negative connection. Feeder sizes may be 1,000,000-, 1,500,000-, or 2,000,000-cm cables. The dc system is an underground system, and it is not economical to transmit dc power at low voltage more than about ½ mile.

The copper bars making up the bus are mounted on porcelain insulators or ebony-asbestos supports. Ebony-asbestos supports lend themselves to ready fabrication for odd bus-bar arrangement. Carbon circuit breakers are used for feeder connections; the circuit breakers may

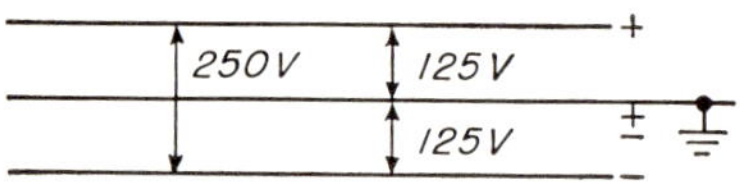

FIG. 2-45. Three-wire dc system has grounded neutral.

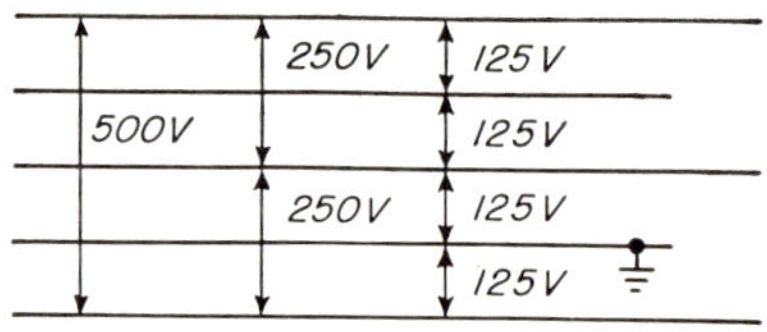

FIG. 2-46. Five-wire 500-volt dc system.

be single pole or two pole. They are mounted on the front of a switchboard panel built of slate or ebony asbestos.

Low-voltage dc systems often have more than one bus, Fig. 2-47. A two-bus system is quite common. One bus is a high-voltage bus (10 percent higher than normal). If the normal bus voltage is 250 volts, the high-voltage bus would have 275 volts. Feeders connected to this bus would supply energy to customers located a little farther away involving higher voltage drop.

The 500-volt dc bus is built in a similar way, Fig. 2-48, but additional insulation is needed and the currents are usually quite small, on the order of 250 to 500 amp. Larger currents, 500 to 1,200 amp, are supplied over 250-volt dc systems. Circuit breakers for the 500-volt system may be one or two pole. If two-pole breakers are used, they are mounted near or at the top of the switchboard panel. If single-pole circuit breakers are used, they are mounted vertically on the switchboard panel, one above the other. The lower circuit breaker has a nonconducting deflector hood over it, so that the hot gases from the circuit breaker arc do not short-circuit the breaker mounted above it.

2-12. DC Railway Bus Structures. The voltage of dc buses supplying power to railway systems may vary from 600 to 3,000 volts. About the upper limit for rotating commutating devices is 3,000 volts. Mercury-arc rectifiers, however, are able to supply power at much higher voltages. The bus structure for railway feeders differs from those for commercial power not only by voltage but also because commercial buses require a separate positive bus and a negative bus with disconnecting and protective devices, while a railway-feeder switchboard has only a positive bus as the rails act as a return circuit for the negative current. Therefore, railway structures are simpler in construction. Railway bus structures may be of the open type or the metal-enclosed type. The

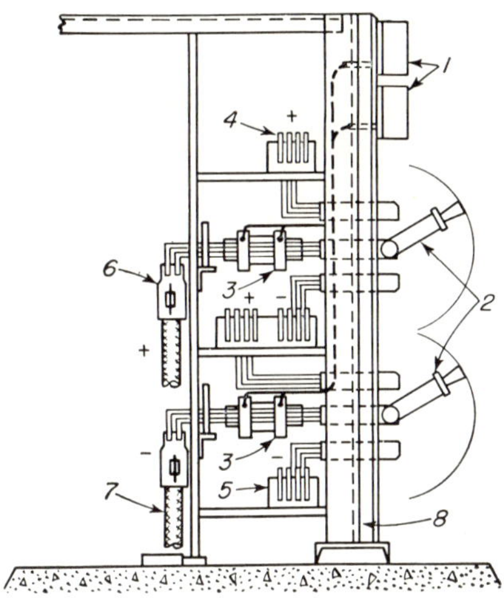

FIG. 2-47. Switchboard for two-bus dc system has one bus at 10 percent higher voltage. 1, dc ammeters; 2, single-pole double disconnecting switches; 3, dc shunts; 4, positive bus, high bus; 5, negative bus, low bus; 6, cable conductor clamps, solderless type; 7, outgoing feeder cables; 8, switchboard panel, ebony asbestos or slate.

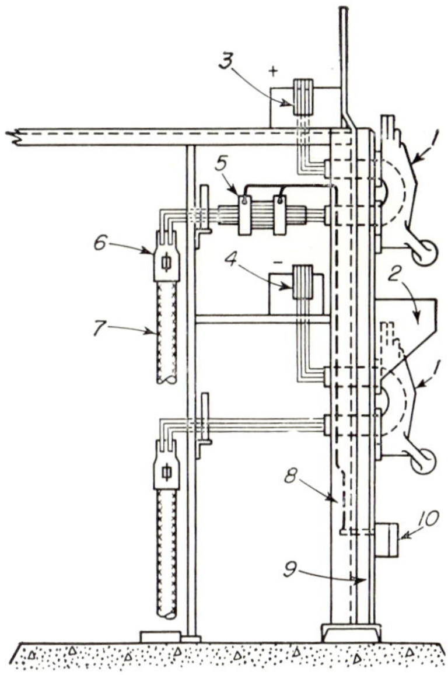

FIG. 2-48. Switchboard for double-bus 500-volt system. 1, single-pole circuit breaker; 2, arc chute; 3, positive bus; 4, negative bus; 5, dc shunt; 6, cable connector, solderless; 7, cable; 8, steel switchboard frame; 9, switchboard frame; 10, horizontal dc ammeter.

railway feeders may be protected by carbon circuit breakers, semi-high-speed circuit breakers.

The simplest railway switchboard is the open type having a slate or ebony-asbestos panel supported by a pipe or angle-iron frame, Fig. 2-49. A single-pole carbon circuit breaker is mounted at the top of panel, with an ammeter mounted directly beneath the circuit breaker. A single-pole disconnecting switch stands below the ammeter. The disconnecting switch is used to disconnect the circuit breaker from the bus *after* the circuit breaker has been *opened*. Six hundred volts is too dangerous for personnel to work on while it is hot or energized.

A semi-high-speed breaker opens its contacts in about $\frac{1}{20}$ sec after the trip coil is energized. The conventional carbon circuit breaker is a little slower, opening about 0.1 sec after the trip coil is energized.

The semi-high-speed breaker is designed with a contactor type of jaws and blade together with an arc chute. This breaker may be mounted at the top of a panel similar to the mounting of the usual circuit breaker. It may also be installed on a panel enclosed in a steel cubicle. The arc chute projects through the top of the enclosure to allow the hot gases to escape.

The high-speed breaker is of entirely different design than the carbon circuit breaker. It has an opening speed of $\frac{1}{60}$ sec. This breaker must

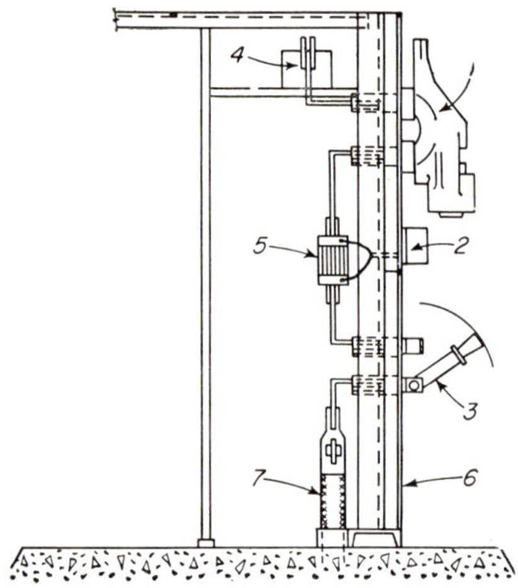

FIG. 2-49. Railway-system dc switchboard. 1, solenoid-operated circuit breaker; 2, ammeter; 3, single-throw disconnecting switch; 4, main bus, positive; 5, dc shunt; 6, slate or ebony-asbestos panel; 7, feeder cable.

be mounted horizontally. It may be mounted on top of an open type of switchboard panel arrangement or within a metal enclosure. It is also built as a truck type, that is, it is mounted on a frame that may be moved in or out of a cell. The contacts are made and broken by bayonet-type contacts. The arc chute projects through the top of the cell. Railway dc buses are generally of the single-bus type. They may be two single buses connected by a bus-tie circuit breaker.

CHAPTER REVIEW TOPICS

1. Name the principal equipment elements in a transmission system.
2. What are the principal steps in generating electrical energy ready for delivery through a transmission system?
3. What requirements must the electrical bus of a generating station meet?
4. Sketch the elements of a modern medium-voltage switchboard arrangement.

5. What are the duties of a circuit breaker?
6. Describe the functions of a disconnecting switch.
7. How can the rating of a bus bar be increased?
8. Sketch some of the arrangements of metal-enclosed switchgear.
9. How do indoor and outdoor buses compare in ground area required?
10. How much greater area must an aluminum conductor have than a copper one to carry the same current?
11. What type of bus is used in outdoor structures?
12. Describe and sketch the various types of insulators used in outdoor buses.
13. What precautions must be taken in providing insulation in areas where the atmosphere is polluted?
14. How are insulators cleaned?
15. How do wooden and steel structures compare in advantages for outdoor buses?
16. Sketch several bus designs used in switching stations and distribution substations.
17. What duties does a transmission-line switching station perform?
18. How are intermediate switching stations controlled?
19. What advantages can be realized from use of a bus-sectionalization switch?
20. What are synchronous condensers and static capacitors used for?
21. What function do regulating transformers perform?
22. What are phase-shifting transformers used for?
23. Describe a typical substation arrangement.
24. What instruments may be found in generating and substations?
25. What are control-board instruments used for?
26. How do recording, indicating, and integrating instruments differ?
27. What is the difference between manual and supervisory control?
28. Name the principal parts of a high-voltage circuit breaker.
29. Describe the arrangement of a bushing.
30. What are the principal features of the tank for a high-voltage circuit breaker?
31. Describe the three principal types of switch contacts.
32. What is the function of arcing contacts?
33. How does a deionizing-type switch work?
34. How does an oil-blast switch work?
35. Describe the three types of switch-operating mechanisms.
36. What is the function of a current transformer?
37. What function is performed by potential devices on modern circuit breakers?
38. Describe the main features of the oil-poor breaker.
39. What are the advantages of single-phase switching?
40. What precautions are observed with an automatic reclosing breaker? Why?
41. What are the maintenance procedures for a circuit breaker?
42. Describe the elements in a circuit-breaker installation.
43. What voltage systems are still in use for dc supply?
44. Are dc systems used for long-distance transmission in the United States?
45. Sketch the general arrangement of a dc bus and switchboard.
46. What are the operating speeds of dc circuit breakers?

CHAPTER 3

TRANSMISSION LINES

By Francis S. Benson

Pacific Gas and Electric Company

3-1. Overhead Systems—Wood Poles. Poles may be of wood or steel. Wood poles for transmission lines are generally quite long, 45 to 70 ft. The poles should be straight and free from knots. The wood used for poles includes douglas fir, western red cedar, cypress lodge pole pine, and western yellow pine. The dimension of the top and butt are specified, a reasonable tolerance in dimensions being allowed. A specification is necessary to obtain duplicate poles of a given size and length. Poles must be treated to discourage wood rot and termites. Termites, small antlike creatures, chew wood. They are the scavengers of the forest, reducing fallen trees to dust.

The most common treatment is creosote. The pole may be treated full length, or it may be butt-treated. Effectiveness of the treatment depends on the penetration of the creosote. When butts of poles are dipped into a creosote tank, the penetration is not deep enough to discourage wood-chewing insects. Several methods give good penetration. One method spikes the pole. The portion of the pole to be treated is indented by sharp spikes and the creosote flows into these indentations and migrates deeply into the wood. Another method places the pole in a tank, extracts the air, and forms a partial vacuum within the tank. This pulls the entrapped air out of the wood cells, allowing the hot creosote to penetrate the cells. Treating a pole increases its life substantially, at slightly increased cost per pole.

Wood poles for transmission-line supports may be used singly, Fig. 3-1, or in pairs. When used singly, one pole can support a 70-kv double-circuit line using relatively short spans, 300 to 500 ft. Single poles may be used for a single line, Fig. 3-2. The conductors may be mounted in horizontal or triangular arrangements. This can be done by mounting one pin-type or post-type insulator on top of the pole, the other two insulators being mounted on a horizontal wooden crossarm, Figs. 3-2 and 3-3.

Another method, the "wishbone" design, Fig. 3-4, may use either steel or wood crossarms. Steel crossarms are shown in Fig. 3-3. In wood crossarms the curved steel angle shown in Fig. 3-3 is replaced by a straight wood arm. This construction uses suspension-type insulators. It is an economical design of 60- to 70-kv line.

FIG. 3-1. This wood-pole line supports double-circuit line of medium voltage, 60 kv, with three-part-type insulators. Lowest crossarm with two wires supports street-lighting circuit. A 120/240-volt circuit held next to the pole by vertical racks. Guy leading across the road is insulated by a wooden guy bar. (*Courtesy of The Ohio Brass Company.*)

When wood poles support voltages of 110-kv or higher, the H-frame construction is used, Fig. 3-5. The name "H frame" describes its construction. Two poles support one long horizontal wooden member mounted near the top. Strings of suspension insulators are supported from the two ends of the long crossarm and in the middle of it. The distance of the conductor from the pole or crossarm brace maintains the required clearance to ground (wooden surface) when the conductor swings up as high as 45° from its vertical "at-rest" position. Line conductors move and are lifted by wind. Wind may also cause sustained mechanical vibrations of the conductors if they are pulled up too tight. The conductors sag when hot and stretch tight when not carrying line current and in cold weather. The action of the wind often causes standing mechanical waves which move the suspension insulator strings from their at-rest positions.

3-2. Steel poles may be the latticed-steel type, the expanded-metal pole, or the tubular-pole type. The latticed-steel type is commonly used, Fig. 3-6. It is a galvanized-steel structure that carries single- or double-circuit lines. The steel tower or pole requires a concrete foundation, while a wood pole does not. Angles imbedded in the concrete foundation act as a support to the tower angles. The tower angles are bolted to foundation-steel members. Often four large machine bolts cast in the concrete foundation anchor the steel pole. In such cases there is a thick plate at the bottom of the pole. The plate bolted tightly against the top of the foundation makes a rigid support for the pole.

FIG. 3-2. Single-circuit 60-kv line built in mountainous country with heavy annual snowfall. Pole in foreground supports the line by a pair of horizontal crossarms. Third pole, in background, supports line in characteristic triangular arrangement. Two guys steady first pole, strain insulator in each guy line. Center pole supports sectionalizing switch covered with snow. Dry snow is a good electrical insulator. (*Courtesy of Pacific Gas and Electric Company.*)

Expanded-metal poles are popular. They are made by cutting the web of a large I beam, then heating the web and mechanically pulling the flanges of the I beam apart. This results in a column of great strength and of relatively light weight. The height of expanded-metal poles is limited by the size of the machine making them and also by the transmission-line designing engineers' specifications.

3-3. Conductors for transmission lines are built differently than those for low-voltage circuits or distribution lines. Transmission lines are usually classified as lines operating at more than 60 kv. Conductors energized at high voltages require a large surface area. The electrostatic field surrounding the surface of a high-voltage conductor may be intense enough to be noisy and also luminous at night when the conductor diameter is too small. Air has a definite electrical breakdown strength, which should not be exceeded. When air is overstressed electrically and breaks down, it causes the phenomena known as corona. Corona results in a continuous loss of energy from the conductor. The line designer must determine the conductor diameter that will prevent corona and its dielectric loss. Transmission lines are now designed for a corona loss of less than 1 kw per mile for a three-phase line. A transmission-line designer must consider conductor spacing, altitude of the line above sea level, as

FIG. 3-3. A 60-kv sectionalizing gang-operated disconnecting switch mounted in triangular arrangement. Switch in closed position. Lower crossarms support two telephone wires. (*Courtesy of Pacific Gas and Electric Company.*)

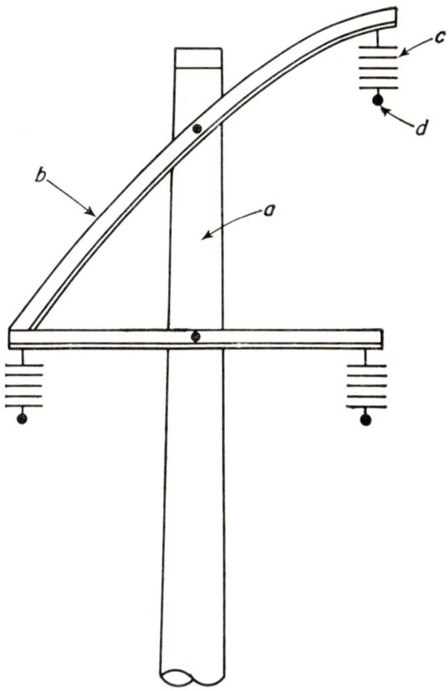

FIG. 3-4. Wishbone design of three-phase transmission line, *a*, pole; *b*, steel angles serving as insulator supports and braces; *c*, suspension insulators; *d*, line conductor.

well as the correct conductor diameter to obtain a design which will give a low corona loss.

Conductor materials may be copper or aluminum. A number of designs are in common use. The hollow copper conductor with an I-beam core is shown in Fig. 3-7.

The I-beam section twists spirally, and the copper strands are wound around the I beam as a center support. This design gives a large outside diameter while using the twisted I beam as mechanical support for long line spans.

Figure 3-8 shows a conductor called type HH. It is made of curved segments tongue and grooved together. Each segment is laid up spirally to make a stiff conductor. The outside of this conductor has a very smooth surface which reduces the corona loss. The electrostatic flux is more intense and concentrates at points on rough surfaces. A stranded conductor cannot present so smooth an outer surface as the type HH.

FIG. 3-5. Two 230-kv transmission lines have larger suspension insulators in lower part of strings above conductor, Lowest disks are grading rings equalizing electrostatic field pattern so each insulator handles proper share of voltage stress. Steel towers at right are snow towers. Right-of-way cleared through forest. (*Courtesy of Pacific Gas and Electric Company.*)

Another conductor in common use is the aluminum conductor steel reinforced (acsr), Fig. 3-9. A core of stranded steel wires acts as a supporting member while the outer strands of aluminum carry the current. This design is used for high-voltage lines with long spans. When steel cable is difficult to obtain, high-strength aluminum-alloy cable is used in its stead. Bare stranded copper wire is used for lower transmission voltages such as 70 and 110 kv. Higher voltages such as 165, 220, and 287 kv use the designs shown in Figs. 3-7 to 3-9.

Another conductor design is the copperweld. Here a thin copper tube is welded over a small steel rod or solid wire. This conductor allows a long span with a small conductor, Fig. 3-10. Copperweld conductors are made in various designs. The copperweld conductors themselves may be stranded to form a three-, five-, or seven-strand cable. Or copper wires may be laid over a copperweld conductor used as a high-strength core. Copperweld conductors find their greatest use at voltages less than 60 kv.

A conductor arrangement has been used experimentally where two or three conductors grouped together form a larger, or bundled, conductor for each phase. Two-, three-, or four-strand cables may be separated by spacers and used in a group to form a large phase conductor. This equivalent conductor design will be used advantageously when heavy currents are carried or transmission voltages are raised above 287 kv.

FIG. 3-6. Latticed steel pole for double-circuit support, operating at 60 kv. Lower crossarm supports telephone circuits. (*Courtesy of Pacific Gas and Electric Company.*)

The joining or splicing of large oversize conductors must be very carefully done. The joints are usually the compression type. Clamped joints are not used under tension. For some concentric stranded conductors, a sleeve is placed over the two conductors to be spliced and a

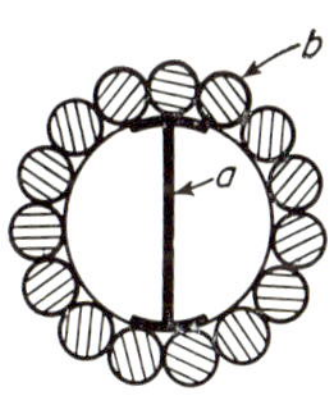

FIG. 3-7.

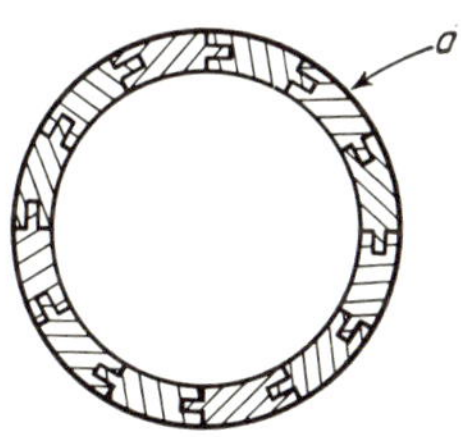

FIG. 3-8.

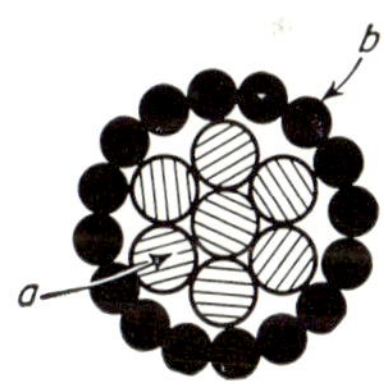

FIG. 3-9.

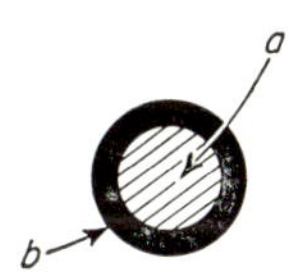

FIG. 3-10.

FIG. 3-7. Hollow copper conductor with an I-beam core. *a,* I beam twisted spirally in a lay opposite to lay of individual wires *b.*

FIG. 3-8. Type HH conductor using curved segments *a* laid in spiral configuration. Individual segments slide back and forth slightly when conductor is wound on reel.

FIG. 3-9. Aluminum conductor steel reinforced commonly called acsr. *a,* steel strands making up the core; *b,* current-carrying aluminum strands.

FIG. 3-10. One form of copperweld conductor. *a,* round steel core over which thin copper cylinder *b* is welded.

special tool twists the two conductors together in the distorted and twisted copper sleeve. Another design also uses a long sleeve, but the conductors and sleeve are squeezed together at short intervals by a hydraulic press, thus forming a very strong joint.

3-4. Line insulators are of two general types, the suspension type and the pin-insulator type, Fig. 3-11 and -12. Suspension insulators are manufactured with clevis-and-pin connections and with ball-and-socket connections. Both styles of connections are commonly used. Suspension insulators have the porcelain *under compression when supporting a conductor*. The weight of the conductor is considerable, as tower spans of high-voltage transmission lines are on the order of 800 ft or more. Pin-type and post-type insulators also have the porcelain under compression. These insulators are used on transmission lines with not more than 70 kv. Above 70 kv, suspension insulators are used.

Fig. 3-11. Three-part insulator has three shells cemented together. This one is designed for 70 kv, pin support, does not cause radio noises. (*Courtesy of Ohio Brass Company.*)

Selection of insulation (insulators) requires careful study. Some areas need considerable overinsulation. Transmission lines built near the sea coast in central California, for example, require considerably greater insulation than do lines built in the interior. Electric light and power companies often make maps showing areas where overinsulation is needed.

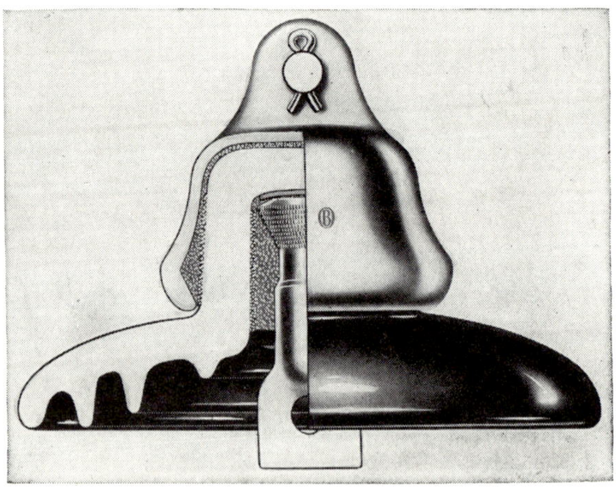

Fig. 3-12. Suspension insulator with pin and clevis for connecting together in string. Corrugations on underside give long electrical leakage distance. Wedge arrangement of cap cemented on top and of cone on pin cemented inside places cement in compression when insulator carries conductor load. (*Courtesy of Ohio Brass Company.*)

Transmission lines along the seashore are subjected to winds blowing in from the ocean. The wind carries a fine salt vapor which deposits salt crystals on the windward side of the insulator. Salt-air vapor may be carried many miles inshore by strong winds. With frequent rains these salt-vapor deposits dissolve and wash off the insulators. In areas where the rain is seasonal, however, the insulator surface leakage resistance may become so low during the dry period that insulators flash over without warning.

Insulators near cement mills, gravel pits, and refineries may become so contaminated that extra insulation is required. Rain falling during windstorms tends to keep insulators clean. When insulators are dirty,

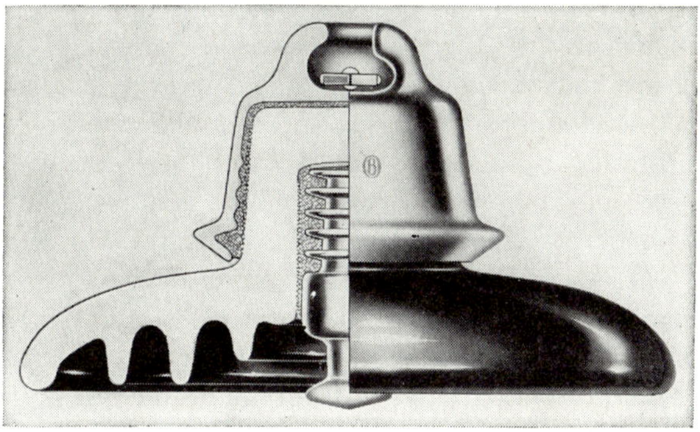

FIG. 3-13. Suspension insulator with ball-and-socket connections. Cement is arranged to be under compression when placed under load. (*Courtesy of Ohio Brass Company.*)

many insulator flashovers occur during light fogs unless arcing horns protect the insulators.

Some insulators may be cracked by dynamic action of the arc, thus overstressing the remaining good insulators in a string. Insulators have been designed, particularly the pin type, with skirts shaped to increase the wind velocity blowing over the underside of the insulator many times. Winds of different velocities and direction tend to deposit dirt on the insulator instead of removing it, unless the insulator parts are shaped properly.

Insulators may be cleaned by high-pressure water jets while the transmission line is energized. A high-pressure pump developing 500 to 850 psi forces water through a nozzle producing a round solid stream with little or no feathering or spraying of the water. This water jet strikes the insulator with a high velocity which literally tears off the dirt and deposits from the insulator surface. The truck carries a 1,000-gal water tank and a telescoping ladder so that the workman can come within 20 to 30 ft of the insulators. The nozzle has a long smooth bore solidly grounded to the truck frame, which in turn is grounded by a rod driven in the earth near the pole. Certain lines may need insulator cleaning several times a year. This can be done at a very low unit cost by this method.

Sometimes a defective insulator has not failed. The water action sometimes causes such insulators to flash over, and the line automatically deenergizes by protective relays opening the breakers at each end of the line. Since the dispatchers know the line is being washed, the trip out does not interfere with normal power flow. Other lines usually have been set to operate in parallel and share the load.

Pollution of insulators is not too serious a problem in areas with frequent rainfall. Lines near gravel pits, cement mills, etc., however, may still require overinsulation and periodic cleaning.

Cleaning insulators on energized lines is a relatively new practice. Conventional methods deenergize the line, ground the conductors at each pole or tower where the insulator cleaning crew is washing and wiping the insulators by hand. This laborious procedure is justified by preventing insulator flashovers with attendant line outages.

3-5. Insulator Protection. Suspension insulators arranged in long strings, Fig. 3-14, sometimes flash over for no apparent reason. Tests made to determine the cause showed that the voltage distribution over each insulator in the string was not equal, suspension insulators nearest the conductor had a much greater percentage of line voltage across them than those on the top of the string at the tower support. Results of these tests initiated the practice of installing larger diameter suspension insulators near the conductor, the remaining insulators being the size originally used in the string, Fig. 3-15. This practice helped to prevent suspension insulators from flashing over. Additional tests showed that adding a metal ring around the outside of the insulator next to the conductor and at the top of the string another metal ring solidly connected or grounded, to the tower, practically equalized the voltage distribution over all insulators in the string, Fig. 3-20. These *grading rings* have reduced the length and weight of the insulator string required. In addition, the grading rings also act as terminals of a long air gap, thus directing the arc of an insulator flashover away from the insulator string. Grading rings have gradually been reduced in size and are now called *arcing rings* or *arcing horns.* These devices have saved many insulator strings from being shattered by the dynamic force of the electric arc.

3-6. Conductor Vibration. Transmission-line conductors are suspended from towers spaced relatively long distances apart, 600 to 1,000 ft. The wind blowing across the line exerts an uplifting force on the conductors, similar to the lift on an airplane wing forced through the air. The force of the wind lifts the conductor a little and places a strain on the conductor. The conductor tries to relieve the strain by going into vibration.

You have undoubtedly heard telephone lines hum and whine when a wind passes them. They are tightly strung, and the wind produces audible vibrations. Transmission lines vibrate similarly, but not rapidly enough to make a sound. The serious effect on the conductor is the continual bending at each insulator-string conductor-clamp connection on each

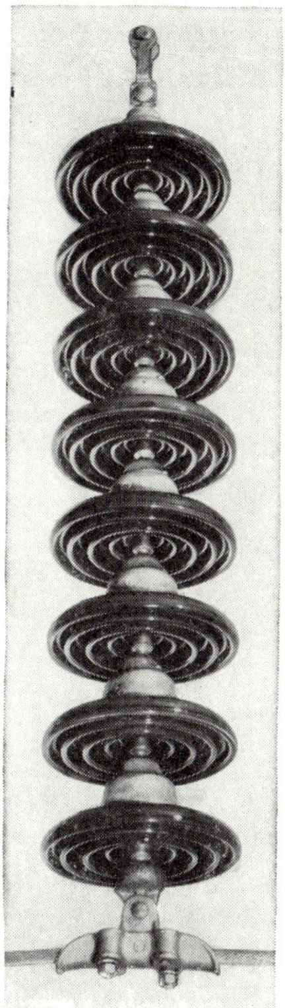

FIG. 3-14. String of ball-and-socket insulators showing fitting to clamp the conductor. This string supports 110-kv line, keeps eddy currents and hysteresis losses to a minimum. (*Courtesy of Ohio Brass Company.*)

FIG. 3-15. A 230-kv insulator string with grading ring at bottom above conductor. This is an early design; modern practice recommends using two rings. (*Courtesy of Pacific Gas and Electric Company.*)

tower, Fig. 3-16. The vibration gradually causes the conductor metal to crystallize and harden, cracking one strand after another. The remaining strands are overstressed mechanically, and they overheat electrically because too much current flows through a smaller cross section. The conductor finally breaks at such a point.

To prevent conductor vibrations, the period of vibration must be changed. This can be done in several ways: (1) changing the spacing between towers, (2) increasing or decreasing the tension of the conductors, that is, increasing or decreasing their sag. When a line is already built these solutions are not practical. A third method places weights on the conductors at or near each insulator string. These weights are called vibration dampers. There are several designs. The weights give the conductor a natural period of vibration that will not respond to the usual range of wind velocities.

It has been found necessary to protect the conductor at each suspension-insulator clamp support from the normal bending of the conductor around the clamp. These long spans require a rather sharp change in direction of the conductor as it passes over the supporting clamp. The conductor is reinforced for a distance of 3 to 5 ft on either side of the suspension insulators by armoring. This consists of additional conduc-

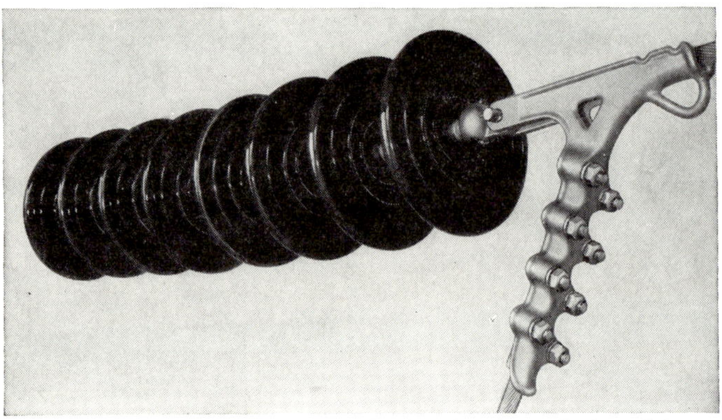

FIG. 3-16. Suspension insulators used as a dead-end string. Left end fastens to steel or wood support. High-strength bolts clamp conductor into strain clamp. These fittings terminate the conductor at a tower.

tors served around the conductor with a long lay, giving a larger conductor diameter at the suspension point. The net effect spreads the strain and stress of the conductor weight at the point of suspension over a much greater area and decreases the sharpness of the conductor bend at the point of suspension. This reduces conductor breakage and line maintenance.

3-7. Conductor hardware is now being made of nonmagnetic material. The conductor when energized has an electrostatic field surrounding it. This field has its greatest strength nearest the conductor. Sharp points cannot be tolerated on conductor hardware, as they cause excessive corona loss. You will recall that corona loss is the power loss from a transmission line which has too small a conductor diameter or too rough a surface. Corona loss is caused by the transmission-line voltage overstressing the surrounding air. The air breakdown or corona shows itself by luminous streamers darting out from the conductor surface. Hardware causing corona may also cause radio background noise; therefore, conductor hardware should have as smooth a surface as possible.

Conductor hardware such as clamps, is also subjected to an electromagnetic field created by the conductor current. As the current in a conductor increases, the electromagnetic field surrounding the conductor increases. Hardware made of nonmagnetic material will have less heating from eddy currents in the material. Eddy currents are small local circulating currents induced in a magnetic material under the influence of a magnetic field. When the conductor hardware does not become hot from corona (electrostatic field losses) or eddy currents (electromagnetic field losses), the conductor will not overheat at the hardware. Conductor hot spots may anneal the conductor and reduce its tensile strength. Hysteresis and eddy-current losses do not become troublesome with conductor currents of less than 150 amp.

A transmission line should be designed to have minimum power loss. Mechanical energy converted into electrical energy losses is money wasted.

3-8. Pole Supports. Wood poles and latticed poles are not stiff enough usually to turn corners or angles in a line unless braced by a guy assembly consisting of an anchor and a guy wire. A guy wire is generally a high-strength stranded-steel cable galvanized to protect it from the weather. Many studies have been made to determine the holding ability of anchors in various kinds of soil. One anchor and a guy or two anchors with guys may be needed to exert the necessary restraining force on the pole line. Guy wires generally make an angle of 45 to 60° with the surface of the earth. There are many types of successful anchors. One that disturbs the earth the least is best. The holding power of earth is much greater when not cut up by too much digging. There are anchors for every possible terrain such as rock or swamp. Anchor designs vary. There are the expanding-leaf, screw, cone, and dead-man type. A dead-man anchor is a 4- to 6-foot length of pole buried in the earth with the guy wire wrapped around it.

When wood poles are used for transmission-line supports, deep holes are needed. Many pole holes are dug by an engine-driven mechanical auger. These augers incline at various angles. They can dig holes at an angle to the surface for anchors of the expanding-leaf type. The anchor is inserted in the bottom of the hole, and the leaves are forced open by pounding or hammering. The hole is backfilled with earth and well tamped. The maximum allowable strain on the anchor is computed. The anchor eye is then ready for attachment of the guy wire. A dynamometer is attached to the guy and a strain put on the guy cable until the dynamometer indicates the desired pull in pounds per square inch.

3-9. Towers. Transmission-line towers are built up of small steel-angle members. The towers are rigid, strong, and self-supporting, Fig. 3-17. They carry transmission-line conductors across fields and hills. They may be especially high and strongly built when used for river or slough crossings, Fig. 3-18. They carry the conductor dead load, that is, the weight of conductors, hardware, and insulator strings, plus the load

Fig. 3-17. Two-circuit tower built of and stiffened with small steel angles. Center conductor extends out farther from tower than upper and lower ones. This prevents arc on lower conductor traveling upward into upper conductor. Overhead ground wire is supported at top of tower. (*Courtesy of Ohio Brass Company.*)

caused by the wind. In addition the tower arms take care of extra loads when ice or sleet forms on the surface conductors. A conductor coated with ice may have sufficient weight to break the steel tower arm if it is not designed to handle such loads.

Steel towers may be single-circuit, Fig. 3-19, or double-circuit designs. In snow country and at high altitudes, a low single-circuit structure called a snow tower is often used, Fig. 3-20. However, most tower lines are double-circuit since the steel towers entail a considerable part of the line cost. The two circuits are spaced far enough apart so that they seldom fly into each other if one conductor breaks.

3-10. Transmission-line Arrangements. Multicircuit steel towers have been built, where the tower supports three to ten 69-kv lines. Such towers are in operation in southern California. Such a special design is used when many circuits leave a generating or large switching station,

FIG. 3-18. Double-circuit high tower carrying 110-kv lines across river. Lines must stand clear of masts of ocean liners passing beneath. Multiple sets of strain insulators form cluster for insulating lines from tower. (*Courtesy of Pacific Gas and Electric Company.*)

all lines running in the same direction for a substantial distance. Transmission line right-of-ways are becoming more and more expensive and difficult to obtain. A multiline transmission tower allows many circuits to pass over a given width of right-of-way, especially if the right-of-way has limited width. One of the main disadvantages of such tower design is the difficulty of working on one circuit while the others remain energized. In fact, so many adjacent circuits make it very difficult to use hot-line tools for changing insulators. If a line burns down, adjacent circuits may be affected and trip out. All circuits must be deenergized when a new conductor must be pulled in to replace an existing conductor which has broken strands or corrosion trouble.

Special towers are built for conductor transpositions, Fig. 3-21. The three individual conductors of each circuit must be transposed or have phase positions interchanged at given intervals to reduce inductive interference on telephone lines paralleling the transmission line. Telephone interference is produced by low voltages induced between a pair of telephone wires and between each wire and ground. The voltages manifest themselves as noise in the receiver of a telephone-line subscriber.

FIG. 3-19. Single-circuit tower with conductors in horizontal position designed to withstand weight and strain of sleet, ice, and snow. (*Courtesy of Pacific Gas and Electric Company.*)

Transposition of line conductors equalizes the voltages on the pairs of wires of telephone lines and also equalizes induced voltages which are out of phase with each other. Telephone lines are two-wire circuit or an equivalent single-phase circuit. Three-phase voltages induced from transposed transmission-line conductors try to cancel each other out and eliminate voltages from appearing on the telephone wires. The result is a great reduction in noise and a much clearer talking circuit. Transposition towers allow one phase of a three-phase circuit to change its position. It may take two or three spans with special towers to effect the complete three-phase transposition. The distance between each transposition varies, but is usually about 15 to 20 miles. The three conductors make a complete twist or barrel in a total distance of 45 to 60 miles.

3-11. Lightning Protection. Tower footings are usually grounded by driving rods into the earth at two points diametrically opposite each other. In areas where an average of five or less lightning storms occur in a year, such tower-footing grounding is satisfactory. Most areas in the United States, however, have many lightning storms per year, calling for an extensive and complete grounding system. Two types of cable serve as ground wires: one, a stranded-steel cable and the other, copperweld conductors. Older steel-tower designs provided only one overhead ground wire. The ground wire is mechanically clamped to the top of each tower at its center point. Tower designs are now arranged for two overhead ground wires placed approximately over each line conductor on a double-circuit tower. The overhead ground wires drain off the induced voltage produced by electrically charged clouds drifting over the line and carry away the current from the direct stroke of a lightning discharge. The overhead ground wires also prevent line insulator strings

FIG. 3-20. Single-circuit snow tower standing on top of ridge acts as dead-end tower. Two strings of insulators support each conductor which spans very long distance to tower on next ridge. Grading rings over pair of insulator strings distribute voltage stress more evenly over insulators. (*Courtesy of Pacific Gas and Electric Company.*)

from flashing over since induced overvoltages are greatly reduced in magnitude.

When the ground resistance is high, it is often necessary to bury copper conductors in the earth radiating out from each tower footing. These buried wires make a counterpoise. A buried conductor may connect from tower to tower; a very expensive counterpoise scheme not used unless absolutely necessary. Other counterpoise arrangements consist of four buried copper conductors, each conductor radiating 100 to 150 ft out from each leg of a tower. The wires or cables should be buried not less than 2 to 3 ft below the surface. Another type of counterpoise includes four buried wires, each connected to a tower footing and running parallel to the axis of the transmission line. This array gives a relatively low equivalent ground resistance, considerably lower than driven ground rods. The latter are used where ground conditions are favorable, such as in farm lands, green pastures, and wooded hills, and in areas where

FIG. 3-21. Transposition tower for single-circuit 165-kv line. Additional cable reinforces each conductor at point of suspension. (*Courtesy of Pacific Gas and Electric Company.*)

ample rain falls. The counterpoise should be used in dry or desert areas when soil conditions are very poor, such as sandy or rocky locations. The counterpoise provides a convenient method of obtaining a low ground resistance in areas having poor earth conditions.

Modern transmission-line protection uses carrier current relaying, supervisory control, telemetering, and communication. Static condensers or capacitors couple a carrier-current transmitter or receiver to the line. However, a separately insulated line wire having a length of one or two tower spans suspended under the tower conductor may also be used as a coupling device. This wire may be compared to the single-wire antenna or the L type. It works, but not so effectively as the modern coupling devices. Since transmission lines carry very large blocks of electrical energy, their relay operation must be fast and positive to maintain system stability. In stable systems, generating stations, loads, and lines remain in synchronism with each other during steady-state and transient conditions. A faulty high-voltage transmission line should be cleared in three cycles when carrying heavy line loads; many high-voltage lines are now cleared of a fault in 5 cycles or less (60-cycle basis).

Lightning is one of the chief causes of line outages. Electrostatic potentials set up in cumulus-cloud formations cause lightning. Violent updrafts or gusts of wind moving thousands of feet into the air at a high velocity create frictional electricity like a giant electrostatic generator. The surface of the earth is usually at a negative potential, although there are times when it abruptly changes from a negative to a positive potential and then back to negative again. The usual voltage gradient at the surface of the earth is about 150 volts per meter or 50 volts per ft. A cloud containing water droplets accumulates an electric charge by the action of wind passing through it. The difference of potential between the lower part of the cloud and the upper part causes voltages to develop of such magnitude that the air between them breaks down, causing a lightning discharge. Sometimes the lower part of a cloud accumulates such a positive charge that lightning discharges between cloud and earth.

As a positive charge on the lower part of a cloud accumulates, it causes an opposite charge (negative) of high magnitude to develop on the surface of the earth and structures beneath the cloud. When lightning discharges between parts of a cloud, the charge on the surface of the earth beneath the cloud (a bound charge) is released. This flows along the surface of the earth until it is equalized by the earth's charge.

If a transmission line operates under a cloud having a high positive charge, it develops a bound charge. When the cloud's charge is equalized, or neutralized, by a lightning discharge within the cloud, the bound charge on the transmission line and ground wires is released. It flows along the ground wires until the charge is absorbed by the resistance of the ground wire and the connections to ground. The released bound charge travels along the ground wires in both directions from its original location, resulting in two steep voltage waves. This phenomena is called a traveling wave. If the line is short and terminates at a substation in a transformer bank, the traveling wave reflects and may easily reach double voltage magnitude. The double voltage at the point of reflection will be impressed across the end turns of transformers. Such end insulation may fail unless protected by lightning arresters or by magnetic shielding, built in modern transformers as a cylinder around the transformer windings. If the traveling waves reach an underground line section, a great portion of the traveling wave is reflected, some part of it continuing through the cable. A cable-terminal connection to an overhead line is also a reflection point and a location where double voltage occurs. This voltage may be high enough to puncture or break down the cable-terminal connection. Lightning arresters should be installed on overhead lines, at the point of junction with underground cable terminals.

Direct strokes of lightning may hit a transmission-line conductor having no overhead ground wires. The resulting high electrostatic voltage (bound charge) on the line travels in both directions and usually flashes over the string of insulators at each of the nearest towers. When a lightning discharge flashes over insulators, it is usually followed by a 60-cycle line-to-ground fault current and relays take the transmission-line out of service. Many times the disruptive or dynamic effect of a lightning discharge breaks the petticoats of suspension insulators, and the line flashes to ground when reenergized. It then remains out of service until linemen locate and replace the insulators.

When a lightning discharge occurs at a tower, the discharge may pass harmlessly off to ground if the ground resistance at the tower footings is very low. If resistance is high, an enormous voltage may build up on the tower and the tower will flash over the insulators to the energized line conductor, again resulting in line-to-ground fault. A counterpoise usually provides a low enough resistance to induced lightning voltages so that they quickly leak off to ground. This reduces the voltage on the tower to less than the normal circuit voltage.

When transmission lines have overhead ground wires and a direct stroke of lightning hits, the overhead ground line will build up a voltage wave having a steep wavefront, traveling in both directions. The induced voltage reduces sharply at each successive tower by flowing harmlessly into the ground until entirely dissipated. Overhead ground wires must be properly centered over the line conductors so that they are protected by a theoretical protection area extending about 30° from the vertical on each side of the ground wire. Outside this angle, a lightning stroke may hit, but inside there is a protected zone that lightning voltages do not disturb. The overhead line wire, together with low tower-footing resistance or a counterpoise connected to the tower footings, results in a transmission line having few if any outages from direct lightning strokes or induced voltages produced by charged clouds passing over the line.

3-12. Protector Tubes. A few years ago protector tubes, a miniature form of lightning arrester, were installed on certain transmission lines in the United States. The tubes are supported by the tower underneath the conductor. They are adjusted to form a gap between conductor and protector less than the flashover impulse value of the insulators. The impulse value of a transmission line is the voltage that just causes an insulator string to flash over. As an example, a 66-kv line may have an impulse insulation of 690-kv based on 1½ by 40 microsecond (μsec) wave, of positive polarity. The 1½ by 40 μsec is a steep wavefront reaching its crest of 690 kv in 1½ μsec and then decreasing to one-half its crest value in 40 μsec. A microsecond equals one-millionth second. A 220-kv line may have impulse insulation of 1,415 kv with a 1½ by 40 μsec wave of positive polarity. The impulse-voltage generators at high-voltage testing laboratories may supply an impulse voltage having either a positive or negative polarity. To allow line-design specifications to be uniform, a positive polarity is usually specified for the impulse voltage.

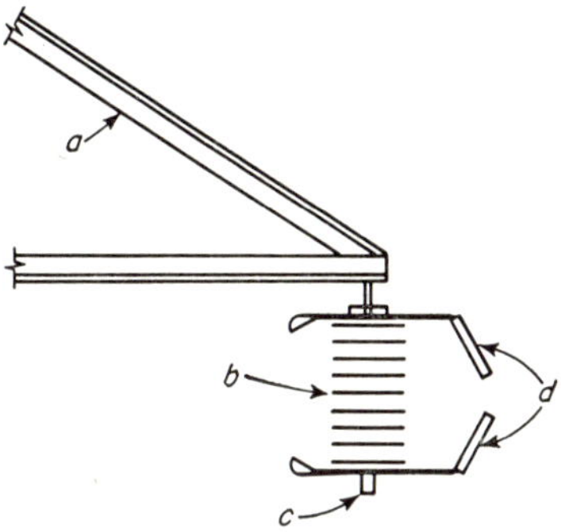

FIG. 3-22. Protector tubes allow high lightning voltage to flash to ground through tower from conductor and prevent normal current from short-circuiting. *a*, tower arm; *b*, suspension insulators; *c*, line conductor; *d*, protector tubes.

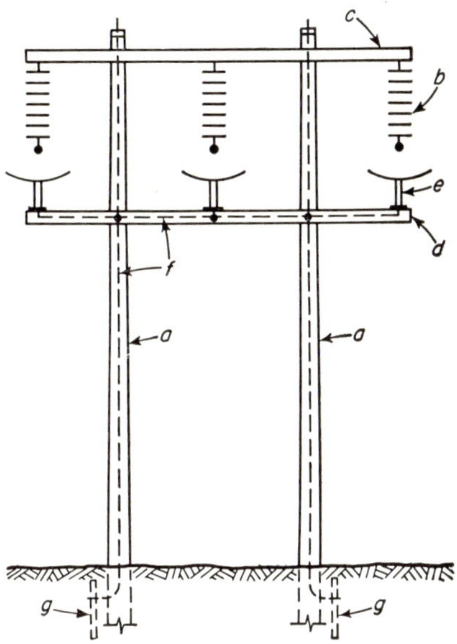

FIG. 3-23. Protective-tube installation on single-circuit wood-pole line. *a*, poles; *b*, insulator string; *c*, crossarm; *d*, crossarm for protector tubes; *e*, protector tube; *f*, ground wire; *g* ground rods.

The protector tubes or ground-fault-neutralizer tubes, as they are called, allow the line conductor to flash to ground during the period of high lightning voltage. After draining the lightning voltage off the conductor, these tubes prevent the 60-cycle current from following to ground. They have an expulsion effect that cuts off the 60-cycle current, allowing the gap to become deenergized after each lightning stroke. These protector tubes have been quite successful in preventing insulator breakages from flashovers.

Protective tubes may be installed across strings of insulators, Fig. 3-22, or they may be mounted on a grounded metal arm underneath the line conductors. They may be used on either wood-pole lines or steel towers. An installation of protective tubes on a wood-pole line is shown in Fig. 3-23. Two or more ground rods are driven into the ground near each pole. The ground rods are connected through a suitable size copper conductor; No. 6 solid is usually the smallest cable size used. These grounding wires extend up to the top of the pole and also connect to each protective tube. The crossarm supporting the protective tubes may be wooden if grounded by a copper ground wire, or it may be a structural-steel shape such as an angle or channel, which also must be grounded.

3-13. Peterson Coil. Since an insulator string flashover caused by lightning voltage is a line-to-ground short circuit, line interruptions may be materially reduced by using a reactance coil (Peterson coil) in the transformer neutral at the station. Its inductive reactance voltage neutralizes the capacitance voltage of the line. The coil connects between a transformer bank neutral and ground, Fig. 3-24. The inductive reactance of the coil must equal the capacitive reactance of the line. No changes, additions, or taps should be added to the system or the neutralizing effect of the Peterson coil will be destroyed. When an insulator string flashes over, the short-circuit current through the Peterson coil is lagging, while the current from the line over the insulators is a capacitive current. The current through the fault, the resultant of these two currents, equals the resistance current at the point of fault if the coil is properly designed. The fault clears itself immediately, and the line is not taken out of service by relay action or opened by circuit breakers. The Peterson coil, very popular in Europe, is being used more and more in this country. Many of the high-tension systems in this country operate with a solidly grounded neutral; therefore the Peterson coil has not been adopted so widely as in other countries where a solidly grounded neutral is not used.

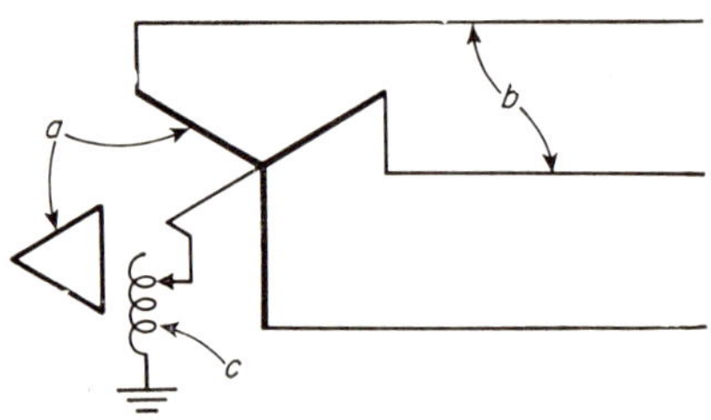

FIG. 3-24. Peterson coil connection in transformer bank. *a*, step-up transformer bank; *b*, line wires; *c*, Peterson coil with taps.

3-14. Rapid Reclosing. Since lightning voltages cause most of the transmission-line outages, methods have been developed to reduce such outages as much as possible. We have mentioned the Peterson coil. Another method is rapid reclosing of line circuit breakers. If lightning voltages cause an insulator string to flash over, the line circuit breakers open by automatic action of high-speed relays. Duration of the short circuit may be as low as three cycles if carrier-current relays and high-speed circuit breakers are used. The line circuit breakers immediately reclose. Since the reclosing time of circuit breakers ranges from 20 to 40 cycles, enough time elapses so that the air surrounding the point of flashover becomes deionized. Deionization means that the molecules of air surrounding the point of flashover have become electrically neutral. If they are ionized the majority of them have a positive charge and there are many free electrons (negative polarity) in the immediate vicinity. It takes a finite length of time for the free electrons to recombine with unstable positive ions or atoms comprising the gas molecules after an electric current has passed through. This recombining process generally requires about 20 cycles. If the circuit is reenergized in less time, the insulator string may flash over again. This immediate reclosing method cuts the line outage to a minimum and restores service quickly. However, if the transmission line carries a very heavy load, compared to the system load, system stability may be lost during the minimum time interval of the 20 cycles required to reclose the line. This method is not so attractive if the load carried by the line is small compared with the system load. Immediate reclosing of the line has a real practical value as it brings the system voltage and frequency back to normal in a much shorter time than any attendant can by manually reclosing a line. This reclosing method opens and closes all three poles of the circuit breakers.

Another method is high-speed single-phase reclosing. Each phase of the line has a separate single-phase circuit breaker with its own operating mechanism. If a string of insulators on one conductor flashes over, that phase is immediately opened by its single-phase circuit breaker through carrier-current relaying operation. The other two phases of the circuit remain in service and transmit power over the two remaining conductors. This is single-phase power, but it keeps the system in synchronism. The faulty phase which has been deenergized is reclosed in 20 to 25 cycles, restoring the circuit to full three-phase operation.

This method of single-phase reclosing has advantages and disadvantages. The advantages are very limited interruption of power, the system does not swing out of synchronism, the system voltage and frequency remain practically normal. The disadvantage: a large block of single-phase power is transmitted to the system tending to single-phase motor driven equipment, a very serious matter. Therefore, every means possible must be taken to restore the line to its three-phase condition as soon as possible. If the single-phase breaker cannot be closed, the other phases of the circuit must be opened. This can be done by a time-delay automatic relay action.

The three methods of reducing outages caused by lightning discharges are: (1) installation of overhead wires and a counterpoise, when necessary, (2) installation of protector tubes, (3) a ground-fault neutralizer or

Peterson coil. The first method seems to give the best and more uniform results.

3-15. Planning. High-voltage transmission lines require careful planning to produce an adequate design and keep maintenance costs to a minimum. First, topographical maps of the route are obtained and the line is tentatively laid out. For important routes an aerial survey is taken over the terrain on which the line will be built. This will produce a series of photographs matched up and pasted together. The composite photographic strip is then carefully scrutinized through a viewer. This stereoscopic instrument blends two pictures into one image, giving a perspective three-dimensional view. It gives the effect of looking at the ground in relief. Hills, valleys, streams, and roads are clearly shown, making it possible to judge steepness of hills and mountains.

The proposed transmission line is laid out on the photograph. Tower locations can be easily spotted and the line sag and clearance roughly estimated. Usually, an aerial survey leaves a doubt at one or two spots along the route. Field parties sent to these locations take additional photographs and map the contour of these areas accurately. Aerial photographs save an immense amount of field surveying.

After the office survey is completed, field parties survey the line. The land and property department contacts the owners of land across which the line will pass for an easement or right-of-way. By the time they have acquired the right-of-way, the route of the proposed line has been carefully surveyed. A profile of the route is sent to the transmission-line engineers. They study the profile together with the aerial photographs and accurately locate the towers and the line, Fig. 3-25. If load-growth forecast indicates the need of a second tower line in a reasonable length of time, the first line is placed along one side of the right-of-way.

The planning engineer determines the load the line will carry and its voltage. The load in kilowatts determines the line current when the voltage is known. The current determines the minimum conductor cross-sectional area, while the voltage fixes the conductor diameter.

The protection engineer, meanwhile, determines the relaying system to be used and whether or not lightning storms are frequent enough to require overhead ground wires. Let us assume that overhead line wires will be needed. From the field notes, the transmission engineer determines which part of the line passes over good moist earth and which part traverses rocky hills with little or no moist earth. In the first part of the line, tower legs would be grounded by ¾-in. ground rods 8 ft long. Resistance of the four rods should average less than 3 to 5 ohms, with a maximum resistance of 10 ohms. The line over the rocky hills would have a crowfoot type counterpoise connected to each tower leg. Such a counterpoise brings effective ground resistance down to an average of 5 to 10 ohms. Maximum resistance at any tower should be kept at 25 ohms, if possible.

A typical line might have the following characteristics: line voltage of 220 kv, conductors of 650,000 cm acsr (aluminum conductor steel reinforced), a capacity of 150,000 kva with line current of about 400 amp, an over-all conductor diameter of $1\frac{3}{4}$ in. Aluminum conductor reduces

FIG. 3-25. Double-circuit 110-kv line passes over rights-of-way through ranches and farms. Right-of-way includes right to enter and leave property with equipment. (*Courtesy of Pacific Gas and Electric Company.*)

weight, while the steel core provides strength for the long spans. Span length varies from 650 to 900 ft. Overhead ground wires could be stranded steel cables or a copperweld conductor. For double-circuit towers, two overhead ground wires would be used at a height above the conductors so that the maximum angle from the ground wire to the top conductor would not exceed 25°.

The counterpoise might be a concentric copper conductor. Copper does not corrode too much when buried in the earth; the outer layer of each strand becomes badly discolored but seems to form a protective coating. Size and length of the counterpoise have a direct relation to the ground resistance. A leakage resistance of 200 ohms is desired and in many cases requires a continuous counterpoise between towers paralleling the line conductors. A crowfoot counterpoise might be required in which the length of the individual wires is limited to 150 ft. This counterpoise would be buried in a trench about 24 in. deep. For rocky

Fig. 3-26. Double-circuit 230-kv conductors have insulator strings with larger insulators next to the conductors. No overhead ground wires are used here because lightning rarely occurs in this region. (*Courtesy of Pacific Gas and Electric Company.*)

Fig. 3-27. Tower turning an angle in single-circuit transmission line. Cable reinforces conductors at suspension point. Conductor at rear of tower needs three insulator strings to hold it in place. (*Courtesy of Pacific Gas and Electric Company.*)

terrain installation of a counterpoise is costly, but it does add considerably to the effectiveness of the overhead ground wires in reducing line outages from lightning voltages flashing over insulator strings.

The steel towers would be designed for a twin circuit, Fig. 3-26. The bottom conductor would be 60 ft above the ground, the conductor spacing (vertical) would be 18 ft, while the separation between phases, of the two circuits, would be 30 ft. Since the line traverses relatively clean country with little or no dust conditions, 18 suspension insulators of the 10-in. type could be used.

Tower design would also include towers for turning angles in the line, Fig. 3-27, and for making transpositions to reduce interference in nearby telephone circuits. Dead-end towers are also required. These towers withstand the strain of dead ending long line spans. Such towers are also needed when the line must turn an abrupt angle. If the terrain is hilly, it may be necessary to use a tower with two strings of insulators. This construction is used to tie down the line, Fig. 3-28, where it may be subject to uplifting action.

FIG. 3-28. Tie-down tower holds each conductor in place, allowing only minimum swing movement. (*Courtesy of Pacific Gas and Electric Company.*)

The insulator strings on a typical transmission line might be equipped with grading rings and arcing horns. Line hardware would be of the low-loss type, and the conductor itself would be armored or reinforced at the point of connection to the insulator-string hardware. Vibration dampers would be installed on the conductor on either side of the insulator string support.

3-16. Maintenance. A transmission line needs periodic inspection and cleaning of insulators. Suspension insulators should be individually meggered to find cracks or flaws. One line of a twin-circuit tower line is taken out of service at a time for insulator cleaning and inspection. Each end of the line is grounded. Grounds are also placed on each conductor on each tower adjacent to the one on which the men are working. Although completely cleared and open from the high-voltage bus at each end, a "dead" ungrounded line can develop an induced voltage high enough to electrocute a lineman in contact with the conductor. This induced voltage is both electromagnetic and electrostatic, originating from the parallel live circuit. When the men move from tower to tower, the line foreman must be careful that the men do not attempt to ground the energized line. Line insulators must be carefully wiped, inspected, and meggered. Cleaning of insulators is done at intervals dictated by operating experience.

If a line has relayed out of service, it must be patrolled immediately to find the fault location and extent of damage. Patrolling is usually done by men walking along the line. If the terrain is mountainous, an aerial patrol is much faster and is equally effective. Airplanes usually fly too fast to make a good survey, but helicopters have proved successful because they can hover above and along the line. Insulator string breakage and broken conductor strands can be seen through field glasses. When the fault is found, ground crews proceed to the location and make repairs.

Usually a transmission line is taken out of service and the conductors are deenergized when insulators must be changed. With present hot-line techniques, however, faulty insulators can be removed and replaced by hot-line tools. Use of these requires a specially trained crew. The work must be done in a definite sequence, in a logical manner, and the men must work as a team.

3-17. Sleet Melting. Freezing rain accumulating on transmission line conductors as coatings of ice can cause considerable trouble and damage. In many sections of the United States, this condition must be guarded against. In one type of storm, thin coatings of ice form during freezing rainfall. When a windy period follows, the force of the wind acting on the changed shape of the conductors causes them to swing up and down or "gallop." This galloping swings the conductors into contact with each other and causes short circuits and trip outs. In cases of severe galloping the transmission towers may be pulled down by the high mechanical whipping forces developed.

In storms of prolonged freezing rainfall, the coatings may build up to such proportions that the conductors may snap or the towers buckle from sheer overweight of ice. If strong winds follow such an ice build up, whole sections of transmission line may be destroyed.

The safest procedure is prevention of formation of ice on conductors. If the cables are kept at a temperature slightly above freezing, ice will not accumulate. The I^2R or line losses of the load current flowing through a conductor are dissipated as heat. This gives us a convenient method of warming the conductors. Simply build up the current flowing through them till the conductor reaches a temperature at which ice can't form. Current magnitude needed depends on (1) air temperature, (2) wind velocity, (3) conductor cross section and resistance. Operating companies prepare tables for sections of line to guide transmission operators in selecting current flows for sleet prevention.

Once ice forms on the conductors, higher currents are needed to melt it. Their magnitude depends on (1) thickness of the ice coating, (2) air temperatures, (3) wind velocity, (4) amount of time available to remove ice, (5) conductor cross section and resistance.

There are several methods of forcing the needed heating current through transmission-line conductors. The preferred method is one in which the transmission line carries its normal load plus additional load from a parallel line cut out of service or plus additional circulating currents developed by adjusting generator fields and voltage regulators. This usually involves a minimum of switching activity and no interruption of service and can usually be done by normal operating personnel.

Where the preferred method doesn't develop enough current, the transmission line being iced up is cut out of service and the load end is short-circuited by jumpers or special shorting switches. The sending end is then connected to an idle isolated generator. The generator is brought up to speed and field is slowly applied until the needed melting current flows through the shorted transmission line.

Another method using the short-circuiting procedure ties enough transmission lines together so that the sending end can be connected to its bus at normal voltage. Reactance of the conductors is just high enough to limit the current to that needed for melting. The shorting methods involve many operations and usually need an extra crew to perform them.

Where lines are built through territory subject to sleet storms, special design precautions should be taken. One company specifies (1) horizontal arrangement of conductors to avoid jump contacts when ice drops off the conductor, (2) ground wires offset from the vertical over power conductors, (3) lower tensions in conductors and ground wires, (4) flexibility in structures by using wooden members.

One utility system uses a heating current producing about 11 watts loss per foot of conductor, or 175 kw per circuit mile of line. A 2/0 copper conductor calls for 350-amp flow when the temperature is near freezing. This clears a line carrying about ½-in. thick ice in about 30 min.

Since during severe ice storms normal communication channels between generating stations and substations may fail, many companies establish procedures that take place on a definite time schedule during storms. This makes it possible to carry on all switching, shorting, and heating operations regardless of communication facilities.

3-18. Underground Transmission Systems. Transmission at high voltage through underground cables is quite different from transmission at low voltages, such as 15 and 23 kv. However, there are many similarities. Each requires a trench, some systems require ducts, and all systems use manholes for sectionalizing, for stop joints, or for splicing cables. Underground transmission of electrical energy divides itself into two general subjects: underground structures and cable facilities.

Underground structures for high-voltage cables (34.5 kv and above) are divided into two general classes: those for pipe-type cables and those for cables requiring a duct line.

The route of a high-voltage underground-transmission line must be selected with great care. Wherever possible it should run along streets not congested with existing underground structures. Other underground structures in common use are sewer lines, water mains, gas pipes, duct lines for telephone and telegraph cables, duct lines for low-voltage distribution cables. Some cities also have underground steam piping systems for heating buildings. It is advisable to keep the route of high-voltage cables away from such steam-heat piping systems. Heat lost through the insulating covering of the steam pipes is sufficient to raise the temperature of the surrounding earth 5 to 10° or more. High-voltage cables traversing suburban areas generally have to contend only with sewer lines and water and gas pipes, which do not heat the earth. A map survey is first made of the proposed route, using city maps. A profile drawing of the subsoil structures along the proposed route helps to locate the proposed transmission-line depth and manholes. In downtown areas of large cities where the streets are filled with subsurface structures, it is often necessary to dig a deep test trench at specified intervals to find the actual subsurface structures. The bottom of the trench should be at a lower level than all the subsurface installations that might cause conflicts with the proposed duct line.

Underground duct lines and manholes for single-conductor or three-conductor oil-filled cables are built in a manner similar to those designed for low-voltage cable (up to 25.0 kv) duct lines. The duct materials in common use include orangeburg fiber, transite, soapstone, and vitrified tile. Fiber and transite ducts are commonly used for duct-line construction all over the country.

Since high-voltage cables carry large amounts of electrical energy, usually one duct line is constructed for each cable. If the cables are the single-conductor oil-filled type, each cable is installed in a separate duct. High-voltage cables are those classed from 34.5 kv up to 230 kv. A 230-kv underground-cable installation has been in successful operation in Paris, France, for over 10 years. High-voltage cable capacities vary from 30,000 to 150,000 kva. It is not considered good practice to install more than one high-voltage cable circuit of large capacity in a single duct line.

The trench in the street for a duct line is usually dug by machine. Much hand digging must still be done as the machine cannot work around laterals from the sewer, water or gas mains in the street to the property line. The trench can be dug quickly in loam or sand. Shoring will be needed if the trench walls tend to cave in.

Four ducts are generally installed for a line using single-conductor oil-filled cables. The ducts are laid in two tiers of two ducts each. Each duct joint is fitted with a close coupling preventing water leaking into the duct line. An envelope of concrete covers the duct bank. After the concrete sets, the trench for the duct line is backfilled with earth. The tamped soil is allowed to settle. Finally the paving material is placed over the trench. Duct lines should rise between manholes so they drain into the manholes. This is not always possible owing to subsurface structures.

3-19. Cable Heating. Since heat liberated by cables when carrying load must be dissipated into the earth, care must be taken that the duct line does not run through earth already heated by some other source. Earth temperature should not be higher than 68 to 77 F in the summertime; otherwise the cable loading must be reduced. Heat liberated from the cable equals the square of the current in amperes flowing in the cable multiplied by the conductor resistance in ohms at normal operating temperatures (watts = amperes × amperes × resistance in ohms) or $W = I^2R$. The heat produced by the current must travel through the cable insulation, the cable protective sheath, the air inside the duct, the duct-wall material, the concrete envelope around the duct, and then into the earth surrounding the duct line.

The temperature of the earth varies from summer to winter, and also with depth. The earth temperature does not vary so rapidly as the temperature of the air above it. It has a long heating and cooling cycle, reaching its maximum temperature in September and its minimum in February. The months in which the maximum and minimum earth temperatures occur may vary somewhat in different parts of the country. In the major portion of the United States, the average maximum ground temperature is on the order of 25 C (77 F) while the average minimum temperature is 5 C (41 F).

Duct lines are usually buried deeper than the frost line except in the northern part of the country. In some parts, such as the Southwest, the maximum temperature of the earth may reach 35 C (95 F). Low-voltage distribution cables often carry their maximum load during the winter months, unless they are industrial feeders. Advantage of the cooler earth temperature can be taken during the winter to give the cables higher ratings. High-voltage transmission cables may operate at their maximum rating during all parts of the year; therefore, the maximum average ground temperature must be taken into account when laying these high-capacity cable systems.

Dry, sandy, or rocky soil does not transmit heat readily. In moist clay or loam, heat will be readily dissipated. The value of heat resistance is called the *thermal ohm.* The temperature drop through a given material will be $T = WR_{th}$; where W is heat dissipated in watts and R_{th} is in thermal ohms. A diagrammatic graph, Fig. 3-29, shows the relative temperature drops or gradient from conductor through the insulation and duct wall to the earth. Metals are good heat conductors; so the temperature gradient through metal drops slightly. Insulating materials, air, and duct materials are not good heat conductors; so the temperature gradients through them are steeper. Thermal conductivity (ability to conduct heat) of duct materials varies greatly. Soapstone duct walls do not conduct heat so easily as standard transite ducts. This

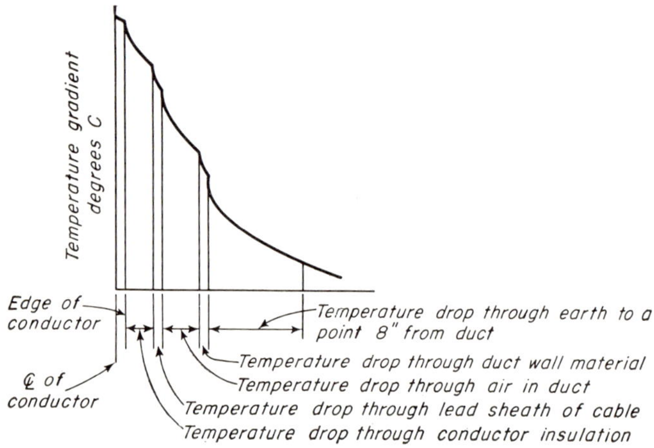

FIG. 3-29. Temperature variation from center of load-carrying underground conductor through insulation, lead sheath, duct air, duct wall, and earth.

is due to (1) thermal resistance of the material, (2) thickness of the material.

Rapidity of heat dissipation also depends on the surface areas of the conductor, insulation, duct wall, and concrete envelope. The larger the surfaces, the greater the heat flow for the same temperature drop. It is possible to increase the spacing between ducts to increase the thermal or heat storage of the concrete envelope. The concrete has a denser mass and heats more slowly. The long walls present a greater surface area to the surrounding earth. Heat from the concrete envelope flows into the earth and attempts to raise its temperature. If the air temperature is less than the soil temperature, the heat will flow to the soil surface and into the air. The quantity of heat that may flow is $W = T/R_{th}$. Heat loss in watts W depends on the temperature difference T between two surfaces and thermal resistivity R_{th} in ohms. Thermal resistance of materials is measured by ohms per centimeter cube or ohms per inch cube. Resistance is measured between any two opposite faces of a cube of the material. It is most important to make a survey of the kinds of soil or rock that a proposed line will traverse.

3-20. Duct Layouts. A duct line section is shown in Fig. 3-30. Three of the ducts carry transmission cables, the fourth duct carries relay-protection connections, pilot wires, or telephone cables. This section is called a four-way duct line. Arrangement of ducts can be changed to meet field obstructions. They may be in a single layer four ducts wide or a single tier four ducts high. While the duct line between manholes should preferably be built with a slight arch with the high point at the center of the duct line, other existing duct lines and pipes in the street often force the duct line to be laid up or down, deviating from a straight line between manholes. Care must be taken that the duct line changes in direction, depth, dips, and offsets be made with a very long radius, or it will be difficult to pull a large cable between manholes.

After the trench is dug to the proper depth, the ducts can be laid by two methods. In the *tier method*, the first layer of ducts is laid and surrounded by concrete in place. After the concrete has set, the second layer of ducts is laid in place and concrete placed around them until the duct line is completed. In the *monolithic method* the first layer of ducts is placed on spaced blocks supported by the ground. Spacer blocks, concrete or wood, are then placed on top of the first layer of ducts and the second layer of ducts is mounted on the blocks. The whole assembly of ducts and spacer blocks is fastened together with stiff wire. If individual ducts must be wired singly, copper wire should be used. Single-phase ac flowing in each duct will cause iron wire to become very hot. This will dry out concrete at that point and cause a hot spot. The entire duct assembly is then covered with a concrete envelope. After the concrete has set, the trench is backfilled with loam and the paving replaced.

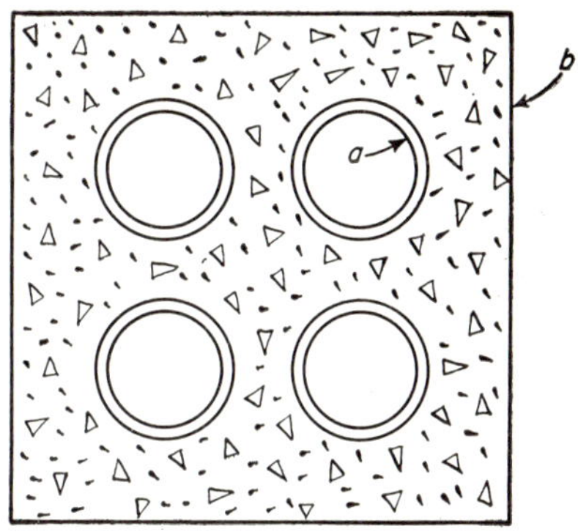

FIG. 3-30. Section through four-way underground duct line. *a*, ducts; *b*, concrete envelope.

3-21. Duct Cleaning. The completed duct line is rodded and brushed out. In rodding, a wooden cylinder, having a diameter slightly less than the duct diameter and about 4 ft long, is pulled through each duct. This cylinder knocks off any water and cement mixture that may have seeped around the duct couplings and hardened. Several circular steel brushes attached to the rope trail the wooden cylinder. These brushes smooth the rough edges on the inside of the duct formed by the cement mixture that has seeped into the duct line. The brushes also thoroughly clean the ducts. The lead ropes or wires are either blown or pulled through during construction, when each length of duct material is joined together.

Rodding also means using a set of duct rods. These are pieces of round wood approximately 5 or 6 ft long, each end with a coupling. A bundle of rods is lowered into a manhole. A duct ram or mandrel and brush assembly are pushed into the duct opening. The rods are coupled to the brush assembly, which is then forced through the duct. After the mandrel is pushed through the duct into the next manhole, it is disconnected and the rods are pulled out and uncoupled. This procedure is repeated for each duct until it is clear and free from interferences; this method is too cumbersome where manholes are more than 400 ft apart.

3-22. Pipe-type cable uses a single pipe. A narrow trench is dug by a trenching machine. The bottom of the trench is deeper than for a conventional duct line. A long easy slope is required for the pipe to pass under or around obstructions. The pipe is usually large (5 to 7 in. diameter), as three single-conductor cables are pulled into it. The pipe is coated with bitumastic material to prevent corrosion. Lengths of coated pipe are approximately 40 ft. These are usually welded together on the surface of the paving adjacent to the pipe trench. The assembled pipe length (sometimes 1,500 ft) is lowered into the trench.

Where many sewer or water-pipe laterals are encountered, the 40-ft lengths must be pushed along the trench until in their proper position before being welded. The trench at the pipe joint must be widened so that a welder may have room to work.

The completed length of pipe between manholes is capped and placed under air pressure of 250 psi. Air is dried by passing it through coils surrounded by dry ice. The moisture condenses out. The dry air should have a temperature not higher than -50 C. There is a negligible amount of moisture in the air at this temperature and below. The pipe welds are tested for leaks, by painting the joints with a soap-bubble solution. If no bubbles occur, the line is left under pressure until the cable is pulled in.

The welded joints are covered with mastic insulation. This is done by placing a form around each joint and heating the mastic material until it flows. It is poured into the form until the entire joint is covered. Before backfilling the trench, the mastic insulation is tested for continuity, by placing a voltage on the iron pipe and passing a grounded metal band over the insulation and taking readings continuously. If leakage occurs at any point on the pipe, a form is placed around the pipe and the heated matsic compound poured in, filling the mold and covering the leak.

A mastic, or bitumastic, covering is subject to deformation or puncture by a phenomena called *cold flow*. When mastic material rests on a hard or sharp object, the weight of the mastic will cause it to sink slowly around the hard object. Weight and pressure cause such a material to deform. This will eventually puncture or form openings in the protective mastic covering, exposing the steel pipe to the soil. The mastic coating over the pipe also acts as a thermal resistance to heat flow. This condition must be provided for, as it is important to maintain a corrosion-free pipe line.

Because of the mastic deforming by cold flow, it is important that the trench bottom be kept free of rocks or hard substances. A layer of sand or loam, 3 or 4 in. thick, should be placed on the trench bottom. The covered pipe is carefully laid on the sand, welded, and tested for air leaks at the joints and for leaks or openings in the mastic coating. After testing, the pipe is covered with sand or loam, for a depth of 6 in. above the top. This loam or sand is settled into place by flooding the trench with water. When the settled material remains about 6 in. above the top of the pipe covering, the rest of the trench is backfilled and the paving replaced.

3-23. Manholes. High-voltage cable installations allow a much greater distance between manholes than is possible with low-voltage installations. Low-voltage (up to 23 kv) installations generally use three-conductor cables. The length between manholes is limited by the weight of the cable that can be pulled into the duct line without deforming the conductor insulation and by the length of the commercial reels available. Manhole spacing for low-voltage cables usually is limited to between 600 and 1,000 ft. A manhole provides adequate space underground to make cable splices. Low-voltage cable installations have many splices in one manhole and often are quite congested. Manholes with high-voltage installations, on the other hand, have few splices, usually only three single-conductor splices or a maximum of six.

Manholes for transmission duct lines are usually long and narrow. They may be 12 to 20 ft long and 5 to 6 ft below the street surface. They may also be deeper, depending on the size of the oil reservoirs (fastened to the manhole ceiling) needed with oil-filled cables.

3-24. Cable Expansion. Manholes must be long enough to take care of expansion and contraction of individual cables caused by temperature changes during load variations. As electrical current through the conductor increases, copper temperature rises, and the cable expands lengthwise and a slight amount in diameter. If the duct is small compared to cable diameter, most of the cable expands along its length, with a large movement at the duct mouth. In large-diameter ducts some movement takes place at the duct mouth, but most of the cable snakes in the duct. That is, the cable moves sideways in the duct, up one side wall and down the other. When the cable snakes, minimum cable movement occurs at the duct mouth, an important advantage.

Cable movement must be carefully considered for lead-covered cables. High-voltage solid and oil-filled cables are paper insulated with a lead-sheath covering. When lead is bent back and forth too often, it tends to crystallize. This eventually cracks the lead at the points of deformation.

When a lead sheath cracks, moisture can enter the paper insulation and cause an electrical breakdown. Cable-sheath cracking occurs in the cable bends at the duct mouth and at the splice. Cable in a manhole should be bent in a long radius not less than twelve to fifteen times the cable over-all diameter. Generally, the duct mouth is rounded so that the cable sheath does not chafe against a sharp edge as it moves.

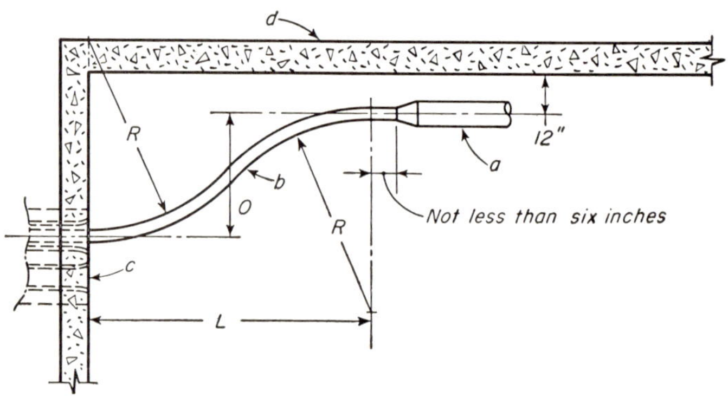

FIG. 3-31. Plan of transmission cable entry into manhole. *a*, cable splice mounted away from wall; *b*, single-conductor oil-filled cable; *c*, duct mouth; *d*, manhole side wall; *R*, cable-bend radius; *O*, cable offset; *L* offset length.

A lead-covered cable with an outside diameter of 2 in. should not be bent in a radius of less than 24 in. Expansion and contraction of electrical cables may be likened to expansion of steam or hot-water pipes (see "Steam-electric Central Stations," chapter on piping). Many have seen the long-radius bends, S-shaped and U-shaped bends that such pipes require. Mechanical expansion of cables is relatively small, but its effect must be allowed for in the design. Cable life may range up to 40 years without sheath cracking or other failures.

Large manholes for high-voltage cable are quite expensive, and it doesn't pay to build the manhole large enough to make two complete right-angle cable bends at each end. The cable is offset in the manhole, forming a flat S bend. The width of the offset should be not less than the radius of the bends, Fig. 3-31. The length of offset *L* depends on the radius of the cable bend and width of the offset *O*. If two 90° bends are used, the width of offset equals the length of the cable offset, Fig. 3-32. The two 90° bends will cause little or no lead-sheath stress with no chance of cracks in the lead sheath at the duct mouth or splice. Usually this arrangement needs a manhole width much larger than necessary for splicing. It is not possible, however, to obtain an optimum cable-training arrangement even if the ducts are brought in and out of the manhole next to the wall opposite the splice. Such an arrangement results in an excessively wide manhole, and the additional cost is not justified. Long cable life can be obtained by cable-training arrangements with smaller offset width and longer lengths.

3-25. Manhole Drainage. Manholes may be built of concrete or brick. The roof must be heavily reinforced with iron rods to withstand the impact of vehicles on the roadway. In manholes built of brick, the floor

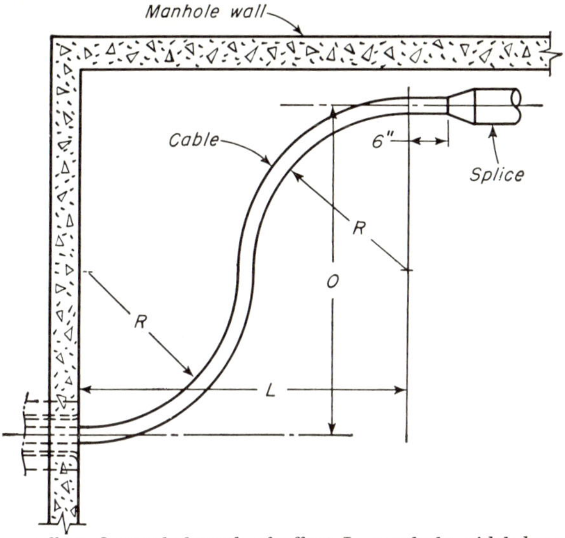

FIG. 3-32. When offset O equals length of offset L, manhole width becomes excessive.

is a reinforced-concrete slab with a sump in one corner, as the manhole floor is often below the sewer line, a motor-driven pump with a float switch keeps the manhole dry. Several feet of water in the manhole usually causes no trouble. If the water rises above the ducts and floods them, an automatic pump should be installed. Water causes chemical corrosion of the lead sheath.

The roof of a manhole is also heavily reinforced concrete. Good practice keeps the manhole width as narrow as possible, probably not more than 5 or 6 ft. Usually one 36-in. manhole opening to the street allows ample ventilation for men working in the manhole. Sometimes, however, two 36-in. manhole openings are needed to provide ventilation for long manholes.

There are several methods of building concrete manholes. The precast manhole is completely formed at some convenient yard or shop. Floor slab, side walls, and roof are all in one piece. Windows are made for the duct entrances. A sump is grouted in the floor slab after the manhole has been lowered in place. A large truck hauls the precast manhole to its location. A derrick or crane lifts the precast manhole off the truck into the prepared hole in the street. The manhole is leveled by use of a surveyor's level and transit. The manhole-cover casting and precast sump are grouted into place, the hole backfilled, and the street is repaved. The ducts enter the manhole through the windows in the walls. This is an economical method of constructing manholes if a power shovel can be used to dig a large deep hole in a street without too much interference with existing underground services.

The prefabricated manhole is built up of precast concrete pieces such as the floor slab, walls, and roof. These separate pieces are hauled to

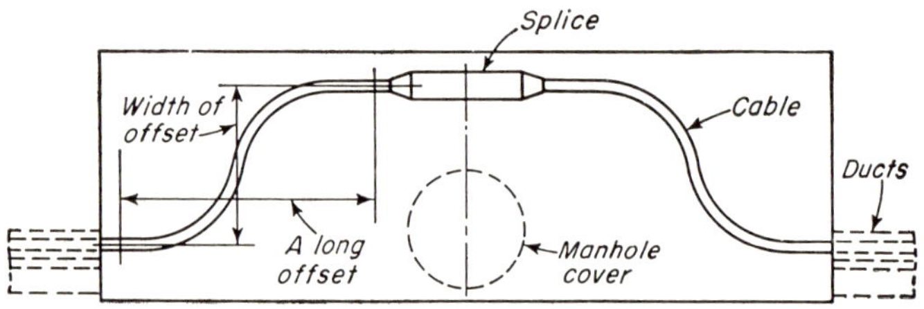

FIG. 3-33. Schematic plan of transmission-cable passage through manhole.

the site and put in place by a crane or derrick. The concrete pieces are assembled and grouted together, forming a strong manhole. Both methods of building manholes require complete standardization of sizes and shapes. Furthermore, an extensive building program is required for them to be economical.

Manholes most often are built in the conventional manner. The street or roadway is excavated for the proper manhole size. Floor slab and roof are formed with lumber, built around various obstructions that may be in the street. The concrete is poured into forms. After the concrete has set, the form boards are removed and the excavation is backfilled around the manhole and paving replaced.

Portable metal forms have been used where many manholes of exactly the same size and shape must be built. These forms are quite costly, but they speed up the construction work.

Manhole accessory equipment includes a built-in ladder. U-bolts cast in the side walls hold block and tackle, for pulling cable through the ducts. Manhole racks cast in the side walls support the cable passing through. Some racks have metal arms with a soapstone block fastened to them upon which the cable rests. The blocks have a circular notch in which the cable rests.

3-26. Manhole Types. There are several types of manholes. Generally they are the straight-through type. The cable runs directly through the hole and out again, Fig. 3-33. A duct line turns a corner in a right-angle manhole, Fig. 3-34. Such a manhole allows the cables to be trained around each wall, giving ample bending radii and large offsets. Sometimes a duct line must be offset, requiring an angle hole,

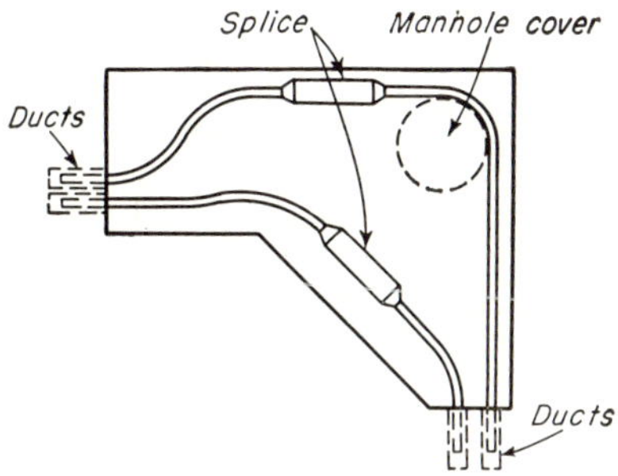

FIG. 3-34. Manhole arrangement for turn in transmission cable.

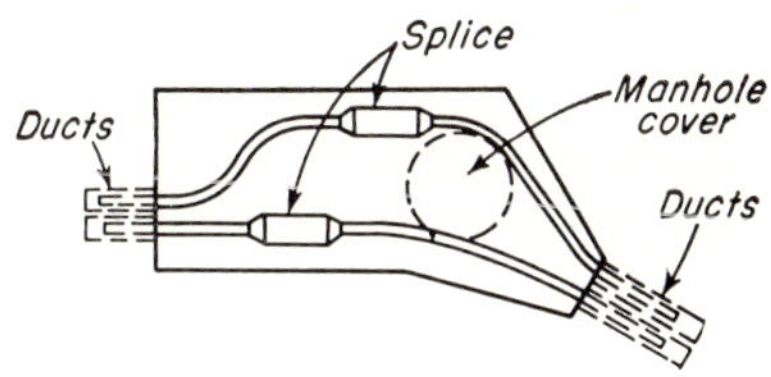

FIG. 3-35. Manhole design varies with the angle to be turned.

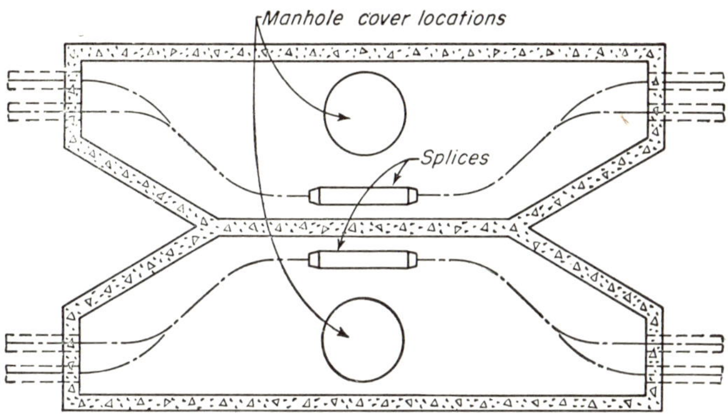

FIG. 3-35*a*. Divided manhole isolates two different transmission cables.

Fig. 3-35. Such a manhole allows a duct line to veer away at an angle and still give long bending radii for the cables.

When two high-voltage cables are pulled in separate duct lines along the same street, it is more economical to use one manhole instead of building separate ones. It is not good practice to place two sets of high-voltage splices in the same manhole. A large manhole is needed for a pair of high-voltage cable circuits. It has a dividing wall separating the two sets of splices and two manhole covers. The two cables are spliced in separate rooms of the same manhole, Fig. 3-35*a*. The two duct lines are well separated so that each can carry its rated load without overheating the soil near the parallel duct line. The individual splices should be separated so that if one fails it will not affect the others.

3-27. Transmission Cables. High-voltage cables (34.5 kv and up) include both solid and oil-filled types. The high-voltage range for solid cables goes up to and includes 69-kv. Oil-filled cables range from 69 up

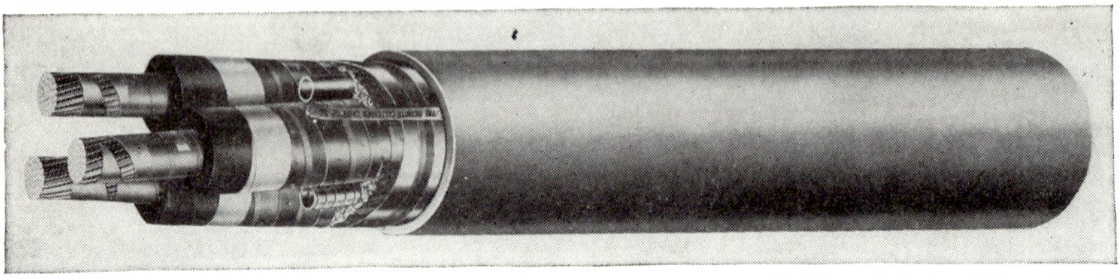

FIG. 3-36. Three-conductor low-voltage low-gas-pressure solid-insulated cable. Each stranded conductor is covered with layers of paper insulating tape saturated with insulating-oil compound. Nonmagnetic shielding tape covering each conductor equalizes the voltage gradient across individual layers of paper insulation. Three tubes in cable interstices act as gas channels. Solid copper tube passes gas from one splice to another. Spiral tubes allow gas to permeate interior of cable length. Nitrogen under 15 psig pressure prevents void formation between paper-insulation layers and prevents moisture from entering the lead sheath. (*Courtesy of Okonite-Callender Company.*)

to 230 kv. These are generally designed for single-conductor operation. Three-conductor oil-filled cables are rather large in diameter and must have a reinforced lead sheath to withstand the internal oil pressure. Figures 3-36 and 3-37 show sections through low-pressure gas-filled cable.

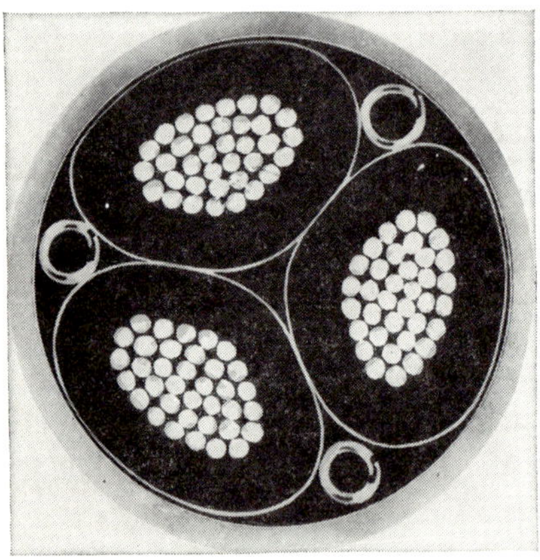

FIG. 3-37. Cross section of cable similar to that in Fig. 3-36, except conductors are sector shaped and not compacted. (*Courtesy of Okonite-Callender Company.*)

3-28. Solid-type 69-kv Single-conductor Cables. The voltage limit for solid insulated cables is 69 kv. In solid insulation the conductor is covered with layers of paper insulation tightly wrapped one over the other. The paper insulation is covered with a nonmagnetic binder tape and the entire assembly encased in a lead sheath. No gas or oil circulates in the cable sheath, Fig. 3-38. Many installations of 69-kv single-conductor cables with solid insulation operate successfully. The single-conductor cable terminals are filled with compound. The solid type of cable is limited in voltage application for the following reasons. (1) As a solid cable carries load, its conductor temperature increases and the cable compound (layers of paper are saturated with an insulating compound) expands. This action stretches the lead sheath, which may rupture or stretch it permanently. (2) As the load on the cable decreases and the conductor cools, a partial vacuum may form within the cable sheath. If pinholes are present in the lead sheath, moist air may be drawn into the cable and be deposited on the insulation, eventually causing failure. (3) As the cable conductor heats, the insulating compound on the layers of insulation become soft and may migrate or flow to low spots in the cable run, especially where the cable sheath has been stretched. This migration causes voids or small air spaces between the layers of paper insulation, which may ionize and break down. The voids, or spaces between layers of paper, become a small spark gap. If the gap or void is small enough, the voltage stress across the layers of paper flashes over during each voltage half wave, that is, during both positive

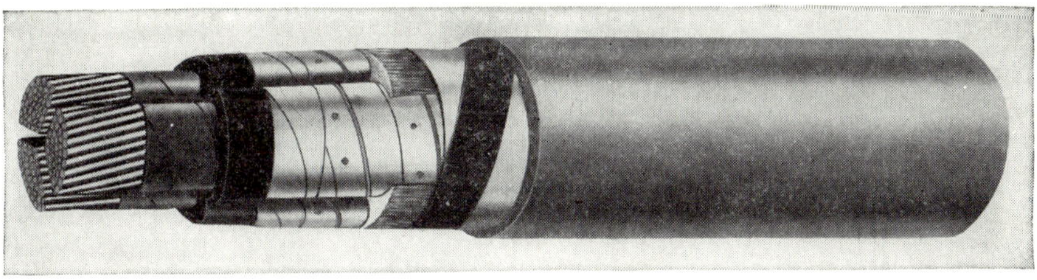

FIG. 3-38. Three-conductor 15-kv solidly insulated lead-sheathed cable. Conductor strands are compacted together. Metallized tape covering each conductor reduces the effect of excessive voltage flux from the irregular conductor surface. Metal binding tape bundles three conductors together before covering with lead sheath. (*Courtesy of Okonite-Callender Company.*)

and negative polarity voltage half wave, a total of 120 flashes per second. This continual arcing eventually creates a hole between the adjacent layers of insulating tapes. This progressive breakdown leads to a complete disruptive cable failure. Solid insulated cables are now limited by technical developments to the 69-kv voltage class, Fig. 3-39.

Solid-type cables are simple to install. They require no auxiliary equipment as for oil-filled cables or high-pressure-pipe oil- and gas-filled cables.

3-29. Oil-filled Cables. The oil-filled cable design was introduced in the United States about 1926. It proved an outstanding success for transmitting large blocks of electrical energy at high voltage through underground cables. Figure 3-40 shows the construction of an oil-filled single-conductor cable. The conductor has a central spiral-steel core around which strands of copper wires lie, making up the needed circular-mil cross section. The steel core has two functions: (1) to form a path for the oil flowing through the cable, (2) to allow the oil pressure to compress the layers of paper insulation. The oil duct or hollow spiral steel core must be large enough to allow the required amount of oil to flow at low pressure, 10 to 20 psi. This is about the limiting pressure a conventional lead sheath can stand without deforming or stretching. If a higher pressure is required, the sheath must be reinforced. Sometimes oil pressures are as high as 30 psi. In such cases, the lead-covered cable sheath must either be reinforced with a second lead sheath extruded over an intermediate layer of jute or be covered with a bedding

FIG. 3-39. A 60-kv solidly insulated single-conductor cable-terminal installation. Conductor terminals are mounted on a steel horizontal support. Connection between underground and overhead circuit is made through oil-filled bushings. (*Courtesy of Pacific Gas and Electric Company.*)

of jute over which is wound a layer of armor wires. These are round galvanized-steel wires or flat steel tapes. The steel tape may be flat and overlapped or may be the interlocked design. Such cable-sheath construction and reinforcement is quite expensive. The usual cable is designed for 20 psi maximum with the operating pressure at 12 psi.

The central core of an oil-filled cable is also made of spiral steel tape, allowing oil to percolate between the copper strands to the wrapped-paper insulation. Oil under pressure compresses the layers of paper and is forced into any voids that may have formed between the layers. Voids or spaces are formed by the temperature cycles of the cable which in turn cause cable expansion and contraction.

One advantage of this cable type is that, if the lead sheath leaks, the oil slowly flows out and the cable loses oil pressure. A pressure gage and relay connect to the oil-piping system. The relay closes its contacts when oil pressure falls to 10 psi or less and sounds an alarm. The cable need not be taken out of service but can be left until convenient to repair the leak. It is necessary, however, to keep the oil reservoirs filled with oil. If the leak is large, the oil pressure drops rapidly, giving an indication of the size of the leak. This gives a supervisory function over the lead sheath. When cables carry large blocks of electrical energy, it is most important that they be left in service as long as possible, even after an oil leak occurs. The oil-filled cable has had wonderful success. A 230-kv oil-filled cable system has been in successful operation in Paris, France, for many years. Many 110- and 132-kv oil-filled cable installations are in successful operation in this country. The oil-filled cable system is made up of lengths connected together through joints in manholes. The joints have small copper tubes connected to oil reservoirs mounted on the ceiling of the manholes. Stop joints between a given number of cable lengths separate the complete cable circuit into smaller sections, Fig. 3-41. Each section has its own set of oil reservoirs. When the cable runs on a slope, a stop joint at the end of each cable length limits the oil pressure. For steep cable slopes it may be necessary to build intermediate manholes. The cables in these manholes are spliced together through stop joints to prevent excessive oil pressure. Separate groups of oil reservoirs serve each section of cable separated by stop joints.

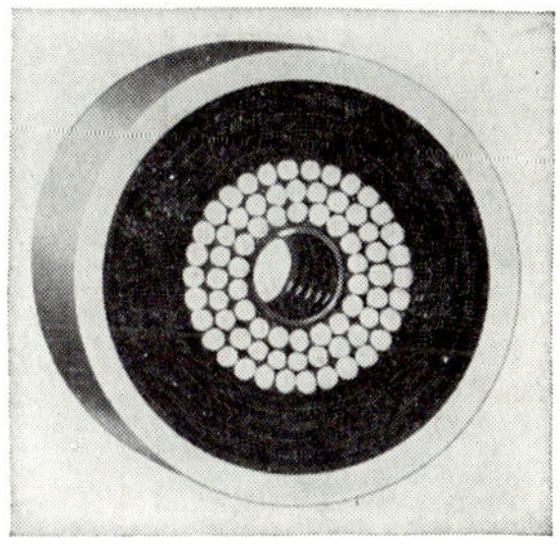

FIG. 3-40. A 120-kv single-conductor oil-filled cable has a stranded-copper conductor surrounding a spiral-steel open core. Oil passes through the open core to permeate the cable at 10 to 20 psig pressure. (*Courtesy of Okonite-Callender Company.*)

3-30. Oil-cable Accessories. An oil-filled cable installation needs accessories, including stop joints, insulating joints, reservoirs, and miscellaneous oil piping, valves, gages, and oil-pressure alarm relays.

Oil reservoirs are built in three general types: (1) the straight gravity-feed type; (2) the balanced-pressure type which exerts a gas pressure on the oil and forces oil to flow into cable lengths that may be at a higher elevation than the reservoir; (3) the straight-pressure type, which has separate cells within the oil reservoir. When the cable thermal cycle changes, oil may be forced into the reservoir, compressing the gas above the oil. As the cable cools, the gas pressure forces the oil to flow into the cable again. The gas in the reservoir acts as a cushioning device, maintaining the oil pressure within the cable.

Stop joints are not used at each cable length except on steep slopes. Normal joints are used to splice lengths of cables together. In a normal splice the conductors are joined by compression-type splicing sleeves. Conductors may be spliced together by soldering the conductor sleeves or by compression joints. A hydraulic press develops a pressure of several tons per square inch, squeezing or indenting the cable conductor and

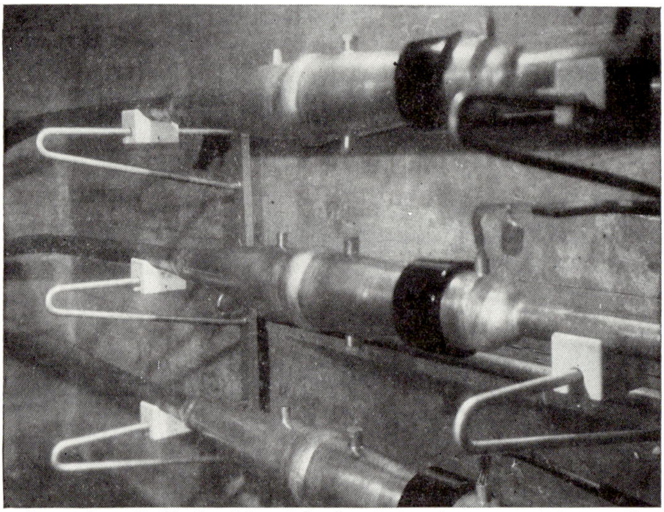

FIG. 3-41. A 110-kv oil-filled cable with sheath insulators on joints. Ground wires connect to each cable on the right-hand side of insulators. (*Courtesy of Okonite-Callender Company.*)

its sleeve together, which makes a very compact tight splice. Furthermore, the electrical resistance of the splice is very low. Some compression joints have been measured for resistance which was on the order of 0.0001 ohm or less. Such low-resistance splices do not develop local heating as a poorly soldered splice may do.

A stop joint is similar to a normal joint except that an insulator or porcelain barrier separates the two halves of the joint. This joint sectionalizes the oil-filled cable into shorter lengths. In case a sheath cracks, all the cable will not lose oil, only one short section. On level ground, the manholes may be spaced as far as 1,800 ft apart. Stop joints may be used every third or fourth manhole. Normal through joints are used in the manholes when stop joints are not installed. A stop joint is similar to a normal joint except that two conductor splices are necessary. Each splice must be fully taped with oiled paper, till the proper insulation thickness is reached. Several layers of varnished cambric are wrapped over the paper insulation. The joints are not filled with compound, as for solid-type cable joints. The cable oil fills the joint at normal pressure, 10 to 12 psi.

The oil reservoirs have valves arranged so that each reservoir connects to its single-conductor cable, but in an emergency, one reservoir may feed all three single-conductor cables.

One contact of the alarm relay closes when excessive pressure occurs; another contact closes when oil pressure is low. One alarm relay attaches to each single-conductor oil reservoir, so that the relays indicate which phase of the single-conductor group of a section is in trouble. Only one set of alarm relays is installed on one section. A multiconductor cable running in the fourth or spare duct of the duct line carries the alarm signal to an annunciator at one of the substations in which the cable terminates. The alarm circuit may connect to a 120-volt ac single-phase system or to a 125-volt dc supply system.

Another auxiliary device is a degasifier. This frees the cable insulating oil from all gas and moisture. It includes a vacuum pump for removing air or entrapped gases from the oil. This assembly is usually mounted on a truck for easy transportation.

3-31. Sheath Currents. Oil-filled cables are generally of the single-conductor type, three-conductor oil-filled cable is used only for low voltage (34.5 kv). Each single-conductor cable runs in a separate duct. The single-phase lead sheaths develop sheath voltages from the single-phase currents flowing in each conductor. These create a field flux cutting the lead sheath and producing single-phase induced voltages. Each sheath voltage must be neutralized or reduced in magnitude; otherwise large currents flow through the lead sheaths and ground connections.

Each joint, whether a normal or stop joint, may have a porcelain insulator mounted in the lead sleeve assembly to separate and insulate the lead sheaths from each other. This insulator limits the induced voltage to one or two lengths of cable. The single-conductor cable positions in a duct line are also transposed (changed in position relative to each other) about every five manhole lengths. Phase A cable takes phase B cable position, phase B cable takes phase C cable position, while phase C cable takes phase A cable position. These transpositions help to neutralize the three induced single-phase voltages. It is still necessary, however, to eliminate, if possible, the induced voltages on each cable length by sheath insulators and bonding the sheath of one cable phase across the sheath insulator to the sheath of another cable phase, this method is called cross bonding.

The same result can be obtained with sheath transformers. Their windings couple inductively between the sheath of one phase connected to another sheath of another phase on the opposite side of the insulating joint. Thus one winding of the transformer may be connected to cable sheaths for A phase and B phase, while the secondary winding is connected to B phase and C phase on the opposite side of the joint insulator.

This method of cross-bonding reduces the voltages induced in the sheath, and they can be practically balanced out. If they could be completely balanced, the sheath voltage would be zero.

Sheath transformers are used only with single-conductor cables. The transformer windings have many taps. It is possible to balance out almost all residual sheath voltages.

3-32. Cable Protection. To protect a lead-sheathed cable from heating by an arc from the electrical failure of an adjacent cable or from burning oil, the cable sheaths in the manholes are covered with asbestos strips soaked in water glass. When the water glass dries, the covering becomes hard and stiff. This covering is known as niagriting. Cable sheaths and joints in each manhole are usually niagrited. This protection covers the cables from duct mouth to duct mouth in the manhole.

Single-conductor oil-filled cable is terminated to a metal wiping sleeve, made up of a large porcelain insulator connected at each substation end by single-conductor terminals. The lead sheath of the cable is wiped directly to the wiping nipple of the cable terminal. Each cable terminal has an oil reservoir. The cable and its terminal are filled with oil from the cable oil system. The oil reservoir absorbs the excess oil caused by the cable load cycles and acts as a regulating pressure reservoir, which tries to keep the oil in the cable near the terminal and the terminal itself at a constant pressure.

Duct lines for oil-filled cable are built to be dry, but water does seep in. If water has a pH value greater than 7.0 it is alkaline, if it has a pH value less than 7.0 it is acid. The letters pH mean "hydrogen ion concentration." Lead cables can usually survive an acid solution, but an alkaline solution attacks and destroys the lead sheath by chemical corrosion. Since cable ducts cannot be kept absolutely dry, the cable sheath must be protected from corrosive influences. This is done in several ways. First, the covering may be a composite lead sheath covered with a neoprene jacket. Neoprene is a synthetic heat-setting compound immune to attack by oil, acid or alkaline solutions, and it is readily extruded over a lead sheath. A disadvantage is the difficulty in pulling such a coated cable through a duct line. A thermosetting compound is rubberlike and can be vulcanized. It retains its solid characteristics when heat is applied.

Cable sheaths may also be covered with a thermoplastic compound. The flamenol type is resistant to oil, acid or alkaline solutions. It has the disadvantage of becoming soft again when heated.

Cable sheaths must be protected against electrolytic action caused by stray dc flowing through the earth. Cathodic protection equipment applies a negative dc voltage to the sheath, making it negative electrically to surrounding metallic structures buried in the ground. The lead sheath is maintained at a negative potential relative to a mass of buried iron which is made positive to the earth. Then an electric current flows off the buried iron and gradually destroys it while the current flows on the protected lead sheath. Cathodic equipment acts as a device to deflect or reroute stray currents and thus protect the cable sheath. It will not protect a cable lead sheath from chemical corrosion caused by an alkaline solution in the earth.

3-33. Pipe-type Cables. Two general pipe-type cable systems are in use: (1) the high-pressure oil system, and (2) the high-pressure gas system. Both systems use pressures on the order of 200 psig.

FIG. 3-42. A 115-kv Oilostatic cable. Three single-conductor paper-insulated shielded cables are pulled into one steel pipe. Copper skid wire wrapped around each conductor separates them and makes it easier to pull the cables into the pipe. Oil at 200 psig fills the pipe. (*Courtesy of Okonite-Callender Company.*)

3-34. Oilostatic Cable Systems. This pipe-type cable has three-single-conductor cables pulled in one duct, Fig. 3-42. The pipe or duct is filled with oil kept at 200 psig. A mastic covering to prevent corrosion coats the pipe exterior. The manholes are spaced far apart, up to 4,000 ft. A separate duct of transite or fiber carries alarm wires, or the alarm wires run from station to station through an existing duct line. Manholes for a pipe-type high-pressure oil cable are similar in shape but smaller in size than those used for oil-filled cables, since only one pipe enters and leaves the manhole. The pipe enters and leaves the manhole in a straight line; so no provision need be made for training the conductors around the manhole. Conductors within the pipe do not have a lead sheath; therefore, sheath cracking or buckling is avoided. All cables are single-conductor type, built of stranded copper conductors of proper over-all diameter to keep corona loss to a minimum. The stranded conductor is covered with layers of paper insulation. Paper insulation in turn is covered with a nonmagnetic shielding tape covered with a paper tape. The over-all cable is covered with copper armor wire, called a skid wire. The whole single-conductor cable assembly is initially covered with a thin lead sheath at the factory. A piano wire placed under the lead sheath and over the skid wire can strip off the lead sheath when the cable is pulled into the pipe.

If the cable circuit is relatively short, no semistop joints are required. Most circuits, however, are rather long, and it is advisable to install a semistop joint at each manhole. A semistop joint requires only one splice instead of the two needed for a stop joint. The copper conductors are joined by a compression-type sleeve connector. A hydraulic press makes the compression splice. The three single-conductor cables pass through a tight stuffing-box assembly at one end of the joint. The stuffing box is oiltight and designed for the particular over-all diameter of the cable installed. Oil at such high pressure will seep in from one side to the other through the layers of paper and between the conductor strands, when a cable section must be taken out of service.

After the three single-conductor cables are spliced and taped for full insulation, shielding tape is wound over each single-conductor insulated joint assembly. A steel cylinder used for covering the joint is pushed over the end bells and the cylinder made oiltight by a clamping ring. Oil is allowed to flow into the semistop joint until it is brought up to a pressure of 200 psig. A small metallic pipe connects each end of the semistop together. This pipe has a valve for shutting off the oil flow from one section to another. A small metal box is often furnished for enclosing pressure gages. A pressure gage attached to each section reads the oil pressure in the pipe on each side of the semistop joint. An oil-pressure relay gives a warning when excessive or low pressures occur.

The three single-conductor lead-covered cables received from the factory are pulled into the pipe. As the cables enter the pipe, the thin lead protecting sheath is stripped off by the piano wire laid under the sheath. While the cables are being pulled in, the pipe is filled with dry carbon dioxide gas. After the cables are in place, a temporary rubber sleeve, called a boot, is fastened to the end of the pipe and covers the exposed cables. The interior of the pipe is then filled with dry degasified oil under a slight positive pressure of 5 or 6 psig.

After all cable lengths are pulled into the duct and capped by boots, they are ready for splicing. Normal joints or semistop joints are used. Generally, semistop joints make it easier to sectionalize a cable in short lengths and find the location of leaks when they occur.

A semistop joint is always installed in the manhole adjacent to the oil-filled terminals at the ends of a cable. Three single copper or brass pipes connect to the end of this joint. Each nonmagnetic tube has a single-conductor cable pulled in it. The terminals are usually mounted on a steel structure, 10 to 12 ft above the ground. The copper or brass tubes bolt to the bottom of the terminals. The whole assembly of single-conductor-cable porcelain terminals, copper tubes, and semistop joints are filled with oil under a pressure of 200 psig. The terminals are huge corrugated-porcelain cylinders designed for the operating voltage of the cable. The conductor terminates in a sleeve or splice connection of the compression type. The sleeve connector is inside the porcelain bushing near the top and covered with paper insulation. The paper insulation and interior of the porcelain bushing is under full oil pressure of 200 psig.

Accessories for an Oilostatic cable include a pumping plant at one terminal of the line. Dry degasified oil fills the cable. The pumping plant stands in a relatively small metal cabinet. A panel with indicating and recording pressure gages gives a graphic record of the oil pressure in the pipe. No reservoirs are required in the manholes. Each manhole may have a semistop joint with shutoff valves and pressure gages.

A pipe-type cable on a hill or long slope has each single-conductor cable anchored to prevent it from sliding downhill during load cycles. One method armors (with copper wires) the upper end of a cable part way. If a cable section is 1,500 ft long between manholes, the cables would be armored for about 150 ft from the top of the slope. A stop joint of special construction with a larger cylindrical section is used. The armor from each of the three cables fastens around clamping rings on each porcelain insulator forming a part of the inside of the enlarged stop joint. This length of armor distributes the strain of the cable downhill movement over a very large cable surface area and prevents it from progressively walking downhill during thermal cycles.

Cable movement inside the pipe is relatively small. The pipe is large enough so that cable conductors can take their expansion and contraction by a snaking movement within the inside of the large steel pipe. The manholes are smaller than those for oil-filled cables.

The pipe-type cable depends on the mastic coating to prevent corrosion. A leak is found by sectionalizing the cable. On finding the faulty section a hole is dug in the street about halfway between the two manholes. The uncovered pipe is surrounded with dry ice and acetone. The oil in the pipe at this point freezes solid at about −82 F. When the oil freezes, the leak can be located in either half section. Another hole is dug at the center of the leaking half section and the oil again frozen. This sectionalizing continues until the remaining pipe section is so short that the oil leak itself can be uncovered. A metal patch can be welded over the leak and additional mastic poured around the repaired section.

This method of finding oil leaks is slow and expensive. Fortunately, leaks occur infrequently.

If the conductor insulation fails and short-circuits to ground, the fault can be found by various Wheatstone-bridge methods. The semistop joints are opened, the compression splices sawed off, and the three single-cable lengths pulled out. New cable lengths are pulled in and respliced. The cable is then tested for voltage breakdown and placed back into operation.

3-35. High-pressure Gas Cables. Two general pipe-type cables use gas pressure at 200 psig. One is the compression type, the other SMD type of cable. Both use a single steel pipe covered with a mastic jacket similar to the pipe assembly for Oilostatic cable. The manholes for SMD cables are larger than for oil-pressure cable, as the joints are longer. The normal and stop joints, large steel flanged cylinders, can be separated at the center by removing bolts from a pair of flanges at the center of the cable joint. The two halves move into recessed openings in the end walls of the manholes. The steel pipe must be laid in the ground and tested for leaks as with Oilostatic installations. The interior of the steel pipe must be clean, dry, and free from sharp metal burrs.

Normal and stop joints are both used. The stop joints sectionalize the cable into short lengths so that it can be tested for gas leaks.

The compression cable system is made up of three single-conductor solid insulated cables covered with a bronze-armored polyethylene sheath. The conductors are a compact stranded conductor having an oval shape. Three of these single conductor cables are pulled into each pipe section. After the cable is pulled in, a temporary end seal is made and dry nitrogen gas admitted at a low pressure, 5 to 10 psig. After the joints are made up and tested, the pipe is filled with dry nitrogen gas under a 200-psig pressure. The sheath remains over each single-conductor cable.

The oval shape of conductor has an advantage in that it can slightly change its shape along the minor axis, during thermal cycles, thus changing the cross-sectional area with but little variation in the outside perimeter of the cable.

The SMD conductor is similar to the Oilostatic type. It has a round concentric stranded conductor, covered with insulating tape. The tape covering is wrapped with a nonmagnetic shielding tape, which in turn is covered with a few paper tapes. Finally a copper armor wire, called a skid wire, is wrapped over the entire single-conductor assembly. A thin lead sheath is extruded over the skid armor. A piano wire underneath the lead sheath strips it off when the cable is pulled into the pipe, Fig. 3-43.

When three single-conductor cables are pulled into a pipe, they are taken off individual reels. They should not be bound together when pulled in. They should be kept loose in the pipe. This allows them to snake during load cycles. As a result there is practically no movement of the cables at the joints.

Terminals for both the compression-type high-pressure gas cable and the SMD type require an oil-filled porcelain terminal fitted to a metal assembly that has an oil-to-gas stop joint made as an integral part. A stress cone, which terminates the shielding tape, must be built within the porcelain terminal. The insulation is built up to about double its thickness, making a slope of approximately $7\frac{1}{2}°$. The shielding tape terminates at the top of the cone. The stress cone relieves the electrostatic voltage stress on the insulation at the point where the shielding tape ends. When a cable is energized, an electrostatic voltage appears on the shielding tape. The tape edge is quite thin, and an excessive electrostatic flux (field lines of force) appears at that point. In time the insulation under the end of the shielding tape may fail owing to concentration of field flux, causing corona cutting. When the shielding tape terminates at top of the cone, the distance of the shielding tape to the conductor has doubled and the electrostatic flux reduced to one-fourth of its original value. This assures long life to the insulation directly under the end of the shielding tape.

FIG. 3-43. Section through high-pressure gas-filled pipe cable for 115 kv. Each conductor has skid wire wrapped around it. Nitrogen at 200 psig fills the pipe interior. (*Courtesy of General Cable Company.*)

If a high-pressure-gas pipe-type cable begins to lose gas, the leak is found by shutting the by-pass valve around the semistop joints and reading the pressure drop in the sectionalized pipes. However, the leak in the faulty pipe is quite difficult to locate. It has been proposed to use audio-frequency amplifiers and a probe laid on the street surface. The noise energy produced by the escaping gas is amplified and appears strongest when the probe is near the leak. This testing would have to be done after midnight when the street-noise level would be at at its lowest value.

Transmission of electrical energy at high voltage over underground cables of various types is well established. Such systems operate satisfactorily with a minimum of maintenance but are more expensive than transmission by overhead lines. High-voltage underground cables can be justified only in large metropolitan areas or around large airports. They have the advantages of not being affected by lightning, windstorms, sleet, snow, fog, and airplanes.

Water-saturated ground does not bother the high-voltage underground cables as they operate with a positive pressure head. The only real enemy of the underground cable system is corrosion. This may be controlled by improved pipe or oil-filled cable coverings.

3-36. Submarine Cables. Submarine cables may be placed under water on the bottom of a lake or river. They carry electrical energy continuously, while submerged. Several types of cables are used. Both low-voltage and high-voltage cables are in operation.

Submarine cables have a much greater carrying capacity than cables installed in duct lines on shore; this is due to the cooling action produced by the water carrying away the heat so much faster than does the soil and earth surrounding a duct line.

3-37. Low-voltage Cables. Low-voltage cables range up to 23-kv; they may be single-conductor or three-conductor. For short lengths crossing small sloughs or rivers, an unarmored cable may be used. The armor provides strength when the cable is lowered into the water or removed to make repairs. This cable uses an ozone-resistant insulation and shielding tape over the insulation. The insulation assembly is covered with a neoprene jacket. Three-conductor cable or three single-conductor cables may be used. When installing the latter, three reels of cable must be mounted on a barge and the cables reeled out simultaneously.

Other types of cable include three-conductor rubber-insulated lead-jacketed cables covered with jute and armor. This type is used for long underwater lengths. The rubber insulation is ozone resistant and covered with a lead sheath. The whole assembly is protected and strengthened by an armor made of galvanized-steel wire. A variation of this cable is a three-conductor cable with ozone-resistant insulation covered with a shielding tape for 5,000 volts and above. The insulated assembly is covered with a galvanized-steel armor wire; no lead sheath is used.

Sometimes, a pair of telephone wires is placed in the interstices of a three-conductor submarine cable. The telephone pairs must be a twisted pair.

Since cables can carry considerably more current when submerged, the shore ends must be a larger size. The shore ends splice to the submarine cable far enough from shore so that the splice stays submerged.

The cables may terminate on an elevated rack for connection to the overhead system, or they may run in a duct line to the substation.

Signs are usually erected on each shore with the wording "Cable Crossing." This warns steamship masters thay they should not anchor their ships in the vicinity, or the ship's anchor may foul the submarine cable and break it.

3-38. High-voltage Cables. High-voltage cables are classed from 34.5 kv up. Three-conductor oil-filled lead-jacketed cables covered with a galvanized-steel-wire armor are used up to 69 kv. The lead sheath protects the paper insulation from the water. Cable oil circulates through the cable and in the insulation, preventing voids from forming between the layers of paper. The head of water over the cable helps to compress the cable insulation which also tends to lengthen the cable life. Oil-filled cable is also used for high voltages, 110 kv, etc. These are single-conductor cables covered with a nonmagnetic wire armor. The oil-filled cables require the usual oil reservoirs and purifiers used with shore cables in duct lines. The oil-filled cable on land is limited to 15 to 20 psig pressure; however, the head of water and addition of armor over the cable assembly gives the lead sheath protection against excessive hoop stresses and allows oil pressures much greater than 20 psig.

CHAPTER REVIEW TOPICS

1. How are wood poles prepared for use on transmission lines?
2. Sketch several different conductor arrangements used on wood-pole lines.
3. Describe the H-frame construction. When is it used?
4. What are the three principal types of steel poles or towers?
5. What is corona? How can it be corrected?
6. What are the features of a hollow-core I-beam conductor?
7. Describe the features of an acsr.
8. How is the copperweld conductor built?
9. How are overhead conductors spliced together?
10. What is the upper voltage limit used with compression-type porcelain insulators?
11. What other conditions beside line voltage determine the amount of insulation needed by a transmission line?
12. Why must insulators be properly shaped?
13. What are the principal features of the water-cleaning method for insulators?
14. Are insulators cleaned while a line is carrying load?
15. What are the functions of a grading ring? Do these rings have other names?
16. What causes a conductor to vibrate? How does vibration affect a conductor?
17. How can conductor vibration be avoided?
18. How is sharp bending of conductors at support points avoided?
19. Why is nonmagnetic material used for conductor hardware?
20. What provisions must be made when a transmission line changes direction?
21. Describe some types of anchors used to brace transmission towers.
22. What conditions require special design of transmission towers?
23. Why are two or more circuits often carried by a line of transmission towers?
24. What are the disadvantages of multicircuit transmission lines?

25. Why must transmission line conductors be transposed?
26. What are the usual distances between conductor transpositions?
27. Why are ground wires used over transmission lines?
28. Why are counterpoises used? What are some of their arrangements?
29. What must be provided to protect a transmission line from lightning damage?
30. Describe the principal actions taking place in a lightning storm.
31. What are the functions of lightning arrestors?
32. What is the protection area under a ground wire?
33. Describe the arrangement of protector tubes.
34. What is the impulse value of a transmission line?
35. What are the functions of a Peterson coil? How is it installed?
36. What are the advantages of automatic rapid reclosing of circuit breakers?
37. Name the three methods for reducing transmission line outages from lightning strokes.
38. What are the principal steps in planning and building a transmission line?
39. What are the steps that must be observed in cleaning insulators on a transmission line?
40. What methods are used to patrol a transmission line?
41. How is sleet prevented from forming on transmission line conductors?
42. Describe three methods of heating transmission conductors for preventing or melting ice accumulations.
43. What design precautions can be taken to prevent icing troubles?
44. What are the principal equipment elements in an underground transmission line?
45. What steps are necessary in planning and building an underground line?
46. What materials are often used in ducts?
47. How many ducts are usually provided for a transmission line underground?
48. What precautions must be observed in regard to cable heating?
49. How is heat dissipated from electrical losses in an underground cable?
50. How should ducts be graded between manholes?
51. How is a duct cleaned by the rodding method?
52. How is pipe-type cable usually installed?
53. What purpose does the bitumastic coating around an underground pipe serve?
54. What limits the distance between manholes?
55. What are the usual proportions of a transmission manhole? Why?
56. What provisions must be made for cable expansion?
57. Why must cable be given a long radius in making turns?
58. What are some of the methods of constructing manholes?
59. How many different types of manholes are used? Describe them.
60. What are common voltage ranges for solid-insulated and oil-filled cables?
61. Describe the features of a solidly-insulated cable.
62. Why are temperature limitations important on solidly-insulated cable?
63. Describe the main features of an oil-filled cable.
64. What auxiliary equipment must be provided with oil-filled cable?
65. What is the usual pressure range for oil-filled cable?
66. Does an oil leak take an oil-filled cable out of service immediately?
67. What are stop joints used for in an underground cable?
68. Why is it important to install good joints with low resistance?
69. How far apart may manholes be spaced for an oil-filled cable?
70. How are sheath currents limited in underground cables?
71. How are sheaths protected from arcing failures on adjacent cables?
72. What is electrolysis? How are cable sheaths protected against it?

73. Describe the features of an Oilostatic cable installation.
74. Describe the semistop joint used with Oilostatic cable.
75. What oil pressure is used with Oilostatic cable?
76. What are the accessories needed with Oilostatic cable?
77. What are the features of a high-pressure gas cable installation?
78. What does a stress cone do in the terminals of high-voltage cables?
79. Why can smaller cables be used for submarine work when carrying a given load, compared to land work?
80. Describe the features of a low-voltage submarine cable.
81. Describe the features of a high-voltage submarine cable.

CHAPTER 4

DISTRIBUTION SYSTEMS—MECHANICAL ARRANGEMENTS

By E. M. ADKINS

Jersey Central Power and Light Company

4-1. Distribution. In its broadest sense *distribution system* refers to the manner in which electrical energy is conveyed from the generators to the many points of use. In its more specialized sense it refers to the lines and circuits over which electrical energy passes along city streets or rural roads to the ultimate consumer.

In general the distribution system is the electrical system between the step-down substation fed by the transmission system and the consumers' meters. There is no definite line between transmission and distribution according to the voltage or bulk capacity.

Energy transmitted at the voltage at which it is used is secondary distribution, for example, the 115–120 volts at which residences are served or the 440–480 volts at which a factory may be served. The system of lines and circuits supplying energy to the secondary distribution system, at a higher voltage, is the primary distribution system. The connecting link between primary and secondary systems is the distribution transformer.

4-2. Overhead or Underground? Distribution systems may be either overhead or underground. For light or medium load densities the distribution system is usually overhead, using poles and open wires mounted on wood crossarms. For greater load densities in the congested areas of cities, the distribution system is usually underground, using ducts and manholes under the surface of streets and sidewalks.

Choice between overhead and underground depends upon a number of widely differing factors. Comparative economics (the annual cost of operation) is the most powerful factor influencing the choice. The capital cost of an underground system may be five to ten times that for an overhead system. The greater capital cost of underground would inevitably mean higher electric rates if it were universally required, as is sometimes suggested.

An overhead distribution conductor has a considerably higher rated safe current-carrying capacity than an underground-cable conductor of the same material and cross-sectional area. On the other hand, the inductive reactance of the underground-cable conductor is considerably less than that of the overhead conductor, because of the closer spacing of conductors. If current-carrying capacity is the important factor influencing the choice of conductor, then the open-wire overhead conductor will be favored. If voltage drop is the important factor, the underground cable would be favored, particularly at low power factors.

In congested city areas the load density, or average peak load demand per 1,000 ft of street length, may become so high that adequate service cannot be supplied from overhead lines. The (1) importance of service continuity, (2) presence of so many wires on the poles, (3) necessity of large transformers on practically every pole, and (4) difficulty of mounting them, together with (5) attendant operating and maintenance difficulties may dictate choosing underground distribution. The exact point at which underground becomes more economical varies with individual cases. With load densities approaching 100 kva per 1,000 ft of street length, the annual cost for comparative overhead and underground distribution may begin to favor certain types of underground. An underground system may show a lower annual cost than a comparative overhead system, even though the capital cost of the underground system is considerably higher. This is influenced by the useful life of the underground system being much longer than the overhead. An overhead system might have a useful life of 25 years, while an underground system might have a useful life of more than 50 years.

Local municipal pressure to remove overhead poles and wires from certain streets to improve appearances may exert considerable influence on choosing underground. Sometimes streets in a certain area are to be torn up and repaved. Under such conditions it may be desirable to change from overhead to underground to take advantage of the cheaper construction costs a few years before the change would have been made ordinarily. An overhead system is susceptible to damage and service interruptions from wind, ice, snow, lightning, and other weather conditions as well as occasional interruptions from traffic hazards. When trouble does occur, however, it is much more difficult to locate and repair on the underground system than on the overhead. The overhead system is much more flexible than the underground. Manholes, duct lines, transformer vaults, etc., are permanently placed once installed. On an overhead system, poles, wires, transformers, etc., can be easily shifted to better take care of changes in load conditions.

Sometimes a combination of overhead and underground is desirable. For example, the primary may be underground while the transformers and secondary are overhead. The entire distribution system may be underground along one street, while overhead laterals from the underground secondary may feed along side streets and alleys together with overhead lateral taps from the underground primary, or numerous other combinations may be desirable. Electrical characteristics of an underground system are similar to those of an overhead system. The final choice between overhead and underground depends upon the comparative annual cost, the influence of noneconomic factors, and the exercise of good judgment.

4-3. Underground Construction. Underground distribution systems are usually used in the most densely built-up portions of cities with high concentration of electric load and where space is at a premium. The

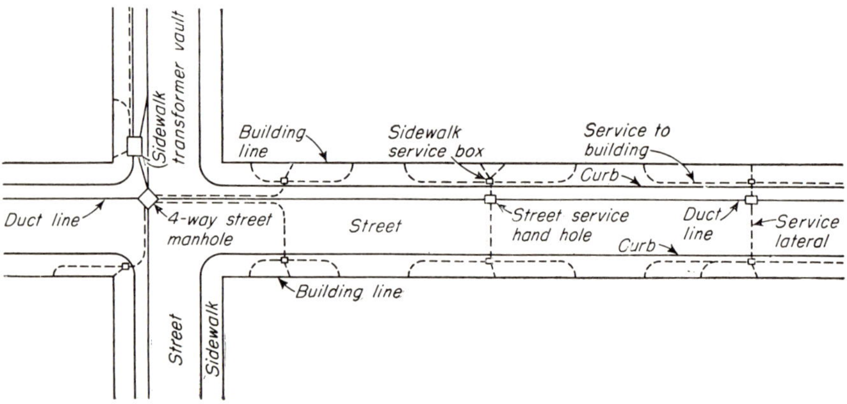

FIG. 4-1. Underground ducting arrangement using service laterals from main duct line under roadway.

loads to be served generally require the greatest possible guarantee of service continuity. A lengthy interruption of electric service to a good portion of the main business section of a city, from failure of the local electric distribution system, would be on the order of a disaster. Frequent interruptions of short duration arouse storms of protest. The underground distribution system, when properly planned and constructed, offers the best chance for continuous, uninterrupted service.

Underground distribution systems in the United States usually use conduit sections carrying insulated cable and varying sizes of underground street manholes. Duct, cable, and manhole location, arrangement, and construction are almost endless in variety. The problem of supplying electric service from an underground system may involve only a few blocks of a small town or a good portion of a large city. In either case the basic principle is the same. Figures 4-1 and 4-2 show a typical underground system section for a town or city. The principal difference between Figs. 4-1 and 4-2 is in the amount of street trenching and pavement replacement. The arrangement shown in Fig. 4-1 requires a maximum of pavement removal and replacement together with a minimum of sidewalk removal and repair. The arrangement in Fig. 4-2 requires a minimum of trenching and pavement repairs but a maximum of sidewalk breakage and repair.

4-4. Construction of Conduit Lines. Underground distribution systems usually have duct lines between manholes, through which insulated cables are pulled. The construction of such duct lines is quite costly. In some cases it is feasible to construct the underground cable system without a conduit duct system. The cable is buried directly in the earth without a duct system. Considerable savings can be realized by using buried cable. When buried directly in the earth, some increase in safe

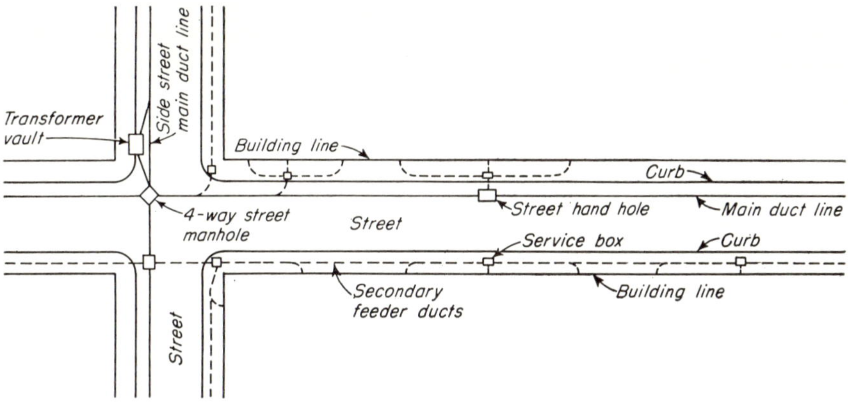

FIG. 4-2. Underground ducting arrangement with secondary feeder ducts under one sidewalk in addition to main duct line under roadway.

current-carrying capacity is gained because of improved heat dissipation. Buried cable installations may be desirable where rate of load growth is low, where the type of load is not likely to change, or where local pressure for removing overhead lines is severe and only a small initial outlay can be warranted.

Buried cable may be especially desirable on streets where repaving is seldom required, where there is a space clear of trees or other obstructions between curb and sidewalk in which the cable may be installed, or where there is an open earth space between the street paving and the curb. The cable is buried in a trench about 18 to 24 in. deep. The trench may be dug by hand, by a special cable-trenching plow, or by a mechanical trench-digging machine. If a long enough cable is to be buried and the route is relatively free of obstructions, the use of the cable plow or trenching machine may be warranted. This is especially true where the trench is dug, the cable laid, and the trench refilled all in a single operation. Where the danger of mechanical damage to the cable from digging is present, the cable may be protected by a metallic armor, by covering the cable in the trench with concrete slabs or creosoted boards, or by similar means.

At street intersections, pavement may have to be broken and repaired. In some cases this can be avoided by pushing a metallic conduit under the paving to the trench on the other side. A hydraulic jack or some other apparatus can be used for this. There is no way to replace or add to existing buried cable except by opening the trench. This same objection also applies to inspection and maintenance of existing facilities.

The first problem in preparing an underground duct system is selection of the route. An accurate map should be made showing location of fire hydrants, street-light standards, curb lines, buildings to be served, areaways, points of entrance for services to buildings, etc. The map should also show the subsurface obstructions in the route of the duct line. The route should seek to avoid all subsurface obstructions such as

1. Telephone and other communication duct lines, manholes, and service laterals.
2. Gas mains and gas service laterals.
3. Water mains, service laterals.
4. Sewer mains, sewer manholes, and house-service connections.
5. Storm sewers and catch basins.
6. Steam pipes.

All such subsurface obstructions should be located as accurately as possible by consulting available maps of the city municipal authorities and data of other utilities. After all surface and subsurface obstructions are located as accurately as possible, a route should be selected with as few curves and bends as possible and which will not have to be placed at too great a depth to avoid crossovers of other pipes and conduits. When the tentative route has been selected, test holes should be sunk at street intersections and intermediate locations to check the location of subsurface obstructions.

The size of the duct depends upon the diameter of the largest cable to be installed, length of conduit run through which the cable must be pulled, number of bends, and present standards of use. Conduit comes in a wide range of sizes, but little is gained by using smaller sizes except a slight reduction in first cost. The 4-in. conduit is probably the most commonly used size and the most satisfactory for general distribution use.

The number of ducts depends upon local distribution requirements, how many ducts will be needed for through circuits, and provision for future requirements. Eight to sixteen ducts for through and local distribution requirements will generally be adequate. For cross streets and adjacent streets where there will be few through-type ducts, four to six ducts will usually be sufficient. The best arrangement of ducts will be a compromise between that which will be lowest in cost and that which permits the highest current-carrying capacity. The safe current-carrying capacity of cables depends upon the maximum temperature at which they can operate without damaging insulation.

The temperature at which a cable operates depends upon the ambient temperature of the earth surrounding the duct bank, the I^2R losses within the duct bank from copper losses in the cable, and the rate at which heat generated by the I^2R losses is conducted away and dissipated. In a duct bank arranged as in Fig. 4-3, the heat-dissipating qualities would not be good, but the average cost of construction would be lower

FIG. 4-3. Duct bank needing minimum construction cost but having poorest heat-dissipating ability. Cables 6, 7, 10, and 11 would run hot.

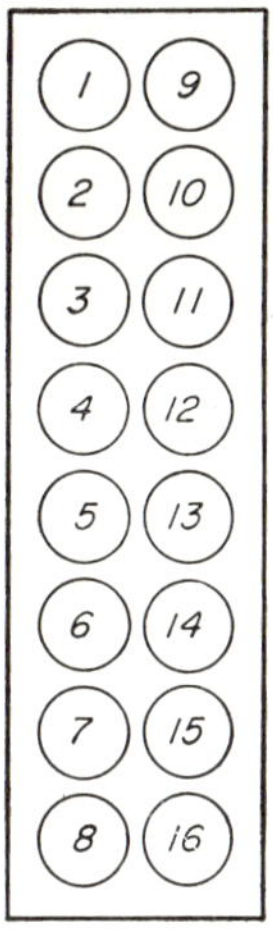

FIG. 4-4. Duct bank with good heat-dissipating ability incurs higher construction cost.

than if the duct bank were arranged as in Fig. 4-4, the best from a heat-dissipating standpoint.

In Fig. 4-3, if all ducts were occupied by cables of the same size and carrying the same load, temperatures in ducts 6, 7, 10, and 11 will be highest, temperatures in ducts 2, 3, 8, 12, 5, 9, 14, and 15 will be next high, and temperatures in ducts 1, 4, 16, and 13 will be lowest. Temperature in ducts 6, 7, 10, and 11 will be highest because the heat generated by I^2R losses will have to be conducted away through the surrounding ducts. The current-carrying capacity of these cables will be lower. If the same cables are arranged in a duct bank as in Fig. 4-4, the least favorable ducts from a heating standpoint will be ducts 4, 5, 12, and 13, but the allowable I^2R losses will be about $1\frac{1}{2}$ times as great as the allowable I^2R losses in ducts 6, 7, 10, 11 of Fig. 4-3 for the same temperature rise. If low duct temperature is needed, the duct bank should be arranged so that at least one side of the duct is in direct contact with the earth. If the lower cost of a duct bank arranged as in Fig. 4-3 is the deciding factor, then cables operating at a low-current density, such as series street-lighting cables, should be located in the inner, less favorable, duct positions. Cables passing through a manhole should be arranged so that they can be efficiently racked to avoid crossovers and to permit adequate working room in the manhole. The arrangement in Fig. 4-4, or some modification, will be the most satisfactory from the standpoint of manhole racking.

There is considerable variety in the types of ducts used. Regardless of the type used, construction of the duct line is approached in the same manner. The route of the duct line is surveyed and clearly marked so the construction forces can easily follow it. Pavement is removed with a minimum of damage. In some cases municipal ordinances specify the manner of breaking and replacing pavement. Duct lines should be graded so that water will not collect in pockets formed by an abrupt change in grade. The line should follow the natural grade of the street and drain into an adjacent manhole. Pockets, or low spots, collect water, which when frozen may do considerable damage to the line. The trench for the duct line should be excavated to as near the depth of grade as possible.

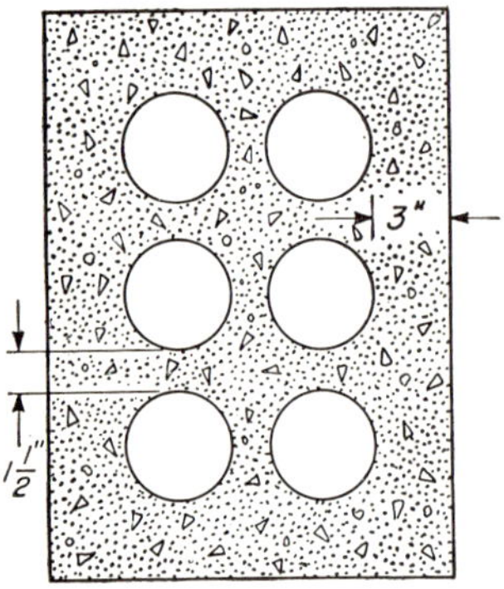

Fig. 4-5. Fiber conduits usually are encased in concrete for protection and strength.

The width of the trench should be kept as close as possible to the width of the duct line and still permit working space for men and materials. The bottom of the trench should be finished to the exact grade required. The slope of the grade should not be less than about 3 in. per 100 ft of duct line. Small amounts of fill required in spots to bring up to grade should be well tamped so that the duct line will lay on undisturbed, or firmly packed, earth. The depth of the trench will be determined by the over-all height of the duct line and the required distance below the top of the pavement. In general, the top of the finished duct line should not be less than 2 ft 6 in. below the top of the pavement.

4-5. Fiber Conduit. Fiber conduit is tubular in shape and made of wood pulp fiber impregnated with a bituminous compound. Nominal sizes of fiber conduit vary in ½-in. steps from an inside diameter of 2 in. to as much as 5 in. The conduit comes in 5- or 8-ft lengths. It has moderate mechanical strength but is somewhat brittle. Fiber conduits are generally enclosed in concrete to protect them from crushing traffic loads. Groups of conduits should be encased in concrete of a thickness similar to that in Fig. 4-5.

The (1) built-up or (2) tier method may be used in constructing a duct bank of fiber conduit. In the built-up method, the full assembly of ducts is mounted on concrete separators, Fig. 4-6*b*, and bound together with baling wire or heavy twine. The complete duct assembly may be laid on a previously poured-concrete base in the trench bottom or laid on concrete separators, Fig. 4-6*a*, spaced about 3 ft apart. After the fiber duct and concrete separators are in place and forms placed along the sides, concrete is poured around the complete assembly. A 1:3:5 concrete mix, consisting of one part of portland cement, 3 parts of sand, and 5 parts of ½-in. gravel, will usually be found satisfactory. The concrete mix should be wet enough so that it will fill all the spaces around the duct and not leave voids or air pockets in the concrete sheath.

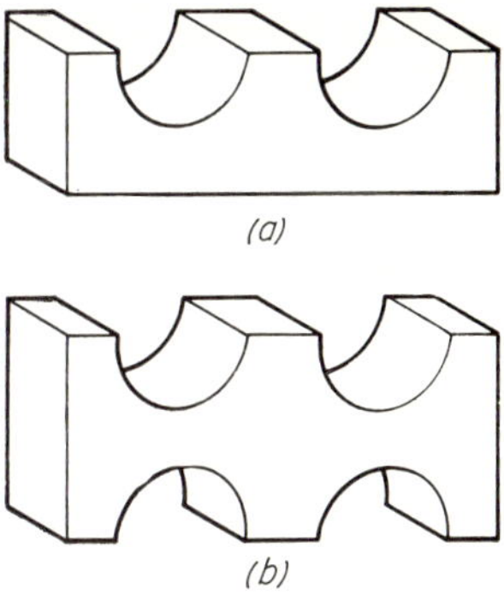

FIG. 4-6. Spacers hold conduits in position while concrete is poured.

In the tier method of construction, concrete separators and forms for the sides are placed in position and the concrete base is poured, using a slightly dry concrete mix. The concrete base is rammed and the first layer of duct placed in position. Concrete separators, Fig. 4-6*b*, are placed on top of the first layer of duct, in readiness for the second layer. Concrete is then poured to a level at the bottom of the second layer of duct. The concrete mix for this and succeeding layers should be wet enough to flow around the duct and fill all voids. The procedure is the same for succeeding tiers of duct until the top tier is reached. Here a top sheath is poured up to grade marks on the side forms to provide a top thickness of concrete sheath similar to that in Fig. 4-5.

The built-up method of duct construction is somewhat lower in cost and has the additional advantage of being somewhat faster. It has the disadvantage of forming unavoidable voids and air pockets. For duct formations more than four or five ducts high, the tier method will generally be more satisfactory. The most commonly used method of connecting conduit lengths is the Harrington joint. It is a fiber sleeve tapered at both ends so that the tapered ends of the conduit can be forced into the sleeve to form a tight fit. The taper is such that the ends of the conduit do not butt. Bends making angles of 45 and 90° allow building turns in the duct line.

4-6. Transite Conduit. Transite conduit is made of a mixture of asbestos fiber and portland cement. It has a smooth bore with a low coefficient of friction, is incombustible, noninductive, and corrosion resistant. Nominal sizes of transite conduit vary in ½-in. steps from an inside diameter of 2 in. to as much as 6 in. The standard length is 5 ft up to 2½ or 3 in. diameter and 10 ft for larger sizes. The mechanical strength is quite high, varying with the wall thickness. Transite duct comes in two wall thicknesses. The thick-walled duct varies from about 0.30 in. for the smallest sizes to 0.35 in. for the larger sizes. The thin-walled variety varies from about 0.20 to 0.25 in.

The thick-walled duct has sufficient mechanical strength to be installed directly in the earth without being encased in concrete for all except the most severe loading conditions. The thin-walled duct is usually encased in concrete except where the crushing loads are light. The duct bank may be concreted similar to fiber conduit. When conduit is installed without concrete, the first layer of duct is laid in the bottom of the trench and backfilled carefully around the ducts to eliminate air pockets, or voids. If the excavated earth is clean of debris and rocks, it may be used as backfill around the duct. If not, a suitable quantity of soft earth or sand should be obtained to provide a well-packed fill around the ducts to a depth 4 or 5 in. above the top layer of duct. In laying the duct, a separator, Fig. 4-6, made from creosoted wood maintains alignment and separation between ducts. In general, a separation of less than 1½ in. should not be used; a greater separation makes it easier to tamp the backfill around the duct and improves the dissipation of heat from the ducts. Transite conduit buried directly in the earth may produce a considerable saving in cost through elimination of forms and concrete.

4-7. Tile Conduit. Tile conduit is made from vitrified clay in single- or multiple-duct sections. Single-duct and multiple-duct tile conduit are both used for underground distribution. The single duct, however, is usually preferred because of its greater mechanical strength. The outside surface of the single-duct conduit is octagonal in shape with a round bore. The nominal inside diameter ranges from about 3¼ to 4¼ in. Duct comes in standard lengths of 18 in. Multiple duct comes in sections of 2, 4, 6, or 9 ducts. The bore is square in shape and may be obtained in sizes having nominal inside dimensions of 3¼ by 3¼ in. or 3½ by 3½ in. with a standard length of 24 or 36 in.

Single-duct conduit is laid similar to fiber conduit with a concrete envelope about 3 in. thick enclosing the duct bank and a lesser thickness of concrete between ducts. In laying the single-duct tile conduit, joints should be staggered for greater mechanical strength. A laying mandrel preserves the alignment and cleans the duct as it is pulled along after each section is completed.

Multiple-tile duct is also laid similar to fiber duct. Multiple-duct sections are bedded on a concrete base equal to the thickness of the concrete envelope encasing the entire duct bank. Succeeding duct sections are joined by a cement mortar. Joints are made with steel dowel pins and then usually wrapped with a thin cloth covered with paint or liquid cement to prevent leakage of the mortar into the ducts. Laying mandrels, in addition to the dowel pins, preserve alignment of the ducts and clean them as the line is laid. The duct bank is encased in concrete to a thickness of at least 3 in. Tile duct for power cables is not used as frequently as fiber or transite conduit.

4-8. Concrete Duct. Concrete conduit, cylindrical in shape with a round bore, is made from a mixture of portland cement and sand. Concrete or stone conduit usually is manufactured near the point of installation because of the care with which it must be handled. Considerable quantities of concrete conduit have been installed by large users of duct. Single-duct concrete conduit is installed similar to fiber conduit, being encased in concrete. A very satisfactory concrete duct line may be built with a patented machine which forms a complete duct bank in place in the trench. The machine consists of a driving motor, a hopper receiving the concrete mix, a set of revolving feeder screws, duct-forming mandrels geared to the driving motor, a molding compartment, and a set of light angle rails on which the machine is mounted. As the concrete mix feeds into the hopper, the feeder screws force the mix into the molding compartment where the duct is formed over the revolving mandrels. The packing action of the revolving feeder screws produces a dense concrete with formed ducts having a smooth bore and propels the machine along the angle rails, laid in the bottom of the trench and along the side of the duct bank, at the same rate that the duct bank is formed. If a considerable amount of duct line is to be built, important savings may be effected by this patented duct-forming machine. Savings result from the elimination of forms, reduction in manpower, and much greater speed with which the duct line is built.

4-9. Wood Duct. Wood conduit is sometimes used for duct lines where the cost of other types is prohibitive, where it is imperative to complete the installation as quickly as possible, and where traffic loading is not heavy enough to crush the duct. Wood duct sometimes serves as distributor ducts and service laterals. Wood duct, made from Southern yellow pine or Norway spruce, is impregnated with creosote as a preservative. It comes in single-duct sections, 6 to 8 ft long, of square cross section. The circular bore varies in ½-in. steps. Mortise-and-tenon joints are used. Conduits lay directly on the earth in the trench bottom. Layers of duct are laid directly on top of each other with the joints staggered both vertically and horizontally. The top layer of duct is generally covered with heavy creosoted wood plank for mechanical protection.

In all types of duct construction, protection against cave-ins must be provided, especially in sandy or loose, porous soil. If the soil will not support itself, or if vibration of heavy traffic causes caving, the trench sides will have to be supported by wood sheathing until part of the backfill is completed. Excavation, backfill, and removal of excess material is generally the largest single cost factor encountered. The principal items of cost in underground duct-line construction include:

1. Pavement breaking and removal.
2. Excavation.
3. Cost of duct-bank materials.
4. Shoring, sheathing, and forms.
5. Labor.
6. Backfill and removal of excess material.
7. Pavement replacement.

4-10. Manhole Construction. In any underground distribution system, manholes are needed. They must be located at points along the duct line to provide (1) terminal points for cable runs, (2) space for installation of auxilliary equipment, (3) space for installation of transformer equipment, (4) junction points where cable runs from cross streets may be joined, (5) points at which service taps may be made, etc. The manhole must be large enough to provide space for the equipment and adequate working space for workmen installing and maintaining it. Since they are usually under street surfaces, they must have sufficient mechanical strength to withstand any crushing loads, such as heavy traffic.

Manhole shapes and sizes vary widely according to the intended use. Rectangular and octagonal shapes are probably the most common. Numerous variations, resulting in odd shapes, may be necessary to avoid subsurface obstructions or to fit unusual requirements. The most commonly used materials are brick, concrete block, and poured concrete. Reinforced poured concrete is probably most frequently used. A typical rectangular type street-cable manhole is shown in Fig. 4-7. Such a manhole routes cables at street intersections or other locations where a terminus for cable runs is needed. Entrance of the duct line should be located to facilitate routing cables through the hole. The manhole should be large enough horizontally to provide ample space for cable bends, space for largest cable splices, mounting space for any auxiliary equipment, and ample working space for installation and maintenance workmen. The rectangular type generally varies from 6 to 9 ft in length or width, according to local need. Depth should be great enough to allow at least 1-ft clearance between the bottom of the duct line and the floor of the manhole and sufficient headroom for working. Headroom clearance of 6 ft or 6 ft 6 in. will usually be adequate.

The floor of the manhole should be reinforced concrete not less than 4 in. thick. If the bottom of the manhole is not below sewer level, a connection to the sewer may be used where a dry hole is required. If the sewer is above manhole floor level, a small sump should be provided and the floor sloped to drain into it. This facilitates pumping when needed. The walls should not be less than 8 in. thick and should be steel

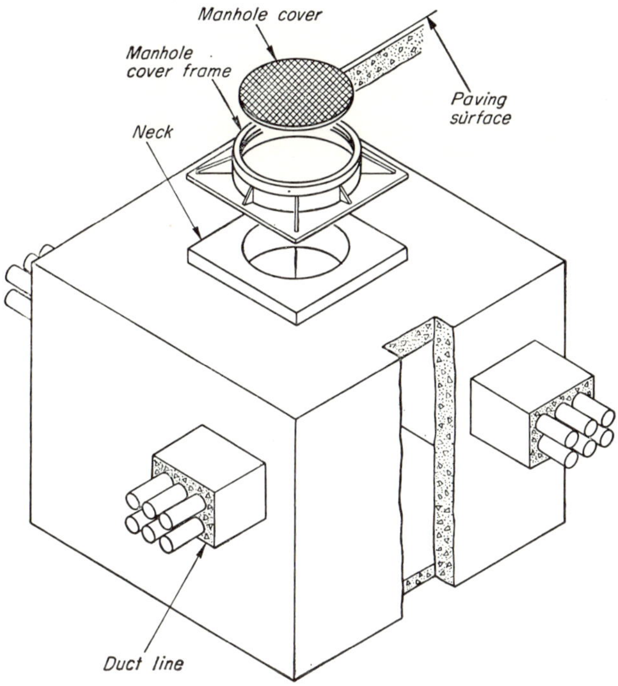

FIG. 4-7. Street-cable manholes must be built to meet several requirements.

reinforced if made of poured concrete. The roof must bear the heaviest traffic loads without cracking or collapsing. If the roof is made of poured concrete, fairly heavy reinforcing will be adequate for smaller manholes. In the larger holes the roof should be supported on sections of steel rails or I beams in addition to steel reinforcing. The roof should be 8 to 10 in. thick. Entrance to the manhole is through an opening in the roof. The entrance consists of a removable cast-iron cover mounted on a cast-iron frame. The frame is mounted on a neck extending above the top of the manhole to such a height that the manhole cover will be level with the pavement. The manhole should not be made any deeper than necessary to avoid unnecessary excavation and overly long necks.

It will generally be desirable to provide vent holes in the manhole cover so that harmful or explosive gases and vapors can escape. The manhole opening is generally 30 to 36 in. in diameter.

Manholes should be located so that water pipes, sewer pipes, gas mains, or duct lines of other utilities will not pass through the manhole. If this cannot be avoided, every effort should be made to offset the conflicting line so that it will by-pass the manhole. If manholes are to be constructed of poured concrete, a 1:3:5 or 1:2:4 mix will generally be satisfactory. Forms represent a considerable item in concrete manhole construction. Some saving can be realized by using collapsible forms which can be used over again. If a standard manhole design can be used, steel sectional forms will be desirable. If any considerable amount of manhole construction is contemplated, important savings can be effected by casting the manholes in the work yard. They are transported to the point of installation and set in place in the previously excavated hole. Precasting is particularly feasible with smaller manholes of the various types. In the larger type, the precast method can sometimes be used by casting the manhole in parts and joining them in the field by grouting or pouring a short joining section. Where the duct line enters the manhole, duct bells should protect the cables from abrasion caused by movement over the rough edges of the conduit ends.

4-11. Transformer vaults are essentially the same as cable manholes except that they are larger. The larger size is necessary because of the larger size of electrical equipment and the need for dissipating heat from transformer energy loss. If the cubical content of the vault and venting means are not large enough, rise in temperature may seriously affect the load-carrying capacity of transformers and cables. Transformer vaults may be under the street surface or under the sidewalk. If the transformer bank is small, or if it is not feasible to locate the vault underneath the sidewalk, the vault may be located in the street and serve as a cable manhole also. A typical transformer vault is shown in Fig. 4-8. In a street-type vault the opening must be large enough to pass the largest size transformer to be installed. Removable openings also must be provided to give workmen access to the vault. If the transformers and other equipment are too large to pass through the opening in the ordinary manhole-cover frame, a removable reinforced-concrete slab is usually located in the vault roof. When transformers or other large equipment is to be removed from the vault, the pavement will have to be broken and replaced, unless the removable slabs are a part of the paving surface. The difficulty of providing adequate ventilation and the traffic along the street interfering with the opening of the vault for access are disadvantages of the street-type vault.

If the transformer vault can be located wholly under the sidewalk, or half under the sidewalk and half under the street, the cost of construction will be somewhat lower. In this case all openings are in the sidewalk. The removable roof slab or slabs are in the sidewalk and form a part of the surface. They are easily removable and give ready access for installation or removal of equipment. The inside physical dimensions of a transformer vault depend upon the physical dimensions of the equipment to be installed and, to a considerable extent, upon personal

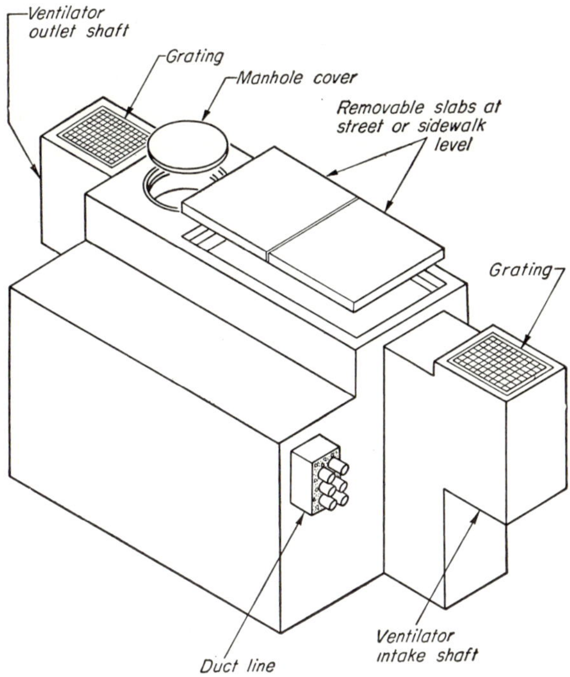

FIG. 4-8. Medium transformer vaults built underground require considerable volume, must provide ventilation for heat dissipation.

opinion. A good rule-of-thumb measure is to provide 2 to 2.5 cu ft of internal volume per kva of installed transformer capacity.

Where transformers are located in a confined space, the temperature in the vault rises because of the heat energy released by the iron and copper losses. The temperature rises until the rate at which heat energy dissipates equals the rate at which heat energy develops by the losses. If the temperature within the vault rises because of poor heat dissipation, the internal temperature of the transformers may exceed the safe operating level and rated load cannot be delivered. On the other hand, if the rate of heat dissipation is high, the transformers may deliver load in excess of their rating for short periods of time, without exceeding their safe operating temperature. Heat dissipates by conduction through the vault walls and by circulating cool air through the vault. A maximum vault temperature of about 40 C should be sought when running under load.

Conduction through the vault walls into the earth dissipates only a small part of the heat. The remainder must be disposed of through ventilation. Cooling air is frequently provided by ventilating shafts, Fig. 4-8. In this case the cool-air intake vent opens into the vault near floor level, while the opening to the hot-air exhaust vent is on the opposite side of the vault near the roof. This method assumes that cool air will be drawn into the vault and expelled by the chimney effect produced in the vault. The vent shafts may be constructed of poured concrete or of concrete or tile pipe sections. Regardless of the method used, the external vent shafts add apprecially to the cost.

The vent opening often is directly in the vault roof instead of in external vent shafts. Each vent opening may serve as either an exhaust or intake, or both, but excessive drainage of surface water into the vault must be prevented. Regardless of the ventilation method, a good rule-of-thumb measure allows a net vent area of 1.0 to 1.75 sq ft per vent per kilowatt of heat-energy loss to be dissipated by ventilation. Protective steel gratings over ventilator openings reduce the opening to only about 60 percent of its gross area. In some cases adequate cooling cannot be secured by natural ventilation; the desired cooling effect may be realized by forced ventilation, using electric fans automatically turned on and off by thermostatic controls.

4-12. Street Service Boxes. These are located at frequent intervals to give access to secondary mains from which low-voltage service taps can be made. The street service box or handhole is considerably smaller than the street-intersection manhole, Fig. 4-7, since space need be provided only for installing low-voltage secondary equipment and making service connections to secondary mains. Since the ducts occupied by the secondary mains and laterals are usually the only ones concerned, the handhole is frequently made only as deep as the top layer of duct where the secondary cables are generally located, and with only an earth bottom. The other ducts, if any, pass through the bottom of the hole without being broken. Services may be tapped from the secondary mains in the service handhole and run directly to the consumer's premises, or lateral taps may be run from the mains in the handhole to sidewalk service boxes, and one or more services then run from the sidewalk box. Sidewalk service boxes are small, shallow boxes, usually about 3 by 3 by 3 ft, set in the sidewalk at the curb to give convenient access to service connections without interference from street traffic.

OVERHEAD DISTRIBUTION

4-13. Mechanical Loading of Pole Lines. Poles supporting overhead wires and other equipment are subjected to mechanical strains. Overhead wires are subjected to strains from the tension with which they are strung. When a force is applied against an object, it produces stress within the object. Stress measures the force distributed internally within the object. Stresses caused by mechanical forces, or loads, are of three general types:

Tensile stress results from application of a force tending to elongate, or stretch, the body beyond its normal length, Fig. 4-9(*a*). A wire strung between two poles, under tension, is subjected to tensile stress.

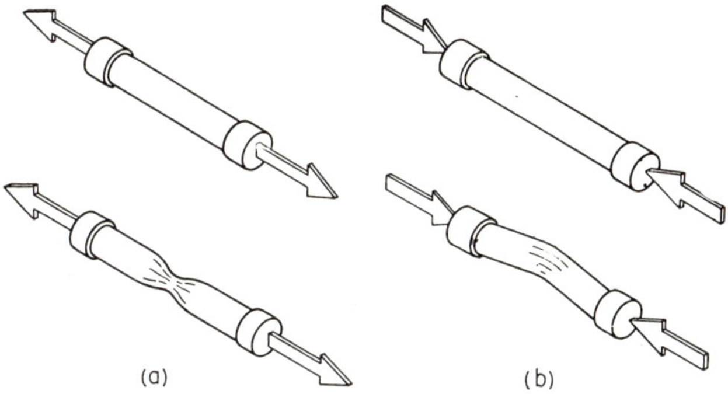

FIG. 4-9. (*a*) A body stretches when forces tend to pull it apart. When the forces exceed its tensile strength, the body fails by "necking down" and then breaking. (*b*) A body compresses when forces squeeze it together. It fails by buckling and then breaking when forces exceed its compressive strength.

Compressive stress results from application of a force tending to shorten the body, the opposite of tensile stress, Fig. 4-9(*b*). A heavy transformer hung on a pole subjects the pole to a compressive stress.

Shearing stress results from the application of a force which tends to separate the body into two parts, sliding the two parts along the surface of a plane separating them, Fig. 4-10. A heavy transformer hung on a wood crossarm attached to a wood pole by a steel bolt sets up a shearing stress in the bolt between the crossarm and the pole. The bolt transfers the force (the weight of the transformer) from the crossarm to the pole and resists the tendency of the crossarm to move along the pole surface where the two are joined by the bolt. The cut through a wire is caused by the shearing force exerted by wire cutters.

The force causing stress in an object acts across each unit of cross-sectional area. When the internal stress is uniformly distributed, the unit stress, or stress per unit of area, equals

$$S = \frac{F}{A}$$

where S = unit stress, psi
F = applied force causing the stress, lb
A = area over which the stress is applied, sq in.

In the case of tensile and compressive stress, the area is perpendicular to the direction of the applied force. In the case of shearing stress, the cross-sectional area is taken parallel to the direction of the force.

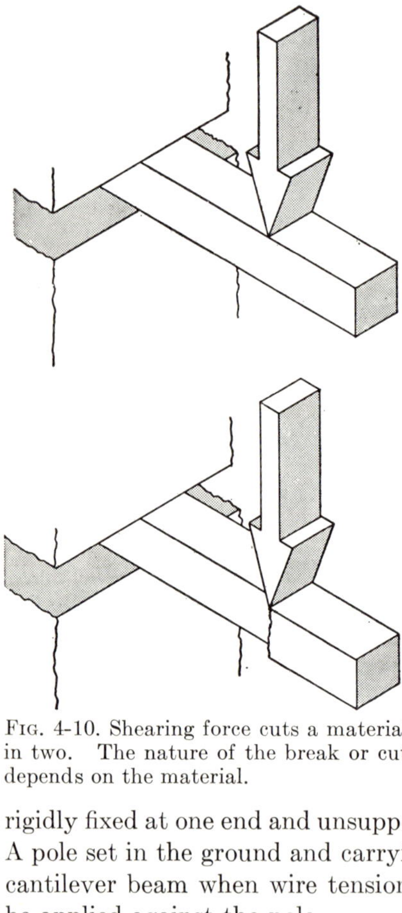

FIG. 4-10. Shearing force cuts a material in two. The nature of the break or cut depends on the material.

For example, suppose that a No. 00 copper conductor, whose cross-sectional area is 0.1045 sq in., is subjected to a tension of 2,500 lb. The unit tensile stress is

$$S_t = \frac{2{,}500 \text{ lb}}{0.1045 \text{ sq in.}} = 23{,}923 \text{ psi}$$

Again, if the transformer previously referred to weighs 500 lb and a ⅝ in. bolt, having a cross-sectional area of 0.2485 sq in., fastens the crossarm to the pole, the shearing stress in the bolt will be

$$S_s = \frac{500 \text{ lb}}{0.2485 \text{ sq in.}} = 2{,}012 \text{ psi}$$

Tensile, compressive, and shearing stresses may all be present in a body at the same time.

A cross member supported at the ends and loaded at some point between the supports, Fig. 4-11, is a simple beam. A beam, Fig. 4-12, rigidly fixed at one end and unsupported at the other is a cantilever beam. A pole set in the ground and carrying wires at the top is an example of a cantilever beam when wire tension or wind causes a horizontal force to be applied against the pole.

4-14. Elasticity and Ultimate Strength. Practically every solid body possesses elasticity. Elasticity permits a body to stretch under tension or to shrink under compression and then return to its original condition when the stress is removed. A rubber band is an example of great elasticity, while a strip of lead has only slight elasticity. Up to a certain limit, stress applied to a body produces deformation which disappears when the stress is removed. Beyond this stress limit the material will not return to its original dimension but receives a permanent deformation. This limit is the elastic limit of the material. When the stress is less than the elastic limit of the material, the deformation is directly proportional to the unit stress. When the stress exceeds the elastic limit, the deformation increases until failure occurs. The stress which causes failure or rupture is the ultimate strength of the material.

A structure must be built to withstand the greatest stress it is likely to experience. The maximum stress at which a structure is designed to operate normally is the working stress. The working stress must generally be low enough so that (1) the elastic limit will not be exceeded, which would cause permanent deformation of the structure, (2) it allows

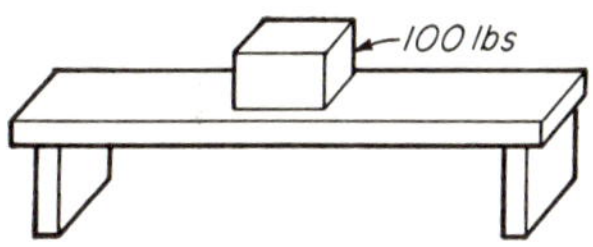

FIG. 4-11. A simple beam carries a load which it supports by resting on two points of a bearing at its ends.

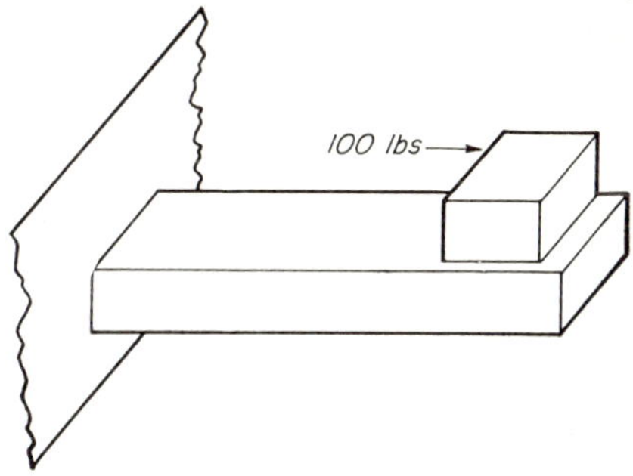

FIG. 4-12. A cantilever beam rigidly supported at one end only usually carries a load at the other end.

for maximum variation in strength and quality of material likely to be encountered, (3) it makes reasonable provision for unusual and unforeseen conditions and hazards to which the structure may be subjected. The ratio of working stress to the ultimate strength of the material is the *design safety factor*.

If a load or force is applied to a cantilever beam, Fig. 4-12, at a certain distance from the fixed end, the beam will deflect or bend. If the distance from the fixed end to the point of application increases, the deflection of the beam increases. A point will be reached at which the beam fails or breaks, even though the distance between the force and the fixed end is the only thing that has changed. The amount of deflection at any point along a beam depends upon the force causing the bending and the distance at which the force is applied. The distance from the fixed end to the point of force application is the moment arm. The product of the moment arm and the applied force is the bending moment. The bending moment measures the effect the bending force has on fiber stress along the beam.

Thus, suppose that a 35-ft wood pole is set 6 ft deep in the ground, so that its base is fixed, and a 1,000-lb pull is applied at a point 2 ft from the top of the pole. The distance from the ground line to the point at which the pull is applied is 27 ft. The pole, then, is a cantilever beam with a moment arm of 27 ft, and the bending moment at the ground line is equal to

$$27 \text{ ft} \times 1{,}000 \text{ lb} = 27{,}000 \text{ lb-ft}$$

At a point 10 ft above the ground line the moment arm is 10 ft shorter and the bending moment is only 17,000 lb-ft. The amount of bending will be reduced, and the fiber stress caused by bending will be less.

4-15. Elongation and Compression. If a small, uniform strip of wood is held in the hands and forcibly bent until broken, the failure of the wood fibers is first seen by cracks across the fibers on the convex outside surface of the strip and by wrinkles indicating forcible shortening along the

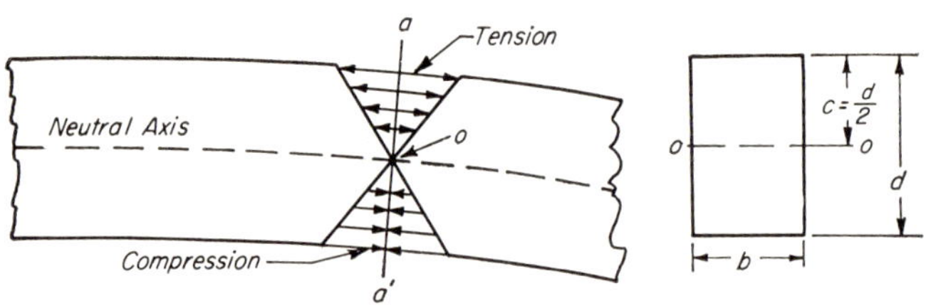

FIG. 4-13. In a cantilever-beam section the upper fibers are under tension, the lower fibers under compression.

concave inside surface. The bent piece of wood represents a beam under bending load. Failure of the outer fibers indicates that the distribution of internal stress is not uniform but is a maximum at the outer surfaces. Failure of the fibers on the convex surface due to elongation and that on the concave surface due to shortening indicate the presence of both tensile and compressive stresses as a result of the bending load.

Figure 4-13 shows a section aa' through the beam of Fig. 4-12, at right angles to its length. The maximum fiber stress in tension occurs in the outer surface near a. The tensile stress decreases, within the beam, from a maximum at a to zero at o. At o the stress changes from tensile to compressive and increases from zero at o to a maximum in compression at a'. Stress in the extreme outer fibers is a maximum because they are subjected to the maximum elongation and shortening under bending. The stress along a plane surface through o equals zero because there is no change in length under bending. The plane surface through o is the neutral surface, and a line marking the neutral surface through the beam is the neutral axis. The neutral surface passes through the center of gravity.

The fiber stress within the beam, at any section, depends upon the bending moment at the section caused by the external bending force.

The internal forces within the beam represent opposition which the beam offers to bending or deformation. Since the internal stresses depend upon the external bending, the internal stresses may be thought of as setting up an equal and opposing resisting moment within the beam. Up to the point of failure the resisting moment equals the bending moment. The internal resisting moment may be figured as

$$M = S\frac{I}{c}$$

where M = internal resisting moment = external bending moment lb-ft
S = maximum fiber stress psi
I = moment of inertia of the beam section
c = distance from the neutral axis to the extreme fibers (for a rectangular beam section see Fig. 4-13), ft
$\frac{I}{c}$ = section modulus of the beam section

This equation determines the maximum fiber stress at any point along a beam if the external force, moment arm, and section modulus of the beam section are known or it determines the allowable external force or moment arm if the maximum fiber stress and one of the other quantities are known. The value of I, c, or I/c for the various shapes of beam sections is given in various structural handbooks on beam sections.

4-16. Pole Loading. In overhead distribution systems the poles supporting the wires and cables must have sufficient mechanical strength to withstand all expected loads. Poles are normally subjected to vertical and horizontal loads. Vertical loading results from the dead weight of equipment such as crossarms, insulators, conductors, transformers, etc. Vertical loading also includes ice sometimes forming on conductors in certain regions and at times weighing more than the conductor itself. The pole acts as a column in supporting vertical loads. Heavy loads such as large transformers eccentrically mounted on the pole produce bending stresses which may exceed the safe value and cause failure, if the installation is not designed properly.

Horizontal loads on poles develop (1) from conductor tension at angles in the line, (2) at line dead ends, and (3) from transverse forces set up by wind blowing on poles and conductors. These forces also increase if ice forms on the conductors. Side pulls from conductor tension can be more or less accurately determined and compensated by guying or bracing the pole. It would be impractical, however, to guy every pole in a distribution line to compensate for transverse loading of poles by the wind. The logical procedure uses poles with sufficient mechanical strength in bending to withstand the maximum load likely to be encountered.

4-17. Safety Factor. A safety factor must be selected so that expected maximum working stresses will be safely below the elastic limit of the materials in a structure. In determining the maximum fiber stress which the structure must withstand, the maximum loading to which a line is likely to be subjected must be known. Part 2 of the National Electrical Safety Code—Safety rules for installation and maintenance of electric supply and communication lines—gives the nearest approach to a nationally accepted standard for strength requirements and probable loading of electrical-line structures. The National Electrical Safety Code is published by the National Bureau of Standards and represents the consensus of such varied interests as power, telephone, telegraph, and others as to minimum requirements for safety in the construction, operation, and maintenance of electric lines. Recommendations of the NESC represent minimum safety factors in structures. In some states the NESC has been adopted outright; in others local codes may apply in determining minimum requirements.

In determining maximum loads which may be imposed on structures supporting overhead lines, the NESC divides the country into three zones—light, medium, and heavy loading—based on weather conditions. There probably are, however, localities in each zone where the prescribed loading may not be applicable. In such a case design loading may be adjusted to local requirements by local authority. The NESC shows dimensions of seven different classes of wood poles taken from the standards of the American Standards Association and gives data on the ultimate fiber strength of the various wood poles. The code also gives vertical, horizontal, and resultant loading for various types of conductors under conditions of light, medium, and heavy loading. The Code recognizes that different degrees of hazard exist and that structures of greater strength are required at those points where the hazard is the greatest. The various grades of construction specified are determined by the safety factors recommended by the Code. The NESC applies the safety factor by establishing the maximum fiber stress permitted. If NESC limits the maximum fiber stress to 50 percent of ultimate strength, the safety factor equals 2. In determining maximum fiber stresses in unguyed poles supporting distribution lines, it is assumed that the ground holds the pole rigidly and that the earth around the pole base will not give and relieve bending stresses in the pole. This condition may be realized if the maximum horizontal load occurs when the earth is frozen or if the pole is set in solid material such as concrete.

The NESC gives the maximum transverse, horizontal wind load on the different types and kinds of conductors, in pounds per foot of conductor length, for the specified wind loading under light, medium, and heavy loading conditions. For other types of conductors such as aerial-supply and communication cables, open communication conductors, etc., which are not included in the code loading tables, the transverse wind load per foot can usually be obtained from the manufacturer or from the utility concerned. The wind blowing on the pole also causes a bending moment which is a maximum at the ground line of an unguyed pole. The bending moment from wind on the pole may be calculated as

$$M_p = \frac{PH^2}{72\pi}(C_g + 2C_t)$$

where M_p = bending moment at the ground line, due to wind, lb-ft
P = wind pressure, lb per sq ft
H = height of pole above ground, ft
C_g = pole circumference at ground line, in.
C_t = pole circumference at the top, in.

The design of a wood pole line, providing an adequate factor of safety to withstand bending load, is illustrated in the following example.

Example: A section of line is to be built on 40-ft wood poles set 6 ft deep in the ground. Each pole will carry three crossarms. The top arm will be 1 ft below the pole top, the middle arm will be 2 ft below the top arm, and the bottom arm 2 ft below the middle arm. The top and middle arms will carry three conductors each, on which the transverse wind load equals 0.623 lb per foot of length. The bottom arm will carry three conductors on which the transverse wind load equals 0.503 lb per foot of length. Span length will be 200 ft. The wood poles have a rated ultimate fiber strength of 7,400 psi, and the transverse wind load will not exceed 9 lb per sq ft. The NESC requires that the line be built with a minimum safety factor of 2. What must be the minimum pole circumference at the ground line when installed?

Solution:

1. Moment arm – top arm = 40 – 6 – 1 = 33 ft
   Moment arm – middle arm = 40 – 6 – 1 – 2 = 31 ft
   Moment arm – bottom arm = 40 – 6 – 1 – 2 – 2 = 29 ft
2. Bending moment from transverse wind on conductors

   Top arm $= 3 \times 0.623 \frac{\text{lb}}{\text{ft}} \times 200 \text{ ft} \times 33 \text{ ft} = 12{,}335$ lb-ft

   Middle arm $= 3 \times 0.623 \frac{\text{lb}}{\text{ft}} \times 200 \text{ ft} \times 31 \text{ ft} = 11{,}588$ lb-ft

   Bottom arm $= 3 \times 0.503 \frac{\text{lb}}{\text{ft}} \times 200 \text{ ft} \times 29 \text{ ft} = 8{,}752$ lb-ft

   Total 32,675 lb-ft
3. The internal resisting moment of the pole

$$M = \frac{SI}{c} = \frac{SC^3}{3{,}790}$$

where M = bending moment = 32,675 lb-ft at the ground line
S = allowable maximum fiber stress = 7,400/2 = 3,700 psi
C = circumference of the pole at the ground line

4. Minimum circumference of the pole at the ground line to withstand transverse wind load on conductors without exceeding fiber stress of 3,700 psi

$$C^3 = \frac{3{,}790M}{S} = \frac{3{,}790 \times 32{,}675 \text{ lb-ft}}{3{,}700 \text{ psi}} = 33{,}470$$
$$C = (33{,}470)^{1/3} = 32.23 \text{ in.}$$

5. Bending moment from wind on the pole

$$M_p = \frac{PH^2}{72\pi}(C_g + 2C_t)$$

From the above minimum pole circumference at the ground line, necessary to withstand transverse wind load on the conductors, the minimum pole top circumference can be found in tables in the NESC for the kind of pole corresponding to an ultimate fiber stress of 7,400 psi. Using these values, the additional bending moment from wind on the pole can be determined. For this case assume that the pole top circumference is 20 in. The bending moment due to wind on the pole, then, is equal to

$$M_p = \frac{PH^2}{72\pi}(C_g + 2C_t) = \frac{9 \text{ lb} \times 34 \text{ ft} \times 34 \text{ ft } (32.23 + 20 + 20 \text{ in.})}{72 \times \pi} = 3{,}325 \text{ lb-ft}$$

6. The total bending moment from transverse wind load on conductors and pole is

$$M = M_c + M_p = 32{,}675 + 3{,}325 = 36{,}000 \text{ lb-ft}$$

7. The minimum ground-line circumference of the pole to withstand the total bending moment due to transverse wind load on conductors and pole without exceeding a fiber stress of 3,700 psi, is equal to

$$C = \left(\frac{3{,}790 \times 36{,}000 \text{ lb-ft}}{3{,}700 \text{ psi}}\right)^{1/3} = 33.3 \text{ in.}$$

The nearest standard size pole whose minimum ground-line circumference is *not less* than 33.3 in. should be used.

4-18. Guying. Loading on poles results from the combined action of vertical and horizontal forces. Vertical forces are the dead weight of the equipment supported and the vertical component of guy tensions. Poles generally have considerably more strength under direct compression than needed to support their dead-weight loads. Wood poles, in particular, are relatively low in strength under bending as compared to their strength under compression.

Horizontal loads result from transverse wind load on the poles and wires, angular change in the direction of the line, change in the number or size of wires, or dead-ending or terminating a line. Poles are not subjected to continuous side pull from transverse wind load on poles and conductors. Unbalanced side pull from wind varies with wind velocity, span length, and projected area of pole and conductors against which the wind blows. Such side pulls are intermittent and vary. Pole lines are generally designed to withstand these side pulls without extra support such as guys. At an angle in the line, however, an additional side pull exerted on the pole depends upon the number of wires, the angle turned, and the wire tension. Minimum side pull will never be zero but depends upon the minimum wire tension and the number of wires just as the maximum side pull depends upon the maximum tension and the number of wires. An angle pole is subjected to a continuous side pull which varies less widely between minimum and maximum than the side pull from wind on a straight section of line. Such side pulls may be much greater than the resisting force of the pole requiring bracing or guying. Guys may be one of the following types:

Arm guys connect the ends of crossarms and the *next* pole, generally at a lower level. They counteract unbalanced tensions which tend to twist the arm about the pole or turn the pole.

Pole guy or head guy uses the same principle as the arm guy. It attaches at or near the level of the horizontal pull and connects to the next pole at a lower level transferring the pull to the next pole, but with a shorter moment arm. The unbalanced pull can be further distributed by extending another pole guy from the second pole, etc. For instance, suppose a pole line is built with two or more crossarms carrying wires and that the line on the bottom arm is terminated or dead-ended on a certain pole while the wires on the other arms continue. Tension in the wires on the dead-end arm causes a bending stress in the arm and pole and tends to rotate the pole or pull the arm out of position if the wire tension on each side of the pole is not symmetrical. If the dead-end load is great enough to overstress the arm under bending, or if the arm tends to turn or twist out of position, an arm guy would support both the crossarm and the pole against bending. If the arm is not overstressed in bending, a head guy would be used to counterbalance bending in the pole from dead-end load.

Anchor guys attach to the pole at, or near, the point where the horizontal pull occurs, and the other end connects to one of numerous types of mechanical anchors set in the earth.

Stub guy is a combination of pole guy and anchor guy. If it is impossible to drop an anchor guy against side pull because the anchor would be in the roadway, a pole guy would be run across the roadway to a shorter or stub pole and an anchor guy dropped from the stub, if needed, to balance the side pull.

4-19. Angle Supports. At an angle pole where the line changes direction, wire tensions combine to form a resultant side pull on the pole. *B* of Fig. 4-14 represents an angle pole where the line turns an angle a, and AB and BC represent wire tension or pull in the direction of the line. If the line were straight, BA and BC would be opposite in direction and their resultant would be zero if they are equal, no side pull would result. When the line turns through an angle, however, the forces BA and BC combine vectorially as shown by the dotted lines to form a side pull on the pole equal to BD. If forces BA and BC are equal, their resultant BD will bisect angle ABC. BA and BC can generally be considered equal and BD considered to bisect the angle with slight error. The component of BA along BD is equal to

$$BD' = BA \sin \frac{a}{2} = D'D$$

Tension BC being equal to BA contributes an equal component to the resultant side pull. If t is the maximum tension per conductor and n the number of conductors, the total side pull due to conductor tension on pole B equals

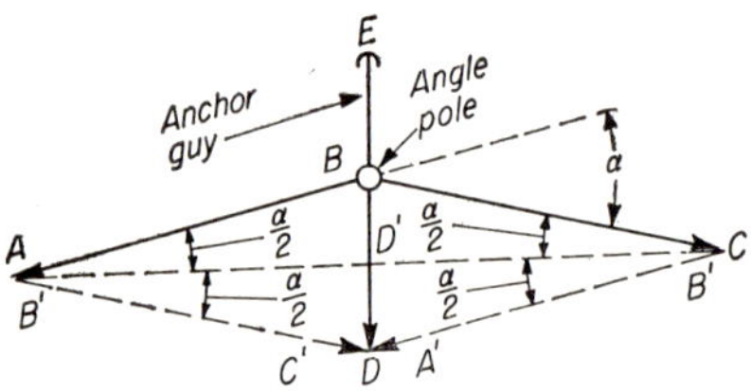

FIG. 4-14. Wire tension on an angle pole causes a force in direction D which must be balanced by guy wire BE.

$$P_t = BD = 2nt \sin \frac{a}{2} = 2T \sin \frac{a}{2} \quad (1)$$

where $T = nt$ = total conductor tension in the direction of BA or BC

If the side pull on the angle pole is great enough, the bending stress may exceed the allowable working stress or even the ultimate strength. In this case a guy installed in the direction BE will balance the side pull of force BD and prevent bending of the pole. The fifth edition of the NESC (1941) specifies an allowable percentage of ultimate fiber stress due to longitudinal loading resulting from conductor tension and an allowable percentage of ultimate fiber stress from transverse wind loading on pole and conductors at angle poles. In determining the permissible fiber stress in an angle pole at the ground line, if not guyed, or at the point of guy attachment, if guyed, the NESC assumes that the side pull on the angle pole results from

1. Wind pressure on the pole surface
2. Wind pressure on conductors
3. Resultant side pull from conductor tension

even though wind pressure on conductors is twice taken into account, since the wind load of b also contributes to conductor tension of c. The adequacy of pole strength at an angle in the line, to meet code requirements, can be determined by the following equation:

$$\frac{(M_p + M_c)100}{\% S_1} + \frac{M_t \times 100}{\% S_2} = M_r \quad (2)$$

where M_r = required resisting moment of the pole, lb-ft

M_p = bending moment due to wind on the pole $= \dfrac{PH^2}{72\pi}(C_g + 2C_t)$

M_c = bending moment due to wind on conductors

$$= \frac{(\text{span } 1 + \text{span } 2)wnH}{2}$$

M_t = bending moment due to conductor tension $= 2HT \sin \dfrac{a}{2}$

$\% S_1$ = allowable fiber stress in pole, in percent of ultimate strength, for transverse loading

$\% S_2$ = allowable fiber stress in pole, in percent of ultimate strength, for longitudinal loading at dead ends

w = wind pressure, lb per foot of conductor length—from the NESC

In the above equation, for bending moment due to wind on conductors, the value of M_c will be slightly high because the assumption is made that the wind is perpendicular to the direction of the line on both sides of the angle pole. For strict accuracy M_c should be multiplied by $\cos a/2$.

If M_r exceeds the ultimate resisting moment of the pole, a guy will be needed to prevent excessive fiber stress under maximum loading. If M_r does not exceed the ultimate resisting moment of the pole, a guy, theoretically, will not be required. This, however, is based upon the assumption that the pole stands in a rigid setting. This is not strictly true. At an angle pole the constant side pull from conductor tension is liable gradually to displace the pole from the vertical position because of the earth giving way around the base. This displacement eases bending stress but is objectionable in appearance and may increase wire sags enough to affect clearances. In other words, an angle pole, without a guy, is no stronger than the earth in which it is set.

4-20. Soil Types. Whether a side guy should be installed on an angle pole depends upon the type of earth in which the pole stands and its condition at the time of maximum loading as well as upon the fiber stress of the pole under bending. In areas subject to heavy loading, the ground will likely be frozen at the time of maximum loading so that poles approach a rigid setting. On the other hand, if the pole stands in loose or soft soil, the setting is far from rigid and guying will be required at side pulls considerably below those which produce a bending moment equal to the value of M_r.

In determining whether or not a guy can be safely omitted, consider that the maximum loads on the pole cannot be determined exactly, the strength of wood poles vary considerably, and the rigidity of an angle pole depends on the type of soil in which it stands at the time of its maximum load. For these reasons, general solution of safe loading for all conditions must be very approximate and liberal safety factors must be used. Resistance that the soil around the pole base offers to displacement depends upon the soil type, depth of pole setting, moisture content of the soil, size of the pole butt, how well the backfill was tamped, and how long the pole has been set. In general, the earth in which poles stand may be grouped as follows:

Class 1. Hard rock.

Class 2. Shale, sandstone, and soft rock. May be solid or in layers.

Class 3. Hard, dry. Hardpan. Usually found under a Class 4 stratum. Resembles soft rock.

Class 4. Crumbly, damp. Not enough moisture to pack into a firm ball when squeezed in the hand. Clay may predominate.

Class 5. Firm, moist. Contains enough moisture to form a firm ball when squeezed in the hand. Most soils in well-drained areas, other than hillsides, fall into this classification. Clay may predominate, but other soils—including stones, gravel, and sand—are usually present.

Class 6. Plastic, wet. Similar to class 5 but with greater moisture content. Found in low areas where surface water collects, where drainage is poor, in flat terrain, or in close proximity to rivers or marshes. Plastic when squeezed and readily assumes any shape.

Class 7. Loose, dry. Low in moisture content. May be found in arid regions, along or near beaches, in filled-in or built-up areas when very dry. Consists of sand or gravel with slight bond between particles.

Loose, wet. High in sand, gravel, loam, or silt content. In general soil resembles mud and generally anchor hole will fill with water. Has holding power equal to loose, dry.

Class 8. Swamps and marshes.

While exact determination of the resistance of soils to failure cannot be made, a reasonable approximation can be used and adjusted upward or downward as experience and judgment dictate. The values in Table 4-1

TABLE 4-1*

	S_e = % *of Pole Ultimate Resisting Moment*
Class 1, 2, and 3—Solid rock; shale, sandstone, and soft rock; hard, dry	50
Class 4—Crumbly, damp	40
Class 5—Firm, moist	35
Class 6—Plastic, wet	30
Class 7—Loose, dry or loose, wet	25
Class 8—Swamps and marshes	20

* From Hubbard and Watkins, *Elec. World*, Oct. 11, 1947.

are percentages of the ultimate resisting moment of the pole at the ground line.

The values are arbitrary, based on the assumption the pole will be set to standard depth, the diameter of the hole will be a minimum when setting the pole, and the backfill will be well tamped.

Values in Table 4-1 for several of the soil classifications may permit pole stresses in excess of the allowable fiber stress for the highest grade of construction as defined by the NESC. The resistance the earth setting offers to pole displacement at angles can be compared with pole loading by the following equation:

$$M_p + M_c + M_t = S_e M_r$$

or

$$M_p + M_c + 2HT \sin \frac{a}{2} = S_e M_r \qquad (3)$$

If Eq. (3) is solved for angle a, the result will be the maximum permissible angle of turn in the line without side guying. If the maximum permissible angle of turn by Eq. (3) exceeds the maximum angle by Eq. (2), then the latter will govern, since it depends upon the allowable fiber stress in the pole.

4-21. Alternate Methods. If the soil around the pole base is not firm enough to withstand side pull at an angle, a side guy can sometimes be avoided by installing a crib brace, or similar underbrace, Fig. 4-15. Such a brace increases the rigidity of the pole setting by enlarging the bearing surface of the pole against the earth. If a crib brace, Fig. 4-15, is used, the values in Table 4-1 may be increased by about 15 percent for each square foot of projected brace area, provided the brace bears against undisturbed earth and has the proper relation to the forces exerted against the pole. Thus, a crib brace consisting of an old pole section 4 ft long and 10 in. in diameter has a projected area of 3.33 sq ft. Properly installed it raises the earth resistance taken from Table 4-1 by 3.33 × 15 percent = 50 percent. Crib braces are sometimes used when the side pull slightly exceeds the minimum value, requiring a guy where the right to install an anchor guy is difficult to obtain.

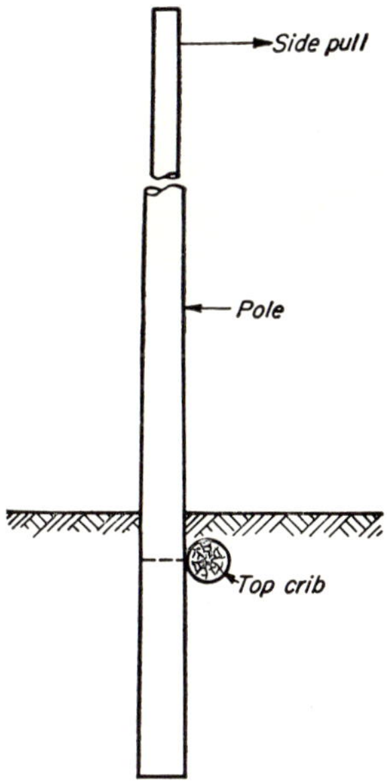

FIG. 4-15. A crib brace will strengthen a pole setting so that a side guy may not be needed to balance side pull.

Greater resistance to overturning can be obtained by increasing the setting depth of the pole rather than using a crib brace, provided the pole is long enough. For each foot of increased setting depth above normal, the values in Table 4-1 may be increased by one-third, or 33.3 percent.

The angle through which the line may turn or the resultant side pull acting against the pole can be calculated as indicated in Fig. 4-16. B is an angle pole on which the line changes direction by a degrees. If 100 ft is marked off in the direction of the line, as from B to A and from B to C, and points A and C are joined by a straight line, the length of line BD', perpendicular to AC, is a measure of either the angle of turn or the side pull against pole B. Thus,

$$\frac{BD' \text{ ft}}{100 \text{ ft}} = \frac{d}{100} = \sin\frac{a}{2}$$

but distance BD' is also representative of one half the resultant side pull against the pole as in Fig. 4-14, from which

$$\frac{BD' \text{ lb}}{T \text{ lb}} = \frac{BD}{2T} = \frac{T_r}{2T} = \sin\frac{a}{2} \tag{4}$$

where T_r = resultant side pull in pounds in the direction of BD
T = maximum tension in the direction of AB or BC due to all the conductors

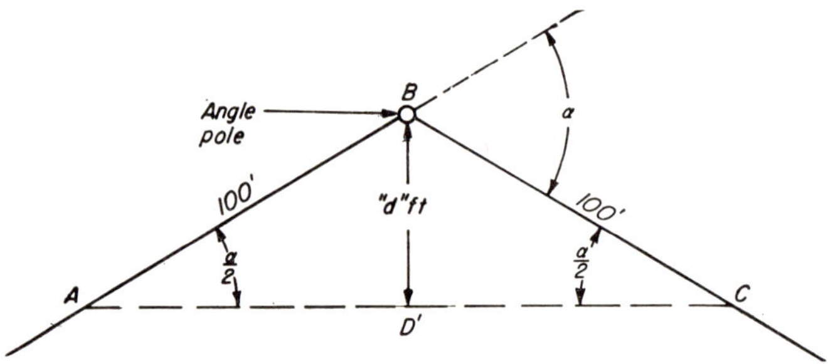

FIG. 4-16. Side pull on an angle pole may also be calculated trigonometrically.

These equations are equal because

$$\sin\frac{a}{2} = \frac{T_r}{2T} = \frac{d}{100} \qquad T_r = \frac{2 \times T \text{ lb} \times d \text{ ft}}{100} \qquad a = 2\sin^{-1}\frac{d}{100}$$

Figuring guying requirements is speeded up by Table 4-2. It shows the ultimate resisting moment of wood poles (M_r) of different species and for different circumferences (taken from Appendix 3B of Part 2 of the NESC).

At angles less than 60° turned by the line, the side pull from conductor tension is less than total conductor tension in the line direction. For such angles a single guy in the direction BE, Fig. 4-14, bisecting the angle ABC will be ample. When turning an angle more than 60°, the side pull will exceed line tension in the direction of AB or BC, Fig. 4-14. For such large angles, guys should be installed in the direction of the line, Fig. 4-17, to prevent a tendency to displace the pole if the angle does not exactly bisect the line angle and to avoid guys of extreme strength.

Example 1: At an angle in a distribution line the distance d, Fig. 4-16, is 30 ft. The angle pole set to standard depth in class 5 earth is a 40-ft creosoted southern pine pole with a ground-line circumference of 34 in. and a top circumference of 21 in. The line consists of three No. 4 copper wires mounted 33 ft aboveground. Maximum conductor tension will be 1,140 lb, span length from angle pole is 120 ft and 140 ft, maximum transverse wind pressure on the pole is 4 lb per sq ft, and transverse wind load on wires is 0.463 lb per foot of conductor (NESC heavy loading). Allowable fiber stress in pole from transverse load is 50 percent and from longitudinal dead-end load is 75 percent. Ultimate strength of the pole is 7,400 psi. (1) What is the angle of turn in the line? (2) What is the resultant side pull on the pole? (3) What is the total ground-line bending moment? (4) Will side guying be required? (5) What is the maximum angle the line can turn without side guying?

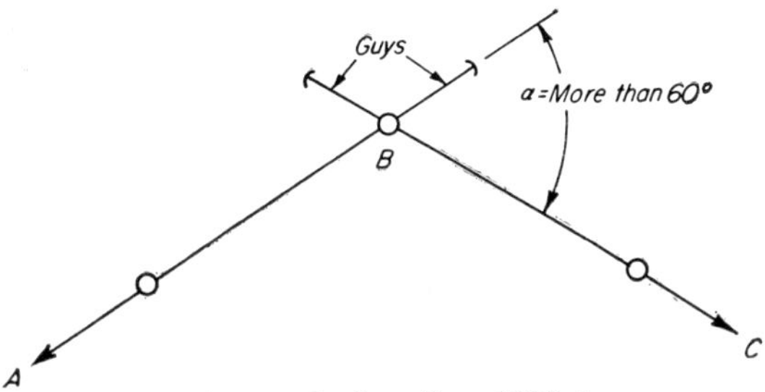

FIG. 4-17. For turning angles less than 120°, two guys are used.

Solution:

1. Angle of turn $= 2 \sin^{-1} {}^{30}\!/_{100} = 2 \times 17°30' = 35°0'$
2. Side pull on pole:

From wind on the pole at 33 ft $= \dfrac{M_p}{33 \text{ ft}} = \dfrac{PH^2(C_g + 2C_t)}{33 \text{ ft} \times 72\pi}$

$$= \frac{4 \text{ lb} \times 34 \text{ ft} \times 34 \text{ ft } (34 \text{ in.} + 42 \text{ in.})}{33 \text{ ft} \times 72\pi} = 47 \text{ lb}$$

From wind on the wires at 33 ft $= \dfrac{M_c}{33 \text{ ft}}$

$$= \frac{3(120 \text{ ft} + 140 \text{ ft})0.463 \text{ lb}}{2} = 181 \text{ lb}$$

From tension in conductors at 33 ft $= \dfrac{M_t}{33 \text{ ft}}$

$$= 2T \sin \frac{a}{2} = \frac{2 \times T \times d}{100} = \frac{2 \times 3 \times 1{,}140 \times 30}{100} = 2{,}052 \text{ lb}$$

Total side pull $= 2{,}280$ lb

3. Ground line bending moment $= HT_r = 33 \text{ ft} \times 2{,}280 \text{ lb} = 75{,}240$ lb-ft
4. From Table 4-2, the resisting moment of a 40-ft csp pole with a circumference of 34 in. at the ground line is 76,000 lb-ft. From Table 4-1 the earth resisting moment for class 5 earth setting is $S_eM_r = 0.35 \times 76{,}000$ lb-ft $= 26{,}600$ lb-ft. A side guy will be required since the bending moment is greater than the earth's resisting moment.
5. Solving Eq. (3) for a,

$$\sin \frac{a}{2} = \frac{1}{2HT}(S_eM_r - M_p - M_c) = \frac{26{,}600 - (47 + 181)33}{2 \times 33 \times 1{,}140 \times 3} = 0.0845$$

$$a = 2 \sin^{-1} 0.0845 = 9°42'$$

If the side pull on a pole at an angle or at a dead end is great enough to overstress the pole or the earth, these stresses are relieved by a guy.

TABLE 4-2. ULTIMATE RESISTING MOMENT OF WOOD POLES AT GROUND LINE—IN POUND-FEET

Pole ground line circum., in.	Douglas fir and creosoted southern yellow pine	Lodge pole pine	Chestnut	Western red cedar	Cypress	Northern white cedar and redwood
20	15,625	13,975	12,700	11,800	10,700	7,600
22	20,800	18,600	16,870	15,750	14,070	10,120
24	27,000	24,150	21,910	20,450	18,240	13,130
26	34,350	30,700	27,850	25,950	23,200	16,700
28	42,850	36,600	34,800	32,440	29,000	20,850
30	52,800	47,200	42,800	39,900	35,640	25,630
32	64,000	57,250	51,900	48,450	43,250	31,100
34	76,800	68,700	62,300	58,100	51,900	37,320
36	91,300	81,500	74,000	69,000	61,600	44,300
38	107,100	95,800	87,000	81,100	72,400	52,100
40	125,000	111,700	101,300	94,550	84,500	60,800
42	144,800	129,200	117,300	109,500	97,800	70,450
44	166,300	148,600	135,000	125,900	112,400	80,900
46	190,000	169,800	154,000	144,000	128,500	92,450
48	216,000	193,200	183,700	171,300	146,000	105,000
50	244,100	218,500	198,000	184,800	165,000	118,700
52	274,600	245,400	222,600	207,700	185,500	133,450
54	307,500	275,000	249,400	232,600	207,800	149,500
56	343,000	306,500	278,200	259,400	231,600	166,600
58	381,400	341,000	309,300	288,400	257,500	185,300
60	422,000	377,400	342,400	319,100	285,000	205,000
62	465,500	416,300	378,000	352,100	314,600	226,300
64	513,000	458,200	415,400	387,300	346,200	249,000
66	561,500	502,000	455,400	425,000	379,600	273,000
68	614,800	549,900	498,500	465,000	415,300	298,500
70	670,000	600,000	544,000	507,000	453,000	325,500

The guy should be attached as near as possible to the point where the resultant of all horizontal loads is applied. A heavily loaded pole may have conductors mounted on crossarms at several different elevations. In such a case, the load at the different elevations should be converted to an equivalent single load at the level of the guy attachment, Fig. 4-18.

4-22. Guy Tension. Figure 4-18 shows a pole with a horizontal or dead-end load T_1, at elevation H_1, and a load T_2, at elevation H_2. An anchor guy is to be attached at elevation H_r to counteract the forces T_1 and T_2. The guy will contain a component of horizontal tension T_h

which will be equal and opposite to the equivalent horizontal pull of T_1 and T_2 at elevation H_r or T_r. Since the bending moment caused by T_h must be equal and opposite to the bending moments caused by T_1 and T_2, then

$$T_h H_g = T_r H_r = T_1 H_1 + T_2 H_2$$

and

$$T_h = \frac{T_1 H_1 + T_2 H_2}{H_g} = \text{horizontal component of tension in guy} \quad (5)$$

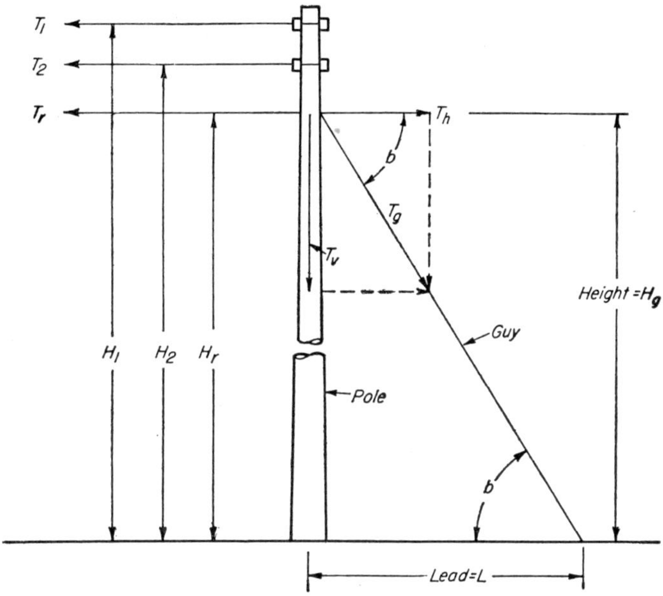

FIG. 4-18. Diagram for calculating the guy tension needed to strengthen a pole carrying several lines.

The guy forms an angle b with the horizontal, Figs. 4-18 and 4-19, such that

$$b = \tan^{-1} \frac{H_g}{L} \qquad \text{or} \qquad \tan b = \frac{H_g}{L}$$

The tension in the guy, then, is equal to

$$T_g = \frac{T_h}{\cos b} = T_h \sec b \quad (6)$$

If the guy in Fig. 4-18 forms an angle of less than 90° with the vertical, the tension in the guy T_g will be greater than the resultant horizontal tension T_h and a vertical component of load T_v will act on the pole so that

$$\frac{T_v}{T_h} = \frac{H_g}{L} = \tan b \qquad \text{and} \qquad T_v = \frac{T_h H_g}{L} = T_h \tan b$$

The total vertical load on the pole is

$$W_v = \frac{T_h H_g}{L} + W + W_1$$

where W = weight of all equipment mounted on the pole
W_1 = weight of the pole itself

As the angle b increases, the total guy tension T_g and its vertical component T_v also increase, even though the horizontal pull against the guy

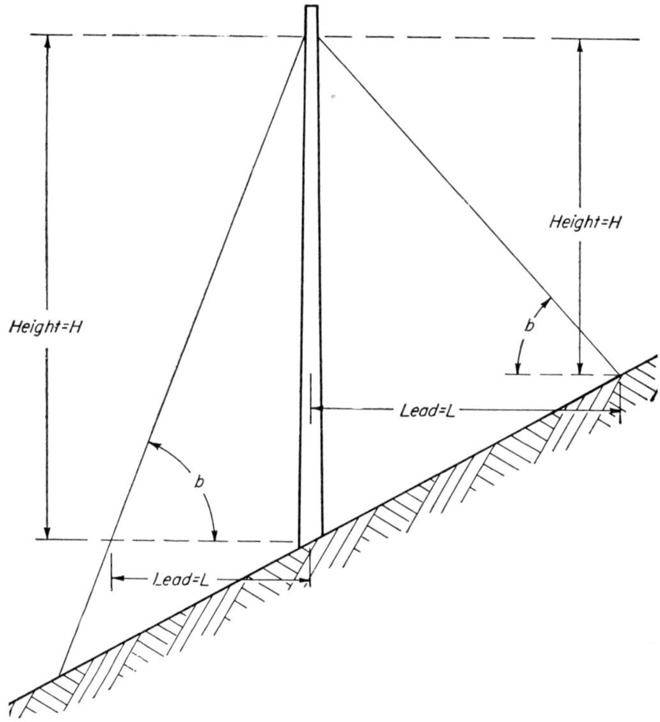

FIG. 4-19. When a pole stands on a slope, account must be taken of the horizontal distances to put the guy under proper tension.

remains constant. For this reason the ratio H_g/L should be kept as small as practicable to limit guy tension to the lowest practicable value.

The NESC specifies different safety factors for transverse loading from wind than for loading from conductor tension in determining allowable tension in guys, as was indicated in Eq. (2). The total ultimate or breaking strength, of guy strand, required at an angle or dead end is

$$U = \frac{(M_p + M_c)100}{\% \, S_w \times H_g \times \cos b} + \frac{M_t \times 100}{\% \, S_t \times H_g \cos b} \tag{7}$$

where M_p = bending moment due to wind on the pole, lb-ft
M_c = bending moment due to wind on the conductors, lb-ft
M_t = bending moment due to conductor tension, lb-ft
H_g = effective height of guy attachment, ft
$\% S_w$ = allowable stress in guy, in percent of ultimate strength, caused by transverse wind load, from NESC
$\% S_t$ = allowable stress in guy, in percent of ultimate strength, caused by conductor tension, from NESC
U = ultimate, or breaking, strength required in the guy, lb
$b = \tan^{-1} \dfrac{H_g}{L}$

In figuring strength of guy for a dead end, there will be no transverse wind loading to consider, and consequently M_p and M_c, in Eq. (7), will equal zero.

A guy is considered to take all the horizontal load which might cause bending or deflection of the pole. Deflection of a pole under bending load is, relatively, much greater than stretch of a steel guy wire under load. The guy wire would break under tension before the pole could deflect enough so that the resisting moment within the pole under bending could assume part of the load borne by the guy. The pole is considered to act as a strut, taking only the vertical component of guy tension.

If the difference in elevation between T_1, T_2, and the point of attachment of the guy, Fig. 4-18, is great enough, the bending moment in the pole may exceed the allowable pole stress at the point of guy attachment. In such a case, the pole bending moment at the point of guy attachment is

$$M = T_1(H_1 - H_g) + T(H_2 - H_g)$$

This bending moment can be checked against the required minimum pole resisting moment at the point of guy attachment as determined by Eq. (2).

Example 2: A guy is to be installed on the pole of Example 1. The guy will be attached at an elevation of 30 ft with a lead of 20 ft so that the guy bisects the angle of the line. (1) What resultant horizontal pull must the guy support? (2) What is the tension in the guy? (3) What will be the ultimate strength of the guy required to support the pole, if the NESC allowable stress in the guy is 50 percent of its ultimate for transverse loading and 87.5 percent of its strength for loading due to conductor tension at angles or dead ends?

Solution:

1. From Example 1, total ground line bending moment = 75,240 lb-ft

Equivalent side pull at level of guy attachment $= \dfrac{75{,}240}{30 \text{ ft}} = 2{,}508$ lb

2. Angle of guy $= b = \tan^{-1}\frac{30\text{ ft}}{20\text{ ft}} = 56°19'$

$\cos 56°19' = 0.5546$

Tension in guy $= T_g = \frac{T_r}{\cos b} = \frac{2{,}508\text{ lb}}{0.5546} = 4{,}530\text{ lb}$

3. $M_p = 1{,}554$ lb-ft $\quad M_c = 5{,}973$ lb-ft $\quad M_t = 67{,}716$ lb-ft

$$U = \frac{(M_p + M_c)100}{\%\, S_w \times H_g \times \cos b} + \frac{M_t \times 100}{\%\, S_t \times H_g \times \cos b}$$

$$= \frac{(1{,}554 + 5{,}973)100}{50\,\% \times 30\text{ ft} \times 0.5546} + \frac{67{,}716 \times 100}{87.5\,\% \times 30\text{ ft} \times 0.5546} = 5{,}560\text{ lb}$$

4-23. Anchor Holding Power. Holding power of an earth anchor attached to a guy depends upon several variables. The principal factors are (1) area of the anchor, (2) depth to which the anchor is set, (3) type

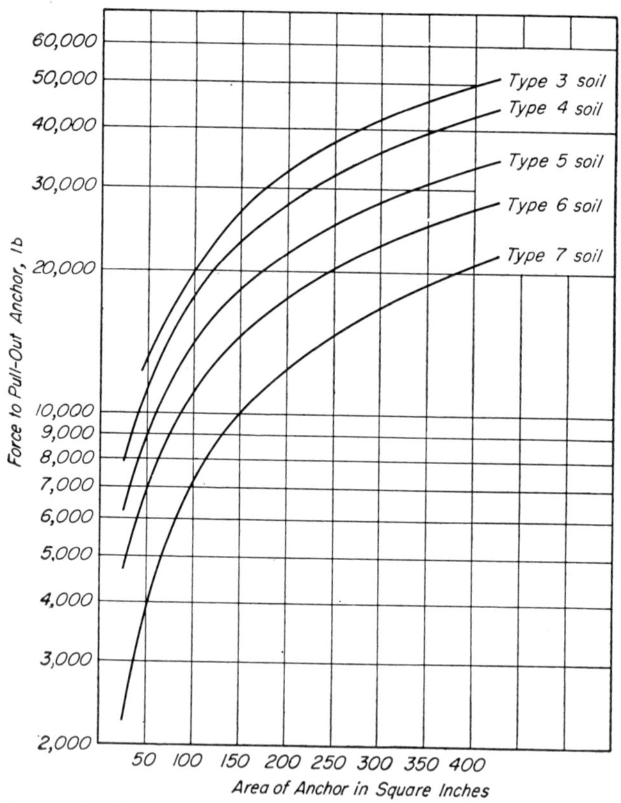

FIG. 4-20. Ultimate holding strength of pole anchors for different soil types and anchor areas.

of soil. Figure 4-20 gives holding power for anchors of different areas installed in soils of the different classifications. The curves are based upon the assumption that an anchor set at an angle of 45° with the horizontal is deep enough so that a rod of standard length for the size and type of anchor will not project more than 6 in. aboveground. The values from the curves should be regarded as the ultimate strength of the anchorage and a suitable factor of safety applied. The safety factor applied to the rod and anchor should not be less than that applied to the guy strand.

If an anchor is selected whose holding power is not less than the required ultimate strength of guy strand figured by Eq. (7), the same safety factor will be applied to both the guy and the anchor. Choosing an anchor should be based upon the most unfavorable combination of loading and soil type likely to be encountered.

4-24. Conductor Sag and Tension. This is an important consideration in an overhead distribution system. Tension in the conductor governs the mechanical load on structures at angles in the line and at dead ends. Excessive tension may cause mechanical failure of the conductor itself. Conductor sag should be kept to a minimum (1) so that excessive horizontal spacing between conductors will not be needed to prevent wires swinging together in mid-span, (2) to avoid extra pole height for sufficient clearance above ground level.

Low conductor tension and minimum sag are not compatible—low sag means a tight wire and high tension, low tension means a loose wire and increased sag. A compromise between the two is needed.

The sag of a wire strung between supports varies with (1) load per unit length of wire, (2) distance between supports, (3) temperature, (4) tension. Loading on the wire includes (1) dead weight of the wire itself, (2) weight of ice or snow clinging to the wire, (3) wind blowing against the wire. The wind is considered to be blowing at right angles to the line and to act against the projected area of the wire, including the projected area of ice or snow which may be clinging to it. The maximum effective weight of the wire is the vector sum of the vertical weight and the horizontal wind pressure. This vector sum, which is a resultant, acts at an angle between the vertical and horizontal planes.

Practically all solid matter possesses elasticity. Tension pulls wire up and decreases its sag. At the same time tension elongates the wire, from elastic stretching, which tends to relieve tension and increase sag. The elastic property of a metallic wire is measured by its modulus of elasticity. The modulus of elasticity of a material equals the stress per unit of area divided by the deformation per unit of length. For example, suppose that a No. 00 hard-drawn copper conductor which has an area of 0.1045 sq in. and a modulus of elasticity of 17,000,000 is subjected to a tension of 3,500 lb. The stress per unit area is

$$S = \frac{\text{tension}}{\text{area}} = \frac{3{,}500 \text{ lb}}{0.1045 \text{ sq in.}} = 33{,}500 \text{ psi}$$

and the elongation of the conductor, due to tension, is equal to

$$e = \frac{\text{stress}}{\text{modulus of elasticity}} = \frac{33{,}500 \text{ psi}}{17{,}000{,}000} = 0.00197 \text{ unit per unit length}$$

The elongation for a tension of 3,500 lb equals 0.00197 ft per foot of wire length, or 0.00197 in. per inch, or it may be 0.00197 per unit of any other length measure. If the modulus of elasticity is low, the elongation is high, and vice versa. A hard-drawn conductor with a load of 1.68 lb per foot of length, a span of 500 ft, and strung to a tension of 5,400 lb will have a sag of about 9 ft 8 in., and the conductor length will be about 500.5 ft between supports. If the conductor length is increased by 1 in., the conductor sag will increase by about 10 in. while the conductor tension decreases by about 400 lb. A small change in conductor length has a comparatively large effect on conductor sag and tension.

Conductor materials are also affected by temperature. A rise increases conductor length, with resulting increase in sag and decrease in tension. Temperature drop causes reverse effects. The change in length per unit of conductor length per degree F temperature change is the temperature coefficient of linear expansion. For hard-drawn copper conductor the temperature coefficient of expansion is 0.0000094. If the temperature rises 60 F, the increase in conductor length for a 500 ft span will be

$$60 \times 0.0000094 \times 500 \text{ ft} = 0.0282 \text{ ft} = 0.3384 \text{ in.}$$

The NESC lists the maximum recommended stress for conductors. The maximum tension in the conductor controls the sag. The temperature varies throughout the year. The loading on the wire also varies from time to time. The problem is to find the sag and tension to which the conductor should be strung at the prevailing temperature, so that it will not be overstressed under the maximum loading expected. The procedure is too long to explain in detail here. The sag for relatively short spans, however, can be closely approximated by the following equation for a specific condition of loading and temperature and for a known tension.

$$d = \frac{wS^2}{8H}$$

where d = sag in plane of resultant load, ft
w = resultant conductor load, lb per foot of span length
S = span length, ft
H = horizontal tension in the wire, lb

The above equation does not provide for determining change in sag and tension from loading or temperature changes.

A line tangent to a wire strung between supports is horizontal at the point of maximum sag and has the greatest angle from the horizontal at the supports. The resultant weight of the conductor acts as a vertical force. If the supports are at the same level, the weight of conductor in one half span on each side is supported at each pole. At mid-span, or point of maximum sag, the vertical component of tension equals zero. The minimum tension occurs at the point of maximum sag, and the tension at this point acts in a horizontal direction and equals the horizontal component of total tension. The maximum tension occurs at the point of support and equals the vector sum of horizontal and vertical components, or,

$$\text{Maximum tension} = T = \sqrt{H^2 + V^2}$$

where V = weight of wire per foot of span times distance from point of maximum sag to support

Distribution lines usually have comparatively short spans with small sag. The difference between the maximum tension T and the horizontal tension H is relatively small because of short spans and small sags. Under such conditions, slight error will result if T is substituted for H in the equation for sag.

Determining accurate values of sag is not so important for distribution lines as for long-span high-voltage transmission lines. Distribution lines, however, are often sagged in a haphazard manner. If the line is strung with too low a tension, the sag will be excessive, with the likelihood of wires swinging together and short-circuiting.

The usual tendency is to pull the wire up too tight. In such a case the conductor may be seriously overstressed and stretched when the heaviest loading occurs and the normal sag after loading becomes excessive. Too often the excess sag is pulled out of the wire, again overstressing the wire on heaviest loading. This cycle of overstressing and pulling up frequently results in broken wires, particularly the smaller ones. Measuring the tension or more accurately sagging the wire reduces unnecessary line breaks. The extra effort is easily justified.

4-25. Conductors. Copper and aluminum are most frequently used for distribution-system conductors. The selection generally depends upon conductivity, weight, mechanical strength, cost, and relative effect upon voltage regulation. Copper best combines these requisites of a suitable conductor and is the reference for conductivity with which other conductor materials are compared. Aluminum is the next best electrical conductor, from the standpoint of conductivity and availability. In comparison, aluminum has a conductivity about 61 percent and a weight about 30 percent that of annealed copper, together with a breaking strength about 43 percent that of hard-drawn copper. In urban overhead distribution the factors affecting voltage drop and mechanical strength (to prevent excessive sag), are likely to be most important. With lower load densities and longer spans of rural overhead distribution, mechanical strength may become the dominant factor. With the high-load densities of urban underground distribution, current-carrying capacity and voltage regulation are generally the dominant factors in selecting the type of conductor.

Conductor sizes are based upon the American wire gage (AWG), which is, also, sometimes called the Brown and Sharpe wire gage. Gage sizes decrease as the wire increases in size. The larger the gage size, the smaller the wire. The largest size is number 0000 written 4/0. Above 4/0 the size is determined by cross-sectional area in circular mils. A circular mil is the area of a circle one mil in diameter. A mil equals one thousandth of an inch. The area of any circle varies as the square of its diameter. The area of a circle in circular mils, thus, equals the square of its diameter in mils, and the diameter of a wire in mils equals the square root of its cross-sectional area in circular mils. The cross-sectional area of a wire in square inches equals its area in circular mils multiplied by 0.7854/1,000,000.

An individual conductor may consist of a single strand, or several strands. If a single strand it is *solid*, if of more than one strand it is *stranded*. A solid conductor is often called a wire, while a stranded conductor is called a cable. If both types are called wires, they should be classified as *solid wire* or *stranded wire* to clearly define the type intended. Solid conductors are generally confined to the smaller sizes because of their stiffness and difficult handling. Distribution conductors larger than No. 2 AWG are usually stranded. Insulated conductors for underground distribution or aerial cable lines are classified as cables and usually are stranded. In the following, data are given on the most commonly used conductors for overhead and underground distribution.

4-26. Bare and Weatherproof-covered Conductors. Conductors for overhead distribution are generally classified as either bare or weatherproof-covered except for insulated aerial cable. Bare wire or cable has no covering of any kind. Weatherproof-covered conductors, widely used in overhead distribution, are formed by placing a closely woven cotton-braid covering saturated with an asphalt compound over the wire. This covering may consist of two thicknesses (double braid) or three thicknesses (triple braid) of braid. Triple-braid weatherproof is the more frequently used. Such covered conductors are classified as weather resistant and not as insulated since the insulating value depends upon its age and dryness. New and unimpaired covering in extremely dry weather has considerable insulating strength. When wet, even if new, the weatherproof covering has slight dependable value as insulation. Weatherproof covering should not be considered as insulation.

Extensive use of weatherproof wire in overhead distribution probably stems from early days when local or municipal codes frequently specified insulation for high-voltage conductors. Authorities were better acquainted with requirements of low-voltage dc distribution than with ac. Covering of low dielectric strength, such as weatherproof, came to be accepted as satisfactory. Some experienced authorities contend that under the best conditions weatherproof covering has sufficient insulating qualities to warrant its use on circuits of 2,400 volts and below because it reduces the number of circuit trip outs from wires swinging together, etc. Others contend that weatherproof covering is not justified because of its higher cost and because it constitutes a safety hazard. Workmen tend to depend upon its unstable dielectric strength and expose themselves to shock, which they would not do with a bare conductor. The number of wires burned down from line short circuits is increased with a covered conductor. It seems that there is slight justification for weatherproof covering except for secondary distribution with low voltages (120 to 240), close spacing of conductors, or where local codes require its use. Recently developed compounds such as neoprene and polyethylene give promise of a better covering for overhead conductors similar to weatherproofing. Neoprene is resistant to oil, heat, sunlight, and chemicals and is very resistant to physical wear and tear. Polyethylene also has good physical properties and in addition has excellent dielectric properties. Both are excellent weatherproof covering.

4-27. Copper Conductors. Copper wires or cables are made in three degrees of hardness: soft-drawn, medium hard-drawn, and hard-drawn. The three grades differ principally in their degrees of ductility and tensile strength. Hard-drawn copper has greatest tensile strength, medium hard-drawn has less, and soft-drawn the least. Hard-drawn copper wire is cold-drawn to size from a stock copper bar. Cold-drawing the copper increases its tensile strength, hardens it, and slightly lowers its electrical conductivity. Hard-drawn copper wire raised to the proper temperature for a long enough time loses its higher tensile strength, becomes softer and more ductile, and is said to be annealed. Medium hard-drawn, made by cold-drawing almost to size, is annealed and then cold-drawn to size and strength.

Soft-drawn wire is cold-drawn to size and then annealed. Because of its lower tensile strength, excessive elongation under tension, and the large sag at which it must be strung, this type is little used as an overhead distribution conductor.

4-28. ACSR Conductors. Aluminum is lower in mechanical strength, lower in weight, and lower in conductivity than copper. A 4/0 stranded aluminum cable has an ultimate strength and a conductivity about equal to a 2/0 soft-drawn copper and weighs about one half as much. Aluminum conductors are usually of a size (cm) based on the equivalent copper size. Resistance and reactance values of all-aluminum conductors are close enough to those of their copper equivalents so that one can be substituted for the other without excessive error. In many cases the greater conductor sag, because of lower tension, will be considered excessive and the all-aluminum conductor cannot be used.

To take advantage of its good conductivity, strands of aluminum are formed about a central core of one or more strands of high-strength steel, making a composite conductor known as acsr (aluminum cable steel-reinforced) of high tensile strength and good conductivity. Such a conductor, equivalent in conductivity to a copper conductor of smaller cross section, will be high in tensile strength suitable for distribution construction, particularly for long-span rural lines.

In an acsr conductor the aluminum strands are assumed to carry all the current flowing. Steel, which has magnetic properties, in the core of the cable affects the apparent resistance of the conductor from iron losses caused by the current flow. This iron loss, adding to the I^2R loss of the conductor, gives the effect of an increased resistance. This apparent increase of resistance is appreciable with high current densities and with a single layer of aluminum strands laid about the steel core. The magnetic effect is practically neutralized by two or more layers of aluminum strands since the direction of the spiral of the second layer is opposite to the first.

4-29. Copperweld-copper Conductors. Copperweld wire consists of a high-strength steel strand covered with a thin sheath of copper welded on. The result, as a conductor, is a compromise between steel and copper. It has much higher mechanical strength than a copper conductor of the same cross-sectional area but somewhat lower conductivity. The strength and conductivity vary with the relative amount of steel and copper forming the strand. When a conductor of high conductivity, comparable to copper, together with high tensile strength is required, copper strands combined with copperweld strands form a composite conductor called copperweld-copper. Current-carrying capacity and tensile strength of such a conductor can be varied by changing the proportions of copper and copperweld strand making up the composite conductor.

4-30. Electrical Characteristics. Characteristics of some of the most frequently used types of bare and weatherproof-covered overhead conductors are given in Tables 4-3 to 4-5.

The current-carrying capacities in these tables are for 60-cycle ac which will raise conductor temperature from ambient air temperature of 25 to

TABLE 4-3. BARE AND WEATHERPROOF-COVERED COPPER CONDUCTORS
TBWP = triple-braid weatherproof-covered

Conductor size		Number of strands	Cond. cross-sectional area, sq in.	Over-all diameter in.		Weight, lb per 1,000 ft		Minimum breaking strength, lb				Amp* approx. current-carrying capacity for 50° C temp. rise	60-cycle resistance, ohms per 1,000 ft	
								Medium hard drawn		Hard drawn				
Circular mils	Gage			Bare	TBWP	Bare	TBWP	Bare	TBWP	Bare	TBWP		25 C	40 C
16,510	8	1	0.01297	0.1285	0.280	49.97	75	644	612	826	785	92	0.6570	0.6932
26,250	6	1	0.02062	0.1620	0.330	79.46	112	1,010	960	1,280	1,216	125	0.4134	0.4365
41,740	4	1	0.03278	0.2043	0.370	126.4	164	1,584	1,505	1,970	1,872	165	0.2602	0.2748
66,370	2	1	0.05213	0.2576	0.460	200.9	260	2,450	2,330	3,003	2,853	225	0.1637	0.1729
66,370	2	7	0.05213	0.292	0.500	204.9	270	2,361	2,243	3,045	2,893	230	0.1670	0.1762
83,690	1	7	0.06573	0.328	0.540	258.4	328	2,958	2,810	3,804	3,614	265	0.1325	0.1399
105,500	1/0	7	0.08286	0.368	0.620	325.7	424	3,703	3,518	4,750	4,514	310	0.1051	0.1110
133,100	2/0	7	0.1045	0.414	0.680	410.9	522	4,641	4,409	5,927	5,630	360	0.0834	0.08796
167,800	3/0	7	0.1318	0.464	0.750	518.1	653	5,812	5,521	7,366	6,998	415	0.0662	0.06990
211,600	4/0	7	0.1662	0.522	0.810	653.3	800	7,269	6,906	9,154	8,696	485	0.05258	0.05551
250,000	...	19	0.1964	0.600	0.930	771.9	985	8,836	8,394	11,360	10,792	540	0.04458	0.04702
300,000	...	19	0.2356	0.629	1.01	926.3	1,174	10,530	10,003	13,510	12,835	605	0.03724	0.03930
350,000	...	19	0.2749	0.679	1.06	1081.0	1,345	12,200	11,590	15,590	14,810	670	0.03201	0.03377
400,000	...	19	0.3149	0.726	1.15	1235.0	1,553	13,850	13,157	17,560	16,682	730	0.02810	0.02963
450,000	...	37	0.3534	0.772	1.20	1389.0	1,724	15,900	15,105	20,450	19,428	775	0.02508	0.02644
500,000	...	37	0.3927	0.813	1.24	1544.0	1,894	17,550	16,672	22,510	21,385	830	0.02266	0.02387
750,000	...	61	0.5890	0.998	1.50	2316.0	2,822	26,510	25,185	34,090	32,386	1,080	0.01552	0.01631
1,000,000	...	61	0.7854	1.152	1.67	3088.0	3,674	35,100	33,345	43,830	41,636	1,285	0.01204	0.01261

* Based upon current sufficient to give a temperature rise of 50 C above an ambient of 25 C with a crosswise wind of 2 fps.

TABLE 4-4. ACSR (ALUMINUM CABLE STEEL REINFORCED) CONDUCTORS
TBWP = triple-braid weatherproof-covered

Circular mils (alum.)	Gage	Equiv. copper size	Stranding	Cross-sectional area, sq in.			Over-all diameter, in.		Breaking strength, lb		Weight per 1,000 ft, lb		Amp* approx. current-carrying capacity for 50 C temp. rise	60-cycle ac resistance per 1,000 ft, ohms	
				Alum.	Steel	Total	Bare	TBWP	Bare	TBWP	Bare	TBWP		25 C	40 C
26,251	6	8	6al–1st	0.0206	0.0034	0.0240	0.198	0.353	1,170	1,110	36.2	69.7	97	0.675	0.715
41,741	4	6	6al–1st	0.0328	0.0055	0.0383	0.250	0.416	1,830	1,740	57.6	101.1	129	0.425	0.451
66,371	2	4	6al–1st	0.0521	0.0087	0.0608	0.316	0.484	2,790	2,655	91.6	144.3	177	0.275	0.291
83,693	1	3	6al–1st	0.0657	0.0110	0.0767	0.355	0.515	3,480	3,312	115.5	170.0	205	0.222	0.235
105,535	0	2	6al–1st	0.0829	0.0138	0.0967	0.398	0.606	4,280	4,073	145.6	227.6	240	0.178	0.189
133,077	2/0	1	6al–1st	0.1045	0.0174	0.1219	0.447	0.660	5,345	5,090	183.7	283.7	280	0.147	0.156
167,806	3/0	1/0	6al–1st	0.1318	0.0220	0.1538	0.502	0.749	6,675	6,360	231.6	352.6	320	0.123	0.130
211,600	4/0	2/0	6al–1st	0.1662	0.0277	0.1939	0.564	0.824	8,420	8,020	292.1	434.1	375	0.104	0.110
266,800	...	3/0	26al–7st	0.2095	0.0341	0.2436	0.642	0.922	11,250	10,705	365.6	538.6	435	0.0663	0.0702
336,400	...	4/0	26al–7st	0.2642	0.0430	0.3072	0.721	1.001	14,050	13,375	462.4	651.4	500	0.0526	0.0558
397,500	...	250,000	26al–7st	0.3122	0.0508	0.3630	0.783	.....	16,190	......	546.4	.....	565	0.0445	0.0472
477,000	...	300,000	26al–7st	0.3746	0.0610	0.4356	0.858	.....	19,430	......	655.7	.....	630	0.0371	0.0393
556,500	...	350,000	26al–7st	0.4371	0.0712	0.5083	0.927	.....	22,400	......	765.0	.....	690	0.0318	0.0337
636,000	...	400,000	54al–7st	0.4995	0.0648	0.5643	0.977	.....	23,600	......	818.0	.....	760	0.0295	0.0313
715,500	...	450,000	54al–7st	0.5620	0.0728	0.6348	1.036	.....	26,300	......	921.0	.....	805	0.0263	0.0279
795,000	...	500,000	54al–7st	0.6244	0.0809	0.7053	1.093	.....	28,500	......	1,023.0	.....	860	0.0237	0.0251

* Based upon current sufficient to give a temperature rise of 50 C above an ambient of 25 C with a crosswise wind of 2 fps.

75 C, with a cross wind of 2 fps. Current-carrying capacity depends, generally, upon the temperature at which the conductor becomes dangerously annealed or softened. The current at which heat generated by the I^2R loss will raise conductor temperature to the annealing point depends upon surrounding air temperature, wind velocity, rain, sunlight, etc.

A wire will have a higher current-carrying capacity on a cold winter day than on a hot summer day. A copper or aluminum conductor can

TABLE 4-5. COPPERWELD-COPPER CONDUCTORS

Type	Equivalent copper size	Strands		Over-all diameter	Bare weight, lb per 1,000 ft	Lb min. breaking strength	Amp* approx. current carrying capacity for 50 C temp. rise	60-cycle ac resistance, ohms per 1,000 ft	
		Copperweld	Copper					At 25 C	At 40 C
8A	8	1	2	0.199	74.27	2,233	100	0.674	0.7200
6A	6	1	2	0.230	101.6	2,585	130	0.426	0.4535
4A	4	1	2	0.290	161.5	3,938	180	0.268	0.2875
2A	2	1	2	0.366	256.8	5,876	240	0.170	0.1820
2F	2	1	6	0.308	222.8	4,233	230	0.172	0.1847
1F	1	1	6	0.346	280.9	5,266	260	0.1367	0.1470
1/0 F	0	1	6	0.388	354.1	6,536	300	0.109	0.1173
2/0 F	2/0	1	6	0.436	446.8	8,094	350	0.0869	0.0935
3/0 F	3/0	1	6	0.490	563.2	9,980	410	0.0695	0.0750
4/0 F	4/0	1	6	0.550	710.2	12,290	470	0.0556	0.0600

* Based upon current sufficient to give a temperature rise of 50 C above an ambient of 25 C with a crosswise wind of 2 fps.

generally operate at a temperature of 75 C, over a long period of time, without danger of annealing. The current-carrying capacity of an overhead distribution conductor, however, is seldom its controlling factor, except for very short lines. Usually the allowable voltage drop will prevent full utilization of a distribution conductor's current-carrying capacity, except for emergency loading usually of short duration. In many cases the maximum current will not exceed 50 or 60 percent of the capacities shown in Tables 4-3 to 4-5. Conductor resistance at 25 C corresponds to light-load conditions, at 40 C it corresponds to normal loaded conditions, or to current values about 50 percent of normal current-carrying capacity. Resistance with ac increases slightly above its value with dc. This is due to the internal inductance of the wire which forces the current out from the center of the wire toward its surface, producing a greater current density near the surface than at the center. This skin effect, negligible with small conductors at 60 cycles, may be appreciable with large conductors.

AC resistance for 4/0 acsr conductors, or smaller, having a single layer of aluminum strands over the steel core, shown in Table 4-4, is based upon a current density of 1,400 amp per sq in. of aluminum while reactance is based upon a current density of 1,200 amp per sq in. Some error results if current density is greater than these values because of the magnetic effect of the current upon the steel core. If there are two or more layers of aluminum strand over the steel core, as in acsr sizes larger than 4/0, this error will be negligible because the direction of the spiral of each

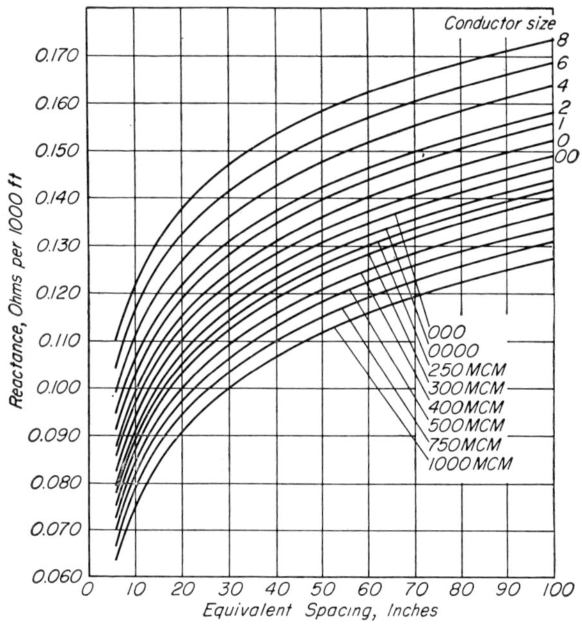

FIG. 4-21. Inductive reactance of copper conductors for different conductor sizes and spacings.

succeeding layer of aluminum strand is reversed. This error in resistance and reactance is not likely to be of importance except for smaller conductors at higher current densities.

Inductive reactance depends upon conductor size and the spacing between conductors. The inductive reactance, in ohms per 1,000 ft, for different wire sizes and for different spacings between conductors is given in Figs. 4-21 to 4-23.

If two conductors of a single-phase ac circuit are separated by a distance d, Fig. 4-24a, the inductive reactance of each depends upon the number of lines of magnetic flux cutting across it when 1 amp flows. If conductors A and B are equal in size, their reactance will be equal. If they move closer together, their inductive reactance will be reduced. This decrease in the reactance of A, for example, is due to a greater number of lines of magnetic flux set up by current flowing in B (opposite in direction to current in A) which cuts across A and reduces the net number of lines of flux linking with A. Moving conductors A and B closer together reduces their inductance and, consequently, their induc-

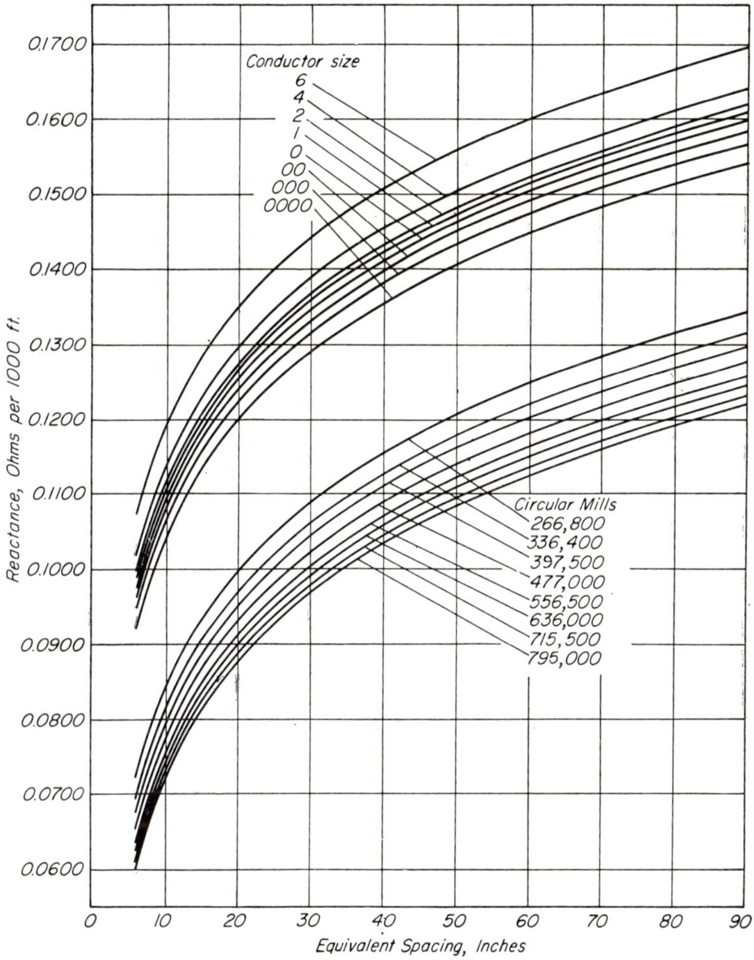

FIG. 4-22. Inductive reactance of acsr conductors for different conductor sizes and spacings.

tive reactance, Fig. 4-25. Similarly, an increase in spacing between A and B increases inductive reactance for each conductor because current in the other conductor has less effect in decreasing the net strength of the magnetic field encircling it.

If the three conductors of a three-phase circuit are arranged as in Fig. 4-24b an equal distance d apart, the inductive reactance of conductors

A, B, and C will be equal, provided they are the same size. The resultant flux from currents in B and C decreases the net flux cutting across A by the same amount that the resultant flux from currents in A and C decreases the net flux cutting across B and that the resultant flux from currents in A and B decreases the net flux cutting across C. Decrease

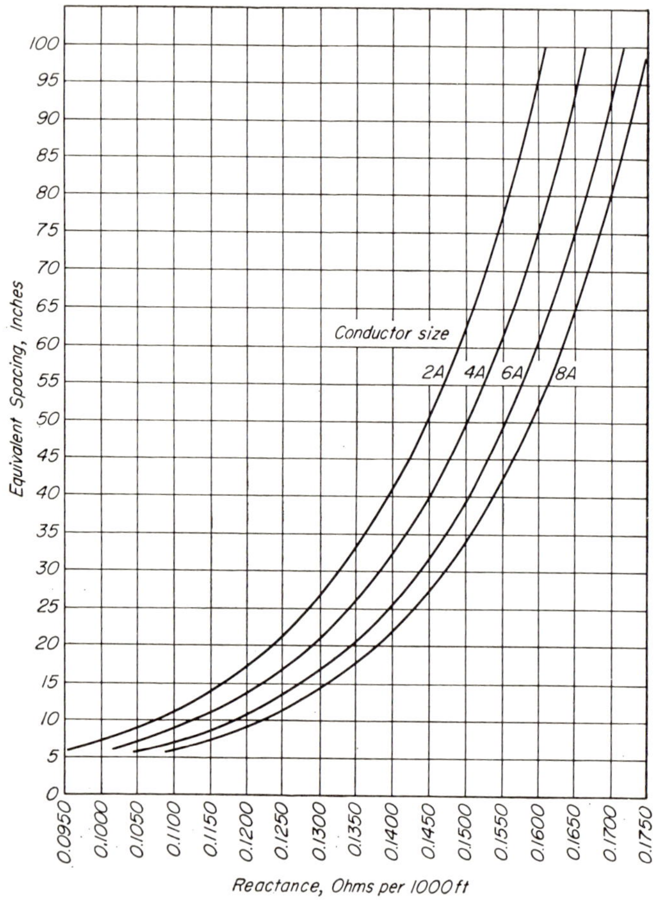

FIG. 4-23. Inductive reactance of type A copperweld-copper conductors for different conductor sizes and equivalent spacings.

in spacing d decreases the reactance, while increase in spacing increases reactance.

If the conductors of a three-phase circuit are arranged as in Fig. 4-24c so that distances d_1, d_2, and d_3 are all different, the inductive reactance of A, B, and C will all be different because a different value of resultant flux cuts across each because of their different spacings. If the three conductors are transposed so that each conductor occupies each position for one-third the line length, the reactance of A, B, and C for the full line length will be the same. This is the same as though the reactance of A, B, and C for any third of the line were averaged and then multiplied by three. Average value of reactance per conductor will be the same as though the conductors were symmetrically spaced as in Fig. 4-24*b* but with a spacing such that

$$d = \sqrt[3]{d_1 \times d_2 \times d_3}$$

This value is the equivalent spacing used in determining inductive reactance of any three-phase line with unsymmetrical spacing. Distribution lines are most frequently arranged with unsymmetrical spacing, Fig. 4-24*c*. For comparatively short lines the average reactance, correspond-

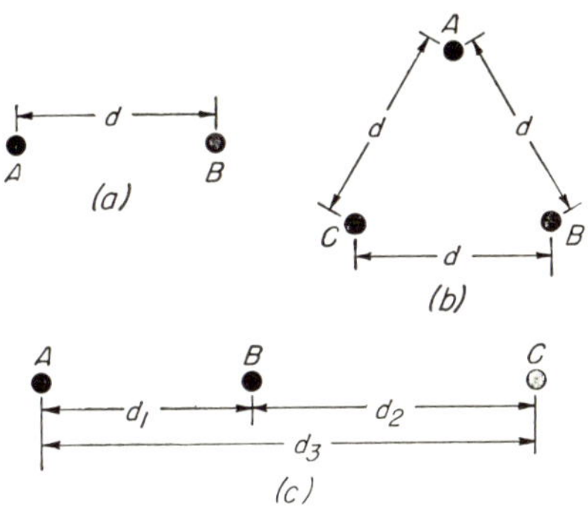

FIG. 4-24. Conductor spacing arrangements.

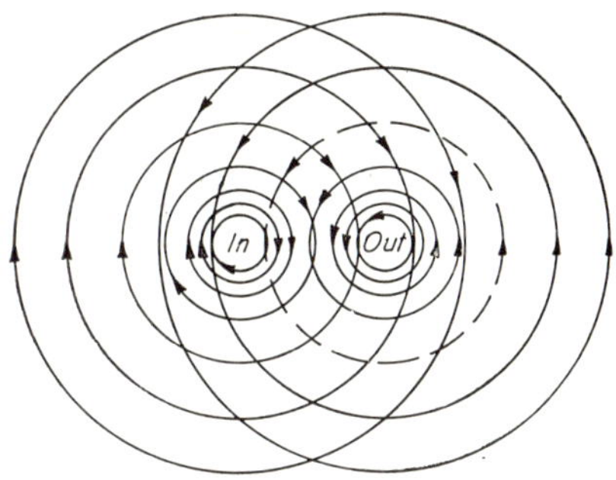

FIG. 4-25. Inductive reactance of a pair of conductors increases with increased spacing because fields have less effect in counteracting each other.

ing to reactance for equivalent spacing, will be sufficiently accurate for normal use, and transpositions may be omitted. For long lines, however, as in rural distribution, there may be considerable difference between highest and lowest value of total conductor reactance, particularly if there is a large difference in spacing. In this case, transpositions to balance and equalize the reactances of the three conductors may be advisable.

4-31. Insulated Conductors. Insulated cables usually consist of a stranded conductor, for flexibility, with an insulating covering of high dielectric strength. This insulating covering frequently consists of a rubber compound, varnished cambric, or paper. Other synthetic compounds also possess high insulating qualities.

Insulation for a rubber-insulated cable consists of one of several grades of rubber compounds covered with an outer protective sheath. The sheath, or jacket, may consist of braid, lead, or neoprene. The outer sheath provides a protective covering for the insulation proper and does not contribute anything to the insulation of the conductor. Where cables are installed in wet or exposed locations, lead or neoprene sheaths are used.

The insulating covering for varnished-cambric insulated cable consists of a cloth-fabric tape impregnated with an insulating varnish and spirally wound around the wire. The cable insulation is covered by a protective outer sheath of braided fabric, lead, or neoprene as for rubber-insulated cables. Varnished cambric is not water resistant. If the cable is to be used in exposed locations the insulation must be covered by a weatherproof and watertight outer sheath. Lead sheath is most frequently used with varnished-cambric insulated cables for underground and overhead distribution.

The insulating covering for paper-insulated cables consists of paper tape impregnated with oil and spirally wound about the conductor to the required thickness of insulation for the rated voltage. Paper for cable insulation has quite high dielectric strength but is particularly susceptible to water absorption. The outer protective sheath of paper-insulated cables must also be watertight and moisturetight. Lead sheathing is most frequently used.

Varnished-cambric and paper-insulated cables are not quite so flexible as rubber-insulated ones because of the danger of separating the layers of insulation at the point where the tape laps. For this reason, sharp bends with varnished-cambric or paper-insulated cables should be avoided. Insulated cables may be either single-conductor or multiconductor. Multiconductor cables are less flexible but are lower in cost than a group of comparable single-conductor cables. Single-conductor cables are generally used for large sizes, lateral cable runs in underground systems, for overhead aerial cable runs, for transformer-bank connections in underground vaults, for secondary low-voltage mains and laterals in underground radial or network systems, etc. Three-conductor cables may be used for three-phase main-feeder cable runs from the power source at the substation to a connection with the outgoing overhead distribution feeders, for primary distribution feeder mains in underground distribution systems, etc.

Aerial cables for distribution-line construction is general practice where trees prevent using open wires or where cables along the streets in business areas prevent unsightly congestion prior to conversion to underground. Aerial cables are insulated conductors suspended from a high-strength messenger wire. The cables may be suspended from the messenger by cable rings or cable ties, or they may be bound to the messenger wire by a flat-copper binding strap as in preassembled self-supported aerial cable. The messenger wire supports the cable weight and withstands the stringing tension. The aerial-type cable may be single- or multiple-conductor. For three-phase service it may consist

TABLE 4-6. 5,000 AND 8,000-VOLT PREASSEMBLED RUBBER-INSULATED AERIAL CABLE

Cond. size	Covering, 64th in.			Copperweld messenger size		Messenger strength, lb		60-cycle ac resistance ohms per 1,000 ft			60-cycle react. to neutral, ohms 1,000 ft		Amp capacity at 75 C from 40 C ambient	
	Insulation		Sheath					Cond. at 75 C	Messenger at 40 C					
	5,000 volts	8,000 volts		5,000 volts	8,000 volts	5,000 volts	8,000 volts		5,000 volts	8,000 volts	5,000 volts	8,000 volts	5,000 volts	8,000 volts
8	10	12	4–5	3-#9	3-#9	5,129	5,129	0.811	0.9733	0.9733	0.0535	0.0570	56	65
6	10	12	4–5	3-#9	3-#9	5,129	5,129	0.508	0.9733	0.9733	0.0495	0.0530	76	85
4	10	12	4–5	3-#8	3-#8	6,282	6,282	0.3206	0.7718	0.7718	0.0475	0.0494	100	110
2	10	12	4–5	3-#7	3-#7	7,922	7,922	0.2014	0.6121	0.6121	0.0438	0.0456	137	149
1	10	12	4–5	3-#7	3-#6	7,922	9,754	0.1596	0.6121	0.4854	0.0420	0.0436	157	171
1/0	10	12	4–5	3-#6	11/32″	9,754	11,280	0.1263	0.4854	0.4180	0.0410	0.0420	184	195
2/0	10	12	5–6	11/32″	3/8″	11,280	13,890	0.1005	0.4180	0.3314	0.0392	0.0413	214	225
3/0	10	12	5–6	3/8″	3/8″	13,890	13,890	0.0796	0.3314	0.3314	0.0380	0.0397	249	259
4/0	10	12	5–6	7/16″	7/16″	16,890	16,890	0.0626	0.2629	0.2629	0.0370	0.0383	287	300
250 mcm	11	12	5–6	1/2″	1/2″	20,460	20,460	0.0530	0.2085	0.2085	0.0368	0.0374	322	330
350 mcm	11	12	5–6	9/16″	9/16″	24,650	24,650	0.0379	0.1653	0.1653	0.0349	0.0355	405	406
500 mcm	11	12	5–6	9/16″	9/16″	24,650	24,650	0.0269	0.1653	0.1653	0.0332	0.0338	500	505

either of a three-conductor cable, rubber, varnished-cambric, or paper insulated with a lead sheath covering and suspended in rings, or of three single-conductor cables assembled in one of several methods. For the lower distribution primary voltages, however, rubber-insulated single-conductor cables are most frequently used.

The safe current-carrying capacity of an insulated cable depends upon many variables. The relative importance of these variables depends upon the conditions under which the cable operates. The safe current-carrying capacity of a cable is, itself, a variable. It can be determined for a specified set of conditions but will change if the conditions change. Temperature is the usual specific condition determining the ultimate limit of safe current-carrying capacity, although, in distribution, voltage regulation frequently limits current-carrying capacity to a value lower than the maximum permitted by temperature.

TABLE 4-7. SINGLE-CONDUCTOR 600-VOLT RUBBER-INSULATED NEOPRENE-SHEATHED COPPER CABLES

Conductor size	Conductor diameter, in.	Thickness of covering, 64th in.		Over-all diameter, in. rubber and neoprene	Approx. current-carrying capacity at 75 C, amp	Ohms per 1,000 ft	
		Insulation rubber	Sheath neoprene			AC resistance at 50 C	60-cycle* reactance
6	0.1836	4	2	0.405	81	0.4671	0.0455
4	0.2316	4	2	0.455	104	0.2946	0.0423
2	0.2922	4	2	0.515	135	0.1850	0.0396
1	0.3320	5	3	0.615	153	0.1469	0.0414
0	0.3725	5	3	0.655	175	0.1162	0.0400
2/0	0.4185	5	3	0.700	197	0.0925	0.0389
3/0	0.4700	5	3	0.755	225	0.0732	0.0377
4/0	0.5275	5	3	0.810	254	0.0574	0.0366
250 mcm	0.5754	6	4	0.920	280	0.0490	0.0377
350 mcm	0.6811	6	4	1.025	337	0.0351	0.0364
500 mcm	0.8134	6	4	1.160	407	0.0251	0.0348

* The inductive reactance at a spacing equal to the over-all diameter of the cable is increased by 20 percent to compensate for random lay in the duct. For installation of two or three cables in an iron conduit the values in the table should be further increased by 25 percent to compensate for the magnetic effect of the iron.

From a temperature standpoint, the current-carrying capacity is limited by the safe maximum temperature for the insulation. Insulation temperature depends on temperature of the conductor and the surrounding medium. The temperature rise of the assembled cable depends on the heat generated by the I^2R loss in the conductor, the i^2r loss from induced currents in the metallic sheath or armor surrounding the cable, the dielectric losses induced in the insulation itself by voltage stresses. Heat from these losses must flow outward through the thermal resistance of the various surrounding mediums, such as the insulation, the duct bank, the earth, the air, etc. With any given value of current flowing in the conductor, the temperature will continue to rise until the rate at which heat dissipates equals the rate at which heat generates. The maximum safe current-carrying capacity, then, is determined by the ambient temperature of the surrounding medium (the air in the case of

TABLE 4-8. SINGLE-CONDUCTOR RUBBER-INSULATED NEOPRENE-SHEATHED SHIELDED COPPER CABLES. 4,000 TO 5,000 VOLTS GROUNDED OR UNGROUNDED AND 7,000 TO 8,000 VOLTS GROUNDED

Conductor size	Insulation thickness, 64th in.		Sheath thickness, 64th in.	Approx. over-all diameter, in.		Approx. current-carrying capacity		AC ohms per 1,000 ft		
								Resistance, 50 C	60-cycle* reactance	
	5,000-volts	8,000-volts		5,000-volts	8,000-volts	5,000-volts	8,000-volts		5,000-volts	8,000-volts
6	10	12	2–4	0.73	0.82	81	84	0.4671	0.0567	0.0623
4	10	12	3–4	0.81	0.90	104	108	0.2946	0.0539	0.0577
2	10	12	3–4	0.87	0.96	135	139	0.1850	0.0501	0.0535
1	10	12	3–5	0.91	1.00	153	157	0.1469	0.0481	0.0524
0	10	12	3–5	0.95	1.04	175	179	0.1162	0.0464	0.0504
2/0	10	12	3–5	1.00	1.08	197	203	0.0925	0.0446	0.0486
3/0	10	12	3–5	1.04	1.13	225	231	0.0732	0.0433	0.0470
4/0	10	12	4–5	1.10	1.19	254	262	0.0574	0.0426	0.0452
250 mcm	11	12	4–5	1.21	1.27	280	288	0.0490	0.0424	0.0440
350 mcm	11	12	4–6	1.29	1.38	337	342	0.0351	0.0406	0.0426
500 mcm	11	12	4–6	1.45	1.51	407	412	0.0251	0.0385	0.0404

* The inductive reactance at a spacing equal to the over-all diameter of the cable is increased by 20 percent to compensate for random lay in the duct. For installation of two or three cables in an iron conduit the values in the table should be further increased by 25 percent to compensate for the magnetic effect of the iron.

aerial cable and the earth in the case of cables in underground ducts) and the maximum temperature at which the insulation can be operated.

It will not be possible to cover all the many types of cables used, their voltage ratings, etc., nor their characteristics under all conditions of use. However, approximate physical and electrical characteristics of some of the most frequently used types of distribution cables are given in Tables 4-6 to 4-10.

The current-carrying capacity shown in Table 4-6 for aerial cable is based on the current which will raise the temperature of the copper conductor from 40 C to 75 C, the safe operating temperature of the insula-

TABLE 4-9. SINGLE- AND THREE-CONDUCTOR VARNISHED-CAMBRIC INSULATED LEAD-SHEATHED COPPER CABLES, 4,000 TO 5,000-VOLT NEUTRAL GROUNDED OR UNGROUNDED, AND 7,000 TO 8,000-VOLT NEUTRAL GROUNDED

Characteristics	Conductor size										
	6	4	2	1	0	2/0	3/0	4/0	250,000	350,000	500,000
A. 5,000-volt single-conductor:											
1. Conductor insulation, 64th in.	9	9	9	9	9	9	9	9	10	10	10
2. Over-all diameter, in.	0.59	0.64	0.70	0.74	0.78	0.86	0.91	0.97	1.05	1.15	1.32
3. AC resistance at 65 C, ohms per 1,000 ft	0.4730	0.2987	0.1888	0.1506	0.1200	0.0965	0.0771	0.0616	0.0528	0.0384	0.0281
4. Reactance, 60 cycle, ohms per 1,000 ft*	0.0581	0.0539	0.0501	0.0480	0.0463	0.0458	0.0441	0.0427	0.0425	0.0405	0.0391
5. Current-carrying capacity, 85 C	97	126	163	185	211	240	270	307	337	407	488
B. 5,000-volt three-conductor:											
1. Conductor insulation, 64th in.	6	6	6	6	6	6	6	6	7	7	7
2. Belt insulation, 64th in.	4	4	4	4	4	4	4	4	4	4	4
3. Over-all diameter, in.	1.08	1.19	1.35	1.43	1.52	1.62	1.78	1.88	2.04	2.33	2.61
4. AC resistance at 65 C, ohms per 1,000 ft	0.4730	0.2987	0.1888	0.1506	0.1200	0.0965	0.0771	0.0616	0.0528	0.0384	0.0281
5. Reactance, 60 cycle, ohms per 1,000 ft*	0.0378	0.0353	0.0332	0.0321	0.0311	0.0303	0.0295	0.0286	0.0291	0.0281	0.0270
6. Current-carrying capacity, 85 C	73	96	124	141	162	185	211	240	263	319	386
C. 8,000-volt single-conductor:											
1. Conductor insulation, 64th in.	11	11	11	11	11	11	11	11	11	11	11
2. Over-all diameter, in.	0.65	0.70	0.77	0.80	0.87	0.92	0.97	1.03	1.08	1.18	1.35
3. AC resistance at 65 C, ohms per 1,000 ft	0.4730	0.2987	0.1888	0.1506	0.1200	0.0965	0.0771	0.0616	0.0528	0.0384	0.0281
4. Reactance, 60 cycle, ohms per 1,000 ft*	0.0610	0.0567	0.0526	0.0504	0.0495	0.0476	0.0460	0.0444	0.0432	0.0411	0.0398
5. Current-carrying capacity, 84 C	95	121	158	178	200	230	258	293	323	386	463
D. 8,000-volt three-conductor:											
1. Conductor insulation, 64th in.	7	7	7	7	7	7	7	7	7	7	7
2. Belt insulation, 64th in.	6	6	6	6	6	6	6	6	6	6	6
3. Over-all diameter, in.	1.21	1.34	1.47	1.56	1.65	1.78	1.89	2.01	2.11	2.39	2.67
4. AC resistance at 65 C, ohms per 1,000 ft	0.4730	0.2987	0.1888	0.1506	0.1200	0.0965	0.0771	0.0616	0.0528	0.0384	0.0281
5. 60-cycle reactance, ohms per 1,000 ft*	0.0397	0.0369	0.0344	0.0332	0.0323	0.0314	0.0304	0.0298	0.0291	0.0281	0.0270
6. Current-carrying capacity, 81 C	72	92	120	136	154	176	202	229	252	303	366

* If cables are installed in iron conduit, the reactance values in the table should be increased by 25 percent to compensate for the magnetic effect of the iron.

TABLE 4-10. SINGLE AND THREE-CONDUCTOR PAPER-INSULATED LEAD-SHEATHED COPPER CABLES. 4,000 TO 5,000-VOLT NEUTRAL GROUNDED OR UNGROUNDED, AND 7,000 TO 8,000-VOLT NEUTRAL GROUNDED

Characteristics	Conductor size										
	6	4	2	1	0	2/0	3/0	4/0	250,000	350,000	500,000
A. 5,000-volt single-conductor:											
1. Conductor insulation, 64th in.	6	6	6	6	6	6	6	6	6	6	6
2. Over-all diameter, in.	0.53	0.58	0.64	0.68	0.72	0.76	0.83	0.89	0.94	1.04	1.19
3. AC resistance at 65 C, ohms per 1,000 ft	0.4730	0.2987	0.1888	0.1506	0.1200	0.0965	0.0771	0.0616	0.0528	0.0384	0.0281
4. 60-cycle reactance, ohms per 1,000 ft*	0.0551	0.0500	0.0475	0.0454	0.0438	0.0425	0.0415	0.0401	0.0392	0.0377	0.0364
5. Current-carrying capacity, 85 C	85	111	143	161	183	209	238	269	294	356	433
B. 5,000-volt three-conductor:											
1. Conductor insulation, 64th in.	6	6	6	6	6	6	6	6	6	6	6
2. Belt insulation, 64th in.	4	4	4	4	4	4	4	4	4	4	4
3. Over-all diameter, in.	1.13	1.27	1.42	1.43	1.38	1.47	1.56	1.66	1.76	1.95	2.20
4. AC resistance at 65 C, ohms per 1,000 ft	0.4730	0.2987	0.1888	0.1506	0.1200	0.0965	0.0771	0.0616	0.0528	0.0384	0.0281
5. 60-cycle reactance, ohms per 1,000 ft*	0.0368	0.0344	0.0323	0.0312	0.0303	0.0295	0.0287	0.0278	0.0275	0.0262	0.0251
6. Current-carrying capacity, 85 C	73	96	124	141	162	185	211	240	263	320	390
C. 8,000-volt, single-conductor:											
1. Conductor insulation, 64th in.	9	9	9	9	9	9	9	9	9	9	9
2. Over-all diameter, in.	0.62	0.67	0.75	0.79	0.83	0.87	0.92	0.98	1.03	1.15	1.28
3. AC resistance at 65 C, ohms per 1,000 ft	0.4730	0.2987	0.1888	0.1506	0.1200	0.0965	0.0771	0.0616	0.0528	0.0384	0.0281
4. 60-cycle reactance, ohms per 1,000 ft*	0.0593	0.0552	0.0517	0.0497	0.0480	0.0461	0.0446	0.0430	0.0418	0.0405	0.0384
5. Current-carrying capacity, 85 C	84	109	140	158	176	204	232	263	290	350	423
D. 8,000-volt three-conductor:											
1. Conductor insulation, 64th in.	7	7	7	7	7	7	7	7	7	7	7
2. Belt insulation, 64th in.	4	4	4	4	4	4	4	4	4	4	4
3. Over-all diameter, in.	1.19	1.34	1.48	1.50	1.45	1.53	1.62	1.74	1.82	2.01	2.26
4. AC resistance at 65 C, ohms per 1,000 ft	0.4730	0.2987	0.1888	0.1506	0.1200	0.0965	0.0771	0.0616	0.0528	0.0384	0.0281
5. 60-cycle reactance, ohms per 1,000 ft*	0.0387	0.0360	0.0335	0.0323	0.0314	0.0306	0.0296	0.0290	0.0283	0.0271	0.056
6. Current-carrying capacity, 85 C	72	94	122	138	156	181	206	235	259	315	381

* If cables are installed in iron conduit, the reactance values in the table should be increased by 25 percent to compensate for the magnetic effect of the iron.

tion. The current-carrying capacities in Tables 4-7 to 4-10 are for cables installed in nonmetallic underground ducts and represents the current necessary to raise the conductor temperature from an ambient earth temperature of 20 C to the maximum safe operating temperature of the particular type of insulation. These safe current-carrying capacities are also based on the assumption that three outside ducts in the duct bank are occupied by three cables per duct, in the case of single-conductor cables, or by one cable per duct, in the case of three-conductor cables, and that the cables are all the same size, equally loaded and operate at an average load of 75 percent of the rated maximum load. The resistance and reactance shown in the tables for lead-sheathed cables assume the lead sheaths are bonded together and grounded. The inductive reactance for single-conductor cables, shown in the tables, are increased 20 percent above the values which would hold if the cables were symmetrically spaced throughout their length to account for random lay of the cables in the duct. If cables are installed in iron conduit, the inductive reactance in the tables should also be increased by 20 to 25 percent to account for the magnetic effect of the iron conduit. For this reason single-conductor cables carrying heavy current loads should not be installed in iron conduit when one cable per duct is installed.

CHAPTER REVIEW TOPICS

1. What is generally meant by the distribution system?
2. How do underground and overhead conductors compare in respect to capacity, inductive reactance, and voltage drop?
3. What factors dictate the selection of underground systems rather than overhead?
4. How do overhead and underground systems compare in service reliability?
5. Describe the essential elements of construction in an underground system.
6. What are the advantages of buried cable?
7. What steps are followed in planning an underground installation?
8. How many ducts may be used for through and local distribution routes?
9. What determines the safe current-carrying capacity of cables?
10. How should underground ducts be arranged for minimum installation cost?
11. How should underground ducts be arranged for best heat dissipation?
12. What is meant by racking of cables in manholes?
13. Outline the precautions to be taken in installing ducts.
14. Name five different types of ducts.
15. Describe the built-up and tier methods of duct installation.
16. How are fiber ducts joined to form a continuous tube?
17. What are transite ducts made of and what are their characteristics?
18. Can transite ducts be buried directly in the earth?
19. What are duct separators used for?
20. Does tile conduit have a round bore?
21. How are tile ducts joined?
22. How are concrete ducts usually manufactured and installed?

23. Describe the operation of a duct-forming machine.
24. What are the principal items of cost in laying underground ducts?
25. What are manholes used for?
26. What factors must be considered in designing manholes?
27. Describe various methods of constructing manholes.
28. What conditions must the design of transformer vaults meet?
29. How does heat dissipate from a transformer vault?
30. Describe the function of street-service handholes.
31. Define the three types of stress to which overhead construction is subject.
32. What is elasticity?
33. What is the ultimate strength of a material?
34. Define the term moment arm.
35. What do elongation and compression do to a material?
36. Define and show how to calculate internal and external moments.
37. Enumerate the loadings to which a pole may be subjected.
38. What is the National Electrical Safety Code?
39. Does the NESC apply throughout the United States?
40. What is a safety factor? Give examples.
41. When must a pole be guyed?
42. Describe four types of guys.
43. How does soil affect the strength of a pole?
44. Describe the eight types of earth in which poles may be set.
45. What is a crib brace?
46. Describe various types of anchors used with guys.
47. What is the relation between conductor sag and tension?
48. What does the sag of a conductor depend on?
49. What metals are usually used for distribution conductors?
50. What factors must be considered in choosing distribution conductors?
51. Define a circular mil.
52. Describe stranded and solid conductors.
53. How many kinds of copper conductor are available? Describe them.
54. Describe the construction of an acsr cable.
55. What is a copperweld conductor?
56. Name the factors that determine the current-carrying capacity of a conductor.
57. What does inductive reactance of a conductor depend upon?
58. How can inductive reactance of a three-phase line be minimized?
59. What types of insulation are used on distribution conductors?
60. What types of sheaths are used on distribution conductors?
61. What function does a messenger wire perform?
62. What conditions fix the current-carrying capacity of a cable?

PROBLEMS

Problem 4-1: The maximum fiber stress from bending under transverse wind load will not exceed 33 percent of the ultimate in a distribution pole line constructed in accord with the NESC. The minimum ground-line circumference of the poles in the line was 39.5 in. when the line was built. If the NESC recommends that the poles be replaced when the maximum fiber stress, under the same wind loading, reaches 75 percent of the rated ultimate strength, what will be the permissible average depth of decayed wood at the ground line when the poles are due to be replaced?

Solution:

1. The internal resisting moment of the pole $= M_p = \dfrac{SI}{c}$

The cross section of the pole is circular. The resisting moment of a circular section can be written

$$M_p = S\frac{I}{c} = \frac{SC^3}{3{,}790} \qquad \text{lb-ft}$$

where S = allowable working stress, psi
C = circumference of the pole, at the ground line, in.

2. If S, the allowable working stress, is written in percent of U, the ultimate strength, the resisting moment for the average pole can then be written as follows:

$$M_p = \frac{SC_1^3}{3{,}790} = \frac{0.33U(39.5)^3}{3{,}790} \qquad \text{when the poles were installed}$$

$$M_p = \frac{SC_2^3}{3{,}790} = \frac{0.75UC_2^3}{3{,}790} \qquad \text{when the poles are ready for replacement}$$

Equating these two equal values and solving for C_2, the ground-line circumference at the time of replacement, we have

$$\frac{0.75UC_2^3}{3{,}790} = \frac{0.33U(39.5)^3}{3{,}790} \qquad \text{from which}$$

$$C_2^3 = \frac{0.33 \times (39.5)^3}{0.75} \qquad \text{and}$$

$$C_2 = 30 \text{ in.}$$

3. The diameter of the poles at the ground line

$$D = \frac{C}{\pi} = \frac{39.5}{\pi} = 12.56 \text{ in. when installed}$$

$$D_1 = \frac{30}{\pi} = 9.55 \text{ in. at replacement}$$

$$\text{Average depth of decay} = \frac{12.56 \text{ in.} - 9.55 \text{ in.}}{2} = 1.5 \text{ in.}$$

Problem 4-2: A rural line consisting of four primary conductors mounted on 40-ft wood poles is contemplated. The moment arm for transverse wind load on the conductors will be 33 ft. The conductors will be subject to a transverse wind load of 0.43 lb per ft. The ultimate resisting moment of the poles at the ground line is 48,000 lb-ft. The bending moment due to wind on the pole will be equal to 1,400 lb-ft. What will be the maximum permissible span length if the maximum fiber stress is limited to 50 percent of the ultimate strength of the poles?

Solution:

1. Allowable bending moment on poles = 50% of 48,000 lb-ft = 24,000 lb-ft
2. Maximum allowable bending moment due to wind on the conductors
= 24,000 lb-ft − 1,400 lb-ft = 22,600 lb-ft
3. 22,600 lb-ft = 4 cond. × 0.43 lb × 33 ft × span length

$$\text{Span length} = \frac{22{,}600\text{-lb ft}}{4 \text{ cond.} \times 0.43 \text{ lb} \times 33 \text{ ft}} = 400 \text{ ft}$$

Problem 4-3: A 35-ft Western red cedar pole has a ground-line circumference of 35.5 in. (1) What total dead-end pull, at a point 28 ft aboveground, can the pole withstand without guying if it is set in class 4 soil? (2) If the pole is supported by an anchor guy attached at a point 24 ft aboveground with a lead of 15 ft away from the pole, what will be the allowable maximum dead-end pull against the pole if the guy wire has a breaking strength of 11,500 lb and the allowable working stress in the guy strand is 87.5 percent of its breaking strength? (3) What dead-end pull would the guy support if the lead were increased to 30 ft?

Solution:

1. From Table 4-2, the resisting moment of the pole at the ground line = 65,000 lb-ft

$$M_p + M_c + M_t = S_e M_r \qquad \text{or}$$
$$0 + 0 + M_t = 40\% \text{ of } 65{,}000 \text{ lb-ft} = 26{,}000 \text{ lb-ft} \qquad \text{also}$$
$$M_t = \text{dead-end pull} \times \text{moment arm} \qquad \text{from which}$$
$$\text{Dead-end pull} = \frac{26{,}000 \text{ lb ft}}{28 \text{ ft}} = 930 \text{ lb}$$

2. $$\frac{M_t \times 100}{\% \, S_t \times H_g \times \cos b} = u$$

$$\cos b = \cos\left(\tan^{-1}\frac{H_g}{L}\right) = 0.53$$
$$M_t = \frac{11{,}500 \times 87.5 \times 24 \times 0.53}{100} = 127{,}700 \text{ lb ft}$$
$$\text{Dead-end pull} = \frac{127{,}700 \text{ lb ft}}{28 \text{ ft}} = 4{,}560 \text{ lb}$$

3. $$M_t = \frac{11{,}500 \times 87.5 \times 24}{100} \times \cos\left(\tan^{-1}\frac{24}{30}\right) = 188{,}000 \text{ lb ft}$$

$$\text{Dead-end pull} = \frac{188{,}000 \text{ lb ft}}{28 \text{ ft}} = 6{,}720 \text{ lb}$$

Problem 4-4: A line consisting of three No. 2/0, three No. 2, and three No. 4 copper wires mounted at 28 ft, 26 ft, and 23 ft, respectively, turns an angle of 45° on a 35-ft pole. The bending moment from wind on the pole is equal to 2,500 lb-ft and 11,500 lb-ft from wind on the conductors at the heaviest loading. The maximum conductor tension will be 2,640 lb for the No. 2/0, 1,340 lb for the No. 2, and 900 lb for the No. 4 wires. An anchor guy is to be installed with a height and lead of 20 ft. The allowable stress in the guy is not to exceed 50 percent of ultimate for transverse wind loading nor 87.5 percent of ultimate due to conductor tension. (1) What is the resultant side pull at the level of the guy attachment? (2) What is the minimum breaking strength of guy required? (3) What will be the maximum stress in the guy expressed in percent of its breaking strength?

Solution:

1. Bending moment due to wire tension

$$\begin{aligned}
&= 2 \times 3 \times 2{,}640 \times \sin\frac{45^\circ}{2} \times 28 \text{ ft} \\
&= 170{,}000 \text{ lb ft due to No. 2/0 wire} \\
&= 2 \times 3 \times 1{,}340 \times \sin\frac{45^\circ}{2} \times 26 \text{ ft} \\
&= 80{,}000 \text{ lb ft due to No. 2 wire} \\
&= 2 \times 3 \times 900 \sin\frac{45^\circ}{2} \times 23 \text{ ft} \\
&= 47{,}500 \text{ lb ft due to No. 4 wire} \\
\text{Total} &= 297{,}500 \text{ lb ft}
\end{aligned}$$

$$\text{Total bending moment at the ground line} = 2{,}500 + 11{,}500 + 297{,}500 = 311{,}500 \text{ lb-ft}$$

$$\text{Equivalent side pull at point of guy attachment} = \frac{311{,}500 \text{ lb ft}}{20 \text{ ft}} = 15{,}575 \text{ lb}$$

2. From Eq. (7):

$$U = \frac{(2{,}500 + 11{,}500)100}{50 \times 20 \times \cos(\tan^{-1} {}^{20}\!/_{20})} + \frac{297{,}500 \times 100}{87.5 \times 20 \cos(\tan^{-1} {}^{20}\!/_{20})} = 26{,}000 \text{ lb}$$

3. Stress in guy $= \left(\frac{15{,}575}{\cos 45°} \div 26{,}000 \text{ lb}\right) 100 = 84.7\%$ of ultimate

Problem 4-5: Referring to accompanying Fig. 4-26, suppose that

$L_1 = 3$ kw at 120 volts
$L_2 = 1$ kw at 120 volts
$E_s = 120$ volts
Power factor of loads L_1 and $L_2 = 90\%$
Distance from E_s to L_1 and $L_2 = 800$ ft
Size of wires No. 1, No. 2, and $N =$ No. 4 AWG copper
Spacing between wires = 8 in.

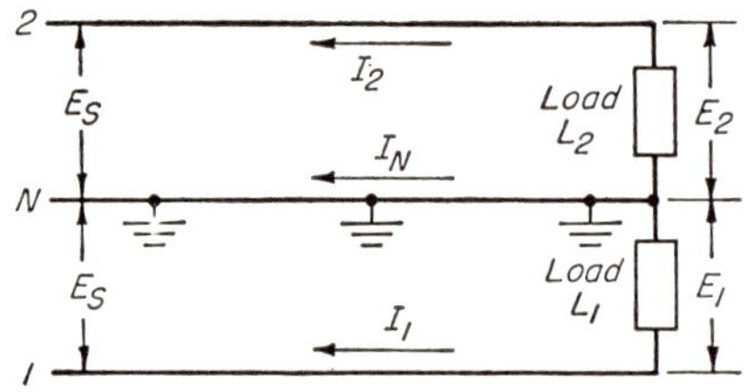

FIG. 4-26. Diagram for Prob. 4-5.

(1) What is the voltage drop in each wire? (2) What is the voltage impressed across loads L_1 and L_2? (3) What is the total voltage drop in the circuit including wires 1 and N? (4) In wires 2 and N?

Solution:

1. Current in wire No. 1 $= I_1 = \frac{3 \text{ kw} \times 1{,}000}{120 \text{ volts} \times 0.90} = 27.77$ amp

Current in wire No. 2 $= I_2 = \frac{1 \text{ kw} \times 1{,}000}{120 \text{ volts} \times 0.90} = 9.26$ amp

Current in neutral wire $= I_N = I_1 - I_2 = 27.77 - 9.26 = 18.51$ amp

$\sin\theta = \sqrt{1 - (\cos\theta)^2} = \sqrt{1 - (.90)^2} = 0.436$

Resistance per wire $= \frac{0.275}{1{,}000} \times 800 = 0.220$ ohm

Reactance per wire $= \frac{0.106}{1{,}000} \times 800 = 0.0848$ ohm

Effective Z per wire $= R\cos\theta + X\sin\theta$
$= (0.220 \times 0.90) + (0.0848 \times 0.436) = 0.235$ ohm

Voltage drop in wire No. 1 $= I_1(R\cos\theta + X\sin\theta)$
$= 27.77 \times 0.235 = 6.53$ volts

Voltage drop in wire No. 2 $= 9.26 \times 0.235 = 2.18$ volts

Voltage drop in neutral wire $= 18.51 \times 0.235 = 4.35$ volts

2. Voltage across $L_1 = E_s$ − volts drop in wire No. 1 − volts drop in neutral
= 120 volts − 6.53 volts − 4.35 volts = 109.1 volts

Voltage across $L_2 = E_s$ − volts drop in wire No. 2 + volts drop in neutral
= 120 volts − 2.18 volts + 4.35 volts = 122.2 volts

3. Total voltage drop in 1 − N = 120 volts − 109.1 volts = 10.9 volts
Total voltage drop in 2 − N = 120 volts − 122.2 volts = −2.2 volts

CHAPTER 5

DISTRIBUTION SYSTEMS—PRIMARY AND SECONDARY

By E. M. Adkins

Jersey Central Power and Light Company

5-1. The primary distribution system includes circuits at voltages somewhat higher than general utilization levels and handles larger blocks of electrical energy than the average low-voltage customer uses. The primary system is generally classified as a delta or a wye system. This refers to the connection of power transformers in the substation supplying primary circuits. The most commonly used nominal primary voltages are

Delta Systems, Volts	*Wye Systems, Volts*
2,400	2,400/4,160
4,800	4,800/8,314
7,200	7,200/12,470
7,620	7,620/13,200
13,800	

The great majority of primary systems use potentials below 5,000 volts, although there are instances of voltages as high as 26,000 for distribution primary circuits. Load densities in urban and rural areas as well as the average annual consumption of electrical energy are steadily increasing. The tendency is toward higher primary voltages since an appreciable saving in over-all economy can be achieved with the higher load densities. A primary circuit operating at 7,200 volts can deliver nine times as much energy as a 2,400-volt circuit over the same conductors with the same percent regulation. Similarly, the 7,200-volt circuit could deliver three times as much energy as the 2,400-volt circuit with the same copper loss.

The vast majority of primary circuits are three phase. Three phase is chosen over single phase because one additional wire, a 50 percent increase in conductor material, increases circuit capacity by 73 percent with the same current and voltage between wires. Also, with the same voltage regulation the three-phase circuit can handle twice as much load as the single-phase circuit. Three-phase primary circuits, however, may have single-phase branches tapped from them. The greater economy of the symmetrical three-phase distribution system has practically eliminated the unsymmetrical two-phase system.

Primary systems are classified as radial or network. A radial system is supplied from a single source, Fig. 5-1. A single substation may supply several primary circuits. If each circuit radiates out from the substation and is separate from the other and does not terminate away from the station by connecting to a common bus, it is a radial circuit.

A primary network system, Fig. 5-4, is supplied from more than one source. If a primary system is supplied from two different substations or by two or more primary feeders from the same substation, the primary is a network system and the feeders are primary network feeders.

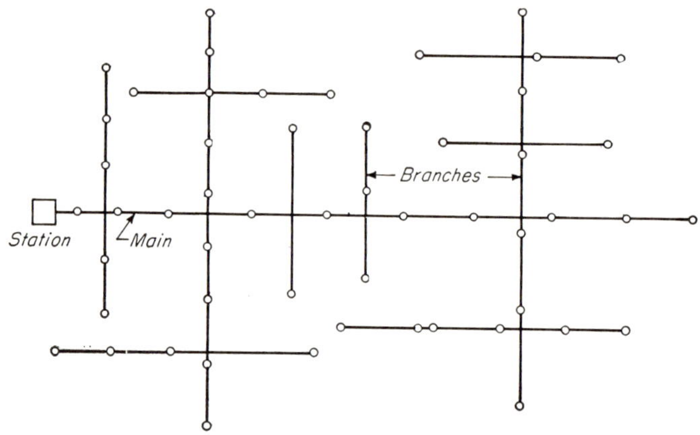

FIG. 5-1. Single-line diagram of simple radial or tree-type primary circuit.

5-2. Radial Tree-type Primary Circuit. The most simple and probably most frequently used type of primary circuit is the tree or trunk-and-branch type as it is sometimes called. This circuit is so named because of its fancied resemblance to the trunk and branches of a tree, Fig. 5-1. A main feeder leaves the substation and traverses the area to be supplied. Transformers are tapped off the feeder to supply various loads along the route. In addition, branch lines are tapped off from the main feeder to supply more distant loads routed along side streets or branch roads. These lateral branch lines may be either single-phase or polyphase. If the branch lines are short, or if the load is small, the lateral taps may be single-phase. Single-phase lines should be tapped from alternate phases to keep the load balanced on the feeder. If the area supplied by the branch line contains polyphase load, or if the number and size of single-phase transformers seriously affect the load balance and voltage drop of the feeder, the branch tap should be polyphase also.

Sometimes it is advantageous to divide the total area supplied by the primary circuit into phase areas and supply all transformers in one section from the same phase. The phase areas or sections are selected, if possible, so that load in each is about the same. This keeps the load balanced on the three-phase circuit. This is of particular advantage where the secondary circuits fed by the transformers connect to form a low-voltage single-phase banked secondary network. Load on the circuit should be kept reasonably well balanced to make best use of substation and feeder facilities. If the circuit is badly unbalanced, voltage regulation on the heavily loaded phase will be excessive and substation facilities, such as voltage regulators, may be overloaded, with the result that circuit performance is much poorer than it should be.

Greatest current density in the circuit will be between the substation and the first transformer connection, or where the first lateral tap to the main feeder is made. From here the current in the line progressively decreases to the end of the feeder. Theoretically, successively smaller sizes of conductor could be used on the main because of decreasing current density. Practically, however, this can seldom be realized. Current-carrying capacity is seldom the governing factor for such a primary circuit. Line regulation or voltage drop is more often the controlling factor. The circuit usually consists of a larger conductor size on the substation end with a somewhat smaller size on the lateral branch lines, and perhaps toward the end of the feeder main.

The maximum load which can be distributed from a primary circuit, if not limited by the maximum allowable current rating of conductors and substation equipment, is determined by the maximum voltage which can be permitted at the point of utilization of the first customer on the circuit and the minimum voltage permitted at the last point of utilization on the circuit, at the time of maximum load demand for the circuit. When loads served by the distribution substation grow, they are taken care of by extending the main or branches until the load limit of the circuit (as determined by current or regulation) is reached and then by subdividing the circuit to form two circuits. This type of radial primary circuit, usually, is not confined to a planned or well-defined area. It is well adapted to areas where the loads are small and scattered and the average load density is low. It is generally used to serve small towns or rural areas, although it is sometimes used in high-density load areas to serve areas immediately surrounding the distribution substation. In this latter case the circuit boundaries are well defined, and the source is in the center of the area to be served.

5-3. The Load-center Radial, or Feeder-and-main, Type. With the load-center radial, or feeder-and-main, type circuit the area to be served is selected with well-defined boundaries. A main feeder runs from the source to the center of the load area, over the most direct route without having any load or transformers tapped from it, Fig. 5-2. From this center of load, simple radial branches radiate in all directions to serve loads along the lateral and cross streets. The feeder-and-main system is a further development of the simple radial system.

The load-center radial system is most advantageously used in the higher load-density areas of large towns and cities. Automatic voltage regulators are necessary with this type of circuit. The conductor size from substation to load center must be large enough to enable voltage regulators to maintain a sufficiently high voltage at the load center so that potential at the ends of the longest and most heavily loaded branch does not drop below the allowable minimum. The maximum capacity of the circuit, as for the tree-type radial, is determined by the allowable highest voltage at the first customer at the load center and the minimum

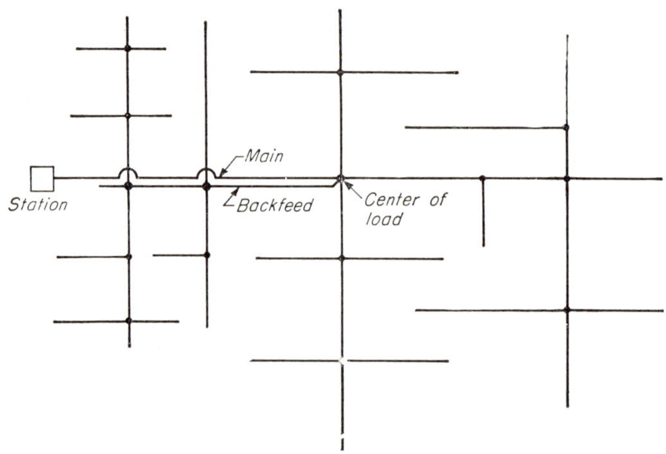

FIG. 5-2. Radial feeder supplied through main-type feeder has better voltage regulation.

voltage at the last most-distant customer. The load capacity of such a circuit may be several times as great as the tree-type radial, even though load density, conductor size, etc., are the same. Radial branches leading out from the load center are three phase, and generally of somewhat smaller conductor size than the main feeder because of much lower current density. Conductor sizes on the branch taps from these radial mains are the smallest in size and may be either three-phase or single-phase as the load requires. These features generally result in materially lower investment per unit of load served.

5-4. Loop Primary Circuit. The loop circuit starts from the substation bus, makes a loop through the area to be served, and returns to the substation, Fig. 5-3. It is the equivalent of two radial circuits with their ends tied together. The loop circuit is not often used because of its high cost. It is sometimes used for special situations, in high load-density areas, where there are fairly large commercial or industrial loads, such as department stores and small factories, which cannot be supplied from an available low-voltage network system, and where demand for uninterrupted service warrants the high investment.

The lateral taps, Fig. 5-3, may supply transformer banks located on individual customer's premises, or they may form short laterals supplying several transformer banks. The circuit is broken up into short sections, to provide automatic sectionalizing in case of line fault. Automatic circuit breakers are inserted on each end of the circuit at the substation and on each side of lateral taps. In addition, an automatic circuit breaker is installed on each lateral tap; making three circuit breakers at each tap. The circuit breaker on the tap opens under a predetermined overload current, while the circuit breakers in series with the loop are controlled by a pilot-wire scheme of operation.

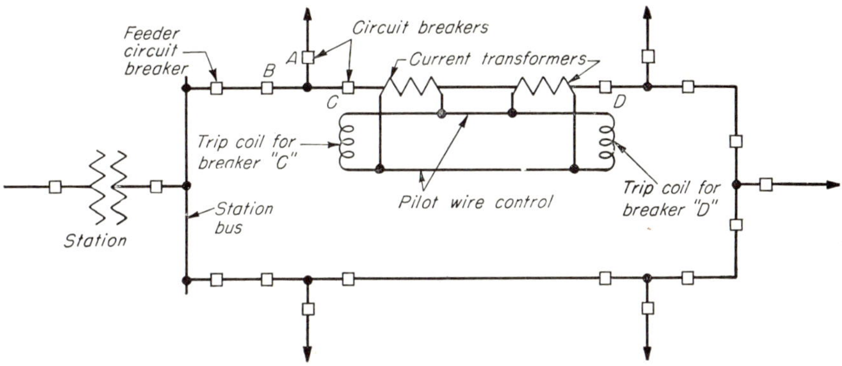

FIG. 5-3. Loop circuit may be used in high-load density areas, achieves higher service continuity.

In Fig. 5-3, suppose a fault develops in the transformer bank or on the line supplied by the tap between circuit breakers *B* and *C*. Circuit breaker *A* will open and disconnect the faulted tap without interrupting the remainder of the circuit.

Circuit breakers *C* and *D* will open if a fault develops on the loop between them without interrupting the remainder of the circuit. Under normal conditions, current flows through the current transformers, controlling circuit breakers *C* and *D*, in the same direction. Secondary current of the current transformers circulates through the pilot wire without passing through the trip coils of breakers *C* and *D*. When the line fault develops between *C* and *D*, current flows from *C* and *D* to the fault. Line current of much higher magnitude now flows through the current transformers in opposing directions. The secondary currents of the current transformers now in opposition, in the pilot-wire system, must each return through the trip coil of the circuit breakers to complete its circuit. As a result, breakers *C* and *D* open, disconnecting the faulted portion and leaving the remainder of the circuit operating but separated into two portions.

Numerous modifications of the loop circuit lower its cost and still maintain a portion of its service continuity. One modification supplies a low-voltage network system. Transformer banks at each tap feed into a low-voltage secondary network. Line circuit breakers in the position of breaker *B* at each tap, Fig. 5-3, are omitted, and the circuit breaker on each tap, such as *A*, is omitted if the network transformer bank has a low-voltage breaker in the network protector. Operation of the circuit will be the same, except that one transformer bank disconnects with each section of line.

5-5. Primary network system for metropolitan areas is suited to a wide range of load density. It sometimes shows considerable economy for

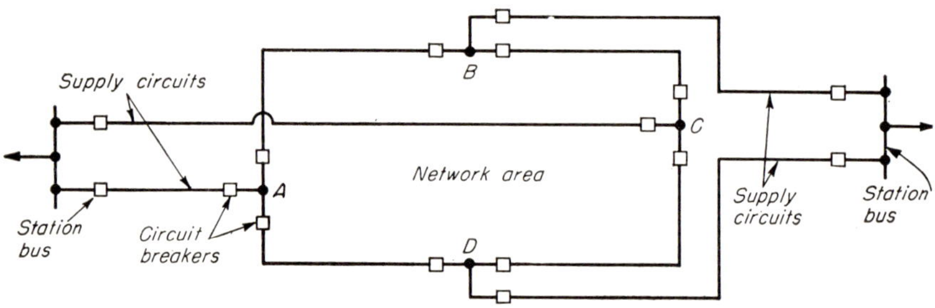

FIG. 5-4. Primary network has similarity to loop circuit but is fed from several different sources.

use in suburban and urban areas of large cities for loads ranging from the high density of commercial and industrial areas to the medium or low-load density of residential areas. A simplified version of the primary network system is shown in Fig. 5-4. Each feeder goes to a selected load center, similar to the feeder-and-main radial circuit, but the mains of all feeders interconnect to form a mesh or primary network. Each feeder is protected by circuit breakers at the substation and at the tie into the network mains. In addition, circuit breakers are installed on the network mains on each side of the main feeder taps. The network is supplied from two or more sources as a general rule. The sectionalizing circuit breakers and the multiple substation supply provide maximum continuity of service. If a line fault developed between *A* and *B*, Fig. 5-4, the circuit breakers on that section would open, isolating the faulted section and leaving the remainder of the circuit undisturbed. If a line fault should develop between *A* and the substation, the circuit breakers at the substation and at *A* would open, isolating the faulted main feeder but leaving the entire network uninterrupted and being supplied from the remaining main feeders. If one of the substations should go out of service, the remaining substation, or substations, would continue to supply the load without service interruption. Such primary networks are generally supplied from several substations.

The networked mains may have single-phase or three-phase lateral branches tapped from them, just as in the radial-type circuit. The transformers supplying load in the network area may be connected as single-phase or three-phase banks, as the load requires. The load, however, should be kept balanced between phases. Disadvantages of having large substations some distance away from the load centers together with the high cost and difficulty of regulation can sometimes be avoided by locating small substations at the load center. In the case of a 2,400-volt or a 2,400/4,160-volt primary network the substations at the area load center would probably be of 1,500- to perhaps 3,000-kva capacity and would be supplied by subtransmission feeders of perhaps 11 to 33 kv. Thus, in Fig. 5-4, substations consisting of 1,500 to 3,000 kva tap-changing underload self-regulating transformers would be located at A, B, C, and D and be supplied by one or more lines operating at a somewhat higher voltage.

The primary network system has extremely good regulation and is very reliable from the standpoint of service continuity. It provides for load growth without the need of frequent system redesign. Overloaded facilities can be relieved by adding substation capacity at or near the center of the overloaded area. It frequently shows a saving in economy of operation and over-all cost in comparison with other types of primary supply.

5-6. Secondary distribution system is the last link in the chain between the generating station and the individual consumer. The distribution transformer links the primary and the secondary systems. The secondary system includes the range of voltages at which the ultimate consumer utilizes the electrical energy delivered to him, namely, 120 or 240 volts. Large industrial plants and factories may find it more economical to utilize the energy at, say, 440 or 600 volts. In such a case, the energy would be distributed through the secondary system and the problem would be the same as for the lower voltages. Frequently energy will be needed in such quantity that economy requires its use at primary voltages. In such a case the user is a primary consumer and distribution-transformer and secondary-system links are omitted.

Essentially the circuits for secondary distribution are the same as for primary distribution, except that they are on a far smaller scale. It might be suggested that since the distribution transformer links the primary and utilization voltages the secondary system is not essential. It is true that each consumer could be supplied without a low-voltage secondary system and with a great reduction in the amount of conductor required. In built-up sections, however, with high load density, this would require a transformer on practically every pole of an overhead system or at practically every service take-off of an underground system. Such a method of secondary supply would have certain advantages, but its higher first cost and lack of operating economy makes it undesirable.

In the ideal situation transformers are located to feed into a secondary system so that the total annual cost is the lowest possible. This depends upon the load density and consumer density of the area to be served. In thinly developed rural areas individual consumers are so far apart that a single transformer rarely supplies more than one or two consumers despite high average load per consumer. No secondary circuit will be required, and the consumer will be served as shown in Fig. 5-5.

Where secondary circuits supply a group of consumers from the same transformer, they may be radial or network in type, as with primary circuits. The basic circuits most common to secondary distribution are shown in Fig. 5-6. Figure 5-6*a* shows a straight two-way feed, the next logical step after a transformer for each individual consumer. It is used where load density and consumer density is somewhat greater but not great enough so that secondary circuits are continuous. The straight two-way feed may be used for load densities and consumer densities requiring a continuous secondary feed along the streets, but in such a case, the circuits, Fig. 5-6*b* and *c*, are generally lower in investment cost and over-all annual cost per unit of load served.

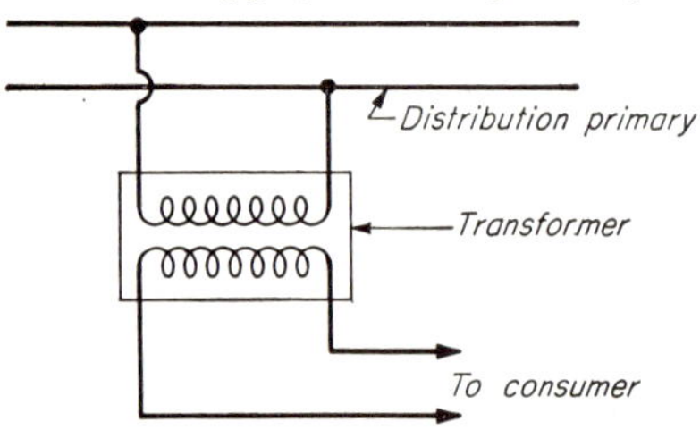

FIG. 5-5. Customer with high load demand may be fed through individual transformer from distribution primary circuit.

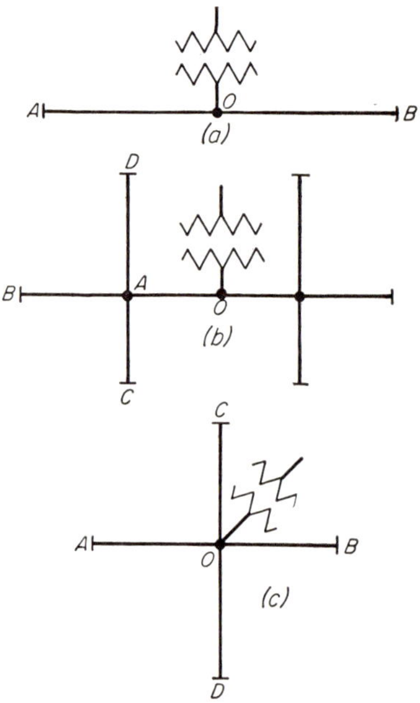

FIG. 5-6. Layout of secondary distribution lines depends on loads in area and allowable voltage drops from transformer.

Figure 5-6*b* shows a two-way feed with laterals. Figure 5-6*c* shows a four-way straight-line feed. The total length of secondary circuit is governed by the maximum allowable voltage drop between the transformer and the end of the circuit at maximum load for a given conductor size. A longer total length of secondary can be served by the circuit of Fig. 5-6*b* with the same load density, consumer density, conductor size, and voltage drop than by the circuit of Fig. 5-6*a*. Similarly, a greater length of secondary can be served from a single transformer with the circuit of Fig. 5-6*c* than with either *a* or *b*. With all conditions equal, an evenly distributed load, and a symmetrical circuit on each side of the transformer, the total length of secondary with circuit *b* will be $\sqrt{2}$ times and with circuit *c* will be two times the length of circuit *a*.

The investment and the total annual cost, for the same load density and wire size, will be lower for a straight-line four-way circuit than for a two-way circuit with laterals. In turn the latter is lower than a straight two-way feed circuit. Further, all three of the circuits in Fig. 5-6 are lower in first and annual costs than a separate transformer for each consumer or small group of consumers. The advantages of the four-way straight-line feed is difficult to achieve with overhead distribution because installation of transformers at corners or street intersections is not easy. This does not arise with underground distribution. The advantages of a four-way feed in the location of the transformer and design of the secondary circuit can be closely approached by placing the transformer one span away from the corner pole and using a larger size conductor to the corner pole. Numerous variations of the three circuit types shown in Fig. 5-6 will be found necessary to fit the local conditions as to block length, load density, customer density, etc.

5-7. Transformer Banking. Where radial-type secondary circuits, Fig. 5-6, are separate and supplied by a single transformer, instantaneous dips in voltage caused by inrush of high starting current to motors may cause highly objectionable lamp flicker on the entire circuit. The increasing use of motor-driven appliances (particularly in residential areas) and the necessity of supplying combined lighting and power from the same circuit have made the problem of overcoming lamp flicker an important one. One of the most effective and economical means of controlling lamp flicker is the banking of distribution transformers. Transformers are banked when two or more separate transformer installations, supplied from the same primary circuit, are paralleled to feed into the same secondary mains, Fig. 5-7. This provides more than one path over which the high inrush of starting current can be supplied. In the circuit of Fig. 5-7 the maximum dip in voltage occurs when a motor connects at a point halfway between two of the banked transformers. With the motor halfway between, about one half the starting current will be supplied by the transformers on each side. The voltage dip from motor starting, in this case, won't be more than one half as great as it would be if each transformer supplied its own individual secondary circuit.

The banking of transformers also lowers the normal voltage drop and saves installed transformer capacity by the effects of *diversity*. All consumers do not call for their individual maximum demands for power at the same time. The maximum demand on a transformer will be considerably less than the sum of all individual consumer maximum demands. This difference between the combined maximum demand and the sum of individual maximums is called diversity. It increases as the number of consumers increases. Thus, if three transformers of the same size with about the same demand are banked together, the combined maximum demand of the three will be somewhat less than the sum of the individual demands. As a result, the maximum demand on each is lower, the maximum voltage drop in the secondary circuit is lower, and sharp variations in secondary voltage due to current inrush and load shifts are largely

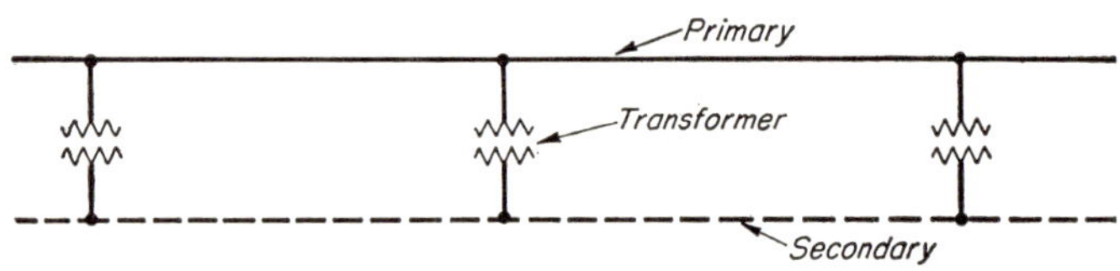

FIG. 5-7. Transformer banking helps to decrease disturbance from voltage dips, takes advantage of individual customer demand diversity.

eliminated or smoothed out. This provides a smoother voltage characteristic and a saving in transformer capacity.

5-8. Line and Transformer Protection. Any number of transformers, from 2 to as many 10 or 12, can be banked together. If the transformers are single-phase, they must all be supplied from the same phase of the primary feeder. Normally, transformers are not solidly banked together, as in Fig. 5-7, since a secondary-line or transformer short circuit would operate all the transformer protective fuses and interrupt service to all consumers. The usual method sectionalizes the banked transformers and secondary circuit by fuses, or other overload protective device, in a manner similar to that shown in Fig. 5-8.

In Fig. 5-8*a* fuses will automatically clear a line fault without removing any transformer from service and without interrupting service to any but those consumers on the faulted section of line. If a fault develops within one of the transformers, it will be automatically disconnected from the line by blowing the two secondary-line fuses and the primary transformer fuse without interrupting service to any consumer. Secondary-line fuses must be large enough to pass the total load between transformers if the adjacent transformer is disconnected. At the same time these fuses should be small enough so they will operate under minimum short-circuit current.

Transformers are available with built-in protective devices for the type of banking shown in Fig. 5-8*a*. They have thermal-actuated low-voltage circuit breakers with contacts in the transformer low-voltage leads performing the same function as the line fuses. Transformers with circuit-breaker protection in the secondary have considerable advantage over fuses for this type of banking. The size of transformer is limited by the fuse size available. In addition, the secondary circuit breaker protects the transformer from overload failure, which is impossible to provide satisfactorily with fuses. For this type of banking, the circuits of Fig. 5-6*a* and *c* are most commonly used.

If transformers are banked as shown in Fig. 5-8*b*, the fuses are located at some point away from the transformers. Either a secondary-line

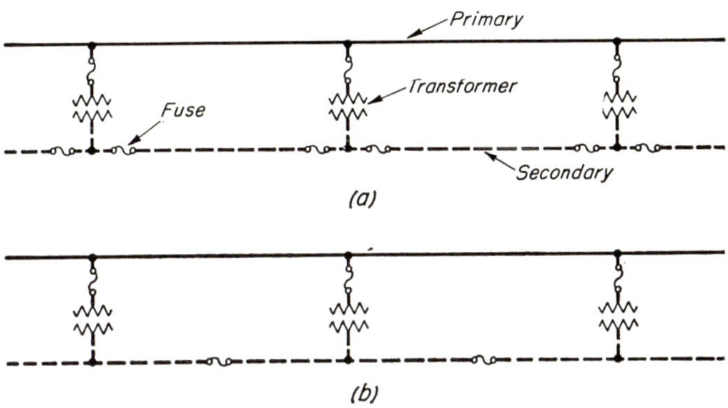

FIG. 5-8. Transformer banks must be protected by fusing or cutouts in the event of accidental short circuits or overloads.

fault or a fault within the transformer will remove transformer and secondary circuit between fuses from service, resulting in interruption of service to all consumers between fuses. The banking fuses should be located so that the voltage drop from fuse to transformer will be the same in all directions, or as nearly the same as pole or manhole spacing permits. If banking fuses can be located in this manner, only low values of current, caused by shifting of the neutral point by diversity of load demand on adjacent transformers, will flow through them. This permits smaller banking fuses and results in lower frequency of fuse operation. Banking fuses should be about equal in size to one half the rated full-load secondary current of the smaller transformer adjacent to the fuse. If these sizes operate too frequently, the fuse size can be progressively increased until the rate of operation reduces to a satisfactory level. This method of banking may be satisfactorily used with any of the circuit types shown in Fig. 5-6. If a transformer protective fuse operates without opening of adjacent banking fuses, as sometimes happens, cascading, the successive operation of adjacent transformer fuses, occurs with interruption of service to the entire banked network. This is usually caused by overloaded transformers and too large banking fuses.

Figure 5-9 shows a combination of the banking methods shown in Fig. 5-8 using the circuit shown in Fig. 5-6*b*. This method of banking is particularly useful with transformers having built-in secondary circuit breakers and if local topography will permit. Built-in secondary circuit-breaker protection combined with smaller intermediate banking fuses reduces the hazard of cascading, reduces the service interruption from a secondary line fault to one half the customers served by the transformer, and increases the permissible size of transformer bank at each installation.

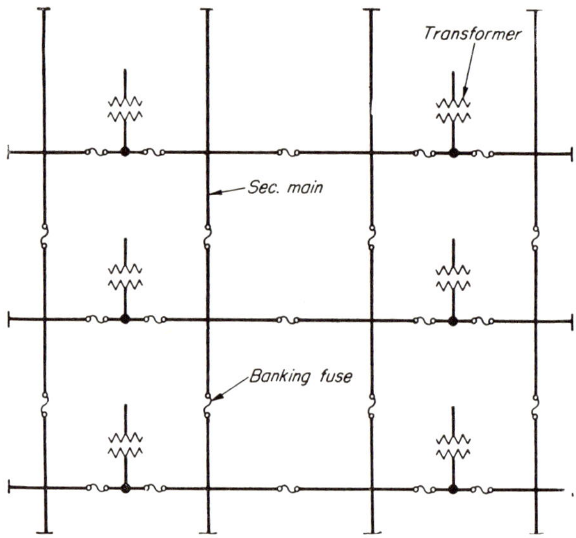

FIG. 5-9. Banked secondary network should be protected by fuses

The best results will be achieved if the same or nearly the same size transformers are used throughout the banked group. If this is not feasible, the ratio between the largest and smallest sizes should be kept to not more than 1.67 if possible. Fuses and thermal-actuated circuit breakers should be coordinated from the banking fuse farthest from the particular transformer through the transformer primary fuse to assure that protective devices will operate in their proper sequence.

Banked secondary systems may be either single-phase or three-phase, but with all conditions equal three-phase will generally be slightly higher in cost. For single-phase circuits the number of single-phase transformers banked together is generally limited to four or five to keep the load balanced on the three primary phases. When an appreciable number of consumers requiring three-phase service are scattered over the area, three-phase secondary circuits, using three-phase transformer banks, banked together will often be lower in first cost and more economical than single-phase. This is particularly true when the three-phase power demand of the different consumers is comparatively small. In any case the comparative over-all cost for banked secondaries will be lower than for indi-

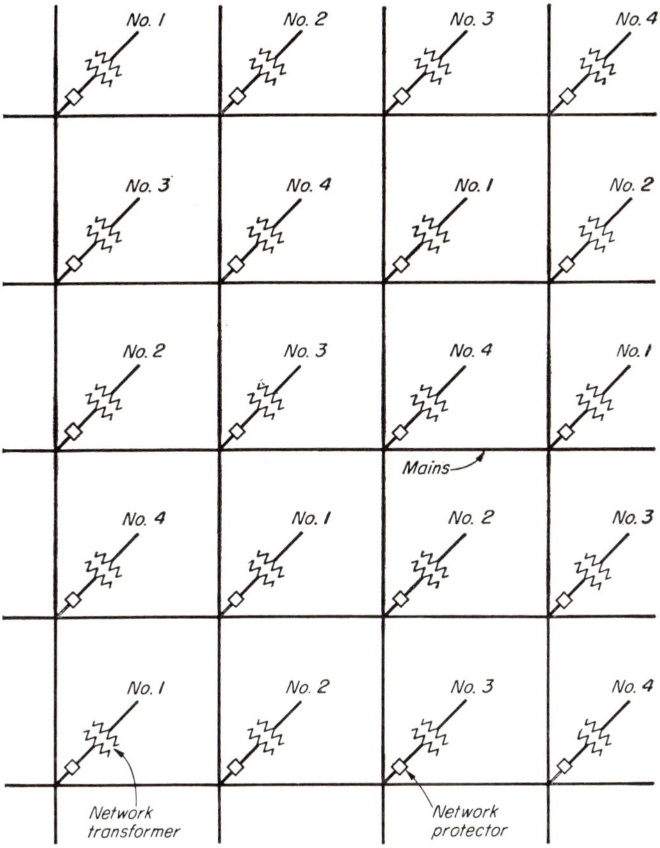

FIG. 5-10. Low-voltage network may be fed from four different primary feeders, may be protected by network protectors or circuit breakers.

vidual radial-type secondary circuits for medium load densities and for either single- or three-phase circuits.

5-9. The low-voltage network system is similar in principle to the banked-secondary system except that it is larger in capacity and more extensive in its circuit ramifications. In the low-voltage network system the secondary mains form a solidly connected secondary mesh or grid, Fig. 5-10. The network system of low-voltage distribution is used in high load-density areas such as the downtown business section of cities and large towns where tall office buildings, hotels, theaters, stores, etc., have a high average load density per block or per unit of street length. The network system of low-voltage distribution is particularly well adapted to areas where underground distribution avoids overhead congestion. In medium load-density areas where the high cost of underground distribution is not economically feasible, an overhead low-voltage network system, similar to an underground network, may be found desirable.

The low-voltage grid forming the network system is supplied by two or more primary feeders. These primary feeders, usually radial in type, feed into the low-voltage network through network-type transformers. The primary feeders interlace so that no two adjacent transformer banks are supplied from the same feeder. This is done so that if one primary feeder goes out of service the remaining feeders can supply the load without overloading and without excessive voltage drop. The network must be designed so that at least one of its primary feeders, together with its transformers, is in the nature of spare capacity. The tie points at which the transformers feed into the network should be located so that the load is divided evenly between banks and so that maximum voltage drop between transformer and the electrical mid-point between banks will not be excessive. Figure 5-10 shows a small network grid supplied by four primary feeders. The network is symmetrical, and the transformers feed into it at street intersections where the current divides in four directions. The number at each transformer indicates the feeder, No. 1, 2, 3, or 4, supplying the transformer bank. If the network is properly designed, it will maintain adequate and satisfactory service when any one of the four primary feeders are out of service. In this case the spare capacity equals 25 percent of the total. In the case of a large network system, with many primary circuits, several of the primary circuits together with their transformer banks provide spare capacity.

5-10. Network Protectors. Connection between low-voltage winding of the transformer and the grid is made through a network protector. The latter consists of a low-voltage air circuit breaker and the relay controlling it. It may be mounted on the transformer or be separately mounted. The relay equipment consists of a reverse-power relay which opens the breaker and a phasing relay to close the breaker. The master relay must have its closing contacts in the closed position as well as the phasing relay before the circuit can close.

The primary feeders supplying the network are protected by breakers at the substation. A line short circuit on one of the primary feeders causes high current flow from the substation bus to the fault. This opens the substation feeder circuit breaker. Simultaneously, current flows to the feeder fault from the network through the transformers normally supplied by the faulted feeder. This reversal of power flow causes the network protector circuit breakers to open. This completely disconnects and isolates the faulted feeder without interrupting service to any of the consumers on the network. When repairs are made, the faulted feeder can be restored to service by closing the substation feeder circuit breaker, and the protector relays will close their circuit breakers at the instant the feeder synchronizes with the network. The transformer voltage must be about 2 volts higher than the network voltage so that power flow will be from the transformer to the network and not the reverse. The reverse power relay is sufficiently sensitive so that it operates the protector circuit breaker with currents as low as the transformer exciting current.

A three-position switch on the high-voltage side of the network transformer assures greater safety in working on primary cables or transformers by disconnecting a transformer from its feeder. This switch, manually operated, may be thrown to closed, open, or grounded position. The grounded position protects workmen against the hazard of reenergized equipment through accidental reclosing of the feeder circuit breaker or the network protectors.

There is no overcurrent protection on the high-voltage side of the network transformer, as for transformers supplying radial secondary circuits, to automatically disconnect a single transformer under fault conditions. The network protector, however, contains fuses in addition to its relay-controlled circuit breaker. The fuses act as back-up protection to disconnect the transformer from the network in case of failure of the network protector under fault conditions. In case of a low-voltage phase-to-phase or phase-to-ground short circuit, the network is usually designed to burn itself clear without using sectionalizing fuses or other overload protective devices. Burning clear of a faulted secondary-network cable refers to burning away of the metal forming the contact between phases or from phase to ground until the low voltage of the network system can no longer maintain the arc. The system must be capable of delivering high current values to the fault to achieve rapid clearing. The larger the cable, the higher the short-circuit current required to burn the faulted cable clear. For this reason cables of about 500 mcm are usually about the largest used for secondary-network mains. Burning clear of the low-voltage fault damages the cable, and if conditions are not just right a considerable length of cable beyond the fault may be damaged or destroyed. To assure rapid clearing of such faults without excessive destruction of cable, current limiters are sometimes used. The limiter, a high-capacity fuse, is installed in each phase conductor at junction points in manholes. It opens and isolates the faulted cable before any appreciable amount of cable is destroyed. The limiter should be large enough to pass the normal network load current without melting and small enough to operate and clear the faulted section of cable quickly before the network protector fuses blow.

Primary feeder voltages supplying the network vary from 2,300 to 33,000 volts. In large cities with an extensive network system, the higher primary voltages are advantageous because a smaller number of feeders will be needed and one substation transformation may be avoided by taking feeders directly from the subtransmission system. The transformers are usually three-phase units designed for network service. They vary in size from 150 to 1,000 kva on underground networks. The 300- and 500-kva units are probably the most frequently used sizes in street-type vaults, the smaller sized units being used for lower load-density areas, and the larger units being used where the concentrated load is too large for service from street mains and to avoid an excessive number of transformer banks at some points. The normal three-phase or single-phase pole-type transformer may be used for overhead networks in sizes ranging from about 75- to 150-kva banks. The transformers and network protectors used in underground systems may be obtained in submersible or nonsubmersible types.

Load in high-density network areas will be composed of both single- and three-phase loads. Practically all network systems are three-phase four-wire wye-connected and consist of three insulated phase conductors and a bare, solidly grounded neutral conductor. The secondary mains usually consist of single-conductor cables which may be either metallic or nonmetallic sheathed. In recent years nonmetallic-sheathed rubber-insulated cables have gained favor because of their improved insulating value and because there is less chance of phase-to-ground faults developing with nonmetallic sheath. The most frequent size of cable for secondary mains is in the neighborhood of 4/0 or 250 mcm. With these sizes, clearing faults is considerably improved, but in heavy load-density areas two or more sets of paralleled mains may be needed.

Secondary networks are in operation at nominal wye voltages of 115/199, 120/208, 122/211, and 125/216. The great majority, however, are designed to operate at four-wire 120/208 volts wye. There is an established trend, however, to operate networks at 125/216 volts which will probably supply the consumer's meter at about 123/213 to 120/208 volts. Some systems supply the network from more than a single substation or transmission source if available, particularly for an extensive network, so that the loss of a single supply source will not cause a complete loss of service to the network area. In an extensive network system it is generally best to provide nonautomatic sectionalizing so that it can be separated into sections to facilitate restoration of service after complete service interruption. It would be impossible to reclose all feeder circuit breakers and network protectors simultaneously in restoring service to a large network system, without tripping feeder circuit breakers and blowing network protector fuses. Service can be restored to each small section and the sections rejoined at the sectionalizing points to restore the entire network to normal. Load growth is accommodated by adding primary feeder and transformer facilities without appreciably changing the system of secondary mains.

5-11. Distribution-system Voltage Spread. There is no established and inflexible limit as to what the secondary-service voltage to supply lights and small power shall be. For residential and commercial lighting service, the nominal secondary-voltage supply is usually kept within 115 to 125 volts. A majority of utility systems base their supply facilities upon a nominal secondary lighting voltage of 120 volts to neutral or a voltage of 240 volts between the outside wires of a three-wire single-phase secondary circuit. Lights and appliances are usually given a rating of 120 volts, although nominal ratings of 115 or 125 volts may be obtained for special conditions. Small single-phase motors are generally given a nominal rating of 115/230 volts. Polyphase motors to be supplied from the mains of a low-voltage secondary system are generally nominally rated at 208, 220, or 440 volts.

Too wide a variation between nominal voltage rating of lights, appliances, or motors and actual service voltage is not desirable. For ordinary incandescent lamps, for instance, a reduction in voltage to 95 percent of lamp rating reduces lumen output to about 85 percent and the wattage to 91 percent of normal. If the supply voltage increases to 105 percent of nominal rating, a corresponding increase in lumen output and lamp wattage occurs, together with a reduction in average lamp life of about 50 percent. The effect of voltage variation from rated values on fluorescent lamps and on appliances and motors will be less severe, but such differences should be kept within reasonable limits.

The range of voltage variation normally allowable in a distribution system is determined by the highest voltage permissible at the service-entrance switch of the first consumer, off the first transformer on the feeder, and the lowest voltage permissible at the service-entrance switch of the last consumer, on the end of the secondary circuit fed by the last transformer on the farthest end of the primary feeder. These comparative voltages are expressed in terms of the nominal lighting voltage. A variation of about ± 3 percent from nominal rating, or a spread of about 6 percent in secondary voltage, is the normal maximum variation in secondary voltage sought in good distribution design. Thus, if the nominal distribution lighting voltage is 120, the highest voltage of the first consumer should not exceed 124, while the lowest voltage of the last consumer should not be less than 116. Another way of expressing the allowable voltage spread would be to give the allowable voltage drop in distribution transformers, in the distribution primary between the first and the last transformer, and in the distribution secondary circuit in terms of the nominal secondary voltage. Thus, if a voltage drop of 3 percent were allowed in the transformer, 3 percent in the secondary circuit, and 4 percent in the primary circuit between the first and the last transformer, the total voltage spread, in terms of nominal secondary supply voltage, would be 10 percent. More than 124 volts at the first consumer on the line will cause serious reduction in normal rated life of 120-volt lights, while less than 116 volts will cause serious reduction in operating efficiency and revenue. Motors are not so seriously affected by voltage variation as lamps, being designed to operate satisfactorily within a ± 10 percent range of nominal voltage rating. Even with motors, however, trouble will frequently arise in the operation of 220-volt single-phase or three-phase motors on a three-phase four-wire 120/208-volt wye secondary supply if phase-to-neutral potential drops much below 120 volts.

5-12. Distribution-line Voltage Drop and Regulation. Resistance, inductance, and capacitance are present in all electric circuits. Because of the comparative short length of distribution primary and secondary circuits, however, line capacitance is negligible. In comparison, line resistance and inductance, even in the short lengths of secondary circuits, exert a powerful influence on the relation of line currents and voltages.

A simple ac circuit can be represented as in Fig. 5-11*a* in which

R_L = resistance of the load, ohms
R = resistance of the line conductors supplying the load, ohms
X_L = inductive reactance of the load, ohms
X = inductive reactance of the line conductors supplying the load, ohms
E_1 = voltage impressed across the circuit at the source, volts
E = voltage impressed across the load at the end of the line, volts
I = current flow through the circuit, amp

The time-phase relation of currents and voltages in the circuit is shown in Fig. 5-11*b*. Here, IR_L and IR represent components of impressed voltage, in phase with the current, necessary to overcome the resistance in the load and line conductors, respectively, while IX_L and IX represent components of impressed voltage, 90° out of phase with the current, necessary to overcome inductive reactance in the load and line conductors, respectively. The voltage impressed across the load equals the vector sum (Art. 1-25) of IR_L plus IX_L, while the voltage at the source E_1 equals the vector sum of all the component parts.

Voltage drops across resistances or reactances are in opposition to the impressed voltage consumed in overcoming their effects. Thus, a component of impressed voltage consumed in forcing a current through a resistance is shown as being in phase with the current, while a voltage drop across a resistance is shown as being in opposition to the current, or 180° out of phase. Likewise, a component of impressed voltage consumed in forcing a current through an inductive reactance is shown as leading the current by 90°, while a voltage drop across an inductive reactance is shown as lagging the current by 90°. The arithmetic sum of the voltage drops across the resistances and inductive reactances of an ac

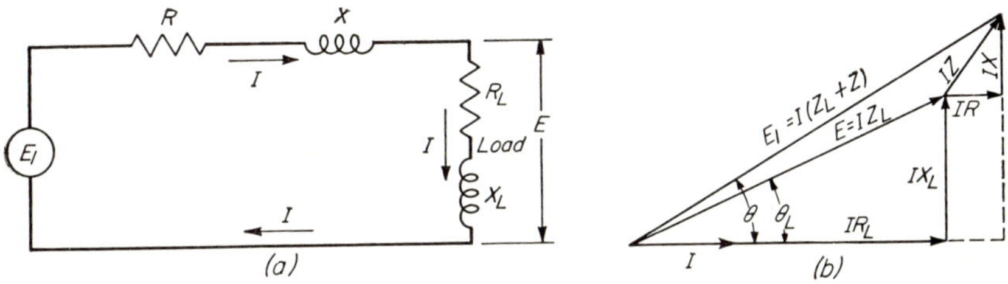

FIG. 5-11. Voltage and current relations in a simple ac circuit.

circuit will exceed the value of impressed voltage due to the 90° phase-angle difference. Figure 5-11*b* shows the voltage impressed across the load equal to

$$E = \sqrt{I^2R_L{}^2 + I^2X_L{}^2} = I\sqrt{R_L{}^2 + X_L{}^2} = IZ_L$$

and the voltage at the source equal to the vector sum of the voltage at the load plus the voltage across the line impedance, or

$$E_1 = \bar{E} + \overline{IR} + \overline{IX} = \bar{E} + \overline{IZ}$$

The vector sum of E plus IZ differs from the arithmetic sum except when the ratio of R/X equals ratio of R_L/X_L. The power-factor angle at the source may be greater or less than the power-factor angle at the load, depending upon the ratio of reactance to resistance in the line as compared to the ratio of X to R at the load. If the current vector is taken as the reference axis, Fig. 5-11*b*, the component of voltage in phase with the current equals $E \cos \theta$ while the component of voltage 90° out of phase with the current equals $E \sin \theta$.

The arithmetic difference between E_1 and E, Fig. 5-11*b*, is the voltage drop in the circuit because of the resistance and inductive reactance of the line conductors between source and load. The voltage regulation of a distribution line, at any point, is the difference between maximum and minimum voltage at that point. Regulation is ordinarily considered to be the difference between the no-load and full-load voltages at the end of the circuit. For the ordinary distribution circuit with lagging power factor, voltage drop and voltage regulation are synonomous. Referring to Fig. 5-11*b*, voltage drop in the line equals

$$V = E_1 - E = \sqrt{(E \cos \theta_L + IR)^2 + (E \sin \theta_L + IX)^2} - E \text{ volts}$$

The voltage drop, or regulation, may also be expressed in percent, as follows:

$$\% V_L = \frac{(E_1 - E)100}{E}$$

$$= \left(\frac{\sqrt{(E \cos \theta_L + IR)^2 + (E \sin \theta_L + IX)^2}}{E} - 1\right) 100$$

in percent of the voltage at the load, or

$$\% V_s = \frac{(E_1 - E)100}{E_1}$$

$$= \left(1 - \frac{\sqrt{(E_1 \cos \theta - IR)^2 + (E_1 \sin \theta - IX)^2}}{E_1}\right) 100$$

in percent of the voltage at the source.

Figure 5-11*b* represents the condition when the voltage at the load is known and we must find the voltage at the source needed to deliver the required voltage E at the load. In this case, the IR drop and the IX drop in the line are added vectorially to the load voltage to find the voltage required at the source.

In many cases the voltage, current, and power factor at the source are known and the voltage at the load or the voltage drop in the line needs to be figured. The resistance drop in the line, which lags the current by 180°, is subtracted from the component of source voltage in phase with the current and the reactance drop in the line, which lags the current by 90°, is subtracted from the component of source voltage which leads the current by 90°, as shown in the vector diagram of Fig. 5-12*a*, to determine how much the voltage drop in the line has reduced the voltage at the load. The difference between voltages at the source and the load equals the voltage drop in the line, the line regulation. Figure 5-12*b* shows an enlargement of the impedance triangle of Fig. 5-12*a*. Voltage drop in line resistance and line inductance very closely approaches

$$V = IR \cos \theta + IX \sin \theta = I(R \cos \theta + X \sin \theta)$$

The value $R \cos \theta + X \sin \theta$, sometimes called the effective impedance of the line, represents a constant. When multiplied by the current flowing in the line, it very closely approaches the voltage drop for a given power factor. The error is so small as to be negligible in distribution-line calculations.

5-13. Two-wire Single-phase Circuit. Figure 5-13 shows a two-wire single-phase line supplying loads L_1, L_2, L_3, etc. The voltage drop in the line is the difference between E_1 at the source and E at the end of the line. Voltage drop between source and the first load L_1 equals the voltage loss in line wires AB and A_1B_1. The total voltage drop, from source to the end of the line, equals the sum of the voltage drops in wires AB, A_1B_1, BC, B_1C_1, etc., to GG_1. If the wire from A to G is the same size as the wire from A_1 to G_1, the total voltage drop equals twice the voltage

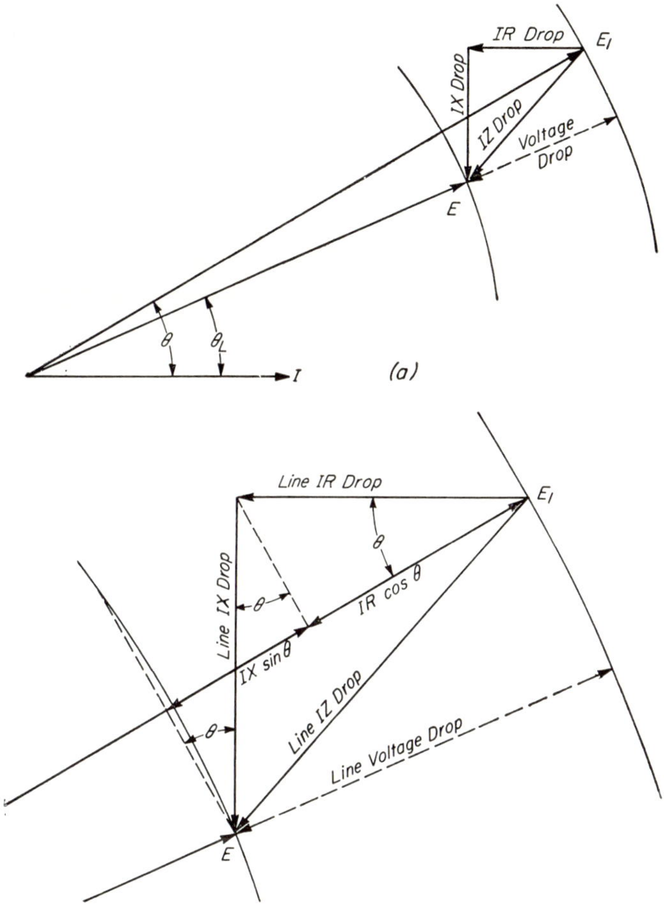

FIG. 5-12. Voltage drop from line impedance can be calculated by vector relations, involves more than simple arithmetic.

drop in one wire to the end of the line. The total current flowing in the line from the source equals the vector sum of all the load currents, or

$$\bar{I} = \bar{I}_1 + \bar{I}_2 + \bar{I}_3 \ . \ . \ .$$

in vector relation. The greatest relative voltage drop, per unit of line length, will occur between AA_1 and BB_1.

If the load, of such a circuit as shown in Fig. 5-13, is evenly distributed, that is, decreases at a constant rate throughout the length of the line, the voltage drop will be the same as if all the load were concentrated at a point one-half the line length. If the product of load L_1 times its distance from the source is added to the product of load L_2 times its distance from the source, etc., to the end of the circuit, the result will equal the equivalent load distance for the entire circuit. If this value is divided by the total circuit load, measured at the source, the result will equal the distance from the source at which the entire load would have to be concentrated for the voltage drop to be equal to the actual drop, assuming the same wire size throughout. With different wire sizes the result will equal the *equivalent length* for each size with a concentrated load at the end of the composite circuit.

Thus, if loads L_1, L_2, L_3, etc., are measured in amperes and the distance between loads is measured in feet, the load-distance value will be in ampere-feet. When divided by the current in amperes, this will give

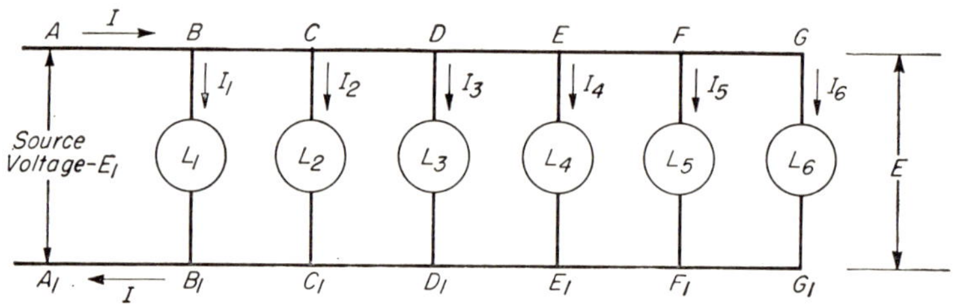

FIG. 5-13. Voltage drops progressively at the loads as they are farther from the source.

the distance from the source at which a load of I amp could be served so that the voltage drop would be equal to that from AA_1 to GG_1 of the circuit shown in Fig. 5-13. The result will be the same if the load-distance value is expressed in ampere-feet, kva-feet, kva-miles, etc. In making calculations of voltage drop, or line regulation, the source may be taken as any point from which the voltage drop is required. In the case of primary circuits the source is usually taken to be the substation supplying the circuit or, sometimes, the first transformer on the circuit. For secondary circuits the source is usually taken to be the distribution transformer supplying the circuit.

Example: A single-phase primary circuit feeds three single-phase loads in Fig. 5-14. The loads take a total of 130 amp, voltage at the source is 2,500 volts, the line wires are No. 2 copper spaced 59 in. apart, the power factor at the source is 85 percent, and the distances in feet as shown. What is the distance to the equivalent load center? What is the voltage at the end of the line? What is the line voltage drop?

Solution:

1. Source to load 1 = 130 amp × 1,000 ft = 130,000 amp-ft
   Load No. 1 to load 2 = 90 amp × 2,220 ft = 198,000 amp-ft
   Load No. 2 to load 3 = 30 amp × 2,500 ft = 75,000 amp-ft
   Total = 403,000 amp-ft
   Distance to equivalent load center = 403,000 amp-ft ÷ 130 amp = 3,100 ft

2. Resistance per 1,000 ft of No. 2 copper (Table 4-3) = 0.176 ohm
   Reactance per 1,000 ft of No. 2 copper, 59 in. spacing (Fig. 4-21) = 0.145 ohm
   $E_1 \cos \theta = 2{,}500 \times 0.85 = 2{,}125$ volts
   $E_1 \sin \theta = 2{,}500 \times 0.527 = 1{,}318$ volts
   $IR = 3.1 \times 0.176 \times 130 \times 2 = 142$ volts
   $IX = 3.1 \times 0.145 \times 130 \times 2 = 117$ volts
   Voltage at end of line $= E = \sqrt{(E_1 \cos \theta - IR)^2 + (E_1 \sin \theta - IX)^2}$
   $= \sqrt{(2{,}125 - 142)^2 + (1{,}318 - 117)^2} = 2{,}318$ volts

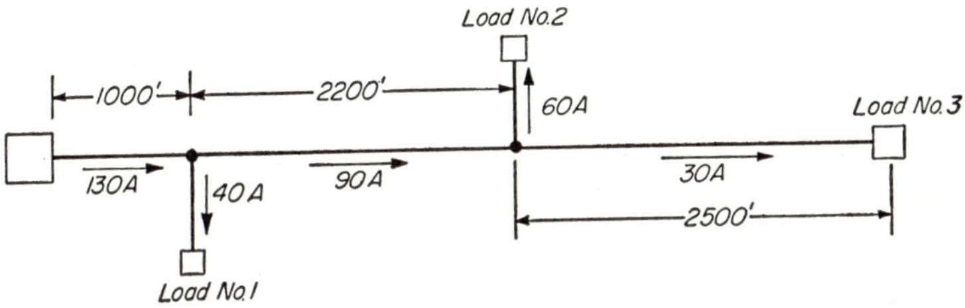

FIG. 5-14. Circuit diagram for example in voltage-drop calculations.

3. Voltage drop in line $= E_1 - E = 2{,}500 - 2{,}318 = 182$ volts, or
   $V = I(R \cos \theta + X \sin \theta)$
   $= 130[(2 \times 3.1 \times 0.176 \times 0.85) + (2 \times 3.1 \times 0.145 \times 0.527)] = 182$ volts
   $\% V_L = (182 \div 2{,}318)100 = 7.85\%$ of voltage at end of line
   $\% V_s = (182 \div 2{,}500)100 = 7.28\%$ of source voltage

5-14. The three-wire single-phase circuit is, in effect, two two-wire single-phase circuits, of the same phase, combined so that one wire is common to both circuits. The combined two-wire circuits also comprise

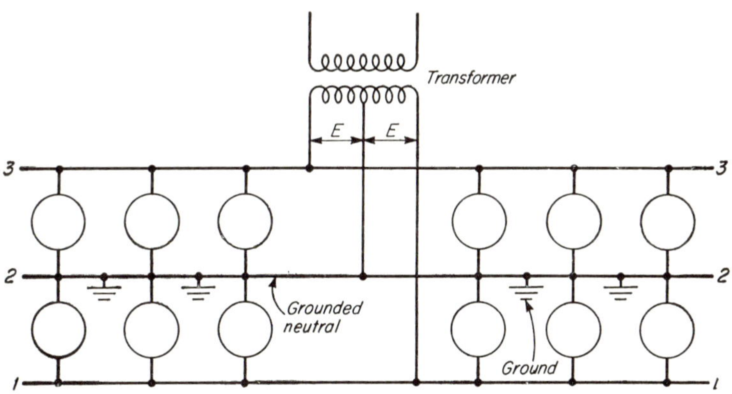

FIG. 5-15. Three-wire single-phase circuit fed from transformer is really two two-wire circuits.

a third two-wire circuit. This type of circuit is common to single phase secondary distribution where the three wires of a 120/240- or 115/230-volt secondary circuit feed from a single-phase transformer to supply lights, appliances, single-phase motors, etc., which have a nominal rating of 115 or 120 volts, Fig. 5-15. Such a circuit permits three two-wire single-phase circuits to be supplied from a single transformer with a total of three wires instead of six. The common wire of such a circuit, the neutral, is usually grounded. The voltage at the transformer, measured between the neutral wire and each of the other two wires, is the same for both of the two 120-volt circuits.

If a load connects between one of the phase legs and the neutral, the current flowing out over the phase leg, at any instant, returns through the neutral. If the current flows into the transformer from the phase leg, it will flow out from the transformer through the neutral wire. Owing to the opposing polarity of the phase legs of a transformer connected for 120/240 volts three-wire, the current will be in opposing directions, at any instant, in the two legs. Thus, when current flows out from the transformer through one of the legs, at any instant, it will flow into the transformer through the other leg. The current flowing in the neutral wire will equal the difference between the currents flowing in the two legs since the neutral wire acts as the return for both 120-volt circuits. The direction of net current flow in the neutral wire depends on the direction of current flow in the phase leg carrying the maximum current at the instant. If loads are connected for 240 volts, across the two phase legs, current which flows to the load over one leg returns over the other leg without passing through the neutral. If the same amount of load connects between each phase leg and the neutral, no current returns through the neutral leg and the load is said to be balanced.

Voltage drop in distribution circuits is frequently expressed in terms of what its effect will be on a 120-volt circuit, or of its effect on lighting voltage. Voltage drop in the three-wire single-phase circuit is expressed in terms of its effect on one of the 120-volt circuits. If the load is balanced, the voltage drop in the two phase legs will be equal, with the same size wire, and no voltage drop will occur in the neutral wire. If the load is unbalanced, the maximum voltage drop occurs in the 120-volt circuit supplying the largest load and consists of the voltage drop in the leg carrying the greatest current plus the voltage drop in the neutral. With unbalanced load, direction of current flow in the neutral wire will be opposite to the flow in one of the phase legs and in the same direction in the other leg, at any instant. The maximum voltage drop occurs in the phase leg and neutral in which the direction of current flow is opposite. Voltage drop in the neutral wire reduces the voltage drop in the lighter loaded 120-volt circuit. If the unbalance of load between two legs of a three-wire circuit is great enough so that the current flowing in the neutral is larger than the current in one of the legs, the voltage measured between the lighter-loaded leg and the neutral, at the end of the circuit, may be greater than the source voltage. This gives the effect of a negative voltage drop.

Where single-phase secondary circuits extend several spans beyond the transformer pole, in overhead circuits, or past several service take-off points, in underground circuits, the load may be expressed in terms of one of several load-distance units convenient for voltage-drop calculations. It may be expressed as kva per pole or as amperes per pole, or it may be expressed in terms of services where the load is reduced to amperes or kva per average service. In any case it is important to keep the load as nearly balanced as possible to provide the best possible regulation and to conserve investment in transformer and secondary-circuit capacity.

5-15. Three-phase Circuits. In Art. 1-27 it was shown that three-phase circuits may be either delta or wye connected. The designation refers to the manner in which substation transformers or the windings of the generator supplying the circuit are connected, in the case of a primary circuit, or to how the distribution transformers are connected in the case of three-phase secondary circuits. Three-phase circuits may be thought of as three separate single-phase circuits. How the separate single phases are connected to use a common wire with one or both of the other phases determines whether the circuit is delta or wye.

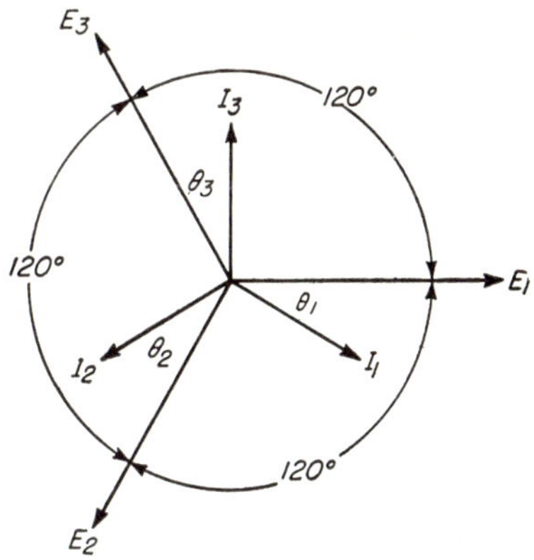

FIG. 5-16. Voltage and current relations in three-phase wye circuit.

5-16. Three-phase Wye Circuits. In the three-phase wye circuit, one terminal of each phase winding of the generator or transformer bank connect together so that all three phases have one common terminal and so that the three-phase voltages are 120 electrical degrees apart, Fig. 5-16. The currents in each phase bear a power factor relation to their single-phase voltage determined by the total resistance and reactance of the single-phase circuit. In Fig. 5-16, phase current I_1 lags θ_1 degrees behind voltage E_1, current I_2 lags θ_2 degrees behind voltage E_2, etc. The phase difference of 120° between voltages E_1, E_2, and E_3 is determined by the time-phase relation of the different phases and the manner of terminal connections at the source, while the phase difference between the currents is controlled by the phase relation of the voltages and by the ratio of reactance to resistance in each single-phase circuit. The currents may differ in phase by more, or less, than 120°, depending upon the impedance characteristics of each circuit. If the loads connected to each phase at a given point are equal, the load is said to be three-phase. If the loads connect from phase wire to phase wire, the three-phase load is said to be delta connected. If the loads connect from the three-phase wires to a common neutral point, the three-phase load is said to be wye connected. The common neutral point of the wye-connected load may, or may not,

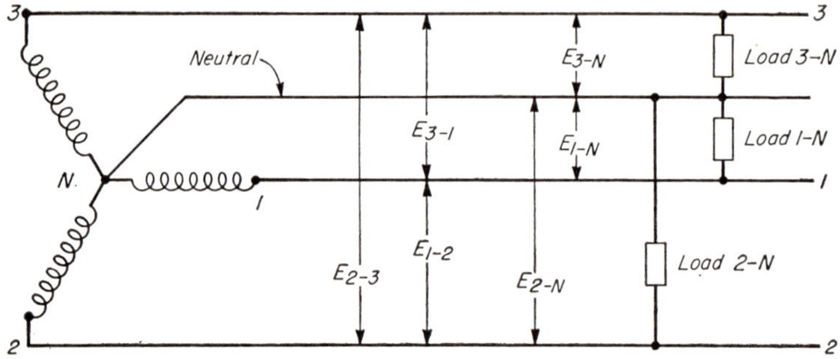

FIG. 5-17. Three-phase wye circuit with neutral brought out to form four-wire distribution circuit.

be connected to the neutral wire of a wye-connected circuit. Polyphase loads are balanced and cause balanced currents to flow in the phase wires with no current flowing in the neutral of a wye-connected circuit regardless of whether the load is wye or delta connected.

Wye-connected distribution circuits usually consist of three-phase wires and a common return neutral wire. Single-phase transformers, or other similar single-phase loads, are usually connected from one of the phase wires to the neutral wire, Fig. 5-17. The voltage rating of transformers or other equipment connected in this manner is lower than when connected between phase wires, being equal to $E_{LL}/\sqrt{3}$. In a three-phase four-wire wye circuit, Fig. 5-17, no current will flow in the neutral wire if loads 1-N, 2-N, and 3-N are equal since the current in phases 1, 2 and 3 is balanced, and hence the vector sum of the return current for all three phases in the neutral wire equals zero. If the loads are not equal, a net return current flows in the neutral wire. Assuming that the current in all the wires, including the neutral wire, flows from the source toward the load at any instant, then, according to Kirchhoff's first law, for ac,

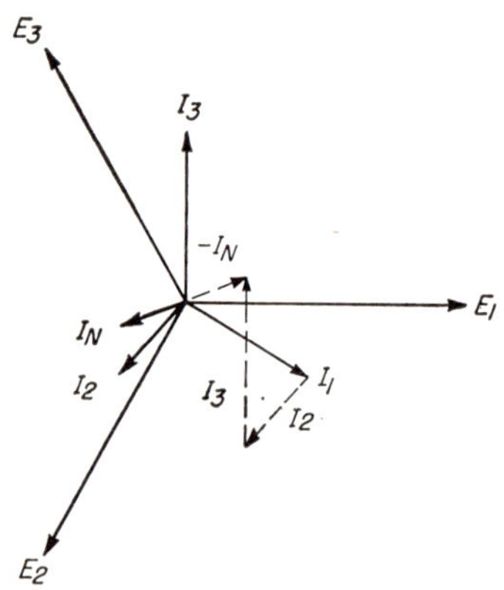

FIG. 5-18. To find current in neutral, the three-phase currents are added vectorially. Closing vector $-I_n$ equals neutral current.

$$\bar{I}_1 + \bar{I}_2 + \bar{I}_3 + \bar{I}_N = \bar{0}$$

in vector form, or

$$\bar{I}_1 + \bar{I}_2 + \bar{I}_3 = -\bar{I}_N$$

as shown in Fig. 5-18.

If the loads are balanced, so that no current flows in the neutral, the voltage regulation of the line will be due to its IR and IX drops. If loads 1-N, 2-N, and 3-N are equal, the current in wires 1, 2, and 3 will be equal, and if the wires are the same size, the voltage drop in each wire will be the same. In this case, the voltage drop is calculated as for a single-phase circuit, except that the line resistance and reactance is for a single wire since the neutral wire, the return of the single-phase circuit, carries no current. The voltage drop in a balanced three-phase four-wire wye

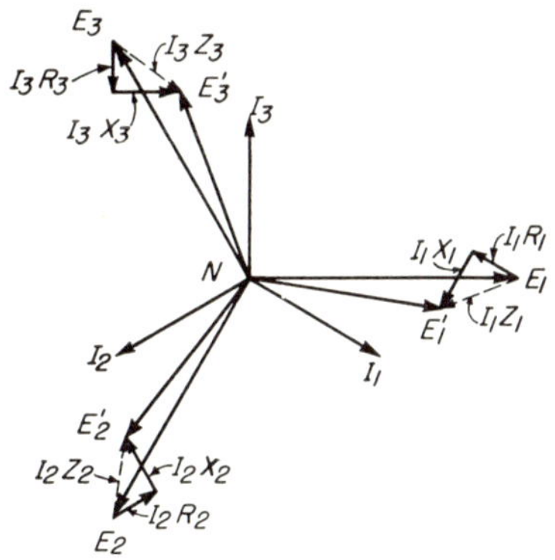

FIG. 5-19. Voltage drop in wye circuit with balanced load.

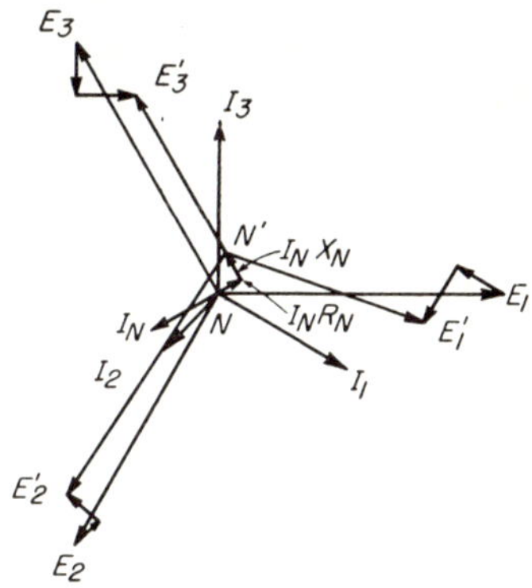

FIG. 5-20. Voltage drop in unbalanced wye circuit.

circuit is shown in Fig. 5-19. The vector diagram for one phase is identical with that of Fig. 5-12 for a two-wire single-phase circuit except that the IR and IX drop of Fig. 5-19 is for a single wire.

If the load is unbalanced, unequal currents flow in the phase wires and current will flow in the neutral wire. Voltage regulation will then be affected by the IR and IX drop in both the phase wires and the neutral, Fig. 5-20. Current flow in the neutral shifts the neutral point from N to N' as shown. Usual practice for a wye-connected primary circuit makes the neutral wire common with the grounded neutral of the various secondary circuits. In other words, the same wire is used as a neutral for both primary and secondary circuits. The neutral wire is grounded in many places, and any current flowing in the neutral returns to the source through both the neutral and the earth since the two are in parallel.

With unbalanced load on a four-wire wye circuit, the voltage drop is the difference between the phase-to-neutral voltage at the source and the phase-to-shifted-neutral point at the load, or for example, the difference between E_1 and E'_1, for the drop in phase 1, of Fig. 5-20. Similarly, the voltage drop in phase 2 equals the difference between E_2 and E'_2, while the voltage drop in phase 3 equals the difference between E_3 and E'_3. The effect of neutral current on voltage drop is not the same for all phases. Under certain conditions the voltage drop in the neutral may cause a negative voltage drop with reference to one of the phases so that there is a *rise* in voltage between source and load. This condition might result, for instance, if current unbalance in the phase wires and the impedance of the neutral-wire circuit is great enough so that the voltage drop in the neutral appreciably exceeds that in one of the phase wires. Voltage drop is difficult to compute for unbalanced load and current flow in the neutral. The voltage drop may be computed graphically by a vector diagram, Fig. 5-20. Usually, satisfactory results can be obtained by averaging the unbalanced line currents and computing the voltage drop as though the loads were balanced and there was no shift in the neutral. Slight error results from this method unless the loads are so out of balance that a relatively large current flow in the neutral causes a large voltage drop.

Example: A primary circuit arranged as shown in Fig. 5-17 has loads 1-N, 2-N, and 3-N draw 220 amp from phase 1, 172 amp from phase 2, and 190 amp from phase 3. The distance from source to load is 5,000 ft. The wire size is No. 00 stranded copper, including the neutral. The source voltage is 2,460 phase-to-neutral and 4,261 volts between phase wires. The power factor at the source is 90 percent. The phase wires are mounted on crossarms, with flat spacing, so that separation between wires 1 and 2 is 14.5 in., between wires 2 and 3 is 44.5 in., and between wires 3 and 1 is 59 in. (1) What is the current flowing in the neutral return? (2) What is the average voltage drop in the circuit? (3) What is the average voltage at the load?

Solution: 1. Since the current in lines 1, 2, and 3 are each at 90 percent power factor, they may be represented by three vectors 120° apart, as shown in Fig. 5-21. If the vector representing current I_1 is made the reference vector along the *x axis*, then the vectors for currents I_2 and I_3 may be represented by components along the x and y axes. Thus, the component of I_2 along the x axis equals *oa*, while the component of I_3 equals *ob*, or

$$oa = I_2 \cos 240° = I_2(-\cos 60°) = 172(-0.5) = -86 \text{ amp}$$
$$ob = I_3 \cos 120° = I_3(-\cos 60°) = 190(-0.5) = -95 \text{ amp}$$

In a similar relation the components of I_2 and I_3 along the y axis, equal to *oc* and *od*, respectively, are

$$oc = I_2 \sin 240° = I_2\,(-\sin 60°) = 172\left(\frac{-\sqrt{3}}{2}\right) = -149 \text{ amp}$$

$$od = I_3 \sin 120° = I_3 \sin 60° = 190 \times \frac{\sqrt{3}}{2} = 165 \text{ amp}$$

Assuming that the flow of current in the neutral is toward the load

$$I_1 + I_2 + I_3 = -I_N$$

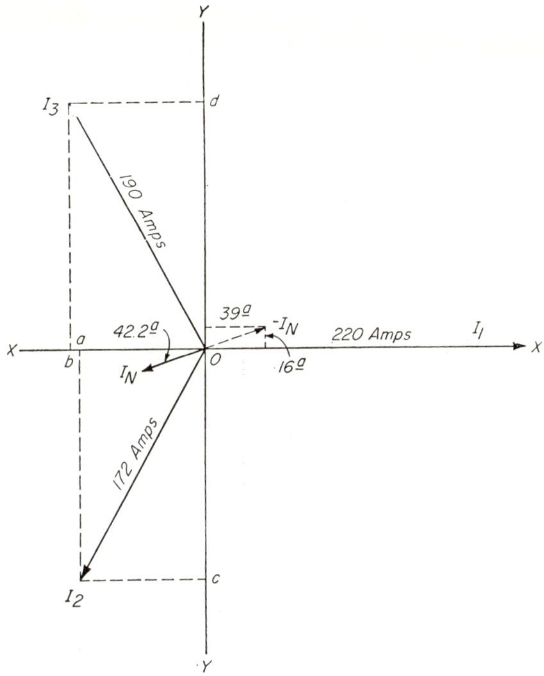

FIG. 5-21. Vector diagram for example in unbalanced three-phase circuit.

then, the components of $-I_N$ are

$$220 \text{ amp} - 86 \text{ amp} - 95 \text{ amp} = 39 \text{ amp in phase with } I_1$$

and

$$0 \text{ amp} - 149 \text{ amp} + 165 \text{ amp} = 16 \text{ amp leading } I_1 \text{ by } 90°$$

Then

$$I_N = \sqrt{(39)^2 + (16)^2} = 42.5 \text{ amp leading } I_1 \text{ by an angle } \theta$$

such that

$$\theta = \tan^{-1}\frac{-16}{39} = \tan^{-1}(-0.4103) = 202°18'$$

2. The equivalent spacing between conductors $= \sqrt[3]{14.5 \times 44.5 \times 59} = 33.64$ in.
Resistance of No. 00 copper wire (Table 4-2) = 0.088 ohm per 1,000 ft
Reactance of No. 00 copper, 34-in. spacing (Fig. 4-21) = 0.125 ohm per 1,000 ft
$\cos\theta = 0.90 \qquad \sin\theta = 0.436$
Effective impedance $= R\cos\theta + X\sin\theta = (0.088 \times 0.90) + (0.125 \times 0.436)$
$= 0.1337$ ohm per 1,000 ft

$$\text{Average current per phase wire} = \frac{220 + 172 + 190}{3} = 194 \text{ amp}$$

$$\text{Average voltage drop} = 5 \times 0.1337 \text{ ohm} \times 194 \text{ amp} = 130 \text{ volts}$$

or

$$(130 \div 2{,}460)100 = 5.3\% \text{ of source voltage}$$

For comparison, if the current in the neutral were taken into account and if the impedance of the neutral and earth return were taken to be equal to the impedance of a No. 00 copper wire, the voltage drop in each phase would be

Phase 1 = 160 volts = 6.56%
Phase 2 = 132 volts = 5.37%
Phase 3 = 88 volts = 3.58%
Average = 127 volts = 5.16%

3. The average voltage at the load = 2,460 − 130 = 2,330 volts phase-to-neutral, or

$$2{,}330 \times \sqrt{3} = 4{,}036 \text{ volts phase-to-phase}$$

5-17. Three-phase Delta Circuits. In the three-phase delta-connected system, transformer and circuit connections are made as shown in Fig.

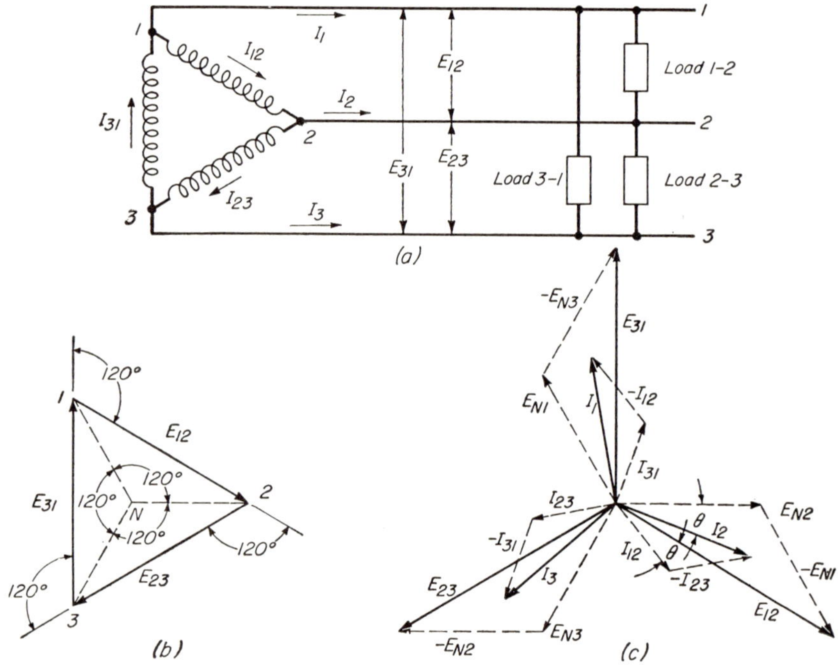

FIG. 5-22. Delta circuit, three phase, has no neutral. Single-phase loads can be connected between any two wires.

5-22. In the delta-connected circuit only the three phase wires are required, loads connect between phase wires, and there is no common terminal connection for all phases. Referring to Fig. 5-22*a*, the current flowing in each phase wire at any instant equals the vector difference of the currents flowing in the two windings connected to it in accord with Kirchhoff's first law. This states that the vector sum of the currents flowing toward a point at any instant is equal to the vector sum of the currents flowing away from the point

$$I_1 + I_{12} = I_{31} \quad \text{or} \quad I_1 = I_{31} - I_{12}$$
$$I_2 + I_{23} = I_{12} \quad \text{or} \quad I_2 = I_{12} - I_{23}$$
$$I_3 + I_{31} = I_{23} \quad \text{or} \quad I_3 = I_{23} - I_{31}$$

The vector relation of the delta, or phase-to-phase, voltages are shown in Fig. 5-22*b*. N is the equivalent neutral point of the delta which, together with the dotted line vectors E_{N1}, E_{N2}, and E_{N3}, represent the equivalent wye-connected system of voltages:

$$E_{12} = E_{N2} - E_{N1}$$
$$E_{23} = E_{N3} - E_{N2}$$
$$E_{31} = E_{N1} - E_{N3}$$

If the loads between phases are balanced, then

$$I_1 = I_2 = I_3 \qquad \text{and} \qquad I_{12} = I_{23} = I_{31}$$

Also, if the phase-to-phase voltages are equal, as shown in Fig. 5-22*b*, then E_{N1}, E_{N2}, and E_{N3} are equal. Thus, with balanced loads and balanced voltages the line current equals $\sqrt{3}$ times the phase current and the phase-to-phase voltage equals $\sqrt{3}$ times the equivalent phase-to-neutral voltage, or

$$I_1 = I_2 = I_3 = \sqrt{3}\, I_{12} = \sqrt{3}\, I_{23} = \sqrt{3}\, I_{31}$$

and

$$E_{12} = E_{23} = E_{31} = \sqrt{3}\, E_{N1} = \sqrt{3}\, E_{N2} = \sqrt{3}\, E_{N3}$$

Vector relations between phase-to-phase and phase-to-neutral current and voltage and the time-phase relation between currents and voltages of a delta-connected circuit are shown in Fig. 5-22*c*. The power-factor angle is measured between the line current and the equivalent phase-to-neutral voltage or between the phase current and the phase-to-phase voltage.

Voltage drop in each phase wire between source and load, due to conductor resistance and reactance, affects the equivalent phase-to-neutral voltage of a three-wire delta circuit in the same way that it affects the phase-to-neutral voltage of a wye circuit, except that there is no neutral voltage drop. The voltage across a load connected between two phase wires of a delta circuit equals the voltage between the two wires at the source minus the vector difference of the voltage drop in the two wires. In other words total loss of voltage, between source and load, if the load is balanced, equals $\sqrt{3}$ times the voltage drop in one wire. If load on the circuit is unbalanced so that unequal currents flow in the phase wires, the voltage drop in each wire will be unequal and voltage between phase wires, at various points along the line, will be unequal. The vector sum of the phase-to-phase voltages and the vector sum of the currents, however, must still equal zero. Unbalanced load shifts the equivalent neutral point of the delta voltages. Under unbalanced load conditions the phase-to-neutral voltages at the load are unequal, the phase-to-phase voltages are unequal, and also not equal to $\sqrt{3}$ times the phase-to-neutral voltage. They equal the vector difference of two of the unequal phase-to-neutral voltages.

Computation of voltage drop in delta-connected circuits, under unbalanced load conditions, must be done by rather cumbersome equations, using complex quantities, or by graphical construction if extreme accuracy is required. For distribution-circuit computations, however, such accuracy is not usually justified. Sufficient accuracy will usually be realized if the unbalanced currents are averaged and the computation is made for a balanced load. This voltage drop is the average drop in voltage from the equivalent phase-to-neutral voltage of the delta circuit, at the source. The drop in phase-to-phase voltage, between source and point of computation, then, equals the computed voltage drop per wire times the $\sqrt{3}$.

FIG. 5-23. Feeder-type induction voltage regulator for automatic control of voltage with load variation. (*Courtesy of General Electric Company.*)

5-18. Voltage Regulators. To hold distribution-circuit voltages within allowable limits, means must be provided to increase the circuit voltage when too low and decrease it when too high. The most widely used method of automatically adjusting line voltage is through induction- or step-type voltage regulators. The induction voltage regulator is a variable-ratio transformer with reversible polarity. Interconnected primary and secondary windings boost or lower the line voltage. Induction voltage regulators are made in either single- or three-phase types. Owing to their greater flexibility, single-phase regulators probably are more frequently used for distribution primary feeders, Fig. 5-23.

The step-type voltage regulator is an autotransformer with taps and can reverse polarity of tap connections to the line so that the line voltage can be either raised or lowered. The voltage varies by changing the tap to which the line is connected, *i.e.*, by tap changing under load.

Similarity between a single-phase induction voltage regulator and a single-phase transformer is shown in Fig. 5-24. It is similar in structure to single-phase transformers, Figs. 1-62 and 1-63, except that the center leg of the core, which the primary and secondary windings cover, is cut and mounted on a shaft. The shaft rotates so that the magnetic relation between the primary and secondary windings can be changed. Any flux linking both primary and secondary windings, however, must cross an air gap.

In Fig. 5-24*a* the regulator is in the maximum boost position. In this position the regulator is similar to an additive-polarity transformer in

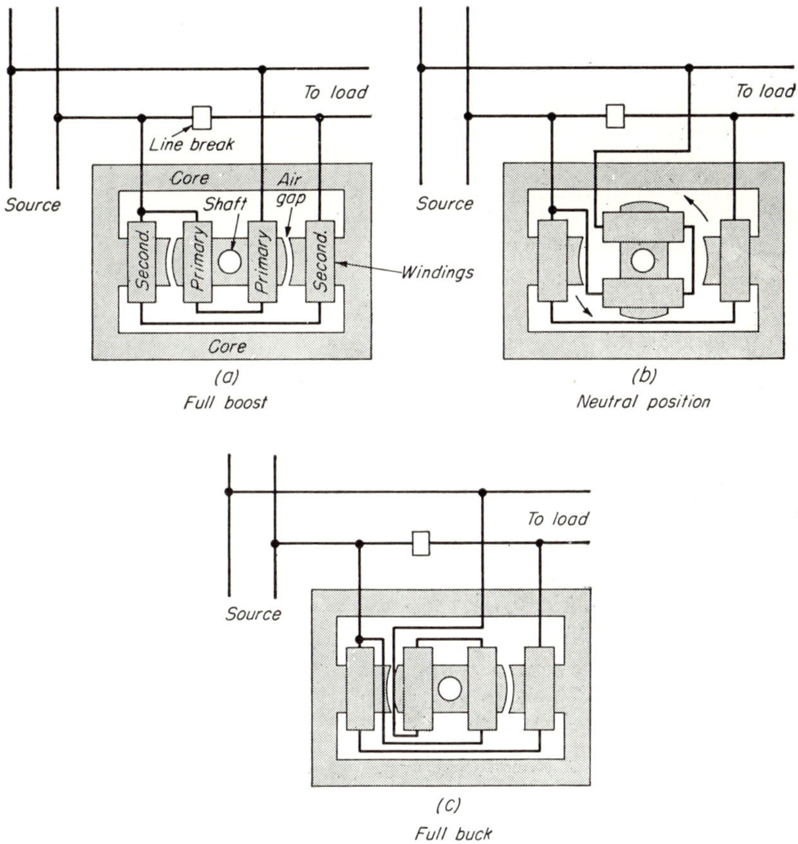

FIG. 5-24. Induction voltage regulator has winding that can be rotated in relation to stationary winding so that line voltage can be raised or lowered with fixed source voltage.

which the secondary winding connects in series with the line so as to add to the primary-line voltage. As the rotor turns to the position in Fig. 5-24*b*, the flux lines linking with the series winding gradually decrease. When the windings are in position *b*, no flux set up by the primary winding links with the secondary, or series winding. When in this position no voltage is induced in the series, or stator, winding and the regulator is in neutral position. When the regulator is "on neutral" it produces no change in line voltage. In the actual regulator the shunt (primary) and series (secondary) windings are in slots on circular cores in a manner similar to rotor and stator windings of a generator. This permits a smooth voltage change over the full range of the regulator.

When the regulator is in neutral position, the shunt and series windings are out of inductive relation to each other. Full line current may still flow through the series coil, and if there is no counter-magnetomotive force the full ampere-turns of the series winding constitute an inductive reactance in series with the line and an excessive reactive voltage drop occurs across the series winding. In the actual regulator a second winding, electrically at right angles to the primary winding, is placed on the rotor. This winding is short-circuited on itself so that, when the regulator rotates toward the neutral position, the coil comes into inductive relation with the series winding. The magnetic flux, set up by the line current in the series winding, passes through it and induces an emf. This induced emf causes current to flow in the short-circuited coil. The magnetizing effect of the ampere-turns in the short-circuited coil counteracts the magnetizing effect of the ampere-turns in the series winding just as in the primary and secondary windings of a transformer. This permits full line current to flow through the series winding with negligible voltage drop. It also equalizes regulator losses and reactance at all rotor positions.

If regulator-shaft rotation continues until the position shown in Fig. 5-24*c*, the inductive relation between shunt and series windings is just reverse of that in *a*. The magnetic flux produced by the magnetizing shunt winding and linking with the series winding is equal in *a* and *c*, but direction of the flux through the series winding has been reversed. The voltage induced in the series winding equals that induced when the windings are in position *a* but opposes the line voltage. In Fig. 5-24*c* the windings are similar to a transformer in which the secondary winding connects in series with the line to buck, or lower, the primary line voltage. In other words, the regulator is in maximum buck position, and while the line connection is unchanged the relative polarity of the series and shunt windings has been reversed by mechanical rotation.

Series windings of induction regulators generally produce a voltage equal to 10 percent of the line or exciting voltage. The regulator can either boost the line voltage by 10 percent or buck it down by 10 percent, giving it a full regulating range of 20 percent, or ± 10 percent. For example, a 2,400-volt regulator can vary the source voltage from a minimum of 2,160 to a maximum of 2,640 volts. Some types of regulators have two secondary windings of 5 percent each. When these two coils are connected in series, the regulator can provide ± 10 percent regulation. When the two coils are connected in parallel, the current rating

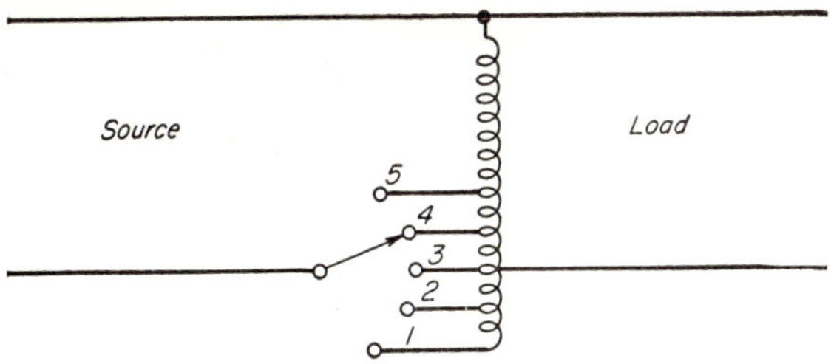

FIG. 5-25. Step voltage regulator uses the autotransformer principle combined with variable taps.

of the regulator is doubled but the regulating range is halved so that it is capable of providing ±5 percent regulation only.

Step voltage regulators perform the same function as the induction regulators except there is no movable core to change inductive relation between shunt and series windings. With the step-type regulator, increase or decrease in circuit voltage is obtained by changing the number of turns in the series winding. This is done through winding taps, as with an autotransformer, Fig. 5-25. If the line on the source side connects to tap 3, line voltage on the load side will be unchanged. If connected to tap 5, the relative number of turns on the load side increases and the voltage on the load side increases in proportion. Conversely, connection to tap 1 decreases the load-side voltage since the number of turns between the load-side terminals is less than between the source-side terminals.

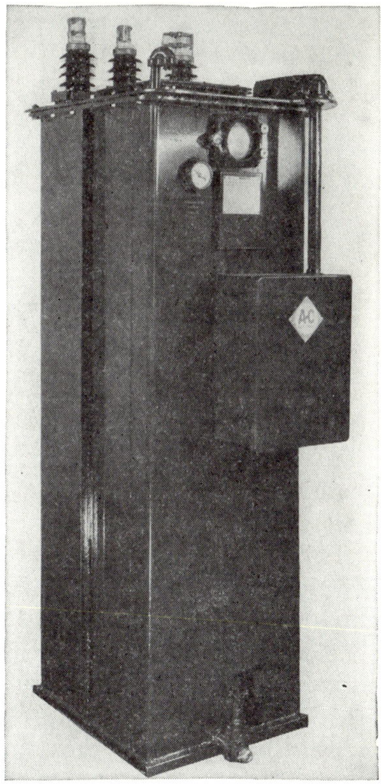

FIG. 5-26. Step-type feeder voltage regulator for automatic control of circuit voltage with varying load. (*Courtesy of Allis-Chalmers Manufacturing Company.*)

Step-type feeder voltage regulators used at substations usually provide the full regulating range of ±10 percent with 32 steps ⅝ percent each. With such small steps the rate of voltage change is gradual and smooth. A typical step-type feeder voltage regulator is shown in Fig. 5-26 and a simplified schematic diagram in Fig. 5-27. There are eight taps in the series winding, numbered 1 to 8.

Each tap represents a voltage change of $1\frac{1}{4}$ percent with a ± 10 percent regulator. The number of taps required is decreased by a reversing switch for reversing connections of the series winding to the line. Reversing the connections reverses the polarity of the series winding.

To control the load-side voltage, the tap on the series winding is changed. This change from one tap to the next under load is done with a small autotransformer having a midtap A. Leads a and b from the autotransformer bridge from one tap to the next so that no gap for arcing exists in moving from one tap to another. When the reversing switch connects to contact R, Fig. 5-27, the polarity of the series winding is additive with respect to the line voltage and boosts the line voltage

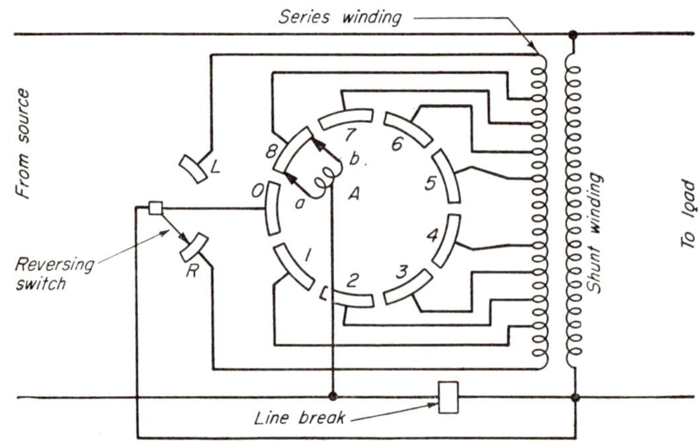

FIG. 5-27. Step voltage regulator circuit diagram shows rotating commutator-type switch and reversing switch to give wide control in small steps.

according to the number of turns in series with the line. When leads a and b are on contact O, the series winding is completely cut out of the line with no difference between source and load-side voltage. If the reversing switch remains on contact R and the rotating tap changer moves so that a is on contact 1 while b remains on contact O, the autotransformer bridges across them. Since the line connects to the midtap of the autotransformer, the voltage is boosted by one-half the voltage difference between contacts O and 1, or by $\frac{5}{8}$ percent. When leads a and b are both on tap 1, the voltage is boosted twice as much, or $1\frac{1}{4}$ percent. When a and b are both on tap 8, the regulator is on the maximum boost position of plus 10 percent. When the reversing switch is on contact L, the polarity of the series winding reverses, becoming negative with respect to the line. The same process of tap changing produces reduction of line voltage until the maximum reduction of minus 10 percent is reached when a and b are both on tap 8. Thus, with 8 taps on the series winding, a total of 16 steps of $\frac{5}{8}$ percent each are available to boost and buck the voltage. This corresponds to a total of 32 steps of $\frac{5}{8}$ percent each across the full regulating range of the regulator.

Induction or step-type voltage regulators are often used in conjunction with shunt-type capacitors (Art. 6-2) for improvement of power factor and voltage regulation. If the capacitors are located between the equivalent center of load, determined as in the example on page 270, and the feeder regulator, the regulator compensator will not represent true voltage conditions in the line and will not cause the regulator to accurately compensate for line voltage drop. This error will be small if the capacitors are closer to the center of load than to the regulators, but, if the capacitors are just beyond the regulators and the line voltage drop to be compensated for to the center of load is large, the error may be large enough to require correction.

With the capacitor just beyond the regulator terminals and with an inductive load being supplied, the 90° leading current going to the capacitor will cancel out an equivalent part of the 90° lagging component of current going to the inductive load supplied by the line. This will result in a lower current and higher power factor in the line up to the capacitor bank, but beyond the capacitor bank the line current, the power factor, and the voltage drop will be the same as though the capacitor were not there. The compensator cannot determine whether the capacitor is adjacent to the regulator terminals or beyond the center of load. Consequently, if the capacitor is near the regulator the current through the compensator will not be in proportion to the line current causing line voltage drop.

To compensate for the capacitor effect, the R and X setting should be increased so that the regulator will produce a voltage boost comparable to the actual voltage drop even though the current through the compensator is lower than equivalent line current. The same effect can also be secured by raising the voltage setting of the voltage-regulating relay and thus increasing the voltage which the regulator is supposed to hold constant at the center of load. The factor by which the compensator or voltage-regulating relay should be increased to compensate for capacitor effect is approximately equal to

$$F = \frac{K}{K_1}\left(1 - \frac{L_1}{L}\right) = \text{multiplying factor}$$

where K = kva to regulator without capacitor bank
K_1 = kva to regulator with capacitor bank
L_1 = distance from regulator to capacitor bank
L = distance from regulator to equivalent center of load

Thus, for the circuit illustrated in Fig. 5-34, page 290, the R and X settings calculated, should be increased by the following amount if a 300 kva capacitor were installed just beyond the regulator, and if the 850 kva of circuit load had a power factor of 85 percent,

$$
\begin{aligned}
K &= 850 \text{ kva} \\
K_1 &= \sqrt{(850 \cos \theta)^2 + (850 \sin \theta - 300)^2} = 737 \text{ kva} \\
L &= 7{,}250 \text{ ft of } \#00 \text{ plus } 650 \text{ ft of } \#2 \\
L_1 &= 0 \\
F &= {}^{850}\!/\!_{737}(1 - 0) = 1.15
\end{aligned}
$$

If the capacitor bank were located half way between the regulator and the center of load, the correction factor would equal

$$F = {}^{850}\!/\!_{737}(1 - \tfrac{1}{2}) = 1.075$$

The error in regulator setting is small except when the capacitor bank is quite close to the regulator and the capacitor bank is large.

5-19. Regulator Controls. A voltage regulator depends upon automatic operation of such controls as the voltage-control relay, or contact-making voltmeter, the motor-control relay, and the line-drop compensator to duplicate conditions of voltage drop in the regulated feeder. Rotor rotation of an induction-type regulator and tap-changer rotation in the step-type regulator are effected by a motor. An auxiliary potential transformer or a transformer winding powers the motor and the various control devices. The voltage supplied from the transformer secondary winding is always proportional to the output, or load-side, voltage of the regulator. A voltage-regulating relay (or contact-making voltmeter) controls the driving motor, Fig. 5-28.

The voltage-regulating relay consists of a small solenoid excited from the auxiliary potential supply. The iron core, or armature, of the solenoid is attached to a contact beam pivoted about its center. The pull on the armature and the pivoted contact beam from the magnetic effect of the solenoid are balanced by a compensating spring. The pull on the armature is proportional to the ampere-turns of the solenoid, which in turn is proportional to the voltage applied. The pivoted contact beam is in mechanical equilibrium when the armature weight, the beam weight, the magnetic pull on the solenoid, and the tension in the compensating spring are balanced. In this condition, the beam is level. If the voltage across the solenoid or the tension in the spring changes, the beam of the voltage-regulating relay tilts. The beam can be balanced either by varying the tension in the compensating spring or by varying the resistance inserted in the potential supply to the solenoid and thus changing the ampere-turns of the solenoid. Both methods are used.

The voltage-regulating relay is shown in Fig. 5-28. There are two contacts on the beam of the voltage-regulating relay. When the mechanical balance of the beam is upset, it tilts and one of two contacts closes. This energizes a motor-control relay which energizes the regulator motor. In Fig. 5-28, if the supply voltage decreases, the pull of the compensating spring exceeds the solenoid pull and the contact beam tilts upward. This closes the contact R and energizes the *raise* coil of the voltage-control relay. The relay in turn closes its contacts and energizes the motor to turn the rotor of an induction regulator, or the tap changer of a step regulator, to bring about an increase in voltage. When the voltage rises enough, the pull of the solenoid again balances the compensating-spring pull. The balanced beam opens the contact, stopping the motor where

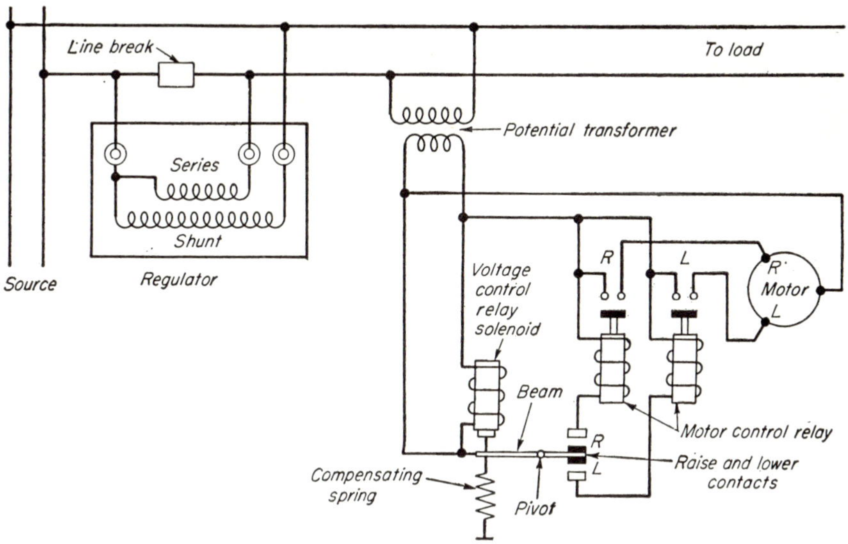

FIG. 5-28. Circuit demonstrating operating principle of voltage-control relay.

the regulator-voltage increase compensates for the decrease in supply voltage.

When the supply voltage increases, the solenoid will pull harder than the spring tension. The beam tilts downward, closing the lower contact L causing the regulator to lower the line voltage to its previous value. The voltage-regulating relay can be used to hold the voltage on the load side of the regulator at a constant value. If the regulator is at, or near, the load, it will hold the load voltage constant at whatever value the contact beam of the voltage-regulating relay is set to level off at.

If regulators are on the primary feeder at the substation and the load is some distance away, a constant voltage at the load cannot be maintained by the voltage-regulating relays alone. Voltage drop in the primary feeder and the transmission supply system may vary appreciably between peak load and light load. When the voltage-regulating relay is set high enough to maintain a satisfactory voltage during peak load, the voltage may be excessive at light load. To maintain a constant voltage at the load, the regulator must raise or lower the feeder voltage in proportion to the primary-line voltage drop and in accord with variations in bus supply voltage. The line-drop compensator performs this duty in conjunction with the voltage-regulating relay.

The line-drop compensator, Fig. 5-29, consists of a resistance R and an inductive reactance X, each can be varied independently. These connect in series with the secondary of a current transformer, the primary

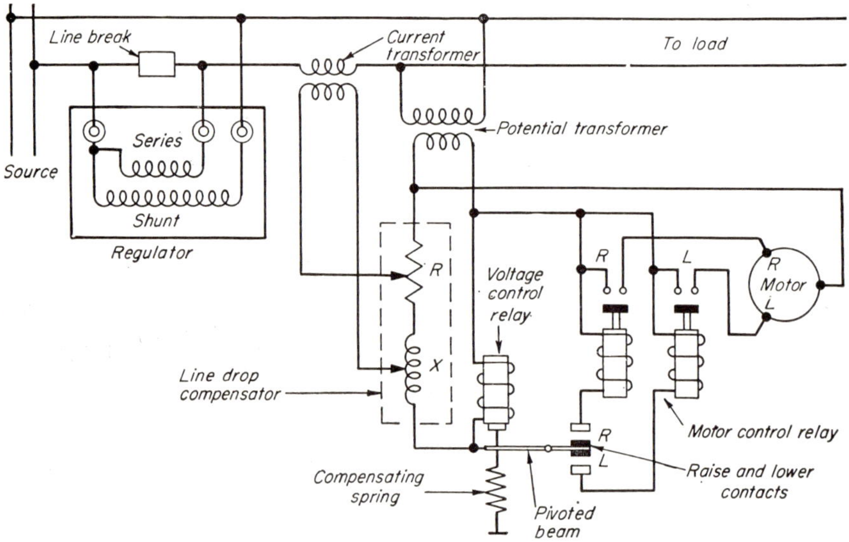

FIG. 5-29. Circuit of regulator controls using line-drop compensator.

of which connects in series with the line to be regulated. They also connect in series with the secondary of a potential transformer whose primary connects across the wires of the circuit to be regulated. The current flowing from the secondary of the current transformer and through R and X is proportional to the current flowing in the line and to the ratio of the current transformer. The potential transformer supplies exciting current to the solenoid winding of the voltage-regulating relay. The current transformer is connected so that its current flows through R and X in the same direction as the negligibly small exciting current flowing to the voltage-regulating relay from the potential transformer. Current from the current transformer causes a voltage drop across R and X which decreases the voltage impressed across the solenoid winding of the voltage-regulating relay. If R and X are adjusted so that the voltage drop across each bears the same ratio to the output voltage of the potential transformer as the actual resistance and reactance voltage drops in the primary line bear to the feeder voltage on the load side of the regulator, the voltage impressed across the solenoid winding of the voltage-regulating relay will be proportional to the voltage at the load at all times. When R and X are properly adjusted, the potential-transformer secondary circuit, including the compensator, is a miniature duplicate of voltage conditions in the primary feeder for all load variations. Any variation in line current produces a like variation of current in the secondary of the current transformer, and any variation in voltage, either at the source or the load, will be faithfully duplicated in the line-drop compensator circuit. The voltage impressed across the coil of the voltage-regulating relay causes the regulator to raise or lower the line voltage enough to keep it constant at a value where the relay beam remains level, a point determined by the R and X setting.

5-20. Application of regulators has so far been concerned with single-phase circuits only. Regulators, however, have their greatest use on three-phase circuits. Three-phase primary circuits are either delta or wye connected and are, in effect, three interconnected single-phase circuits. A three-phase three-wire delta primary circuit may be regulated by two regulators connected as in Fig. 5-30, or by three regulators connected as in Fig. 5-31. When using two regulators they are, in effect, connected in open-delta with the shunt windings connected between phase wires and one phase wire common to both regulators, as in Fig. 5-30*b*. When the two regulators are both on neutral, voltages between phase wires are equal so that $AB = BC = CA$. When the two regulators are on full boost, the voltage between phases A and B will be increased by AA' and the voltage between phases C and B will be increased by an equal amount CC'. When both regulators are in full buck position, the voltage between phases will be decreased by AA' and by CC' so that the line voltage on the load side of the regulators is

$$AB - AA' = BC - CC' = CA - CC' = CA - AA'$$

The full range of regulation provided by the two regulators equals twice the value of AA'. Either regulator may assume any position between full boost and full buck in accord with the setting of the line-drop compensator and the voltage regulating relay. In this case the effect of each regulator will be different, and the voltage between phases, on the load side of the regulator, will not be equal.

If three regulators are connected as in Fig. 5-31 to regulate a three-phase three-wire line, the voltage between any two phase wires will be affected by action of two regulators. If the regulators are delta-connected on a three-wire line as in Fig. 5-31*a* and *b*, the voltage across phase AB will be increased to $A'B'$ when the regulators are in the boost position. This boost is brought about by the addition of both voltages BB' and AA'. The latter is not added directly to voltages AB and

BB' but at an angle of 120°, representing the vector difference between phase-voltages AB and CA. Where single-phase regulators are used, each regulator acts independently. Where the voltage induced in the series winding is 10 percent of the exciting voltage across the shunt winding, the full range of regulation varies from a 10 percent boost above to a 10 percent buck below exciting or bus voltage.

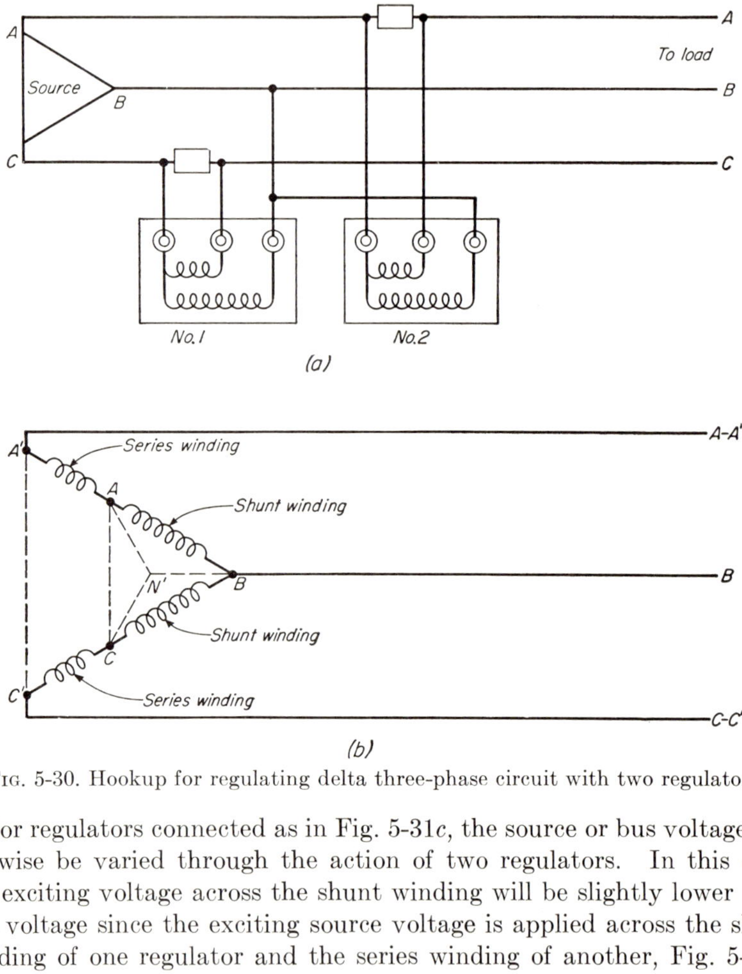

FIG. 5-30. Hookup for regulating delta three-phase circuit with two regulators.

For regulators connected as in Fig. 5-31*c*, the source or bus voltage will likewise be varied through the action of two regulators. In this case, the exciting voltage across the shunt winding will be slightly lower than bus voltage since the exciting source voltage is applied across the shunt winding of one regulator and the series winding of another, Fig. 5-31*d*. In either case, the range of regulation provided by three regulators, connected as in Fig. 5-31, will be approximately 50 percent greater than with two regulators connected as in Fig. 5-30, or about 15 percent above or below normal line voltage.

In a three-phase four-wire wye-connected primary circuit, three regulators connected as in Fig. 5-32 are required. The shunt windings of the

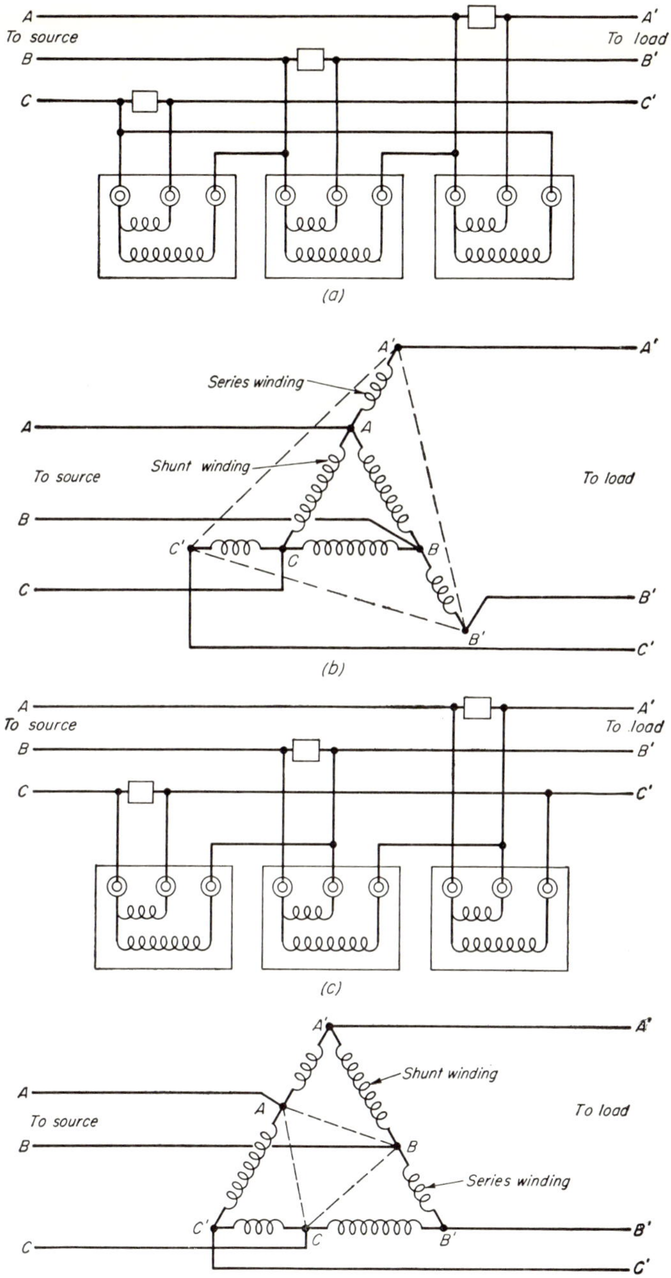

Fig. 5-31. Hookup of three-wire delta circuit using three single-phase regulators.

regulators connect between phase and neutral wire. Each phase is regulated independently by a single regulator. Full range of regulation per phase equals ±10 percent. In a three-phase wye circuit at unity power factor, the current flowing in each phase wire is in phase with the phase-

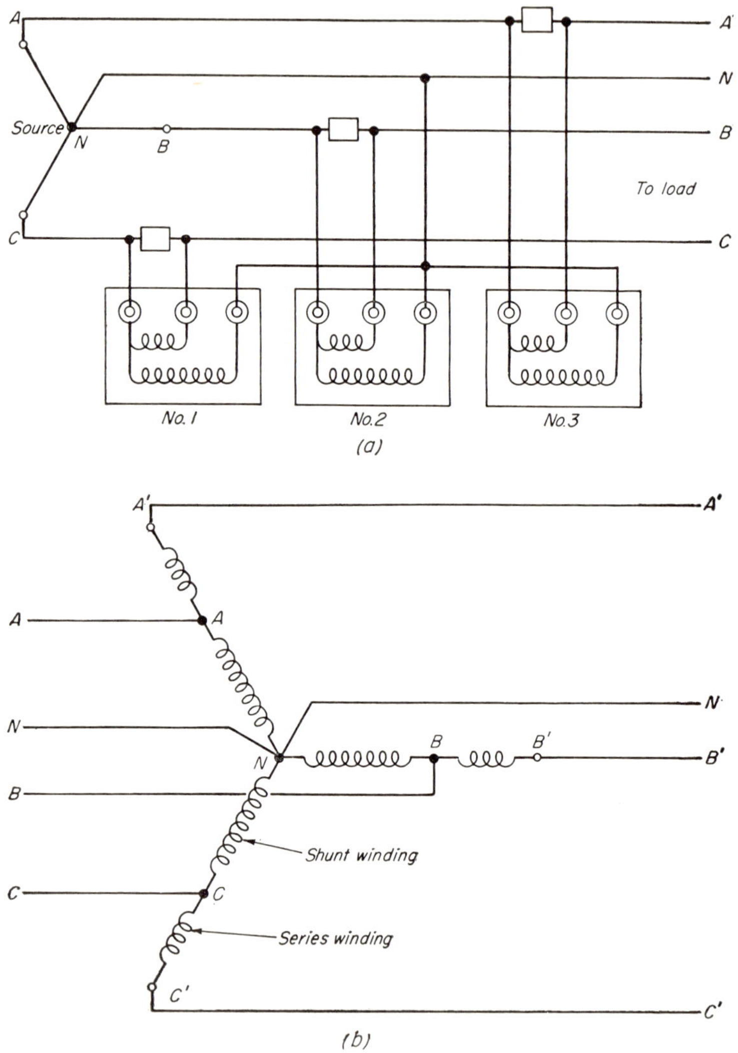

FIG. 5-32. Connections of wye circuit using three single-phase voltage regulators.

to-neutral voltage. In a three-phase three-wire delta circuit the current in each phase wire is in phase with the equivalent phase-to-neutral voltage of the equivalent wye at unity power factor. In a delta circuit, then, the current flowing in the line is 30° out of phase with the phase-to-phase voltage (see Art. 1-27). When regulators connect in a delta primary circuit as in Figs. 5-30 and 5-31, the current flowing through the resistance and the reactance of the compensator is shifted 30° out of phase with the impressed voltage, in addition to the power-factor angle, owing to this phase-angle difference between phase-to-phase voltage and phase current. This gives an incorrect image to the compensator of voltage drop in the primary line since the voltage drop in each primary wire is caused by the current flowing in the wire and is referred to the voltage to neutral of the circuit while the voltage drop in phase-to-phase voltage is a result of current flowing in two separate phase wires.

In the older types of regulators, correction for phase-angle difference between phase-to-phase voltage and line current is provided by three current transformers, one in each phase, with their secondaries connected in delta for installations of either two or three regulators. In the more recent regulator designs, correction for phase-angle difference between current and voltage is provided within the compensator through a phase-shifting impedance network. This permits using either two or three regulators on a delta circuit with only one current transformer per regulator and without interconnecting their secondaries, each regulator being completely independent in operation and control. In the older types of regulators, also, the current and potential supply for the regulator control and motor is provided by current and potential transformers external to the regulator. In the later designs all these are built into the regulator so that each is completely self-contained.

5-21. Size of Regulator Required. The size of a regulator is usually given in terms of its transformer capacity in kva, or in terms of its full-load current capacity and ability to vary line voltage. A regulator which has a nominal rating of 2,400 volts, full-load current of 200 amp, and is capable of varying line voltage from 10 percent below to 10 percent above normal would be classified as a 48-kva regulator, or it might be classified as a ±10 percent 200-amp regulator. The transformer capacity of a regulator is determined by its maximum series voltage rating and maximum current rating. The transformer capacity, or kva size, of the regulator is determined by multiplying the voltage of the series winding at either full buck or full boost by the current through the series winding at rated full load and dividing by 1,000. Thus, in the example cited above,

$$\frac{10\% \text{ of } 2{,}400 \text{ volts} \times 200 \text{ amp}}{1{,}000} = \frac{240 \text{ volts} \times 200 \text{ amp}}{1{,}000} = 48 \text{ kva}$$

5-22. Determination of Compensator Settings. Correct operation of the regulator depends upon the ability of the compensator faithfully to reproduce variations in line voltage from no-load to full load and to impress these same percentage variations on the voltage-regulating relay.

The variable resistance and reactance in the compensator must be adjusted so that current from the current transformer produces a voltage drop across R and X proportional to the drop in the line.

Example: Suppose that a 2,400-volt circuit similar to that shown in Fig. 5-33 is regulated by two 60-kva 250-amp regulators, connected as shown in Fig. 5-30. It is required that 2,400-volts be maintained at the load. Setting the compensator to maintain required voltage at the load might be determined as follows:

1. Line characteristics:
   Resistance, No. 00 copper = 0.08796 ohm per 1,000 ft
   Reactance, No. 00 copper = 0.1246 ohm per 1,000 ft

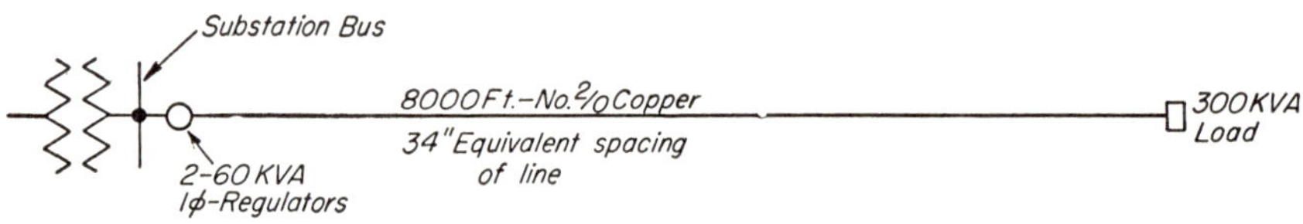

FIG. 5-33. Circuit using voltage regulators.

2. Load on line causing voltage drop is 300 kva over 8,000 ft of line. This is equivalent to 2,400 kva of load over 1,000 ft of line in terms of line voltage drop. The line load, expressed in terms of load and distance, then, is equal to

$$8{,}000 \text{ ft} \times 300 \text{ kva} = 2{,}400{,}000 \text{ kva ft} = 2{,}400 \text{ kva per } 1{,}000 \text{ ft}$$

3. Potential transformer ratio, 2,400 volts to 120 volts = 20:1
   Current transformers, 250 amp primary to 5 amp secondary
4. Actual current flowing in the line is

$$(300 \text{ kva} \times 1{,}000) \div (\sqrt{3} \times 2{,}400 \text{ volts}) = 72 \text{ amp}$$

Equivalent current flowing in the line is

$$(2{,}400 \text{ kva per } 1{,}000 \text{ ft} \times 1{,}000) \div (\sqrt{3} \times 2{,}400 \text{ volts}) = 577 \text{ amp per } 1{,}000 \text{ ft}$$

5. Voltage drop per line wire to neutral is

$$IR = 0.08796 \times 8 \times 72 = 51 \text{ volts} \quad \text{or} \quad IR = 0.08796 \times 577 = 51 \text{ volts}$$
$$IX = 0.1246 \times 8 \times 72 = 72 \text{ volts} \quad \text{or} \quad IX = 0.1246 \times 577 = 72 \text{ volts}$$

6. For the voltage drop in the line to be reproduced in proportion in the circuit of the compensator, a 5-amp current from the regulator current transformers would be required. However, instead of 5 amp only $^{72}/_{250}$ of 5 amp is delivered from the 250-amp current transformers. In order to compensate for this relative difference between the current flowing through the resistance and reactance of the compensator and the current flowing through the resistance and reactance of the line, and so that the voltage drops in the compensator will be proportional to the voltage drops in the line, the amount of resistance and reactance in the compensator is increased. This increase equals the ratio of current transformer rating to current in the line.

7. The compensator is intended to be a miniature duplicate of the line as far as voltage drop is concerned except that the compensator operates at 120 volts while the line operates at 2,400 volts. Consequently, a drop of 1 volt in the compensator is equivalent to a drop of 20 volts in the line. To reduce the IR and IX drops in the line to their corresponding equivalent values in the compensator, the line drops should be divided by the ratio of the potential transformer supplying the compensator circuit.

8. Calculated values of IR and IX voltage drop in 5 represent the voltage drop in a single-phase wire. The potential transformer which supplies voltage to the compensator circuit, however, connects between two phase wires and consequently is affected by the voltage drop in two phase wires. To express the voltage drops in 5 in terms of phase-to-phase voltage these values should be multiplied by the $\sqrt{3}$.

9. Taking into account the current and potential transformer (PT) ratios and the ratio of phase-to-phase and phase-to-neutral voltages, the R and X settings of the two regulators would be equal to the following:

$$IR = \frac{250}{72} \times \frac{120}{2,400} \times \sqrt{3} \times 51 \text{ volts} = 15.3 \text{ volts}$$

$$IX = \frac{250}{72} \times \frac{120}{2,400} \times \sqrt{3} \times 72 \text{ volts} = 21.7 \text{ volts}$$

Since the regulator is to maintain a constant value of 2,400 volts at the load the contact beam of the voltage regulating relay should be adjusted to level off at 120 volts (2,400 volts divided by the PT ratio) and the dials on the compensator set for 15.3 volts of resistance drop and 21.7 volts of reactance drop. The compensator has two dials with graduated scales. A maximum setting of 24 volts resistance drop and 24 volts reactance drop can be made. If the circuit of Fig. 5-33 had been a four-wire 2,400/4,160-volt wye-connected circuit, controlled by three regulators connected as in Fig. 5-32, the 30° phase-angle shift between the current and voltage of the compensator would be eliminated because the potential transformers would be connected phase-to-neutral instead of phase-to-phase. The compensator setting of the regulators would be affected by the lower current in the line, due to the higher line voltage, and by phase-to-neutral voltage supplied to the compensator instead of phase-to-phase voltage.

Determination of the compensator setting for a circuit similar to that in Fig. 5-33 is relatively simple. This type of circuit, however, occurs much less frequently than the type where the feeder supplies a large number of small loads scattered over a relatively large area, such as a general distribution feeder serving lighting and small power load. In this case the regulation limits are governed by the minimum permissible voltage at the last transformer on the line where the line voltage drop is a maximum and the maximum allowable voltage at the first transformer on the line serving the customer nearest the source of supply. The compensator, in controlling the operation of the regulator, cannot distinguish between a circuit in which all of the load is concentrated at a single point and a circuit in which the load is distributed along the line and numerous branches over a large area. Consequently, such a primary circuit must be reduced to an equivalent simple circuit like that in Fig. 5-33 to determine the compensator setting required.

The equivalent simple circuit is found by locating the point at which all of the load of the complex circuit could be concentrated so that the IR and IX voltage drop from the source to the point equals the actual IR and IX voltage drop from the source to the farthest end of the feeder where the voltage drop is a maximum. This point is the center of load of the actual feeder. As an example, suppose that the circuit shown in Fig. 5-33 were extended to supply additional loads, as in Fig. 5-34, and that 118 volts, in terms of lighting voltage, is to be maintained at the end

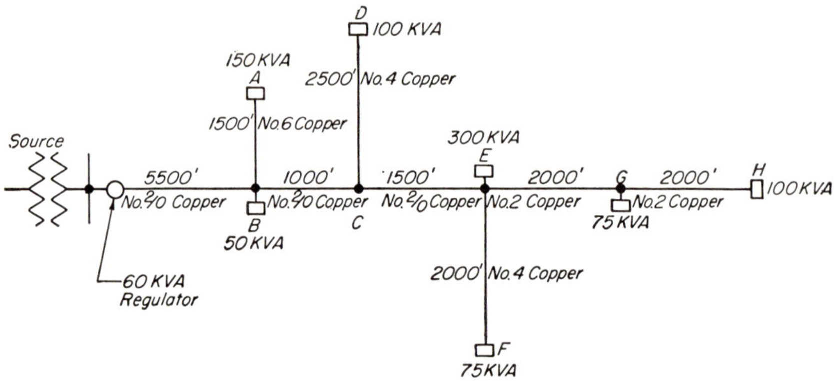

FIG. 5-34. Typical circuit data for calculating voltage-regulator characteristics.

of the line. For such a circuit the compensator setting might be determined as follows:

1. It is required that voltage at the farthest end of the line shall not drop below 2,360 volts primary, or 118 volts in terms of lighting voltage. Calculation indicates the maximum voltage drop will occur between the source and load H. A minimum of 2,360 volts, then, must be maintained at H and the current between H and the source to vary with the load flowing over the line.

2. Line characteristics:

	R per 1,000 ft	X per 1,000 ft
No. 00 copper, 34-in. spacing	0.08796 ohm	0.1246 ohm
No. 2 copper, 34-in. spacing	0.1729 ohm	0.1336 ohm

3. Since the voltage drop from E to H is greater than the voltage drop in branches AB, CD, or EF the voltage drop in these branches may be neglected just as though loads A, D, and F were connected at points B, C, and E. The total circuit load equals 850 kva. The equivalent circuit loading in terms of kva per 1,000 ft, which is a measure of the circuit voltage drop, is

a. Over No. 00 copper:

From source to B	= 850 kva × 5,500 ft	=	4,675 kva per 1,000 ft
From B to C	= (850 − 200) × 1.0	=	650 kva per 1,000 ft
From C to E	= (650 − 100) × 1.5	=	825 kva per 1,000 ft
Total		=	6,150 kva per 1,000 ft

b. Over No. 2 copper:

From E to G	= (550 kva − 375 kva) × 2,000 ft	= 350 kva per 1,000 ft
From G to H	= (175 − 75) × 2.0	= 200 kva per 1,000 ft
Total		= 550 kva per 1,000 ft

From the viewpoint of the compensator the circuit of Fig. 5-34 consists of 7,250 ft of No. 00 copper [(6,150 kva/M ft × 1,000)/850 kva] and 650 ft of No. 2 copper

[(550 × 1,000)/850] with a load of 850 kva concentrated at the end. This point corresponds to the center of load of the circuit.

4. The maximum load current flowing in the line will be equal to about 208 amp. The voltage drop per wire will be

IR, No. 00 copper	= 0.08796 × 7.250 × 208 amp	= 133 volts
IR, No. 2 copper	= 0.1729 × 0.650 × 208 amp	= 23 volts
Total IR drop		= 156 volts
IX, No. 00 copper	= 0.1246 × 7.250 × 208 amp	= 188 volts
IX, No. 2 copper	= 0.1326 × 0.650 × 208 amp	= 18 volts
Total IX drop		= 206 volts

5. Taking into account the transformer ratios and the ratio of phase-to-phase and phase-to-neutral voltages the compensator setting should be about

$$R = \frac{250}{208} \times \frac{120}{2{,}400} \times \sqrt{3} \times 156 = 16.2$$

$$X = \frac{250}{208} \times \frac{120}{2{,}400} \times \sqrt{3} \times 206 = 21.4$$

These settings take into account only the voltage drop in the primary circuit. If the regulators are to compensate for the voltage drop through the transformers serving the loads the R and X should be increased by the resistive and reactive voltage drop in the average transformer, referred to the phase-to-phase voltage of the primary circuit. For example, suppose that the transformers serving the loads in Fig. 5-34 have an average resistance drop of 1.7 percent and reactance drop of 2.2 percent. If the transformers are connected between phases and if the equivalent of 118 volts is to be maintained on the secondary side, the above value of resistance should be increased by 1.7 percent of 118 volts and the reactance by 2.2 percent of 118 volts. In other words, the R setting should be increased by 2.0 volts to 18.2 volts and the X setting should be increased by 2.6 volts to 24 volts. With this setting of R and X the regulator would maintain the equivalent of 118 volts on the secondary side of the transformer bank at H. If the regulator is also to compensate for voltage drop on the secondary side of transformers, the R and X setting of the regulators should be further increased by an amount equivalent to the resistance and reactance voltage drop in the secondary circuit. In this case the X setting cannot be further increased because it cannot exceed 24 volts.

6. Assuming the loads in Fig. 5-34 represent the distribution of load over the feeder at the time of peak load and the voltage-regulating relay is set for 118 volts, 2,360 volts will be maintained on the primary at H for all variations of load and for all power factors. The other limiting factor is that the voltage at B must not be excessive. For the setting given and for a voltage of 2,360 volts to be maintained on the primary at H, the voltage at B, with a power factor of 85 percent, will reach a maximum of 2,500 volts at peak load, as can be shown by calculation. The voltage at B, then, will vary from 2,500 volts at peak load, at 85 percent power factor, to 2,360 volts at no load.

In the ordinary case of distribution feeders serving both power and lighting load, it is usually necessary to include compensation for voltage drop in the step-down transformers on the feeder, and frequently for voltage drop in the secondary circuit. To do this the transformers should be loaded to about the same percentage of rating and the secondary circuits carefully designed to provide about the same range of voltage drops. This avoids excessive voltage at the first transformers on the circuit and low voltage on the secondary side of the last transformers on the circuit. To use regulators to best advantage the primary and secondary circuits must be designed for the best voltage regulation and the transformers on the circuit must be placed and loaded to obtain the best over-all limits of regulation.

CHAPTER REVIEW TOPICS

1. Describe the primary distribution system.
2. What voltages are frequently used for primary distribution systems?
3. How do single- and three-phase circuit capacities compare in respect to number of wires and in respect to voltage regulation?
4. Describe radial and network primary distribution systems.
5. What precaution must be observed in feeding single-phase loads from a three-phase system?
6. Where does the greatest current density occur in a distribution feeder?
7. What is the usual controlling factor in determining conductor size?
8. Describe the radial tree-type primary circuit.
9. Describe the feeder-and-main type primary circuit.
10. Describe the loop primary circuit.
11. How may loop circuits be sectionalized?
12. Detail the features of a primary network system.
13. What are the main features of a secondary distribution system?
14. Describe the advantages of transformer banking.
15. What are the effects of diversity on the maximum load of a transformer?
16. Sketch some of the distribution circuits and show how they are fused.
17. What elements make up a network protector?
18. How does a network protector work?
19. How are secondary networks protected against short circuits?
20. What voltages are used in secondary network systems?
21. What procedure is used to restart a network after complete shutdown?
22. What are the undesirable aspects of too much voltage variation on a distribution circuit?
23. What percentage variation in voltage represents good practice?
24. What is the relation between voltage drop and voltage regulation in a distribution circuit?
25. Draw the circuit and voltage and current relations for a delta-connected three-phase circuit.
26. Draw the circuit and voltage and current relations for a wye-connected three-phase circuit.
27. Describe the principles of the step-type and induction-type voltage regulators.
28. Describe the sequence of actions in voltage regulators.
29. How does a line-drop compensator work in a voltage regulator?
30. Discuss the effect of voltage regulators on the voltages in three-phase circuits, delta- and wye-connected.

PROBLEMS

Problem 5-1: Suppose that a three-phase delta circuit of the same conductor size, circuit length, etc., as that of the example on page 270, supplies an unbalanced load drawing the same current per phase. (1) What is the voltage drop? (2) What is the voltage at the load?

Solution:

1. Effective impedance = 0.1337 ohm per 1,000 ft
   Average current per phase wire = 194 amp
   Equivalent voltage to neutral = 2,460 ÷ $\sqrt{3}$ = 1,420 volts at the source
   Average voltage drop, phase-to-neutral = 0.1337 × 194 = 130 volts
   = 9.15% of source voltage
   Average voltage drop, phase-to-phase = 130 volts × $\sqrt{3}$ = 225 volts
2. Average voltage at the load $= \dfrac{2{,}460}{1.732} - 130$ volts = 1,290 volts phase-to-neutral
   = 2,460 volts − 225 volts
   = 2,235 volts phase-to-phase

Problem 5-2: Suppose that the circuit in Fig. 5-34 is changed to a four-wire 2,400/4,160-volt circuit, which is regulated by three 60-kva ±10 percent 250-amp regulators connected as in Fig. 5-32. What should be the compensator setting to maintain 120 volts (2,400 volts phase-to-neutral) at H, if all other conditions remain unchanged?

Solution:

Line current for 850 kva = 118 amp

IR drop, No. 2/0	= 0.088 × 7.250 × 118 amp =	75 volts
No. 2	= 0.1729 × 0.650 × 118 amp =	13 volts
Total IR drop	=	88 volts
IX drop, No. 2/0	= 0.1246 × 7.250 × 118 amp =	107 volts
	= 0.1336 × 0.650 × 118 amp =	10 volts
Total IX drop	=	117 volts

The potential transformer supplying voltage to the compensator is connected phase-to-neutral so that the voltage is in proper phase relation to the current flowing in the line. Consequently, correction does not have to be made in the setting for the phase-angle difference between phase-to-phase voltage and voltage drop due to phase current. The compensator setting, then, is equal to

$$IR = \frac{250 \text{ amp}}{118 \text{ amp}} \times \frac{120 \text{ volts}}{2{,}400 \text{ volts}} \times 88 \text{ volts} = 9.3 \text{ volts}$$

$$IX = \frac{250 \text{ amp}}{118 \text{ amp}} \times \frac{120 \text{ volts}}{2{,}400 \text{ volts}} \times 117 \text{ volts} = 12.4 \text{ volts}$$

CHAPTER 6

DISTRIBUTION SYSTEMS—CAPACITORS AND TRANSFORMERS

By E. M. Adkins

Jersey Central Power and Light Company

6-1. Capacitors in the Distribution System. Voltage drop in an ac circuit depends upon resistance, reactance, current flow, and the time-phase angle between current and voltage. The voltage drop is approximated by

$$v = E_1 - E = I(R \cos \theta + X \sin \theta)$$

Figures 5-11 and 5-12 together with the above equation show that, for high power factor, resistance causes a greater proportionate part of the voltage drop than reactance. For low power factor, reactance causes a greater proportionate part of the voltage drop than resistance. For instance, a circuit has a resistance and a reactance of 1 ohm each and a power factor of 90 percent. The cosine of the power factor angle is 0.90 and the sine is 0.436. The approximate voltage drop per ampere flowing in the circuit is

$$v = (0.90 \times 1 \text{ amp}) + (0.436 \times 1 \text{ amp}) = 1.336 \text{ volts}$$

The resistance causes 67.3 percent of the voltage drop (0.90 volt), while the inductive reactance causes 32.7 percent (0.436 volt). If the power factor is lowered from 90 percent to 43.6 percent, these proportions will be reversed and the inductive reactance will become the predominate element in causing voltage drop. These actual proportions, of course, vary with the size of and spacing between conductors as well as with the load power factor.

The power factor depends upon the relative proportion of resistance and reactance in the complete circuit. If the reactive component of 60 amp in Fig. 6-1 was eliminated, only 80 amp would serve the useful resistance load and this current would be in phase with the voltage. Voltage drop would be reduced in proportion to the reduction in current, and I^2R losses would be reduced in proportion to the square of the reduction in current. Voltage regulation and load capacity of the circuit would be correspondingly improved.

A capacitor of proper size connected as a load can eliminate the lagging inductive current, see Art. 1-21. Current drawn by a capacitor leads the voltage by 90°. If the capacitor is large enough to draw 60 amp of charging current, the leading component cancels the lagging component, Fig. 6-1, and only 80 amp of current flows between the source and the capacitor, as shown in Fig. 6-2.

6-2. Application of Shunt Capacitors. In any ac circuit the load capacity is determined by voltage, current, and allowable voltage regulation. The load capacity of a circuit is expressed in terms of kva for limiting value of voltage regulation. In any ac circuit the useful component of power in kilowatts equals the kva multiplied by the power factor, or cosine of the power-factor angle. The reactive component of power equals the kva multiplied by the sine of the power-factor angle and is

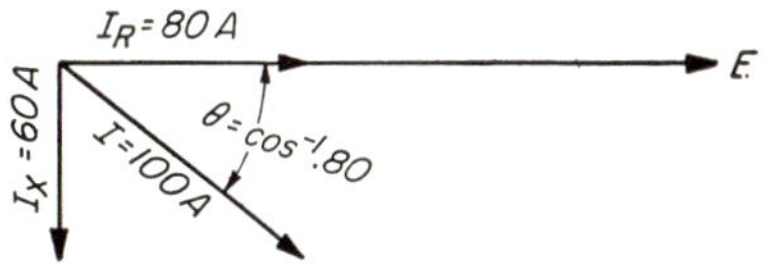

FIG. 6-1. Typical voltage and current relationship with lagging power factor.

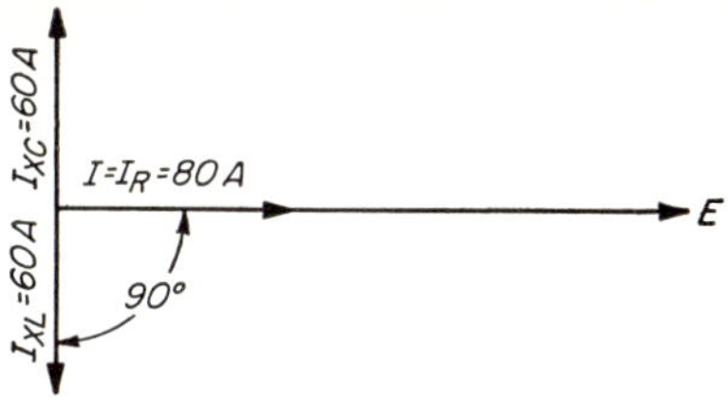

FIG. 6-2. Inductive component of current balanced with capacitive component to make circuit current in phase with voltage.

called kilovars or kvar. If the power-factor angle is lagging, the reactive component in kvar represents the magnetizing power of the circuit alternately stored in the magnetic field and then returned to the circuit twice each cycle, Art. 1-23. If the power-factor angle is leading, the reactive component of power represents the capacitive power alternately stored in the dielectric field and then returned to the circuit. In either case the reactive component of power is wattless and does not represent power to do useful work.

Kilovars are unavoidably present in any ac circuit since they provide excitation for transformers, induction motors, induction furnaces, welding equipments, etc. The undesirable effect upon voltage regulation and energy losses can be reduced or eliminated by connecting capacitors across the phase wires of a circuit. The size of shunt-connected capacitor banks varies from individual units of 5 to 25 kva connected to secondary or primary circuits of a distribution system to a bank of several thousand kva connected at primary voltage to the bus of a step-down substation. In Fig. 6-3 a typical installation, 180 kva of capacitors (four 15-kva units per phase), serves an overhead distribution primary circuit.

Leading capacitive current drawn by capacitors must be supplied by the generators supplying the system. Consequently, shunt capacitors installed on a distribution circuit reduce the current, improve voltage regulation, and reduce energy losses in every part of the system between capacitors and generators. On the other hand, shunt capacitors have no effect upon current or power factor beyond their point of installation. Thus, in Fig. 6-2, current I_R will flow in the circuit up to the point at which the capacitor connects. Beyond this point and up to the load, current and power-factor relations revert to that in Fig. 6-1.

FIG. 6-3. Shunt capacitors for correcting excessive lagging power factor mounted on pole top. (*Courtesy of General Electric Company.*)

Voltage drop in an inductive circuit can be closely approximated as

$$v = I_R R + I_X X_L$$

If a capacitor is connected in shunt with the inductive load, the voltage drop can be approximated as

$$v = I_R R + I_X X_L - I_C X_L$$

where R = resistance of the line, ohms
X_L = phase-to-neutral inductive reactance of the line, ohms
I_R = power component of current in phase with the voltage, amp
I_X = reactive component of current lagging the voltage by 90°, amp
I_C = reactive component of current leading the voltage by 90°, amp

The voltage drop in the line with the capacitor is less than without, as can be seen by the above equation. The effect of the capacitor in raising the voltage equals the difference between the two voltage drops, or

$$\text{Volts rise} = (I_R R + I_X X_L) - (I_R R + I_X X_L - I_C X_L) = I_C X_L = v_r$$

The percent rise in voltage can be similarly expressed as

$$\% \, v_r = \frac{100 \, I_C X_L}{E_N} = \text{percent rise in phase-to-neutral voltage}$$

where E_N = phase-to-neutral voltage of the circuit

Where a shunt capacitor is connected, the effect of R and X_L in producing voltage drop can be neutralized if the capacitor bank is made large

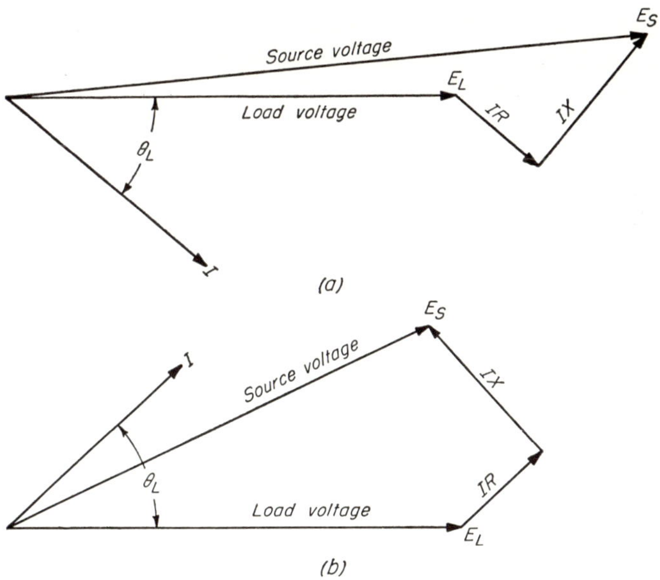

FIG. 6-4. Under leading power-factor conditions, the load voltage can be higher than the source voltage.

enough. The voltage drop will equal zero if I_C, the capacitor current, is large enough so that

$$I_C X_L = I_R R + I_X X_L$$

With this amount of capacitor kva, however, I_C will be larger than I_X and the power factor will be less than unity, leading. If I_C is large enough to produce a low leading power factor, the voltage at the load may be higher than at the source, particularly if X_L is larger than R. The effect of leading power-factor current in raising the voltage of a circuit is shown in Fig. 6-4; in *a* the current lags behind the voltage, in *b* the current leads. The IR and the IX voltage drops in the line are the same in both cases, but in *a* the power factor at the load is higher than at the source and the voltage is lower, while in *b* the power factor is lower at the load than at the source while the voltage is higher. The load is considered to be the resultant of the total resistive, inductive, and capacitive loads in the circuit.

In actual practice it usually doesn't pay to correct fully to unity power factor. In the usual distribution feeder the load is far from constant. During some part of a 24-hr period the feeder load rises to a maximum and varies from this maximum toward a minimum. If the feeder load is predominantly residential, the load will probably rise to a maximum in the early evening, with smaller rises probable in the forenoon. and will drop off to a much smaller minimum from midnight to early morning. The power factor of such a 5,000-volt circuit may be 80 percent to 90 percent during the peak-load period and may then drop to around 60 percent during the minimum-load period because low power-factor transformer exciting current and inductive appliance load, such as refrigerator and oil-burner motors, make up a large part of the total load during this period. If the distribution voltage is in the higher range, such as 11 or 13 kv, the power factor at light load may be near unity, lagging or leading, owing to the capacitive charging current drawn by the line because of its great length. If capacitor kva is installed to raise the power factor at peak load to unity, a low value of leading power factor may result during light load because the leading capacitor kva is larger than the lagging inductive-load kva during this period.

Maximum pf gain per capacitor kva is obtained up to power factors of about 95 percent. As the power factor approaches unity, the amount of capacitor kva required to produce further improvement increases more rapidly. For example, the power factor of a circuit can be increased from 80 percent to 81 percent by a much smaller amount of capacitor kva than will be required for a rise from 99 percent to 100 percent. Figure 6-5 shows the loads, in vector form, of a circuit carrying 800 kw of load at different power factors. In *a* at a power factor of 80 percent the circuit load consists of 800 kw of resistive load and 600 kvar of lagging inductive load which combine to form a total resultant load of 1,000 kva. The addition of 20.85 kvar of capacitive load in *b* raises the power factor from 80 percent to 81 percent, reduces the lagging kvar from 600 to 579.15, and reduces the line load from 1,000 to 987.65 kva. In effect, the installation of 20.85 kva of capacitor load has increased the load capacity of the circuit by 1,000 − 987.65 = 12.35 kva, or approximately 10 kw. Figure 6-5*c* and *d* shows similar relations at 95 percent and 96 percent power factors. Capacitor load of 337 kva raises the power factor of the circuit from 80 percent to 95 percent and lowers the demand from 1,000 to 842.1 kva, while an additional 29.83 kva of capacitor load raises the power factor one more percent, from 95 percent to 96 percent.

From this point the amount of capacitor kva required to produce further improvement in power factor increases rapidly. In Fig. 6-5*e* a total of 486.14 kva of capacitor kva raises the power factor from 80 percent to 99 percent, while an additional capacitor load of 113.86 kva raises the power factor one more percent, to unity. Thus, as the power factor approaches unity the effectiveness of capacitors in improving power factor, lowering line kva transmitted, raising the load capacity, or lowering line copper losses by lowering line current rapidly decreases.

When the effect of shunt capacitors upon the entire electric system is considered, correction to unity power factor is sometimes economically

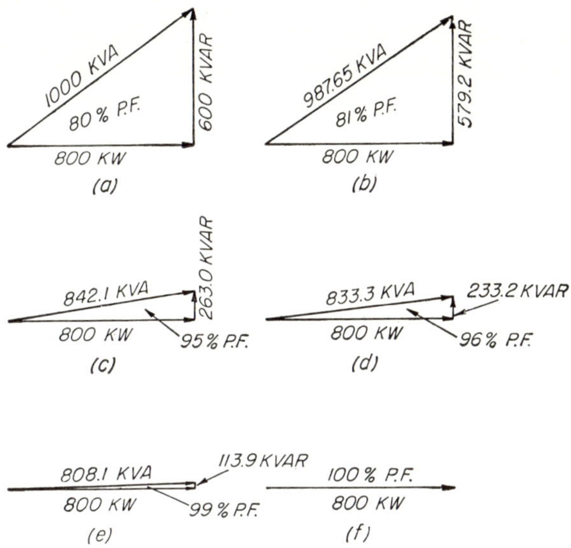

FIG. 6-5. As power factor gets nearer to unity, it takes more kvar to gain 1 percent in power factor. As a result it seldom pays to correct fully to unity.

justified. This is particularly true where the capacitors are fed from the substation bus, the capacitor bank is large, and the principal advantage sought is an increase in substation, transmission-system, and generating-system load capacity together with a decrease in system energy losses. Correction to unity power factor in the distribution system of an industrial plant is also often justified, particularly where the demand charge and the penalty for low power factor are important parts of the cost of electrical energy purchased. Obviously power-factor correction on distribution circuits is economically justified if additional load capacity is required and it can be supplied by installing shunt capacitors at a lower cost than by building new facilities. Power-factor correction in an industrial plant is justified to the point where the annual cost of the capacitors installed is equal to the annual saving in purchased energy.

The maximum reduction in copper loss results when the power factor is corrected to unity at peak load, but the maximum saving in copper energy losses will result only when the power factor is held to unity at *all* loads. The greatest over-all advantage of shunt capacitors is realized if the capacitor kva varies with the load to hold unity power factor at all times. With a load which fluctuates, however, unity power factor can be approached only by switching part of the capacitors out of service as the load decreases and switching them back into service as the load increases toward its maximum.

A switch for shunt capacitors must be capable of interrupting the full charging current of the bank or the maximum current to be encountered at rated voltage. The switching may be done manually or automatically. The numerous methods of automatic control may utilize clock-controlled switches, current relays, voltage relays, or perhaps, relays whose operation depends upon power factor, kilowatts, or kvar, etc. Of the various methods, voltage control is probably most frequently used.

Capacitors will withstand voltages only moderately in excess of their rating. In general they should not be subjected to continuous operation at voltages in excess of 115 percent of rating. Since a capacitor itself raises the circuit voltage and contributes to overvoltage, care should be taken that the charging voltage will not be excessive over a protracted period. Overcurrent protection should be provided to detect and disconnect a faulted unit before the heavy short-circuit current causes the main-feeder circuit breaker to open and also to prevent complete destruction of the faulted unit. Thermal fuses commonly provide short-circuit protection for the small or moderate-sized capacitor banks on distribution primary circuits. The usual practice uses a single fuse for the group of capacitor units connected to each phase. Thus, a three-phase bank of capacitors would be protected by three fuses. The cutout or disconnecting switch in which the fuse is placed should have sufficient capacity to interrupt the full short-circuit current safely if the capacitor bank is delta connected or if it is wye connected and the neutral of the bank solidly connected to the neutral of the wye-connected primary circuit. On wye primary circuits it is usually best if the capacitor bank neutral is solidly connected to the system neutral. If the bank neutral is left floating with no connection to the circuit neutral, noise interference upon paralleled telephone circuits will sometimes be reduced, but other effects may make the floating neutral connection objectionable. In Fig. 6-5*a* is shown a capacitor bank connected with a floating neutral and supplied from a wye primary circuit with one phase open circuited by a blown fuse between the capacitor bank and the source. If it is assumed that current flows from the source to the capacitor bank it will be seen from an examination of Fig. 6-5*a* that the charging current to the units connected to the two live phases must return to the source through the units connected to the dead phase, the dead phase wire, the primary windings of adjacent transformers connected between the dead phase and neutral, and the neutral back to the source. Since this capacitor charging current flowing through the primary windings of transformers connected to the dead phase is not balanced by counter load ampere turns in the secondary winding, it will act something like transformer exciting current. Under proper conditions this may result in the destruction or serious damage to customer's equipment supplied from these transformers because of high voltage. This is particularly true of voltage sensitive equipment such as radio and television tubes. Where shunt-connected capacitor banks are protected by fuses, the choice of the fuse size is influenced by its melting

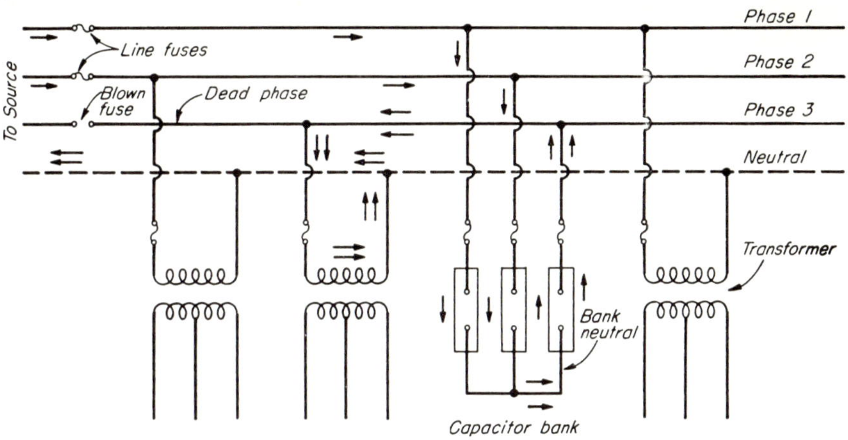

FIG. 6-5a. High voltage due to floating neutral, wye capacitor banks.

characteristic, the kva of the capacitor bank, the circuit voltage, and by local practice. Usually the fuse link size will fall between 150 and 225 percent of the charging current drawn by the capacitor at rated voltage.

Capacitors should be protected by lightning arresters. This is particularly true for capacitors not connected to ground but connected between phases, as in a delta primary circuit. If the capacitors connect from phase wire to a grounded neutral return, lightning arresters are not as essential for protection since the capacitors themselves reduce the magnitude of the surge voltage to ground.

There is capacitance from the terminals of the capacitor to the tank and to ground in addition to the expected capacitance between terminals. Because of this capacitance effect, the metal case or tank of the unit may develop a charge of several hundred volts above ground potential. As a safety precaution, capacitor cases should be solidly grounded. This can usually be done by a connection to the ground lead of the lightning arrester protecting the capacitor.

If a large installation of capacitor kvar is to be made on a primary feeder, it will usually be best to break it up into several moderate-sized banks spread over the feeder rather than as a single installation. To achieve greatest over-all advantage, shunt capacitors should be placed near the equivalent load center of the circuit. This will usually fall between one-half to two-thirds of the distance between first and last points of load take-off on the circuit. If a single large installation is made toward the end of the circuit, a low leading power factor may result in part of the circuit which could have been avoided by spreading the capacitor kva over the circuit in more than one installation. In a radial-feeder-and-main type circuit, Fig. 5-2, for example, the best results will be realized with the maximum number of capacitor kvar, if two or more banks are used and each bank is installed one-half to two-thirds of the

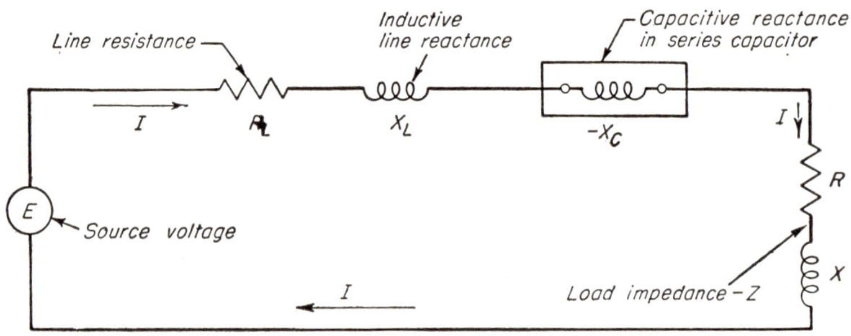

FIG. 6-6. With a series capacitor, change in voltage occurs between load and capacitor, but current flow is the same on both sides of the capacitor.

distance from the first to the last transformer on separate branches radiating from the center of load. In a radial-tree-type circuit, Fig. 5-1, the best results may be obtained by two or more installations at successively greater distances from the substation, depending upon the load density, distance, wire size, power factor, etc.

6-3. Application of Series Capacitors. A series capacitor is connected in series with the line and introduces a lumped value of capacitive reactance in the line impedance. Since capacitive reactance is the exact opposite of inductive reactance (Arts. 1-22 and 1-29), a series capacitor has the effect of negative reactance in the line. A shunt capacitor affects power factor and voltage drop in the line by introducing a component of leading capacitative current in direct opposition to any component of lagging inductive current flowing. A series capacitor produces the same result by introducing into the circuit a negative component of reactive voltage drop in direct opposition to any reactive voltage drop caused by inductance in the circuit.

If a capacitor is connected in series with the line, Fig. 6-6, the amount of current flowing in the line will be the same on either side of the capacitor and no change in voltage or power factor results except between capacitor and load. If a circuit contains only resistance and inductive reactance and serves an inductive load, the vector diagram of current and voltages will be similar to Fig. 6-7*a*. Voltage at the source is *OC*, voltage at the load is *OA*, and voltage drop in the line equals the difference between *OC* and *OA*.

If a series capacitor is added to the circuit, the voltage drop in the line due to the capacitor reactance will oppose and tend to cancel out the voltage drop across the inductive reactance of the line. In other words,

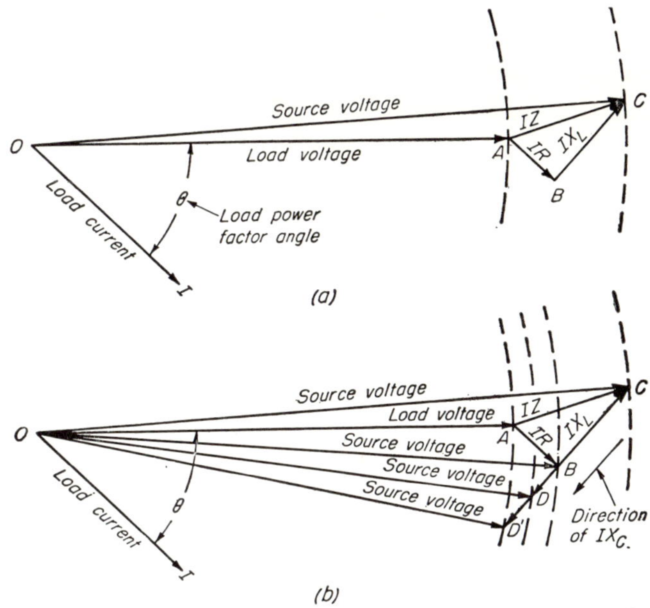

FIG. 6-7. When IX_L equals IX_C, resistance is the only factor causing voltage drop in the circuit of Fig. 6-6. Increasing X_C will eventually cause load voltage to exceed source voltage.

the voltage drop across X_L will be from *C* toward *B* while the voltage drop across X_C will be from *B* toward *C* in Fig. 6-7. The vector diagram of a circuit containing resistance, inductance, and capacitance and serving an inductive load whose power factor is θ will be similar to Fig. 6-7*b*. If the series capacitor is just large enough so that the voltage drop between its terminals is equal to the voltage drop across the inductive reactance of the line, then IX_L is equal to IX_C and the effect will be the same as though there were no reactance in the line but only resistance. With only *IR* drop in the line, the source voltage will have to equal *OB* to maintain voltage *OA* at the load. In this case, the voltage drop in the line equals the difference between *OB* and *OA* which is less than the difference between *OC* and *OA* of Fig. 6-7*a*. Introduction of capacitive reactance into the line impedance by the series capacitor reduces line-voltage drop or raises load voltage if the source voltage remains constant.

If the load power factor is less than unity, lagging and the series capacitor is made large enough so that IX_C is larger than IX_L, CD becomes larger than BC in Fig. 6-7*b*, further reducing the total voltage drop. If X_C is made large enough, the voltage drop across the series capacitor will compensate not only for the voltage drop across the line inductive reactance X_L but also for all or part of the voltage drop across the line resistance, provided the power factor is lagging. If the power factor is unity, voltage drop will be reduced up to where X_C equals X_L. If X_C is made to exceed X_L, the voltage drop again increases.

If the power factor is less than unity leading, the addition of X_C in a series capacitor causes an increase in voltage drop. At high power factors the drop across X_C will have slight effect in compensating for IR drop but will be appreciable at low lagging power factors. Thus, with the power factor and IR drop of Fig. 6-7*b*, if IX_C is made equal to CD' the source voltage will equal the load voltage and the line voltage drop will be zero with both the IR and IX_L line drops being compensated. In general, however, full compensation for line drop is not practical.

Since X_C is concentrated within the series capacitor while R and X_L are distributed along the line, a sharp rise in voltage occurs on the load side of the capacitor which will be reduced by the IR and IX_L drop from the capacitor to the load. The rise in voltage at the load due to the capacitor, or the reduction in voltage drop from the source to the load, can be closely approximated by

$$V_r = IX_C \sin\theta = I_X X_C$$

where V_r = volts rise between the source side and the load side of the capacitor
I = line current, amp
I_X = reactive component of lagging line current, amp
X_C = ohms capacitive reactance in series with the line, ohms
θ = power-factor angle

Excessive and objectionable voltage dips or lamp flicker caused by suddenly applied fluctuating or intermittent loads may be corrected by series capacitors since they are instantaneous in action. Induction or step-voltage regulators cannot be used because they are not sufficiently rapid in action and the dip in voltage must first occur to initiate regulator action. The series capacitor may also be used in place of the induction or step voltage regulator to control normal line regulation when lamp flicker is not present. Voltage regulators, however, are considerably lower in cost. If the series capacitor is used as an automatic voltage regulator, it will be most effective when the inductive reactance of the line is appreciably larger than the line resistance. A series capacitor should be placed so that transformers immediately after the capacitor will not be subjected to excessively high voltage. The capacitor cannot correct for variations in supply voltage at the substation bus as the fully controlled induction or step-voltage regulator does. It may be necessary to use regulators at the substation in addition to series capacitors in the line if there is any appreciable difference between minimum and maximum voltage at the substation bus.

A series capacitor should not be subjected to overvoltage in excess of 200 percent of its rated working voltage and that only for very short intervals of time. Under short-circuit conditions the current flowing in the line may be many times the maximum load current. With a series capacitor in the line, the impedance of the line will be reduced because of the negative reactance in the capacitor itself. This causes a still higher short-circuit current under fault conditions. Under short-circuit conditions the voltage between the capacitor terminals will increase in proportion to the ratio of the current flowing under fault condition to the rated current of the capacitor. Damage to the dielectric can be prevented by a protective device which will by-pass the capacitor with sufficient speed to limit the overvoltage to not more than 200 percent of normal rating within one half cycle. When the line fault is removed and the current flow restored to normal, the automatic protective by-pass must open and replace the capacitor in series with the line. To permit inspection and maintenance, switching should be provided as shown in Fig. 6-8 so that the capacitor can be manually by-passed and disconnected from the line. Current through the line and sustained for any appreciable length of time should not exceed 110 percent, nor should the average current over any 24-hr period exceed 105 percent of the capacitor current rating.

If the capacitor is to correct voltage dips caused by current inrushes of short duration, such as motor starting or welders, it should be of a size so that the largest momentary current inrush will not cause operation of the protective by-pass. The current rating of the capacitor should equal at least two-thirds of the maximum rms value of load-current inrush expected in the line. This limits the voltage rise to not more than 150 percent of normal.

In applying series capacitors, the reactance of the capacitor, in ohms per phase to neutral, can be calculated by

$$X_C = \frac{10 \times (KV_N)^2 \times \% \, V_r}{\text{kva} \times \sin\theta} \qquad \text{or} \qquad X_C = \frac{KV_N \times V_r}{\text{kva} \times \sin\theta}$$

where X_C = reactance of capacitor, ohms per phase to neutral
kva = peak load kva per phase, beyond the capacitor installation

V_r = volts rise to neutral or $\% \ V_r$ = percent volts rise desired
KV_N = line-to-neutral voltage, kilovolts
θ = full-load power-factor angle without the capacitor

The ohms capacitative reactance required to provide a given voltage rise can be calculated by means of the following equation:

$$X_C = \frac{V_r}{I \times \sin \theta} = \frac{V_r}{I_X} \qquad \text{or} \qquad X_C = \frac{10 \times KV_N \times \% \ V_r}{I \times \sin \theta}$$

where I = maximum line current after the capacitor at power factor angle of θ

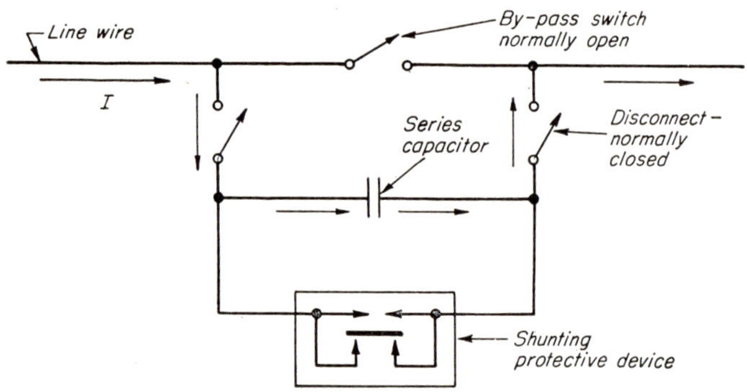

FIG. 6-8. Series capacitors should be suitably switched to permit taking out of service for inspection.

The steps in determining the size of series capacitor are briefly outlined:

1. Determine the resistance and inductive reactance of the circuit.

2. Determine the maximum allowable voltage regulation or voltage dip.

3. Determine the boost in voltage to be supplied by the capacitor. If the capacitor is to act as a voltage regulator under normal sustained load, the minimum voltage rise necessary equals the voltage drop at peak load, without the capacitor, minus the allowable maximum voltage drop. If the capacitor is to control voltage dip or lamp flicker, the minimum voltage rise equals the dip in voltage due to momentary current inrush without the capacitor minus the allowable maximum voltage dip.

4. Determine the ohms of capacitive reactance which must be added to the circuit to confine regulation or lamp flicker within the desired limits. This equals X_C and is determined in accord with the preceding equation for X_C. The value of the current I used in this equation equals the maximum sustained load current if the capacitor is to control line regulation under normal load. If the capacitor is to control voltage dip or lamp flicker, the value of I in the equation equals the momentary current inrush causing lamp flicker which is superimposed upon the normal sustained load current.

5. Determine the minimum current rating required. The rated current should not be less than two-thirds (66.6 percent) of the total rms value of the greatest momentary load current likely in the line and not less than 91 percent of the maximum sustained load current.

6. Determine the minimum voltage rating. The minimum voltage rating equals the continuous current rating from step 5 multiplied by the value of X_C from step 4.

7. Select the nearest standard capacitor unit size whose voltage rating is equal to or greater than the voltage rating determined in step 6.

8. Determine the reactance of a single standard capacitor unit selected in step 7. This equals its rated voltage divided by its rated continuous current.

9. The maximum number of capacitor units per phase which can be used equals the value of X_C per unit from step 8 divided by the total value of X_C per phase from step 4. The minimum number of units per phase which can be used equals the nearest whole number which is equal to or greater than the continuous-current rating from step 5 divided by the rated continuous current per capacitor unit from step 7. In any case the momentary current should not exceed 150 percent nor the sustained current 110 percent of its rated current. The actual value of X_C in the line, then, equals the value of X_C per unit divided by the number of units per phase since the units per phase are in parallel. If the number of units per phase is increased, the equivalent value of X_C per phase will be decreased, the current per capacitor unit will be decreased, the voltage rise due to the capacitor will be decreased, and the voltage drop or voltage dip will increase. If the number of capacitor units per phase is decreased, the reverse effect will take place. Unless the load on the circuit is static and not likely to increase, the minimum number of units per phase should not be used since this will leave no provision for future load increase.

Capacitors are manufactured in several rated kva sizes per unit (such as 10, 15, 25 kva, etc.) and at a wide range of rated terminal voltages from 230 to 8,000 volts or more. The same type of unit connected as a shunt capacitor can also be used as a series capacitor provided it is adequately insulated from ground. A series capacitor is, in effect, a part of the line conductor and requires the same degree of insulation to ground and between phases as the line wires. If a capacitor is used in shunt, it will have a voltage rating equal to the voltage rating of the circuit and the tank may be at ground potential. If a capacitor is used in series, the voltage to ground or between phases may be many times the rated voltage between its terminals. In such cases each unit must have additional insulation to ground. If the additional insulation is built into the capacitor tank, the tank may be solidly grounded as for shunt-connected capacitors. If capacitors insulated for a lower voltage are connected in series with a line of a higher voltage, the additional insulation may be added external to the tank by mounting on insulators. In this case neither the tanks of the individual capacitor units nor any metal cabinet in which the capacitors are mounted can be connected to ground. If capacitors are mounted on insulators, lightning arresters may be omitted. If the capacitor is primarily designed for operation in series, each phase should be protected by lightning arresters on both sides if otherwise unprotected and exposed to lightning surges.

6-4. Voltage Dips and Lamp Flicker. Lamp flicker is a sudden change in the intensity of illumination caused by a corresponding sudden change in the voltage across the lamp. Lamp flicker is usually the result of a suddenly applied heavy load of short duration superimposed on the normal load current. The sudden dip in illumination caused by the inrush of starting current to a motor is an example of this type of lamp flicker.

If the human eye were not sensitive to the variations in the level of illumination with artificial lighting, slight objection would result from lamp flicker on distribution circuits. Under certain conditions, however, lamp flicker can be highly objectionable and may produce great discomfort. The amount of voltage variation producing objectionable lamp flicker varies considerably with different individuals. Flicker objectionable to one may not even be perceptible to another. The degree of objection to lamp flicker depends largely upon the frequency of its occurrence and the rate of change of the voltage dip.

Voltage changes causing lamp flicker may be either cyclic or noncyclic in character. Cyclic lamp flicker is caused by sudden changes in voltage which recur at what approaches definite time intervals. The operation of a spot welder, for example, causes a dip in voltage repeated at nearly definite time intervals for each new weld. Any load causing a periodic variation in the rms value of load current may produce cyclic lamp flicker. Noncyclic lamp flicker results from the infrequent dip in voltage caused by occasional motor starting. Of the two, cyclic lamp flicker is usually the most objectionable. Lamp flicker recurring at regular intervals of six to twelve times per second is the most objectionable. A dip in voltage which occurs only once or twice per day will be less objectionable than a smaller dip repeated at such short time intervals as to be cyclical in form. A somewhat greater variation in lighting voltage can be tolerated with noncyclic than with cyclic dips.

Incandescent lamps of the smaller sizes are more sensitive to voltage dip than the larger sizes because of their lower thermal capacity. Fluorescent and neon lamps also quickly reflect voltage drop by a change in illumination. The annoyance caused by noncyclic lamp flicker is influenced by the frequency of occurrence, amount of dip, and also the rate of change in the rms value of the voltage reflecting the dip. A relatively large dip with a steady decrease in voltage, perhaps spread over several seconds, may be hardly perceptible, while a much smaller voltage dip in a fraction of a second is clearly noticeable and highly objectionable if repeated often enough. In Fig. 6-9 the curve indicates the volts change on a 120-volt lighting circuit and the frequency of occurrence at which lamp flicker will be objectionable to the average individual. Flicker voltages somewhat below these will be objectionable to some individuals, but it is not practical to attempt to limit permissible voltage dip to the point where annoyance to all individuals will be eliminated. These median values will usually provide satisfactory limits. If voltage dip is very infrequent, a flat limit of about 8 volts may be set.

The rate at which voltage dips occur and the types of representative loads causing the flicker may be classified with the frequency of occurrence as follows:

1. Cyclic, high frequency. Dips occurring two to ten times per second may be caused by air compressors, large refrigerators, reciprocating pumps, or automatic spot welders.

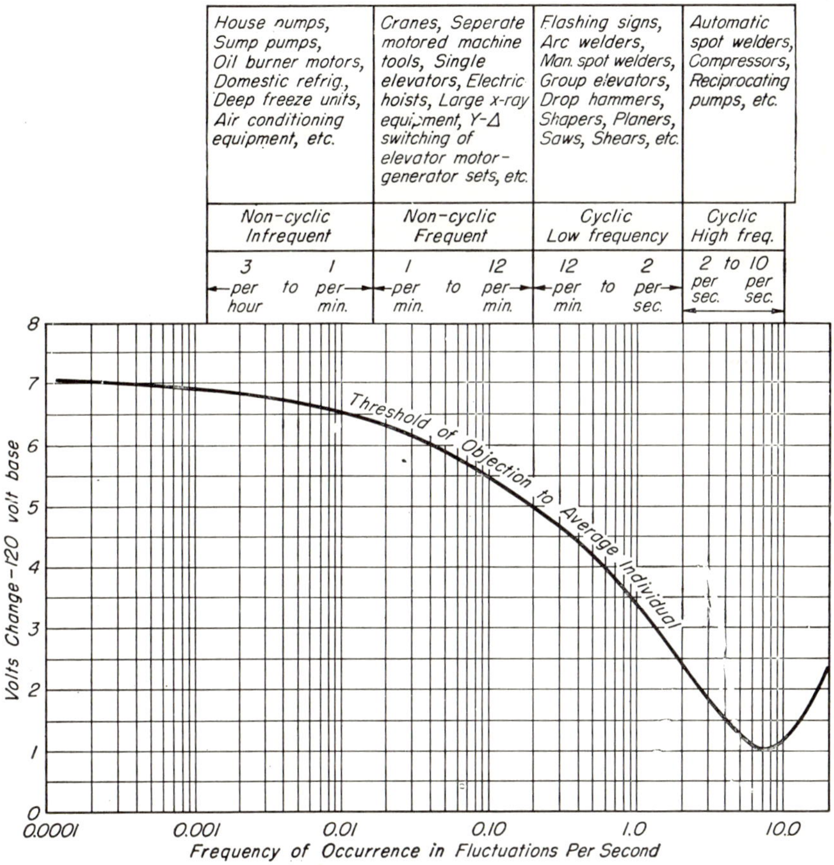

FIG. 6-9. Relation between voltage change and frequency of occurrence at which this effect becomes objectionable on lighting circuits.

2. Cyclic, low frequency. Dips occurring two times per second to twelve times per minute may be caused by motor-driven saws, metal shears, punch presses, planers, drop hammers, group elevators, flashing signs, arc welders, manually controlled spot and seam welders, etc.

3. Noncyclic, frequent. Dips occurring one to twelve times per minute may be caused by electric hoists, cranes, individually motored machine tools, drill presses, single elevators, large X-ray equipments, or wye-delta switching on elevator motor-generator sets, etc.

4. Noncyclic, infrequent. Dips occurring three times or more per hour may be caused by domestic water pumps, cellar sump pumps, domestic refrigerators, oil-burner motors, air-conditioning units, etc.

One of the most frequent sources of lamp flicker is the starting of electric motors. When closing the switch, the effect is similar to short-circuiting the secondary of a high-impedance transformer. During the first few moments the rotor does not turn even though fully energized. During this short interval the current is limited only by the impedance of the motor and its supply circuit. This impedance is relatively low, and the inrush of current may be six or seven times the rated normal full-load running current, as in the case of a squirrel-cage induction motor. When the rotor begins to turn, the effect is somewhat as if additional impedance were gradually inserted into the circuit. Current flowing into the motor decreases to its normal running value as rotation approaches its normal running speed. The starting current nearly equals the locked-rotor current of the motor. It closely approaches the current which would flow if the rotor were locked so that it cannot turn while fully energized.

Typical values of locked-rotor starting current and starting-power factors for general-purpose squirrel-cage induction motors are given in Tables 6-1 and 6-2. Flicker from starting motors operating on secondary voltages results from voltage drop in the secondary-circuit conductors and in the transformer supplying the circuit. The locked-rotor starting current is large enough to cause flicker in lights supplied from the same circuit and transformer but too small to effect lights supplied

TABLE 6-1. APPROXIMATE STARTING CHARACTERISTICS OF SINGLE-PHASE SQUIRREL-CAGE INDUCTION MOTORS

Motor size, hp	Starting power factor, %	Amperes locked-rotor current at starting				
		110 volts	120 volts	208 volts	220 volts	240 volts
Dom. refrig.*	65–85	20–15	18–14			
1/4	50–80	28–16	26–14			
1/3	50–80	33–25	30–22			
1/2	50–80	38–25	34–22			
3/4	50–80	65–55	60–50	35–29	33–28	30–25
1	50–80	.....	.....	37–32	35–30	32–28
1½	50–80	.....	.....	45–36	42–34	39–31
2	50–80	.....	.....	53–45	50–42	46–39
3	50–80	.....	.....	85–64	80–60	74–55
5	50–80	.....	.....	116–101	110–95	101–87
7½	50–80	.....	.....	170–148	160–140	147–128

* Domestic refrigerator.

TABLE 6-2. APPROXIMATE STARTING CHARACTERISTICS OF THREE-PHASE SQUIRREL-CAGE INDUCTION MOTORS

Motor size, hp	Starting power factor, %	Amperes locked-rotor starting current			
		208 volts	220 volts	240 volts	2,400 volts
5	40–60	95	90	83	
7½	40–60	127	120	110	
10	40–60	159	150	138	
15	40–60	232	220	202	
20	40–60	307	290	266	
25	40–60	387	365	335	
30	40–60	460	435	400	
40	40–60	615	580	533	
50	40–60	768	725	667	67
60	40–60	922	870	797	80
75	40–60	1,145	1,085	995	100
100	40–60	1,530	1,450	1,330	133
125	40–60	1,915	1,815	1,665	167
150	40–60	2,295	2,170	1,990	199
200	40–60	3,070	2,900	2,660	266
250	40–60	.....	.....	.....	333
300	40–60	.....	.....	.....	400

from other transformers. Larger sized motors may be served from individual transformer banks to prevent other lighting consumers from being affected. In some cases, with the largest motors, the dip at starting may be great enough to affect lighting supplied through other transformers because of voltage drop in the primary-circuit conductors and in the substation transformers.

Flicker from voltage dip in secondary conductors and the transformer can usually be alleviated by relocating the transformer closer to the source of flicker or by increasing the size of the secondary wires and transformer. This is an expensive means of correction, however, and frequently some other more economical way can be found. The banking together of transformers through their secondary circuits, as discussed in Art. 5-7, is frequently used and is an economical method of controlling secondary lamp flicker. When flicker is caused by starting a small single-phase motor connected between phase and neutral of a three-wire 120/240-volt single-phase circuit as shown in Fig. 6-10*a*, the voltage dip may be reduced about 50 percent by a small 120/240-volt autotransformer connected as shown in Fig. 6-10*b* or *c*.

Flicker can sometimes be satisfactorily controlled by using the proper type of motor starter. If the inrush of locked-rotor current is impeded so that it rises to its maximum value more slowly, the voltage drop may be as great but will not cause an objectionable flicker. A gradually changing current produces no sudden change in illumination level, and the human eye is able to adjust at the same rate that the change occurs. A compression-resistance type motor starter provides this type of flicker control. So does a motor starter consisting of one or more fixed blocks of resistance cut out, or shunted, a block at a time when starting. With resistance added in series with the impedance of the line and the motor, the first inrush of locked rotor current and resulting voltage dip is reduced.

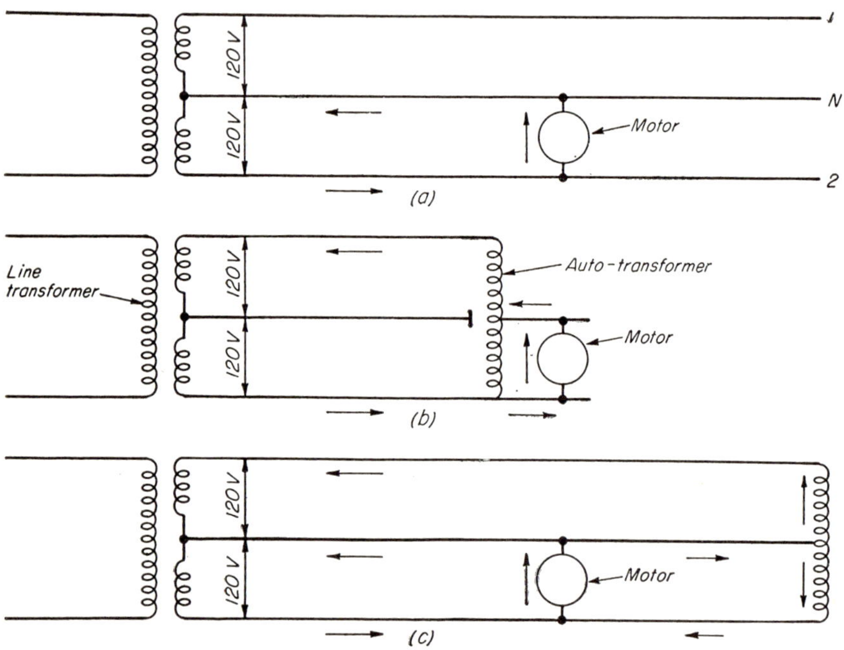

FIG. 6-10. Autotransformers may sometimes reduce lamp flicker caused by motor starting.

If the flicker voltage dip is large enough to exceed satisfactory limits on the distribution primary circuit, it will affect a great many more consumers than if confined to a single secondary circuit. In such a case, flicker may be controlled by providing a separate primary circuit to feed the flicker load or by providing larger primary conductors, substation transformer capacity, etc. Under certain conditions flicker control affecting primary circuits can be best provided by a series capacitor in the primary line.

6-5. Distribution Transformers and Connections. The fundamental theory of the transformer was discussed in Art. 1-30. Single-phase distribution transformers greatly outnumber polyphase because of the normal practice of supplying lighting and the smaller power loads at single-phase from single-phase secondary circuits. Even if the secondary system is polyphase, it may be supplied through single-phase transformers connected in polyphase banks as in the case of a polyphase power load. The usual polyphase distribution system is either two-phase or three-phase. Three-phase distribution has become so predominant, however, as to make two-phase distribution practically extinct in this country.

Transformers may be connected in a number of different ways to supply both single- and polyphase loads at secondary voltages. Since utilization voltages in this country are usually in the range of 120 to 240 volts, the high-voltage winding will be generally referred to as the primary side and the low-voltage winding as the secondary. Transformers to supply three-phase secondary distribution must be connected in either delta or wye on both the primary and secondary side. The most commonly used single- and three-phase bank connections will now be discussed.

Single-phase Connections. Single-phase distribution transformers have one high-voltage primary winding and two low-voltage secondary windings. Secondary coils each rated at a nominal 120 volts may be connected in parallel to supply a two-wire 120-volt circuit, Fig. 6-11*a*, or they may be connected in series to supply a three-wire 120/240-volt single-phase circuit, Fig. 6-11*b*. The latter is most commonly used since it has twice the load capacity of *a* with only 1½ times the amount of wire. Each 120-volt winding represents one-half the total kva rating of the transformer, and if the 120-volt loads connected between each of the two outside wires and the middle wire are equal, the load is balanced and no current flows in the middle wire. Loads on the circuit should be kept as nearly balanced as possible to provide most economical usage of transformer capacity and to hold regulation to a minimum. One leg of the 120-volt two-wire circuit and the middle leg of the 120/240-volt three-wire circuit is normally grounded to limit the voltage to ground on the secondary circuit to a minimum.

Three-phase Delta-delta. The delta-delta connection is formed by tying together single-phase transformers as in Fig. 6-12. In distribution the primary supply is usually an ungrounded delta circuit for this connection. If the bank is connected as in Fig. 6-12*a*, it comprises a three-wire delta circuit with balanced voltages between phases. Such a bank connection may be used where the load consists of balanced three-phase load or where there is no single-phase 120-volt lighting or appliance load. This connection has been used occasionally to supply a balanced three-phase lighting load at 120 volts delta on the secondary side. Figure 6-12*a* shows the bank connected for 240 volts on the secondary side, but this same connection is frequently used at higher secondary distribution voltages, particularly in industrial plants. If the bank is to supply a

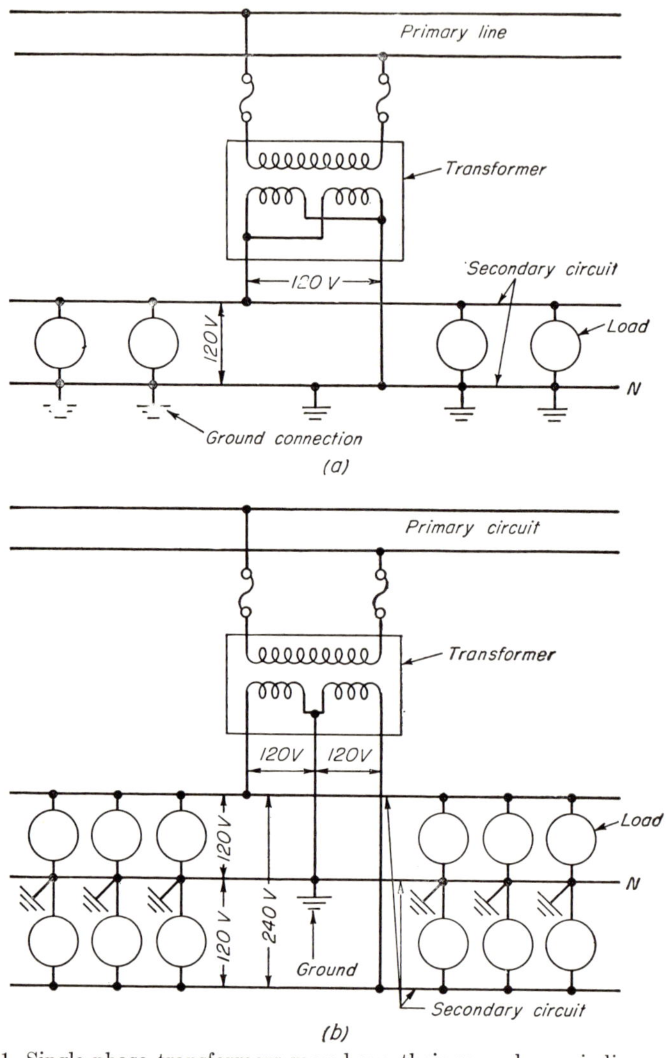

FIG. 6-11. Single-phase transformers may have their secondary windings connected either in parallel, (*a*), or in series to give 120/240-volt service.

balanced three-phase load, equal phase currents will flow in both the primary and secondary and the three transformers should be equal in size with similar characteristics.

If the bank is connected as shown in Fig. 6-12*b*, it will be suitable for supplying both balanced three-phase power load and single-phase lighting and appliance load. With this connection a fourth wire in the secondary circuit connects to a midtap brought out from the middle of the secondary winding of one of the transformers. The transformers are connected in delta on the secondary side to provide 240 volts between phases 1, 2, and 3. The midtap, brought out from transformer 2, therefore, measures 120 volts to phases 2 and 3. The bank is similar to a three-

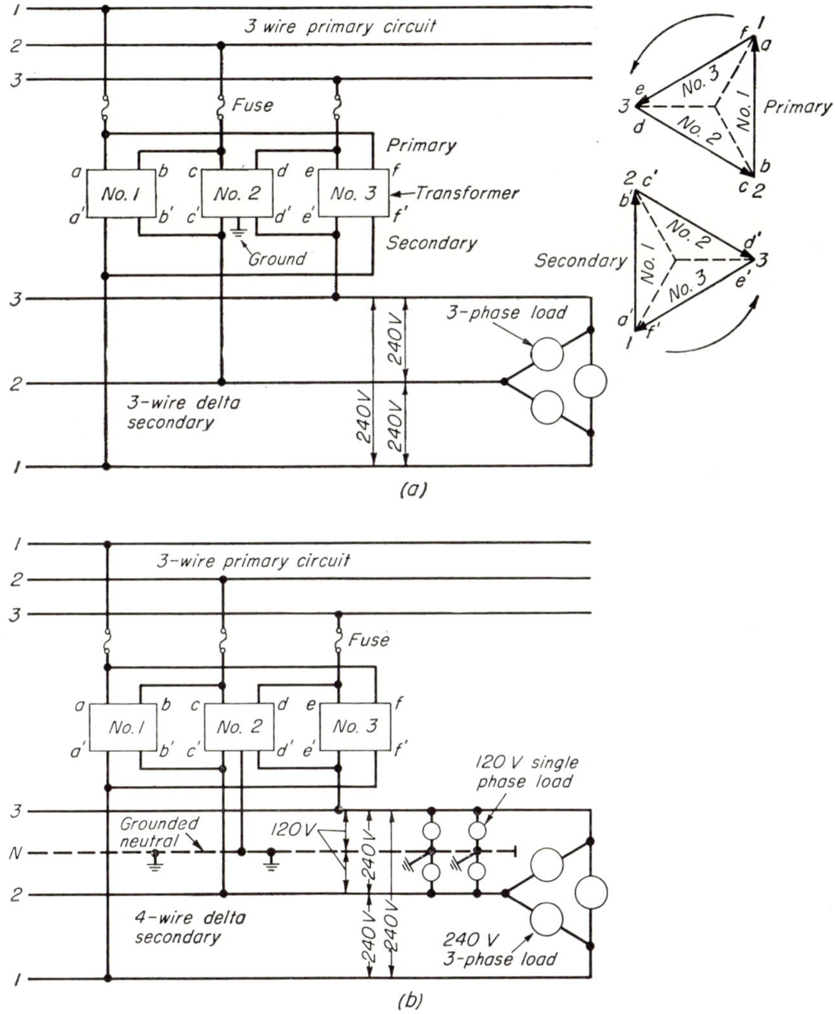

FIG. 6-12. Delta-delta transformer bank connections. By bringing out center-winding connection in (*b*), a 120/240-volt service can be given on one phase.

phase bank, Fig. 6-12*a*, plus a single-phase transformer, superimposed upon the three-phase circuit. The circuit is suitable to serve a balanced three-phase power load plus an unbalanced single-phase lighting and appliance load

Since the single-phase load is not balanced between all three phases, the currents in the phase wires will be unequal and one transformer must be larger to prevent overload with any appreciable amount of single-phase load. Such a circuit may be designated as an unbalanced three-phase four-wire circuit. The midtap is called the neutral of the four-wire delta circuit and should be grounded. In Fig. 6-12*b*, with the midtap of transformer 2 connected to the grounded-neutral wire, the voltage to ground of phases 2 and 3 will be 120 volts while the voltage to ground of phase 1 will be 240 volts × sin 60°, or 208 volts. Even if the bank supplies three-phase power load only, and a three-wire circuit only is required, a midtap from one transformer is usually brought out and grounded as shown in Fig. 6-12*a*.

The vector sum of the voltages across the windings of a transformer bank connected in delta on a three-phase system must equal zero. Since

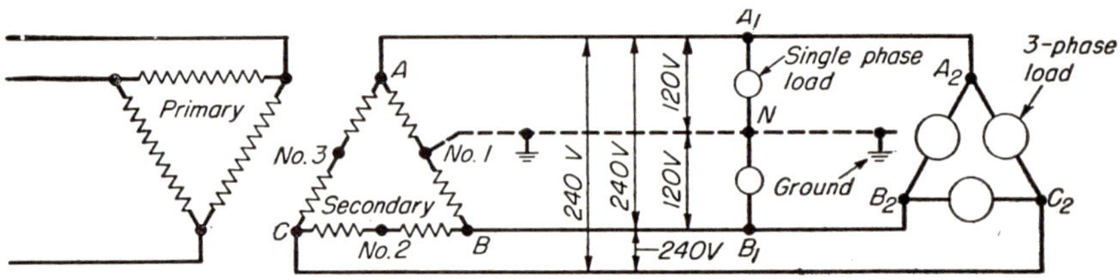

FIG. 6-13. Circuit diagram of single-phase transformers hooked up in delta-delta with 120/240-volt service on one phase.

any symmetrical three-phase system is made up of three equal single-phase voltages 120 electrical degrees apart, any three transformers connected in delta on such a system must have equal voltage ratios. If the vector sum of the delta voltages does not equal zero, it will be reduced to zero, by driving a circulating current through the closed circuit formed by the delta windings. If the difference in voltage ratios is excessive, the circulating current may be large enough to burn out a winding. Similarly, the vector sum of the transformer-impedance voltage drops in a delta-connected bank must equal zero. This feature controls the division of load between transformers connected in a three-phase delta-delta bank.

A "CSP"-type transformer with small secondary breakers in the secondary leads should not be used as the transformer with midtap for 120-volt load in a four-wire delta-delta connected bank. If such a transformer were used as transformer No. 1 in Fig. 6-13, circuit breakers would be located in the secondary leads at points *A* and *B* within the delta. If these switches were to open, the single-phase load connected from *A* to *N* and from *B* to *N* would be supplied from transformers No. 2 and No. 3 acting in open delta, and the neutral would be floating between phases

A and B. If, at any time, the single-phase load connected from A to B through the neutral N becomes unbalanced, the voltage from the higher loaded phase leg to neutral will drop while the voltage from the lighter loaded phase leg to neutral will rise proportionately. This rise in voltage, due to neutral shift, resulting from load unbalance, may be sufficient to damage or destroy consumer's 120-volt equipment.

In Fig. 6-13, the single-phase load A_1B_1 is connected across phases A and B, supplied by transformer AB together with transformers AC and CB, in series with each other and in parallel with transformer AB. If, at a given instant, current flows from A to A_1 to B_1 and back into the transformer bank at B, it will flow through the transformer bank from B to A and through the alternate parallel path from B to C to A. This single-phase load current will be divided between transformer AB, constituting one path, and transformers AC and CB, a parallel path, in inverse proportion to the impedance of the two parallel paths (see Arts. 1-28 and 1-29). The proportion of single-phase load supplied by transformer AB is equal to:

$$\frac{Z_2 + Z_3}{Z_1 + Z_2 + Z_3}$$

where Z_1 = impedance of transformer AB
Z_2 = impedance of transformer BC
Z_3 = impedance of transformer CA

If the single-phase load is an appreciable portion of the total combined load, the transformer across which the single phase load is connected may, of necessity, be somewhat larger than the other two transformers and consequently will have a lower ohmic value of impedance. Ohmic value of impedance varies inversely with the kva rating (see Art. 1-30). If all three transformers have the same ratio of resistance to reactance, the ohmic values of impedance can be added arithmetically since there will be no angular difference between Z_1, Z_2, and Z_3. In this case, the proportion of single-phase load supplied by the two parallel transformer paths can be expressed by the following equations:

$$\frac{(\%\,Z_2K_1K_3 + \%\,Z_3K_1K_2)100}{\%\,Z_1K_2K_3 + \%\,Z_2K_1K_3 + \%\,Z_3K_1K_2} = \%\,L_1 = \text{percent single-phase load supplied by transformer } AB$$

$$\frac{(\%\,Z_1K_2K_3)100}{\%\,Z_1K_2K_3 + \%\,Z_2K_1K_3 + \%\,Z_3K_1K_2} = \%\,L_{2-3} = \text{percent single-phase load supplied by transformers } BC \text{ and } CA \text{ in series}$$

where K_1, K_2, and K_3 are the kva rating of transformers AB, BC, and CA, respectively.

If transformers BC and CA are identical in kva size and percent impedance, the above equations reduce to the following:

$$\frac{2(\% ZK_1)100}{2(\% ZK_1) + \% Z_1K} = \% L_1$$

$$\frac{(\% Z_1 K)100}{2(\% ZK_1) + \% Z_1K} = \% L_{2-3}$$

where $\% Z$ = percent impedance of transformers BC and CA
$\% Z_1$ = percent impedance of transformer AB
K_1 = kva rating of transformer AB
K = kva rating of transformers BC and CA

If all three transformers in the bank have the same percent impedance and the same ratio of resistance to reactance, the above equations for single-phase load division further reduce to the following:

$$\frac{2K_1 \times 100}{2K_1 + K} = \% L_1 \qquad \text{and} \qquad \frac{100K}{2K_1 + K} = \% L_{2-3}$$

The total load on each transformer is approximately equal to one-third the balanced three-phase load plus its portion of the single-phase load.

The above equations provide only an approximate solution for the determination of load on each transformer since both the percent impedance and the ratio of R to X may vary somewhat between different kva sizes. Further error is also introduced in combining single-phase and three-phase power load arithmetically, since there is usually a difference in power factor between the two types of load. In distribution work it is usually satisfactory to make these assumptions, however, since the error will be comparatively slight and on the safe side. In Fig. 6-13, if single-phase current flows from B to A through transformer 1, it will flow from B to C through transformer 2 and from C to A through transformer 3 at the same instant. At the same time three-phase current will be flowing from B to A through transformer 1, from A to C through transformer 3, and from C to B through transformer 2. Thus, the single-phase and three-phase components of current are in the same direction through transformer 1 and in opposite directions through transformers 2 and 3. A similar reverse phase relation also holds true in determining the vector sum of current components through the other two transformers if you want to determine accurately the total current and phase angle through each transformer.

If the two wing transformers are made smaller in size, they will take a smaller proportion of the unbalanced single-phase load since decreasing the transformer size increases the ohms impedance of the parallel path through them. The two wing transformers in an unbalanced delta-delta bank, then, cannot be overloaded by single-phase load on a single-phase without first overloading the other transformer across which the single-phase load is connected. Where single-phase lighting load is combined with three-phase power load and supplied from a delta-delta connected transformer bank, both loads affect the regulation. If the circuit supplies three-phase power load only, the allowable regulation may be somewhat higher than if it supplies lighting load only. Service to a three-phase induction motor load usually will be entirely satisfactory with a line regulation as high as 5 to 10 percent, while satisfactory service to a lighting load may limit regulation to not more than 3 or 4 percent. If combined power and lighting load is supplied over the same secondary wires, a much larger conductor size may be required for the two-phase wires supplying the lighting load than for the power wire, in order to confine line regulation and lamp-flicker voltage dips within satisfactory limits.

With balanced load and a delta-delta connected transformer bank, the line current equals $\sqrt{3}$ times the transformer current. The equivalent phase-to-neutral or equivalent wye voltage is the voltage between phases, or the transformer voltage, divided by the $\sqrt{3}$. If one of the three primary fuses protecting a delta-delta connected transformer bank blows, so that one phase is open circuited, rated voltage will be across one transformer and one-half of rated voltage across the other two. For example, if the fuse connecting transformer 1 to primary phase 1 in Fig. 6-12 should blow, transformer 2 would be connected between the uninterrupted phases 2 and 3 while transformers 1 and 3 are connected in series between the same two phases. As a result, rated single-phase secondary voltage will occur between secondary phases 2 and 3 while one-half of rated secondary voltage will occur between phases 1 and 2, and 3 and 1. The voltage between phases 1 and 2, and 3 and 1 will be in phase with the voltage between phase wires 2 and 3, or single-phase voltage only will be present. Three-phase power load, such as a motor, in attempting to operate single-phase may cause a burned-out winding if not adequately protected against overheating. If the reduction to one-half voltage occurs across the lighting phase, a quick report of service failure will usually result and immediate restoration of the faulted phase can be made.

Three-phase Wye-delta. The wye-delta connection is shown in Fig. 6-14. This connection supplies balanced three-phase load or balanced three-phase load combined with unbalanced single-phase load just as in the case of a delta-delta connected bank. The primary supply, usually a grounded wye circuit, consists of three phase wires and one neutral wire. The neutral wire, running from the neutral of the wye-connected substation transformer bank supplying the primary circuit, may be completely independent of the secondary system, or it may be combined with the neutral of the secondary system. If the primary neutral is independent of the secondary system, it will usually be grounded at the substation only and is an isolated neutral. If the same wire serves as both a primary neutral and a secondary neutral, it will be grounded at many points, including each customer's service and is a multi-grounded common neutral. The common-neutral system is probably the most frequently used because of its greater economy in wire. In either case it is generally considered best not to connect the primary-bank neutral to the primary-circuit neutral if the bank is connected wye-delta. This is because such a connection between the neutrals is not necessary for

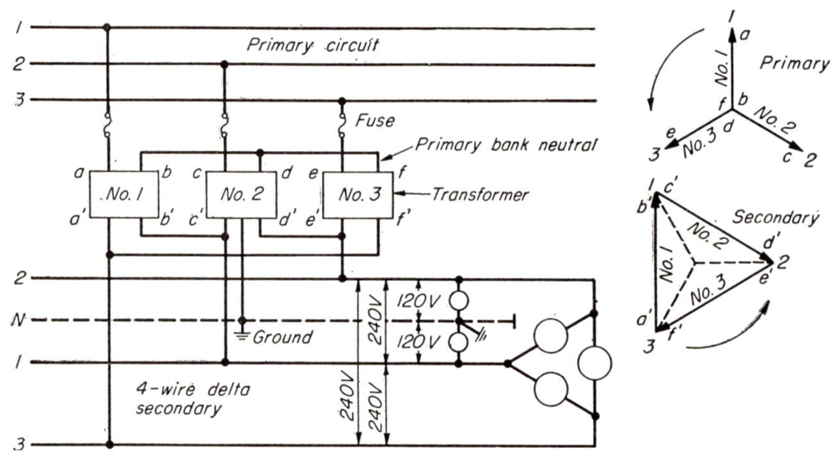

FIG. 6-14. Wye-delta transformer bank connections with four-wire secondary.

satisfactory operation of the bank and prevents a burned-out transformer winding during phase-to-ground faults.

The "CSP"-type transformer with small built-in breakers placed in the secondary winding leads should not be used in any wye-delta connected transformer bank with the primary bank neutral floating. In Fig. 6-14, if such transformers were used, circuit breaker contacts would be located within the transformer, in all secondary leads a' to f'. Referring to the vector diagram of Fig. 6-14, suppose that for the time being the three-phase load is disconnected and only single-phase load is being supplied and that the breaker in lead b' of transformer No. 1 opens. Under this condition the delta is open at b'. Load current will flow in the primary and secondary windings of transformer No. 2, but on the primary side the load current through transformer No. 2 can only find a return path through the windings of transformers No. 1 and No. 3 and phases No. 1 and No. 3. The current through the primary windings of transformers No. 1 and No. 3 must act as magnetizing current, since it is not balanced by any current in the secondary windings. With the secondary delta open at b' the primary bank neutral is free to shift. With this effect and with saturation effects due to high magnetizing currents in transformers No. 1 and No. 3, the primary bank neutral *bdf* will shift toward *c*. The voltage vector *dc*, thus will shorten and the voltages *ba* and *fe* will lengthen. This will produce a proportional distortion in voltage on the secondary side. Serious distortion of the delta and unbalanced secondary voltages will occur with either single-phase or balanced three-phase load or both and regardless of which of the secondary breakers opens. These effects may be serious enough to cause damage to consumer's equipment due to high voltage or overheating.

In Fig. 6-14, if the transformer-bank neutral connects to the grounded neutral wire running back to the substation and a low-resistance primary fault occurs from one phase wire to ground, a high primary current will flow. This large fault current causes an excessive voltage drop in the faulted phase from the substation to the fault. The primary voltage across any transformer connected between the faulted phase and the grounded neutral will be greatly reduced below the voltage across the primary windings of transformers connected between the other two unfaulted phases and the neutral. This is particularly true in the case of a three-phase bank near the fault where the voltage to ground of the faulted phase may approach zero. The connection of the bank neutral to the circuit neutral will stabilize the primary neutral point so that the three induced secondary voltages will always differ by 120 electrical degrees.

Since the vector sum of the three voltages in a delta connection must equal zero, it is seen, by referring to Fig. 6-14, that the voltage between secondary phase wires 1 and 3 will be less than the voltage between phases 2 and 1, and 2 and 3 when the primary voltage across transformer 1 is reduced by a fault to ground on phase 1. Consequently the vector sum of the secondary delta voltages 1 and 3, 3 and 2, and 2 and 1 will not equal zero. If the vector sum of these delta voltages is not equal to zero, it causes a circulating current in the delta-connected windings. If current flows in the secondary windings of the transformers, a corresponding current in proportion to the winding ratio will flow in the primary windings. Thus, the two unfaulted phases will feed current into the faulted phase through the windings of the wye-delta connected bank. This may cause bank fuses to open or one or more of the transformer windings in a wye-delta bank to be burned out or seriously overheated if the fault is near the bank and if the fault resistance is low so that a high fault current flows. If the bank is farther away from the fault, the current through the windings will be reduced by the greater impedance in the circuit or by the voltage impressed across the primary winding of the faulted phase transformer being greater.

If the primary neutral of a wye-delta connected bank is solidly connected, cutting off one transformer from the primary source will not immediately interrupt service or noticeably reduce the quality of service. The remaining two transformers will attempt to supply the entire load. If a wye-delta connected bank is loaded to rated capacity with balanced three-phase load and all three transformers are the same size, the removal of one transformer by the blowing of a sectionalizing line fuse or a protective bank fuse will increase the load on the two remaining transformers to 173 percent of rated capacity (see open-wye open-delta connection). If the bank supplies combined three-phase and single-phase load, disconnection of the large transformer by fuse operation causes an even higher overload on the two wing transformers. With this connection, the blowing of a single protective bank fuse is difficult to detect since no diminution in service quality is immediately apparent, and one of the two remaining transformers may be burned out by the overload. If the bank neutral is not connected to the primary-circuit neutral, but is left isolated as shown in Fig. 6-14, disconnection of one transformer causes a partial service interruption without danger of a transformer burnout. For example, if the protective fuse between phase 1 and transformer 1 blows, transformer 1 will be disconnected from its supply, and its secondary voltage between secondary phases 1 and 3 will collapse to zero. Transformers 2 and 3, however, will now be connected in series between primary phases 2 and 3. The secondary voltage between phases 1 and 3 will be equal to zero, while that between phases 1 and 2, and 2 and 3 will be reduced from 240 to 208 volts and will be in phase with one another instead of having a phase-angle difference of 120°. The two remaining transformers are not likely to burn out, but there is some chance that the three-phase motors supplied may be damaged, unless protected by thermal devices, by attempting to operate single-phase at reduced voltage.

Kirchhoff's first law states that the sum of the currents flowing toward a junction point in a circuit equals the sum of the currents flowing away from the point. In the case of ac this means the vector sum, to take into account angular difference. If the load is unbalanced between phases and the neutral of a wye-delta connected bank is connected to the primary-circuit neutral, the vector sum of the currents flowing in the primary windings toward the neutral point of the bank will not equal zero but will equal the current flowing in the neutral away from the bank. When the bank neutral is isolated from the circuit neutral and the bank supplies combined single-phase and three-phase load, as in Fig. 6-14, the single-phase load must divide so that two-thirds is supplied by the transformer across whose terminals the unbalanced single-phase load is connected on the delta secondary and one-third by the other two transformers in series, regardless of size or impedance, to satisfy the requirements of Kirchhoff's first law. The rated capacity required in a wye-delta connected bank with an isolated bank neutral to supply a combined load can be approximated by

$$\frac{2S + T}{3} = \text{rated transformer capacity across lighting phase}$$

$$\tfrac{1}{3}\sqrt{S^2 + ST + T^2} = \text{rated kva capacity of each wing transformer}$$

where S = single-phase load, kva
T = three-phase load, kva

The above equations assume that both the single-phase and three-phase loads are at unity power factor. For distribution purposes this

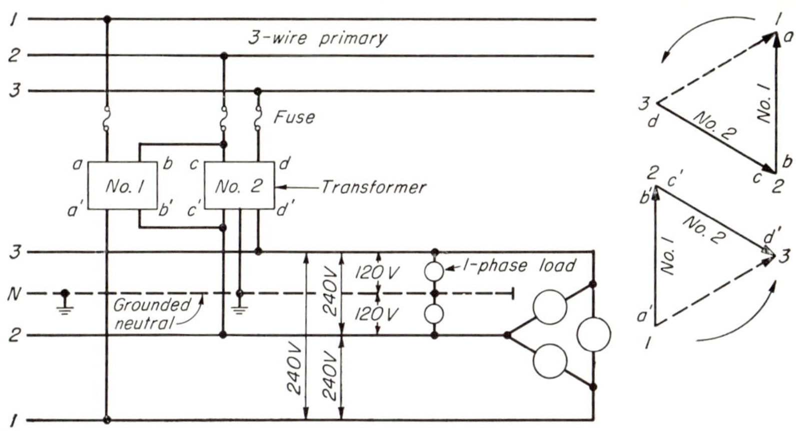

FIG. 6-15. Open-delta primary open-delta secondary bank connections.

approximation is on the safe side and close enough for determining the size of bank required to supply a specific load. The wye-delta connection is used to best advantage when the load is completely balanced three-phase or when the unbalanced single-phase load is small in comparison to the three-phase load. Since one-third of the unbalanced single-phase load current passes through the two wing transformers in series, the total transformer capacity required for the single-phase load is larger than the load itself.

Three-phase Open-delta Open-delta and Open-wye Open-delta. If one of the transformers in the delta-delta connected bank of Fig. 6-12 is disconnected, the two transformers remaining form an open-delta open-delta bank, as in Fig. 6-15. This connection provides a three-phase transformation using two transformers and all three-phases of a three-phase system. The primary supply is usually a three-wire delta circuit, although, a wye circuit can be used if the transformers are connected across phases instead of phase-to-neutral.

If one of the transformers in the wye-delta connected bank of Fig. 6-14 is disconnected and the bank neutral connected to the circuit neutral, the two transformers remaining form an open-wye open-delta bank, Fig. 6-16. This connection provides a three-phase delta secondary from only two phases of a three-phase wye primary circuit. The primary supply must include at least two phase wires and the neutral wire of a wye circuit. If the bank neutral is not connected to the circuit neutral, the two transformers will be in series between the two phases and the secondary voltage will be only single-phase and reduced in value, as in the case of

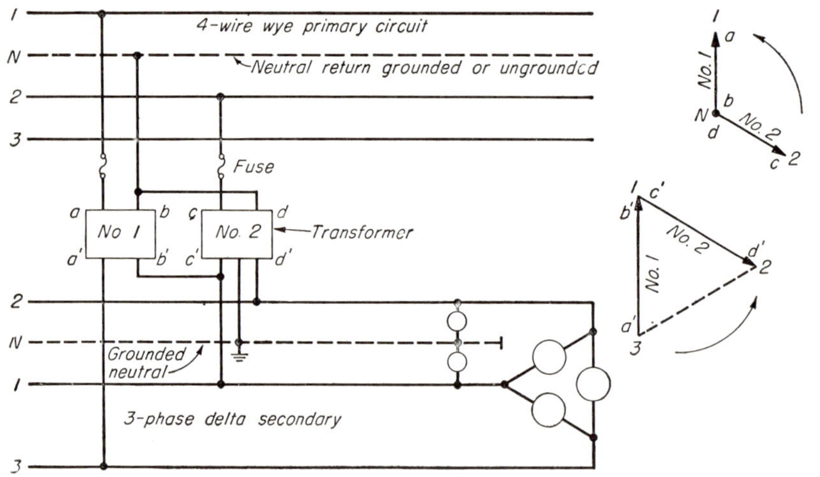

FIG. 6-16. Open-wye primary open-delta secondary transformer-bank connection.

the wye-delta bank when one transformer disconnects by fuse operation. When two transformers are in open delta, the voltage between the two phase wires across which no transformer winding connects equals the voltage between the other phases and is the vector sum of these voltages, as shown by the dotted vectors of Figs. 6-15 and 6-16.

In the delta-delta bank of Fig. 6-12a at the instant the current flow in the secondary of transformer 1 is from terminal a' to b', or from phase 1 to phase 2, the flow of current through the other two transformers will be from terminals c' to d' and from e' to f', as shown by the vector diagram. Assuming that current flows from the bank into each secondary phase wire, the current into phase 1 will equal the vector difference between the secondary currents through transformers 3 and 1, in accord with Kirchhoff's first law, or

$$I_1 = I_{e'f'} - I_{a'b'}$$

Also, in a similar manner,

$$I_2 = I_{a'b'} - I_{c'd'} \qquad \text{and} \qquad I_3 = I_{c'd'} - I_{e'f'}$$

If the bank serves only balanced three-phase load, as indicated in Fig. 6-12a, then

$$I_{a'b'} = I_{c'd'} = I_{e'f'} \qquad \text{and} \qquad I_1 = I_2 = I_3$$

Under this condition the line current (I_1, I_2, or I_3) will be 1.732, or $\sqrt{3}$, times the transformer current ($I_{a'b'}$, $I_{c'd'}$, and $I_{e'f'}$) and will be shifted 30° out of phase with the transformer current as in the vector diagram of Fig. 6-17. If one of the transformers in the delta-delta bank of Fig. 6-12a is removed and the two remaining transformers are left connected open-delta open-delta, the line current flowing to the balanced three-phase load will be unchanged but the transformer current will now equal the line current, or $\sqrt{3}$ times as much as in the delta-delta bank. The capacity of the open-delta bank, thus, will be reduced to $1/\sqrt{3}$ or 57.7 percent of the closed-delta bank although the open-delta bank has a rated transformer capacity equal to 66.7 percent that of the closed-delta bank. In other words, the rated nameplate capacity of the open-delta bank is only 57.7%/66.7%, or 86.7 percent, usable in serving three-phase load. This means that each of the two transformers in an open-delta bank will need to be 1.732 times as large as in a closed-delta bank to supply the same three-phase load. The same relations between line and transformer current will hold on the primary side for open- and closed-delta.

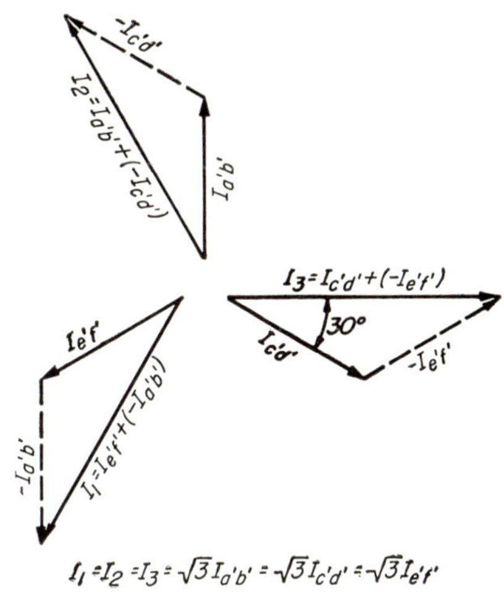

FIG. 6-17. Relation between primary and secondary currents in a delta-delta transformer bank.

If the bank is connected open-wye open-delta, the same conditions apply on the secondary side as for the open-delta open-delta connection. On the primary side the grounded neutral acts as a common return for the primary current through each of the two transformer primary windings. Since the transformer current on the secondary side equals $\sqrt{3}$ times as much as for a closed-delta connection, the current through the primary winding of the open-wye open-delta bank will be $\sqrt{3}$ times as great as for a wye-delta bank. The current in the neutral return equals the vector difference of the two transformer primary currents, or the neutral current equals $\sqrt{3}$ times the transformer current. The neutral current of the open-wye open-delta bank, then, is equal to $\sqrt{3} \times \sqrt{3}$, or three times the current through the primary windings of a wye-delta bank supplying the same balanced three-phase load. If either of the open-delta connections supplies combined single-phase and three-phase load, all the single-phase load is supplied by one of the transformers while the three-phase load is divided between both transformers. The transformer size for an open-wye open-delta bank or an open-delta open-delta bank can be approximated as follows:

(0.577 × three-phase load kva) + single-phase load kva = transformer kva size required across phase supplying single-phase load

(0.577 × three-phase load kva) = transformer kva size required across the other phase to supply three-phase load

The open-delta open-delta bank may be used to advantage in place of the closed-delta bank to serve mixed load when the single-phase load is equal to or greater than the three-phase power load. The open-wye open-delta bank has certain distinct advantages over the wye-delta bank to serve mixed load when the single-phase load is about one-half the three-phase load or larger because of the arbitrary division of single-phase load in the wye-delta bank into components of two-thirds and one-third. Neither of the open-delta connections will usually be found economical for use if balanced three-phase load only is to be supplied, except possibly for quite small three-phase loads in the case of the open-wye open-delta connection.

Three-phase Delta-wye. Combined three-phase power and single-phase lighting and appliance load may be supplied by a three-phase transformer bank connected delta-wye, as shown in Fig. 6-18. The single-phase load is not balanced between two phase wires and a midtap on one transformer as in the case of a four-wire delta-connected secondary but is equally divided between all three transformers, being connected between phase and neutral of all three phases forming the wye secondary. Balanced three-phase load may be thought of as three equal single-phase loads connected between phases or connected between each phase wire and neutral. In either case, the line currents are equal and balanced. If the load is balanced, three single-phase transformers of equal size form the bank. If the load is not balanced, each transformer should be equal to its unbalanced single-phase load combined with one-third of the balanced three-phase load. The single-phase load should be kept in as close balance between phases as possible, to assure most economical use of transformer capacity and to prevent poor regulation from excessive load on one phase.

It is sometimes desirable to supply balanced three-phase load such as 220-volt induction motors from a three-phase four-wire 120/208-volt wye secondary circuit. Since the torque developed by an induction motor is proportional to the square of the voltage, the starting torque will be reduced to about 89 percent of normal when 208 volts from a wye circuit is applied to a 220-volt motor. This reduction in starting torque will sometimes cause unsatisfactory operation of the motor. In such cases it is best to use motors rated at 208 volts and designed to operate on a 208-volt wye circuit, but where this is not feasible the starting and running characteristics of the motor can be corrected by autotransformers to step up the three-phase voltage. The delta-wye connection has distinct

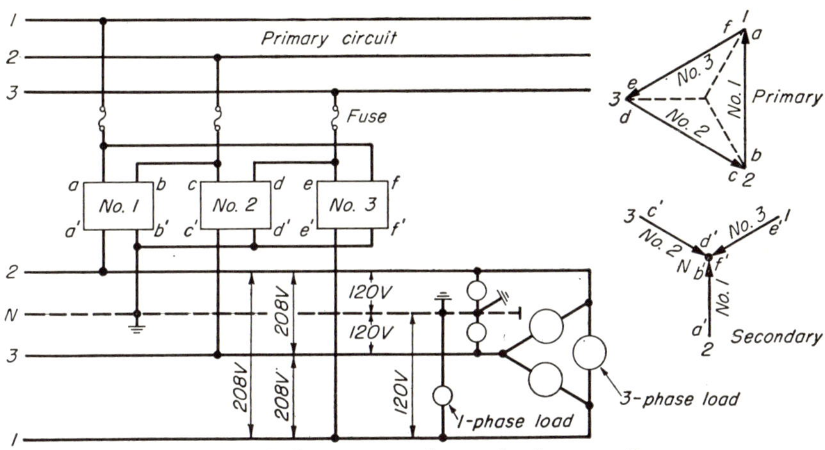

FIG. 6-18. Delta-wye transformer-bank connection.

advantages when supplying combined loads where the load density and the proportion of three-phase load is high. An example of this is found in a congested business section where the load density warrants an automatic low-voltage ac network system or in lower density load areas where single- and three-phase loads warrant the nonautomatic network-type secondary system, such as with banked transformers. The delta-wye bank is also useful where the single-phase load is large enough to cause troublesome unbalancing on the primary or where excessive conductor sizes will be required on the two lighting phases on the secondary side.

Three-phase Wye-wye. The wye-wye connected transformer bank is shown in Fig. 6-19. It is used under the same conditions and to supply the same mixed distribution loads as the delta-wye connection. The division of load between transformers is also similar to that of the delta-wye. Care should be exercised in making the wye-wye connection to make sure that a wire connects primary-bank and primary-circuit neutrals. If not, two primary windings will be in series between each two phase wires. Current into a transformer from its phase wire can return only through the other two phases since there is no neutral return. Thus, the wye-wye without connected primary-bank and primary-source neutrals can operate satisfactorily only when the load is exactly balanced between each phase and neutral on the secondary side with equal phase-to-neutral voltages. Only then will the vector sum of the phase currents add up to zero at the neutral point so that the return current equals zero.

In Fig. 6-19, suppose that the only load on the bank is a single-phase load between phase 1 and neutral. Current will flow out to the load through phase 1, on the secondary side, and return to transformer 1 through the secondary neutral without any current flowing in phase

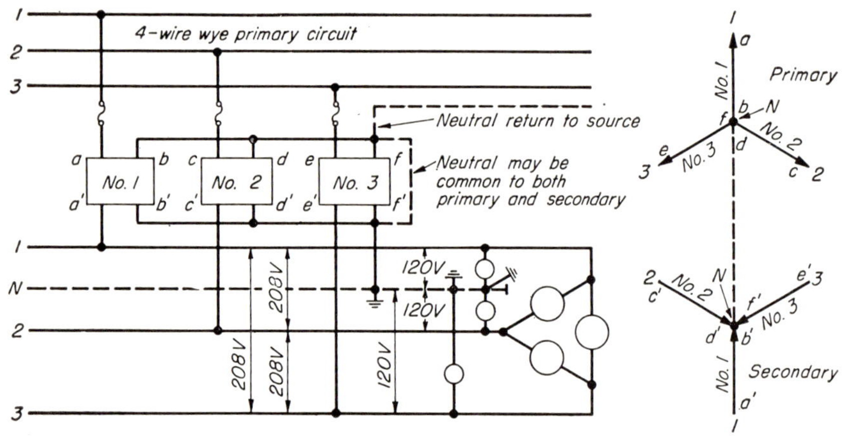

FIG. 6-19. Wye-wye transformer-bank connection.

wires 2 and 3 or through the secondary windings of transformers 2 and 3. On the primary side, current will flow into transformer 1 from phase 1 and return to the source through the primary neutral without affecting either of the other two phases or transformers. Now, if the connection between the bank and source neutrals is removed, the current on the secondary side will be unchanged. But on the primary side current into transformer 1 from phase 1 must flow into the bank neutral through the primary winding of transformer 1 and return through the primary windings of transformers 2 and 3 and their corresponding phases. Since this return current is not balanced by a corresponding current and ampere-turns through the secondary windings of transformers 2 and 3, it must act as a magnetizing current. This overexcitation results in high voltages and a shift in phase on the secondary side just as though the neutral point had violently shifted, producing distortion in the secondary wye. If the primary neutral return is omitted, even a slight unbalance of load can produce an appreciable shift in the neutral with a corresponding distortion of the wye.

The solidly grounded neutral of the various secondary circuits is often used both as a secondary and primary neutral by filling in the gaps in the neutral between adjacent secondary circuits, either single- or three-phase, and running back to the substation. A solid connection is usually made between the primary and secondary bank neutrals and the common neutral return.

In the preceding, the three-phase transformer banks principally used single-phase transformers. Three-phase transformers, however, are frequently used in distribution practice. The transformer has the bank connection already made between windings. One frequently used three-phase transformer is the vertical or triplex type. This transformer has three single-phase core and coil assemblies mounted, usually one above the other, in a single tank with the primary- and secondary-coil leads internally connected and the bank leads brought out through insulator bushings. Standard three-phase transformers consisting of a single laminated core carrying three separate primary and secondary coil assemblies are also frequently used. These transformers have particular advantages where the bank is large or where the space is scarce as in a low-voltage underground network system.

6-6. Voltage Diagrams for Phasing and Paralleling. It is sometimes necessary to construct voltage vector diagrams to (1) correctly connect a three-phase bank in parallel with an adjacent transformer bank, (2) determine the correct phase markings, or (3) determine the correct bank connections. These all depend upon the transformer polarity. The majority of distribution transformers are of additive polarity (see Art. 1-31). In Fig. 1-63*b*, if at a given instant the induced voltage in the primary winding is from H_2 toward H_1, the induced voltage in the secondary winding will be from X_2 toward X_1 so that secondary current will flow out X_1 and return through X_2.

When representing the voltages induced in the primary and secondary windings by vectors, the primary-induced-voltage vector can be given any chosen direction with respect to the primary terminals. The secondary-induced-voltage vector must then be parallel to the primary vector, and in the same direction, since both voltages are induced by the same magnetic flux. If the observer faces the secondary side of the transformer and letters the primary winding leads *ab* and the secondary winding leads *a′b′*, from left to right, the induced-voltage vectors will be as shown in Fig. 6-20 for additive and subtractive polarity and for the assumed direction of induced primary voltage. Note that the letter at the arrow end of the secondary-induced-voltage vector is reversed for additive as compared with subtractive polarity.

The use of arrowheads on vectors is not necessary if the difference between the lettering of vector ends for additive and subtractive polarity is remembered. For three transformers in a three-phase bank the terminals may be successively lettered *abcdef* on the primary side and *a′b′c′d′e′f′* on the secondary side. In a three-phase system, the three

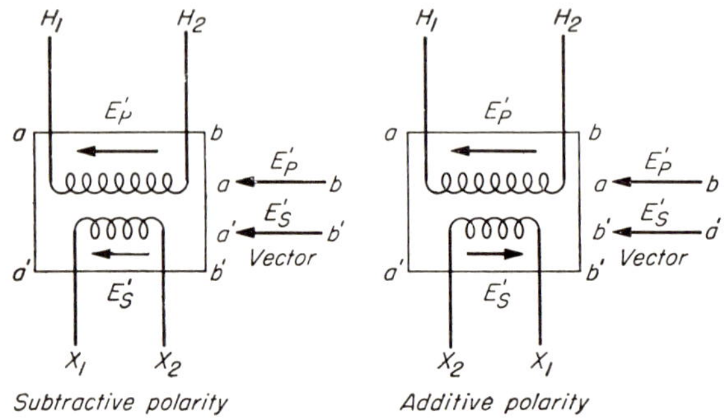

FIG. 6-20. Polarity of transformers can be identified by the lettering given the primary and secondary terminals.

voltages differ in phase by 120 electrical degrees. Vectors representing voltages induced in primary and secondary windings of a three-phase bank, before the transformers are interconnected, are 120° apart, as shown in Fig. 6-21. Letters identify the terminal ends of the transformer winding. The primary is assumed to be the source and the secondary the load, regardless of voltage.

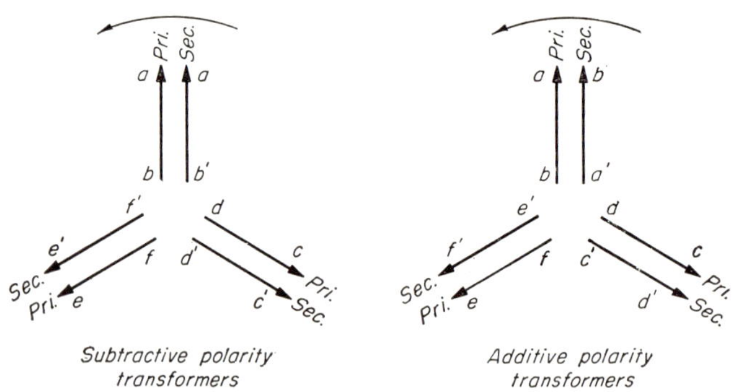

FIG. 6-21. In connecting three-phase transformers, the secondary side must be connected in the same sequence as the primary side.

The actual rotation, or sequence in which phase voltages rise to a maximum, is determined by the generator. In constructing voltage diagrams or checking phasing, however, any convenient sequence of phase identification may be used on the primary side so long as the same sequence and method of phase identification are consistently followed. Counterclockwise rotation of the vectors is assumed, and the identification of phases on the source side is known or assumed. In Fig. 6-21, if vector terminal ends *af*, *bc*, and *de* for primary voltages and *a′f′*, *b′c′*, and *d′e′* for secondary voltages are connected, the voltage-vector diagram for a delta-delta bank will be formed. If the voltage vectors for additive polarity transformers are used, they will be like Fig. 6-12, forming a delta-delta connection. If subtractive-polarity transformers are connected as in Fig. 6-12 or if secondary terminals *a′d′*, *c′f′*, and *e′b′* of additive-polarity transformers are connected, the vector-voltage diagram will be like Fig. 6-22, or the secondary delta diagram of Fig. 6-12 will be rotated through 180°. In a similar manner the bank connections may be checked or the voltage-vector diagram may be constructed for the wye-delta bank of Fig. 6-14, the open-delta banks of Figs. 6-15 and 6-16, the delta-wye bank of Fig. 6-18, or the wye-wye bank of Fig. 6-19 by making the proper terminal connections between the voltage vectors of Fig. 6-21.

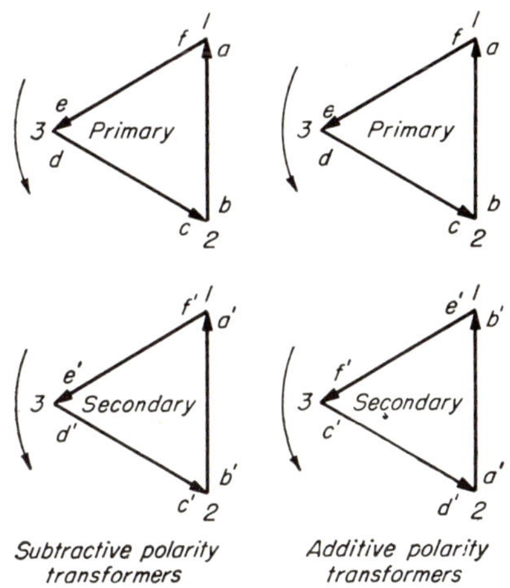

FIG. 6-22. Delta-connected transformers must also be connected in proper sequence on both primary and secondary sides.

After properly connecting the bank it is often necessary to provide phase identification on the secondary side of two adjacent transformer banks for paralleling or banking, or checking motor rotation when an existing primary supply is to be changed from delta to wye. The correct phase identification is usually known, or can be assumed, on the primary or source side. The primary phases may be arbitrarily identified as 1-2-3, *ABC*, *XYZ*, etc., so long as the same method is consistently followed. In the delta vector-voltage diagram of Fig. 6-12 a wye, in dotted lines, is drawn inside the delta for both the primary- and secondary-voltage diagrams. These dotted vectors represent the wye vector-voltage diagram equivalent to the delta vector-voltage diagram.

If the secondary delta is placed over the primary delta so their equivalent neutral points coincide, each equivalent phase-to-neutral voltage on the secondary side is 180° out of phase with the corresponding phase-to-neutral voltage on the primary side. Making the same comparison with the delta diagrams of Fig. 6-22, the equivalent wye voltages are found in phase for this connection. With a delta-delta connection, then, phase identification on the secondary side should correspond to the phase identification of the equivalent primary wye voltage with which its equivalent wye secondary voltage is in phase, or out of phase by 180°.

Thus, in Fig. 6-12, since terminals *af* on the primary connect to phase 1, the wire on the secondary side connected to terminals *a'f'* may also be identified as secondary phase 1 because its equivalent phase-to-neutral voltage is 180° out of phase with the equivalent phase-to-neutral voltage of primary phase 1. If the bank is connected wye-wye and the secondary wye-voltage diagram is placed over the primary wye, their neutral points coinciding, the secondary wye is in phase with the primary wye or out of phase by 180°.

If the equivalent wye of the delta side of a wye-delta or delta-wye bank is superimposed upon the real wye of the wye-connected side so that their neutral points coincide, the two wyes are always out of phase by 30°. There is always a 30° phase-angle difference between a delta and a wye connection. In the connection, Fig. 6-14, the equivalent wye of the secondary delta leads the primary wye by 30°. In the delta-wye of Fig. 6-18, the real secondary wye lags 30° behind the equivalent wye of the primary delta. In the case of wye-delta or delta-wye connections, the secondary wire or bank lead may be given the same phase identification as the primary phase whose real or equivalent wye voltage it leads or lags by 30°. Secondary phase identification may also be determined by a 90° phase-angle difference between primary and secondary phase-to-neutral voltages. Either method is equally satisfactory for delta-wye and wye-delta connections if consistently followed. Modifications of these methods of phase identification may be worked out whereby each phase is identified by its shift from a chosen basic wye. This method is

TABLE 6-3. AVERAGE LOSS AND IMPEDANCE CHARACTERISTICS OF 2,400- AND 4,800-VOLT TO 120/240-VOLT SINGLE-PHASE TRANSFORMERS

Size, kva	Iron loss, watts	Copper loss, watts	Total loss, watts	% R	% X	% Z
1½	22	36	58	2.40	1.20	2.68
3	32	60	92	2.00	1.43	2.46
5	40	102	142	2.04	1.66	2.63
7½	51	151	202	2.01	1.82	2.71
10	62	185	247	1.85	2.06	2.76
15	81	254	335	1.70	2.23	2.80
25	114	394	508	1.58	2.70	3.13
37½	149	530	679	1.41	2.90	3.22
50	186	642	828	1.28	2.95	3.22
75	278	947	1,225	1.26	3.00	3.24
100	265	1,200	1,465	1.20	3.10	3.32
150	550	1,700	2,250	1.13	3.40	3.58
167	620	1,835	2,455	1.10	3.60	3.76

TABLE 6-4. AVERAGE LOSS AND IMPEDANCE CHARACTERISTICS OF 7,200- AND 7,620-VOLT TO 120/240-VOLT SINGLE-PHASE TRANSFORMERS

Size, kva	Iron loss, watts	Copper loss, watts	Total loss, watts	% R	% X	% Z
1½	24	40	64	2.66	1.60	3.10
3	35	70	105	2.33	1.85	2.97
5	52	111	163	2.22	1.84	2.88
7½	64	157	221	2.09	1.90	2.82
10	80	194	274	1.94	2.05	2.82
15	109	263	372	1.75	2.36	2.93
25	155	390	545	1.56	2.64	3.06
37½	213	521	734	1.39	2.85	3.17
50	273	674	947	1.34	2.95	3.24
75	384	946	1,330	1.26	3.25	3.48
100	482	1,243	1,725	1.24	3.52	3.72
150	714	1,725	2,439	1.15	3.80	3.96
167	860	1,882	2,742	1.13	3.94	4.09

useful where phasing must be carried through several transformations involving both delta and wye connections and paralleling.

6-7. Transformer Characteristics. Values of iron loss, copper loss, total loss, percent resistance, percent reactance, and percent impedance for the most frequently used sizes of distribution transformers are given in Tables 6-3 and 6-4. These depend upon design characteristics and may differ slightly for transformers of different manufacture and voltage ratings. They represent average distribution-transformer characteristics. The percent values of resistance, reactance, and impedance of a single-phase transformer can be converted to ohms by

$$Z = \frac{\% Z \times E_T^2}{U_{1\phi} \times 10^5}$$

For a three-phase transformer or three single-phase transformers connected in a three-phase bank, the percent values can be converted to ohms by

$$Z = \frac{3 \times \% Z \times E_N^2}{U_{3\phi} \times 10^5}$$

where Z = resistance, reactance, or impedance, ohms
$\% Z$ = percent resistance, reactance, or impedance
E_T = single-phase transformer winding voltage
E_N = phase-to-neutral voltage of the three-phase connected bank
$U_{1\phi}$ = single-phase transformer kva
$U_{3\phi}$ = three-phase bank kva

CHAPTER REVIEW TOPICS

1. How do resistance and reactance compare in causing voltage drop in a circuit?
2. How do capacitors improve voltage regulation and increase circuit capacity?
3. What is the relation between kw, kva, kvar, and power factor?
4. Does it always pay to raise system power factor to 100 percent with capacitors?
5. How do distribution circuit power factors vary over a 24-hr period? Why?
6. Why should capacitors be switched in and out of a circuit?
7. How are capacitors protected against short circuits?
8. How should capacitors be connected to a distribution feeder for maximum advantage?
9. Compare the effects produced by shunt-connected and series-connected capacitors.
10. Can series capacitors be used instead of voltage regulators?
11. Why are protective by-passes used on series capacitors?
12. What factors must be considered in selecting series capacitors?
13. What causes lamp flicker?
14. What is the difference between cyclic and noncyclic lamp flicker?
15. Why does motor starting cause a dip in voltage?
16. Describe some of the methods of reducing or eliminating lamp flicker?
17. How may a single-phase transformer be connected to a three-phase circuit to supply 120/240-volt service?
18. Draw the circuit diagrams and vector diagrams and discuss the advantages and disadvantages of each of the following transformer circuits: (*a*) three-phase delta-delta, (*b*) three-phase wye-delta, (*c*) three-phase open-delta open-delta and open-delta open-wye, (*d*) three-phase delta-wye, (*e*) three-phase wye-wye.
19. How are absolute and percent values of resistance, reactance, and impedance related?

Problem 6-1: A 2,400/4,160-volt wye-connected distribution feeder carries a peak load of 200 amp at 80 percent power factor. The minimum daily load is approximately 90 amp at about 60 percent power factor. The circuit has a resistance of 0.60 ohm and an inductive reactance of 1.10 ohms to the equivalent feeder load center. (1) How many kva of shunt capacitors will be required, at or near the equivalent feeder load center, to raise the power factor at peak load from 80 to 95 percent? (2) What effect will the capacitors have upon the kva and the amperes in the line? (3) What effect will the capacitors have upon regulation and voltage drop in the line? (4) What will the power factor be at minimum load?

Solution:

1. Maximum load $= 3 \times 2{,}400$ volts $\times 200$ amp/1,000 $= 1{,}440$ kva
   When power factor $= \cos\theta = 0.80$; $\sin\theta = 0.60$
   Kilowatts of load $=$ kva $\times \cos\theta = 1{,}440 \times 0.80 = 1{,}152$ kw
   Kvar of reactive load $=$ kva $\times \sin\theta = 1{,}440 \times 0.60 = 864$ kvar
   $\sin\theta$ at 95% power factor $= 0.3123$
   $$\tan\theta \text{ at } 95\% \text{ power factor} = \frac{\text{kvar}}{\text{kw}} = \frac{\sin\theta}{\cos\theta} = \frac{0.3123}{0.95} = 0.329$$
   Kvar at 95% power factor $=$ kw $\times \tan\theta = 1{,}152 \times 0.329 = 379$
   Kvar of capacitors required to raise power factor from 80% to 95% $= 864 - 379$ $= 485$ kva

The nearest approach to this requirement, using standard single-phase units, is eleven 15-kva units per phase. This will amount to a total of 495 kva which will raise the power factor at peak load to 95.23 percent.

2. $I_r = 200 \text{ amp} \times 0.80 = 160 \text{ amp}$
   Maximum line amperes at 95% power factor $= 160 \div 0.9523 = 168$ amp
   Kva at 95% pf $=$ kw $\div \cos\theta = 1{,}152 \div 0.9523 = 1{,}213$
   Reduction in line amperes at peak load $= 200 - 168 = 32$ amp
   Reduction in demand $= 1{,}440 - 1{,}213 = 227$ kva
3. Voltage drop in line $= I(R\cos\theta + X\sin\theta) = I_rR + I_xX$
   Voltage drop without capacitors $=$
   $$200[(0.60 \times 0.80) + (1.10 \times 0.60)] = 228 \text{ volts}$$
   or $(228/2{,}400)100 = 9.5\%$
   Voltage drop with capacitors $=$
   $$168[(0.60 \times 0.9523) + (1.10 \times 0.3051)] = 152 \text{ volts}$$
   or $(152/2{,}400)100 = 6.33\%$
   Reduction in line voltage drop $= 228$ volts $- 152$ volts $= 76$ volts
   or $(76/2{,}400)100 = 3.17\%$
4. At minimum load, $\cos\theta = 0.60$ and $\sin\theta = 0.80$
   $I_r = I\cos\theta = 90 \times 0.60 = 54$ amp $\qquad I_x = I\sin\theta = 90 \times 0.80 = 72$ amp
   $$\text{Kw} = \frac{3 \times 2{,}400 \times 54}{1{,}000} = 389 \qquad \text{Kvar} = \frac{3 \times 2{,}400 \times 72}{1{,}000} - 495 = 23$$
   $$\text{Power factor} = \cos\left(\tan^{-1}\frac{\text{kvar}}{\text{kw}}\right) = \cos\left(\tan^{-1}\frac{23}{389}\right) = 0.9981 = 99.81\%$$

Problem 6-2: An intermittently operating motor which draws a locked-rotor starting current of 80 amp at 35 percent power factor, and a normal running current of 12 amp at 85 percent power factor, is to be supplied at 2,400/4,160 volts from the circuit of Prob. 6-1. It is proposed to install a series capacitor in the line, in place of the shunt capacitor of Prob. 6-1, to aid regulation and limit lamp flicker from frequent motor starting. (1) What will be the dip in lighting voltage caused by the motor starting without the series capacitor in the line? (2) What size series capacitor will be required to limit voltage dip at motor start to not more than 2.5 percent?

Solution:

1. At motor start, $I_r = I\cos\theta = 80 \times 0.35 = 28$ amp
   $$I_x = I\sin\theta = 80 \times 0.9568 = 75 \text{ amp}$$
   Voltage dip due to motor start, without capacitors,
   $$I_rR + I_xX = (28 \times 0.60) + (75 \times 1.10) = 99 \text{ volts} = 4.12\%$$
2. Required voltage rise due to capacitors $= 4.12\% - 2.5\% = 39$ volts $= V$
   Voltage rise due to series capacitor $= V_r = I_xX_c$
   $$\text{Capacitor reactance} = X_c = \frac{V_r}{I_x} = \frac{39 \text{ volts}}{75 \text{ amp}} = 0.521 \text{ ohm}$$
   Maximum line current without motor $= 200$ amp at 80% pf
   Total $I_r = 160 + 28 = 188$ amp at motor start
   Total $I_x = 120 + 75 = 195$ amp at motor start
   $$\tan\theta = \frac{I_x}{I_r} = \frac{195}{188} = 1.0372 \qquad \cos\theta = 0.694 \qquad \sin\theta = 0.720$$
   $$\text{Momentary line current at motor start} = I = \frac{I_r}{\cos\theta} = \frac{I_x}{\sin\theta} = 271 \text{ amp}$$
   Necessary continuous current rating $= 0.666 \times 271$ amp $= 180$ amp

Minimum rated voltage of capacitor $= I \times X_c = 180 \times 0.521 = 94$ volts
Nearest standard size = 230-volt 32.6-amp 7.5-kva unit
X_c of 7.5-kva unit = 230 volts ÷ 32.6 amp = 7.05 ohm

Maximum number of units per phase which can be used $= \dfrac{7.05}{0.521} = 13$

Minimum number of units per phase which can be used $= \dfrac{180 \text{ amp}}{32.6 \text{ amp}} = 6$

Allowable momentary amperes per unit = 150% of 32.6 amp = 48.9 amp
Normal peak load with motor running = 212 amp

Amperes per unit $= {}^{271}\!/_{13} = 20.85$ amp = 63.8% of rating when motor starts with 13 units per phase
$= {}^{271}\!/_{6} = 45.17$ amp = 138.5% of rating when motor starts with 6 units per phase
$= {}^{212}\!/_{13} = 16.3$ amp = 50% of rating with motor running and with 13 units per phase
$= {}^{212}\!/_{6} = 35.33$ amp = 108.3% of rating with motor running and with 6 units per phase

CHAPTER 7

DISTRIBUTION SYSTEMS—PROTECTION

By E. M. Adkins

Jersey Central Power and Light Company

7-1. Lightning Protection. Under certain meteorological conditions a storm cloud may become heavily charged with static electricity. The charged cloud attracts an equal charge of opposite polarity to the surface of the earth underneath and in any other cloud within range of its electrostatic field. The cloud corresponds to one of the plate electrodes of a capacitor and the earth to the other, while the air between corresponds to the dielectric, see Art. 1-21. If the cloud becomes charged heavily enough, the potential difference between cloud and earth exceeds the air's dielectric strength, and discharge in the form of an electric arc between cloud and earth takes place. The electric arc, a lightning stroke, eliminates the potential difference, restoring electric equilibrium between the two. In the majority of cases the cloud is negatively charged and the earth positively. The lightning stroke may also take place between two oppositely charged clouds without involving a stroke to earth.

The charge of opposite polarity attracted to the earth's surface by the charged cloud is concentrated in conducting and semiconducting bodies directly underneath the cloud. Overhead transmission and distribution lines directly beneath the charged cloud have a heavy charge of opposite polarity concentrated in them. The wires of the electric line are insulated from the conducting surface of the earth by the porcelain insulators and wood poles on which the lines are mounted. Up to the point of lightning discharge, the line insulation is subjected to no potential stress other than the normal stress supplied by the operating voltage of the line. The line conductors and the earth underneath carry a charge of the same polarity induced by the charged cloud. If the lightning stroke from the cloud makes contact with the conductors of the electric line, however, the full charge of the cloud transfers to the line, while the earth underneath retains its charge of opposite polarity. This direct stroke subjects the line insulation to the full potential stress since it is the only remaining dielectric keeping the two opposite charges apart. If the potential stress to ground exceeds the dielectric strength of the line insulation at any point, a breakdown to ground, such as an insulator flashover or puncture of winding insulation, occurs.

If the cloud discharges by a lightning stroke to the earth or to another cloud, instead of by a direct stroke to the electric line, the potential stress between cloud and ground is removed, leaving the original induced charge of opposite polarity to the cloud charge on the line. This induced charge on the line now attracts an equal and opposite charge to the surface of the earth beneath. This places the line insulation under potential stress as the induced line charge seeks a path to ground. Such stressing of the line insulation is the result of an induced lightning stroke.

Many close-by objects, such as trees, buildings, cliffs, and structures, often as high or higher than the conductors of a distribution line, partially shield the line from direct strokes. Fortunately a distribution line is far less subject to a direct stroke than an induced surge. A direct stroke may develop a potential over 10,000,000 volts, while an induced surge probably never exceeds 500,000 volts. The rate of discharge of a lightning stroke creates a wave of surge potential on the line more or less triangular in shape. This wave will have a shape similar to Fig. 7-1. The voltage rises quickly from zero to a maximum and then gradually tapers off to zero as the traveling wave passes the fixed reference point in the line or as it dissipates. A typical lightning voltage wave may rise from zero to its crest in 5 to 10 micro-seconds (millionths of a second) and then taper off to zero in 80 or 100 micro-seconds.

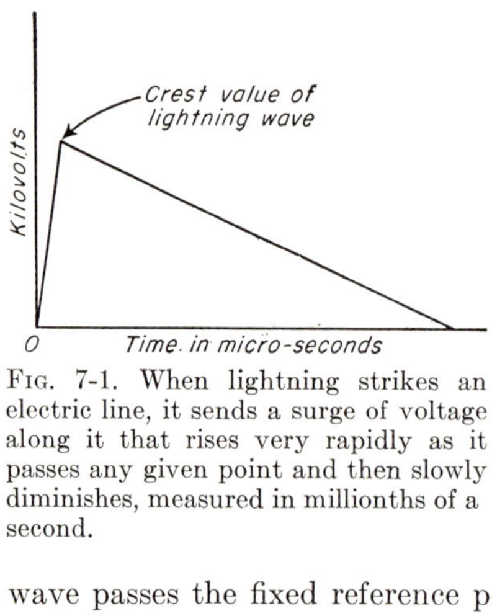

FIG. 7-1. When lightning strikes an electric line, it sends a surge of voltage along it that rises very rapidly as it passes any given point and then slowly diminishes, measured in millionths of a second.

The rate at which a lightning-voltage wave rises to its crest value and then tapers off is of considerable importance in determining the adequacy of line insulation to resist surge voltages. Insulation does not break down instantaneously when subjected to overvoltage but only after a time interval, or time lag. The voltage at which line insulation breaks down depends upon the rate of the surge-voltage rise and the length of time the insulation is subjected to the rising potential. A 60-cycle voltage wave rises from zero to maximum in $\frac{1}{4}$ cycle, or in $\frac{1}{240}$ second. This rate of rise is very slow compared to a lightning surge which rises from zero to its crest in perhaps 5 micro-seconds, or about 833 times as fast. In such a case the insulation would withstand a considerably higher voltage from the transient wave of lightning than from the 60-cycle wave.

The insulation of an overhead distribution line is not constant at every point. The porcelain insulators and the wood poles usually represent the highest degree of insulation against transient overvoltages, while insulated devices such as transformers, overhead and underground insulated cables connected to the overhead line, line capacitor banks, switches, and other pieces of distribution equipment offer a lower degree of protection against overvoltages such as lightning surges. A wave of lightning surge voltage traveling along distribution conductors seeks out the path of least resistance to ground, or in other words, the weakest link in the line insulation to ground. If at any point along the line of travel of the wave, the insulation resistance is low, a breakdown to ground occurs. To prevent the flashover and destruction of insulators and bushings, puncture of winding and cable insulation, a path to ground over which lightning surges can be dissipated must be provided. One of the earliest ways of doing this was by an air gap or spark gap. An open gap between the line and a low-resistance path to ground could effectively isolate the line from ground and still be small enough so that lightning potentials arc across to ground before the insulation or equipment is damaged. The spark gap provides a fair measure of lightning protection, but since the arc across the gap is of low resistance it also provides a low-impedance short-circuit path to ground or from phase to phase for the power circuit. As a result, clearing of the line through breakdown of the spark gap was followed by power-circuit current which operated line-overcurrent protective devices, such as fuses and circuit breakers, or destroyed equipment. Lightning protection now uses arresters which perform the same function as the spark gap but also quickly shut off the following power current before damage can result.

Two forms of lightning arresters are generally used for overhead distribution circuits, the valve type and the expulsion type. The valve type contains an open gap which operates similarly to the spark gap but in addition contains a material acting as a valve in shutting off the flow of power-system follow current, Fig. 7-2. A sealed porcelain or pyrex glass housing protects the arrester assembly from the weather and provides insulation to ground for the line connection. The gap assembly consists of one or more gaps in series. Breakdown of the gaps initiates arrester operation. The gaps isolate the line from ground under normal operating conditions. The lightning-arrester material is in the form of blocks, disks, or pellets and has a resistance which varies approximately in inverse proportion to the voltage impressed across it.

A lightning surge with voltage high enough to endanger the insulation of equipment arcs across the series gap of the arrester and passes through the arrester material to ground. With high surge voltage the resistance through the gap and arrester material will be quite low, practically short circuiting to ground. This removes the voltage stress of the surge from the line. As the surge passes and the voltage impressed across the arrester decreases, the internal resistance of the arrester rises in about inverse proportion to the decrease of impressed voltage until normal line voltage cannot maintain the arc and power current flow through the arrester. The lightning surge lasts only a small part of a half cycle, and the flow of power follow current is usually less than one half cycle.

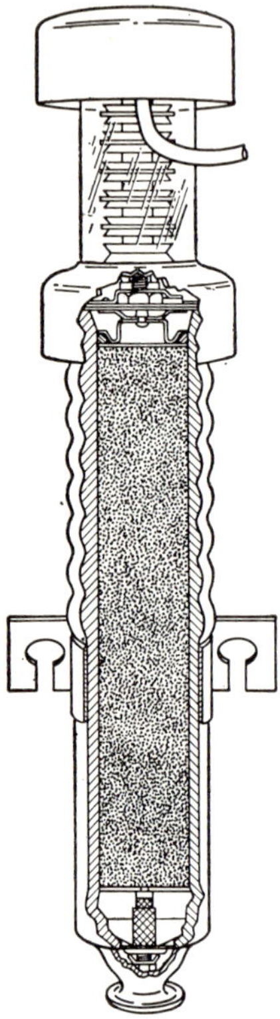

FIG. 7-2. Valve-type lightning arrester lets lightning jump across air gaps at the top to flow through pellet material into the ground through the bottom terminal. Internal line resistance rises with drop in surge voltage until current flow is shut off upon return to normal line voltage. (*Courtesy of Electric Service Manufacturing Company.*)

The expulsion-type arrester consists of one or more internal gaps, between electrodes placed within a tubelike holder made of horn fiber, and an external gap, placed between either the upper tube electrode and the line or the lower tube electrode and the ground connection, Fig. 7-3. The horn-fiber tube is placed within a weather-resisting or porcelain housing. The horn-fiber tube interrupts and prevents the flow of power follow current following lightning discharge. The external gap isolates the line from the arrester ground connection. The length of the external gap plus the internal gap or gaps determine the spark-over voltage at which the arrester breaks down to ground and begins discharge. The arc within the tube dislodges small fiber particles and introduces them into the arc stream. These nonconducting particles tend to deionize or lower the conductivity of the arc. This generates gas under considerable pressure within the arcing chamber of the fiber tube. The vented bottom of the fiber tube permits expulsion of the gas. The deionizing action of the fiber particles together with the expulsion of gas under high pressure literally blow out the power arc and prevent the flow of power follow current beyond one half cycle. When the low-resistance arc path is removed, the normal voltage of the line is not high enough to reestablish the arc across the gaps.

The valve-type arrester sparks over at a lower value of surge voltage than the expulsion arrester because of its smaller total gap distance. The valve-type arrester thus provides a somewhat lower level of impulse protection. Under certain conditions, particularly on distribution circuits of high short-circuit capacity, the power follow current of an expulsion arrester may be large enough to operate fuses between the arrester and the substation. The arrester, which momentarily short-circuits the line during its operation, offers only the small resistance of its internal arc to impede the flow of power follow current. This does not occur with the valve-type arrester because its internal resistance rises so high with normal circuit voltage.

The amount of current which a lightning arrester must discharge varies widely with the severity of the stroke. Evidence indicates that lightning-stroke currents vary from a few hundred amperes with an induced surge to more than 200,000 amp under a direct stroke. Distribution lightning arresters, however, will seldom need to discharge currents in excess of about 40,000 amp, and probably 50 percent of the time discharge will not exceed 1,500 amp. For this reason it is not economical to provide discharge capacity to handle the heaviest direct strokes since the danger on distribution lines is comparatively remote.

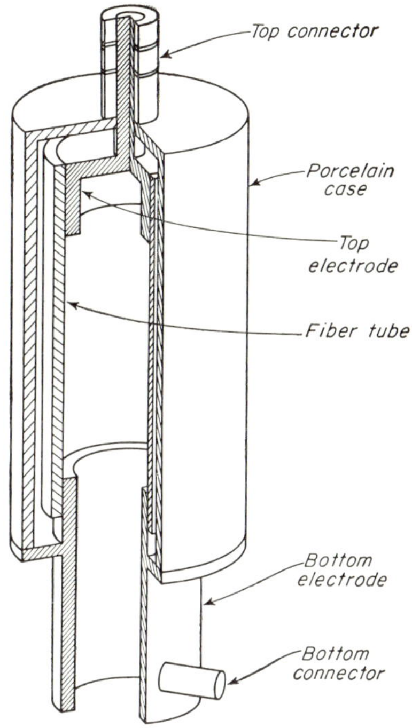

FIG. 7-3. Schematic diagram of expulsion-type lightning arrester with gap between electrodes surrounded by horn-fiber tube. Heat of lightning surge dislodges particles of fiber into arc path decreasing conductivity and raises pressure in tube, causing ionized gas to blow out of the bottom vent and snuff out arc.

The discharge capacity of distribution lightning arresters, under AIEE and NEMA standards, is rated at 65,000 amp for a specified time interval. If the discharge current seriously exceeds the arrester capacity, it will fuse a conducting path through the arrester material of the valve type and destroy its valve action. With excessive discharge current through the expulsion-type arrester, excessive gas pressure builds up within the tube to burst it. The life of the valve type is practically unaffected as long as the moisture seal remains intact and as long as the discharge capacity is not appreciably exceeded. The life of the expulsion type depends upon the rate of erosion of the fiber liner within the expulsion tube, unless the fiber consumed during each arrester operation can be restored. Comparatively little fiber is consumed by the short-duration lightning discharge, but an appreciable amount is removed by power follow current, particularly if the power follow current is heavy. Enough fiber may disappear so that the enlarged bore of the tube wil not develop enough gas pressure to extinguish the power-follow-current arc. The frequency of lightning discharge to be expected at any one location is not very great; so many years of operation can be expected. For this reason the expulsion arrester is sometimes considered to be most suitable for distribution systems of low short-circuit capacity, such as the more lightly loaded rural circuits.

To develop maximum protection, arresters should limit the potential stress across the equipment to be protected to the lowest feasible value. Lightning arresters are usually connected as shown in Fig. 7-4 to provide maximum protection to distribution transformers. This is the interconnection method of impulse protection. The ground terminal of the lightning arresters is connected to the transformer tank and to the grounded-neutral conductor of the secondary circuit. This prevents the rise of surge voltages which might flashover transformer bushings or puncture winding insulation.

The surge-voltage drop in the arrester discharge path to ground is made up of the IR drop through the arrester itself, plus the impedance drop in the arrester leads from line to arrester and from arrester to ground, plus the voltage drop in the resistance of the arrester ground. The drop through the arrester is made up of the voltage drop across the gap or gaps and the drop across the arrester material, in the case of a valve type, and is practically all IR drop. The voltage drop across the arrester ground is also principally IR drop. The voltage drop in the wire leads from the arrester is mostly reactance drop. The IR drop across the arrester material of a valve type is unimportant alone, when compared with the impulse insulation of transformer bushings and windings. The voltage drop across an expulsion arrester is even smaller since it consists of the drop across the gaps only. Reactance voltage drop in the wire leads from the arrester to the ground may be quite high. This is clear when it is considered that a lightning wave rising from zero to its crest value in a few microseconds corresponds to a much higher frequency than a 60-cycle sine wave rising from zero to its maximum in $\frac{1}{240}$ second and that the effect of inductance in the leads will be correspondingly higher (see Art. 1-19). The resistance of short ground rods ordinarily used may vary from a few to several hundred ohms, depending upon local grounding conditions. The total voltage drop across the various elements making up the impedance in the lightning-arrester discharge path may be quite high with a steep-front lightning wave and discharge of thousands of amperes.

If the connection from arrester ground terminal to transformer tank and grounded secondary neutral is omitted, the entire voltage drop in the arrester discharge path will be impressed across the parallel path from the primary line, through the transformer, to the grounded secondary neutral. If the surge voltage drop in the arrester discharge path exceeds the impulse insulation strength of the possible parallel path through the transformer, transformer-bushing flashover or breakdown of

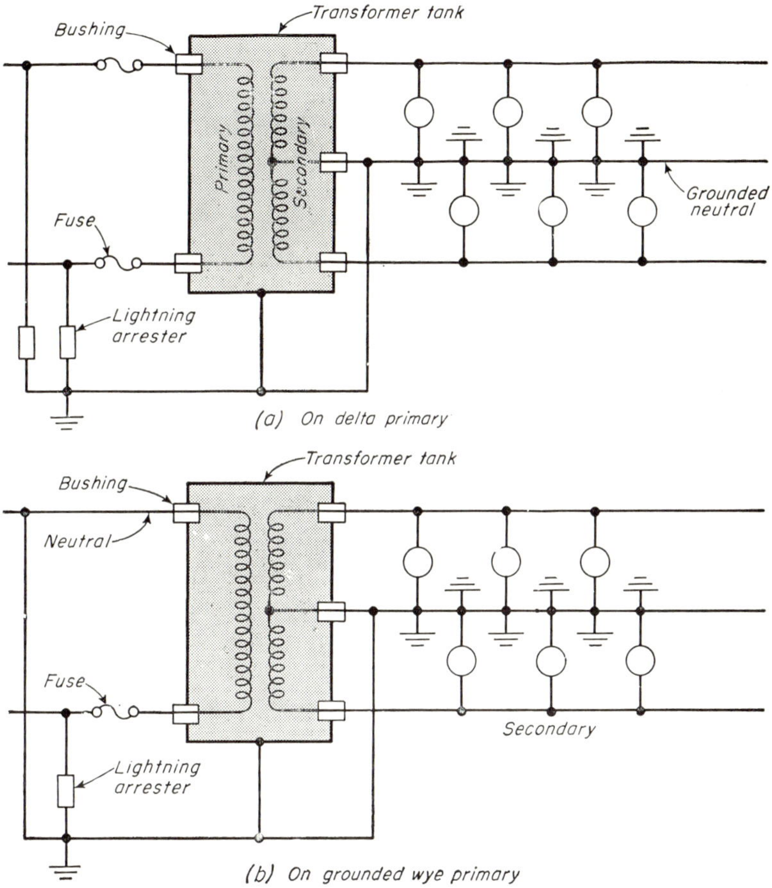

FIG. 7-4. Methods of connecting lightning arresters.

winding insulation to core or tank may occur. If this happens, the protective fuse may blow or the transformer may be seriously damaged. By interconnecting as in Fig. 7-4, the transformer tank and windings will be limited to the potential rise at the ground terminal of the arrester. This rise equals the surge-voltage drop in the arrester itself and the drop in the arrester lead to the line, which usually is negligible. The total surge voltage to which the transformer is subjected, in this case, is safely below its impulse insulation strength.

The arrester connection in Fig. 7-4 grounds the transformer tank. If a grounded tank is undesirable, the connection to the tank may be made through an isolating gap which arcs across or breaks down under surge voltage before transformer bushing or insulation fail. The driven arrester ground at the transformer pole may often be omitted if the impedance to ground of the secondary neutral is quite low, as might be the case with a continuous neutral on an urban distribution system where the customers' service is grounded to a continuous metallic water-pipe system (see rule 2632, National Electrical Code, 1953). Where only a few customers are supplied by the transformer and, particularly, where a continuous metallic water-pipe system is not available for grounding, it is usually advisable to provide a driven ground-rod connection of low resistance at the transformer pole, since the surge-voltage drop across the arrester ground connection will be impressed across the customers' wiring. In case a solid metallic connection between the arrester ground terminal and the grounded secondary neutral is not desired, this connection may also be made by an isolating spark gap. If the transformer bank is three-phase without a grounded secondary conductor, a low-resistance arrester ground should be provided to limit the rise of the customer's equipment above ground potential.

The same principle of interconnection should be used to protect other types of equipment where the impedance drop in the arrester ground lead or ground resistance may subject the insulation to dangerous surge potentials. For example, with an underground or aerial cable, connected to an open-wire overhead distribution line and which has a grounded outer metallic sheath, the ground terminal of the arrester should connect to the grounded metallic cable sheath to limit the voltage stress across the cable insulation or the cable pothead to the IR drop in the arrester. The arrester should be located as near the protected equipment as feasible. In any case, where an underground or aerial cable connects to an open-wire distribution line, lightning arresters should protect the cable from high surge potentials caused by reflection resulting from the change in surge impedance between the open-wire and cable section of line.

7-2. Distribution-system Grounding. Distribution circuits and equipment are grounded for three principal reasons: (1) to stabilize circuit voltages to ground and aid circuit operation, (2) to provide maximum protection by limiting the voltage stress to ground of equipment insulation, (3) to provide maximum safety to persons who operate or come in contact with electrically energized equipment or devices. Satisfactory ground connections may be made in a number of different ways. These may include connections to a continuous metallic-piping system buried in the earth, connections to buried plates or other types of electrodes, or connections to metallic rods or pipes driven into the earth. A ground must have a low resistance, be durable, have adequate current-carrying capacity, and be economical to install and maintain. Of the various types, the water-pipe ground probably satisfies all these requirements to the greatest degree and is most often used for the ground connection on consumers' premises. The water-piping system is best for grounding when the joints are leaded or screwed together, making a solid metallic joint, and when the piping system is extensive and continuous. If the pipe joints are cemented or formed by packing compound so that the run of metal pipe is not continuous, or if the piping system is not extensive even though continuous, as a small water system supplying a single consumer, or if the ground connection is on the house side of the water meter, the advantage of the water-piping system as a ground may not be so pronounced. A ground connection to the best type of water-piping system has a resistance often less than 1 ohm and seldom more than 4 or 5 ohms.

Outside of consumer premises the most commonly used and most satisfactory ground connection is the driven ground. The earth is made up of many different substances of varying chemical composition. Some are excellent conductors and others excellent insulators. The earth at a given location may be made up of successive layers of differing resistivity. Identical grounds a comparatively short distance apart may vary widely in value.

Very little of the resistance to ground is in the rod, pipe, or wire lead to a driven ground. Practically all the ground resistance is in the earth immediately surrounding the rod or pipe. Resistance is inversely proportional to the cross-sectional area, and here the conducting cross-sectional area of the earth is a minimum. Even a relatively poor conductor may pass large currents, with a low resistance opposing flow, if the cross-sectional area of the conductor is large enough. This is true of current flow in the earth. In a driven ground, however, current from rod to earth is confined to a comparatively small contact area with a resulting high current density. It is not practical to attempt to lower ground resistance by driving a ground rod of larger diameter since a prohibitive size would be required to lower the resistance effectively.

Since the greater portion of ground resistance is in the shell of earth immediately surrounding the driven rod or pipe, the choice of material for the conducting rod or pipe is relatively unimportant. The greater conductivity of a copper rod compared with a steel pipe means little since either is a much better conductor than the earth in which it is installed. A steel rod or pipe, however, does have one disadvantage compared with copper in that steel corrodes much faster. An advantageous compromise uses a steel rod with a copper covering or sheath, such as copperweld. Such a rod has the strength and toughness of steel combined with the durability and resistance to corrosion of copper.

Earth resistance usually will be high where the gravel, sand, or stone content in the soil is high with a low content of clay or loam. Resistance will be lowest with a high content of clay or loam in the soil and little or no sand, gravel, or stone. The moisture content of the soil affects its resistance so that it is highest during dry seasons and lowest during rainy seasons. For this reason ground resistance may vary as much as 100 percent. To achieve lowest ground resistance the ground rod should be driven well below the permanent moisture level and also below the permanent frost level in severe climates. Ice in the soil causes a much higher ground resistance. Ground rods should be driven to a depth of at least 6 ft. In distribution practice a slightly longer rod, such as a ⅝-in. by 8-ft rod, is usually most suitable.

The soil in any area will vary widely in type, texture, and resistivity at different depths. A lower ground resistance can often be achieved by driving the rod to a greater depth to reach a strata of soil of lower resistivity. The greater depth will have a higher average moisture content and will proportionately increase the volume of the conducting body of earth away from the rod. In some localities a satisfactorily low ground resistance cannot be realized by driving the ordinary short rod, and a sectional type rod cannot be driven to perhaps a 25 to 40 ft depth because of rock formations. In this case, the ground resistance can be lowered by driving two or more short rods and connecting them in parallel. A greater total length of rod will probably have to be driven to get the same total reduction in resistance of a single deep-driven rod. Parallel-driven ground rods should be separated by at least 6 ft; a greater separation will usually prove advantageous.

If individual low-resistance grounds on the distribution system are required, measurements of ground resistance should be taken. If the resistance at an individual location cannot be satisfactorily lowered by driving to a greater depth or by driving a number of parallel rods, a still lower value can probably be obtained by chemically treating the soil around the rod or rods. Common salt (sodium chloride) is one of the most frequently used chemicals and also one of the most effective in lowering soil resistivity. Magnesium sulfate, copper sulfate, and several other chemicals are also effective. Chemicals mixed with the soil gradually leach out and are carried away by the natural drainage or migration of moisture in the soil. Usually, effects of the treatment last 2 to 5 years, depending upon the soil type and amount of rainfall. The ground resistance gradually rises as the chemical in the soil disappears. Treatment must be periodically renewed to maintain the low value.

One of the most effective methods of realizing low ground resistance is by utilizing the neutral of the secondary-distribution circuits as a ground bus. This is done by connecting the grounded neutral of each secondary circuit to the next to form a continuous and unbroken run of wire throughout the distribution system to which all grounds are connected. This connects consumers' service grounds and all other grounds in parallel, forming an extensive grounding system of low resistance. This method is particularly effective in areas where the run of secondary circuits along the streets is practically continuous and where thousands of individual grounds may be connected in parallel. It is less effective and may not be feasible on rural circuits because of the large gaps between secondary circuits.

7-3. Distribution-circuit Overcurrent Protection. The current in an electric circuit depends upon the impedance of the entire conducting circuit and the impressed voltage. In an electric system the impedance may include complex combinations of series and parallel impedances in the generators, transmission lines, distribution lines, transformers, and loads. The load impedances exert the greatest influence upon normal current flow. Impedance of the conducting network between generator and load is always kept at a minimum to avoid excessive voltage loss. Loads, except in the case of series street-lighting circuits, are connected in parallel. When two exactly equal impedances are in parallel, the effect is the same as though a single impedance of half the ohmic value were substituted for the two (Art. 1-28).

If nothing but load currents ever flowed in the circuit, the problem of protecting distribution equipment would be reduced to the replacement and enlargement needed by normal load growth. The normal hazards of day-to-day operation, however, prevent any such simplicity. It is virtually impossible to construct distribution lines so that the danger of overhead conductors coming together or of insulation failure on underground conductors will be entirely removed. Such unintentional contacts between insulated conductors of a distribution circuit produce results similar to an additional load of nearly zero impedance being suddenly connected to the line. The normal load impedance is effectively short-circuited, causing currents much higher than the normal. All the impedance limiting the flow of current will be in the supply circuit between source and short circuit. When short-circuit currents flow, the faulted section of the circuit must be disconnected to prevent destruction of conductors and equipment. Thermally operated fuses or automatically operated switches, such as oil or air circuit breakers, usually disconnect the faulted section or feeder so quickly that damage is held to a minimum.

7-4. Fuses and Fused Cutouts. In any conductor, if one small section is reduced in cross-sectional area normal continuous full-load current raises the temperature of the reduced section. It will melt, or fuse, and open the circuit if the current is raised sufficiently above normal and maintained for a long enough time. The fuse for overcurrent protection on distribution circuits acts this way and consists of a thin conducting link with a fusible section usually made of silver, tin, zinc, or some metallic alloy having a low but consistently uniform melting point. The fusible section of the link will conduct rated current continuously without the temperature exceeding a predetermined value below the melting point. It will melt and open the circuit if the current exceeds rated value. Fuses carry their rated current continuously but operate within a maximum of

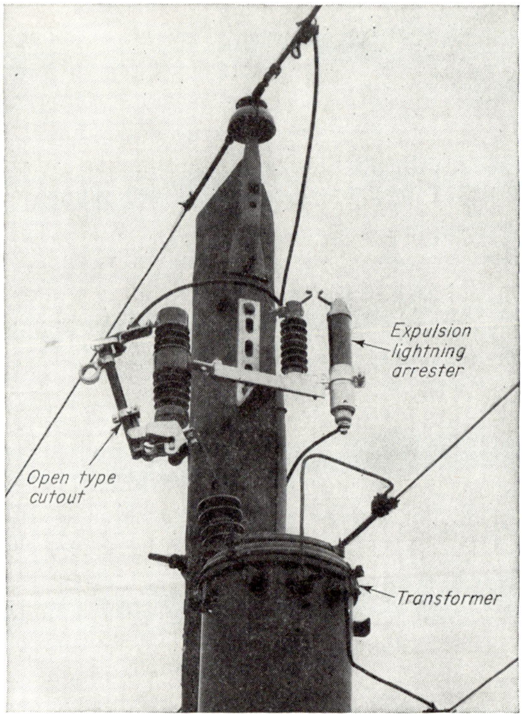

Fig. 7-5. Open-type fused cutouts mounted on pole top, with expulsion-type lightning arrester. (*Courtesy of General Electric Company.*)

5 min if the current rises to 230 percent of their rating. The fuse has an inverse time-current characteristic since the greater the current, the shorter the time in which the fuse melts. Heat develops at a rate proportional to I^2R.

When overcurrent melts a fuse, an arc will be momentarily maintained across the ends of the fused section. At primary voltage the arc has a tendency to hang on and is difficult to break because of the conducting metallic vapors in the arc path and the higher value of voltage. Fuses are generally installed in open- or enclosed-type fuse cutouts, as shown in Fig. 7-5, when used for protection of lines and equipment at distribution primary voltages. The cutout fuse-holder tube has an inner liner of fiber and an upper and lower set of terminals between which the fuse link is connected. The inner fiber tube aids in the extinction of the arc following the melting of the fusible section of the link. The upper metal terminal of the fuse-holder tube contains an inner chamber in which the fusible section of the link is located. When overcurrent flows, the fusible section melts and considerable gas pressure may develop within the tube from vaporizing of the fuse element. The gas is expelled from the open lower end of the tube with an explosive action. This blows the tail end of the fuse link out of the bottom of the tube, stretching out and breaking the arc. Expulsion of hot gases dislodges small fiber particles from the surface of the fiber liner which are blown out in the gas stream. The gas stream being expelled from the tube contains highly conducting metallic vapor. The nonconducting fiber particles introduced into the gaseous arc stream tend to deionize the arc path. This action greatly lowers the conductivity of the arc stream and aids in preventing a sustained power arc through the cutout.

The interrupting capacity of a cutout depends upon the tube-opening diameter and the thickness and mechanical strength of the tube walls. The gas pressure developed within the tube is proportional to the inner diameter. With a small fuse and large-bore tube, a low current may melt the link but fail to develop sufficient gas pressure for satisfactory operation of the cutout. Low gas pressure may fail to expel the fuse pigtail, and arcing between the severed ends of the melted link may char and destroy the fuse-holder tube. On the other hand if the bore of the tube is too small, the gas pressure may rupture the tube. Fuse cutouts as shown in Figs. 7-5 and 7-6 are ordinarily used on overhead distribution lines, in several current and voltage ratings. In general, these cutouts are supplied in 0- to 50-, 50- to 100-, and 100- to 200-amp current ratings for 5 to 15 kv. The rated current-interrupting capacity of overhead-type distribution cutouts, in general, will range from twenty-four to thirty times their rated maximum-normal load current at rated voltage.

Unsealed fused cutouts expelling hot gasses are not suitable for underground distribution circuits where they may be subjected to continual dampness or occasional submersion. An oil-type fuse cutout, in which the fuse link is sealed and immersed under oil, may be used. The oil-fuse cutout consists of a small metal tank to which watertight cable connections may be made. The tank is filled with an insulating oil, and the fuse link, consisting of a metal strap with a fusible section of reduced cross-sectional area, is submerged in oil and connected between the two cable terminals. When the fusible section of the link melts, the oil quenches the arc and interrupts the flow of current. The cutout may be used with expansion chambers to control gas pressure when interrupting heavy overcurrents.

In addition, there are several other types and variations of fuses for the protection of distribution circuits and equipment. Among these are the boric acid fuse and the liquid-type fuse. In the boric acid fuse the fusible element is surrounded by a tube liner of compressed boric acid. The heat of the arc reacts on the boric acid to produce vapor which aids in extinguishing the arc. In the liquid type the fuse element under spring tension is sealed within a glass tube filled with a suitable insulating liquid. When overcurrent melts the fusible element, the spring

FIG. 7-6. Enclosed-type fused cutout and valve-type lightning arrester mounted on pole top. (*Courtesy of General Electric Company.*)

quickly pulls the burned ends of the severed link apart while the liquid extinguishes the arc. There is no expulsion of gas, and the entire assembly including glass tube, fuse link, and liquid, is replaced after a fuse blows. In all these variations the principle of operation is the same, *i.e.*, a fuse section of metal melts to interrupt the current and auxiliary arrangements must break the power arc.

Improved design and factory methods of production have provided fuse links with a high degree of reliability and accuracy. The fuse, however, cannot distinguish between a temporary overcurrent of very short duration, such as inrush current when a motor starts or when wires momentarily swing together in a windstorm, and an overcurrent of long duration which might cause melted conductors and damaged equipment. Each time a fuse operates, the melted or blown link must be replaced. To lessen the number of outages, repeater fuse switches or fuse cutouts have been devised. The repeater fuse consists of an open or enclosed cutout with two or three fuse-holder tubes instead of one. The cutout is mechanically arranged so that when a fuse blows a second fuse-holder tube with fuse link automatically switches into position, after a short time interval, restoring the interrupted service. The two- or three-shot repeater fuse cutout reduces the number of long-time service interruptions, as compared with a single fuse, but is no guarantee against interrupted service from repeated temporary overcurrents or line short circuits.

In general, it is not advisable to attempt to fuse equipment such as transformers or distribution lines with the idea of protecting against overload from normal load growth. Normal practice uses the fuse to protect lines and equipment against abnormal currents. This is because of the character of the fuse and because it is impossible to introduce and vary at will the time delay between the beginning of the overcurrent and melting of the link. Short circuits on distribution lines may require several hundred thousand kva of fault energy to be interrupted if the fault occurs at or near a large distribution substation supplying the distribution system. Fuses on the distribution system most frequently provide fault-current protection on the high-voltage side of distribution transformers, as shown in Figs. 6-12, 6-14, or 6-16. They are also frequently used on primary circuits for branch-circuit protection or to provide automatic sectionalizing where the normal load current is relatively small and the occasional loss of service from temporary faults will not affect too many consumers.

To avoid excessive fuse operation from temporary and short-lived overcurrents in normal load variation, the fuse should be somewhat larger than the sustained or peak-load current. For instance, in fusing transformers on the high-voltage side a fuse smaller than 150 percent of the rated full-load current of the transformer is seldom used. More often transformers are fused at 250 to 500 percent of rated load and are intended to blow only under short-circuit conditions within the transformer itself or on the low-voltage secondary near the transformer. In the case of a fuse used for main or branch-circuit line protection, the fuse size should be appreciably larger than the normal peak-load current likely to flow in the line and through the fuse. On the other hand, a fuse should not be used whose current rating is so high that it may not operate under short-circuit conditions. Where the normal line peak-load current is so large as to require an excessively large fuse, a current-interrupting device of larger capacity, such as an oil switch or circuit breaker, will be required.

7-5. Circuit Breakers. The circuit breaker may be of the oil or air type. In the oil circuit breaker there are two stationary contacts per phase and a movable contact bar which bridges between the two stationary contacts when the breaker closes. Separating and moving apart the stationary and movable contacts break the current. The contacts are immersed in oil contained in a metal tank. The movable bridging contact bar is mounted on an insulated lift rod so that the breaker opens by dropping the lift rod and separating the contacts or closes by raising the lift rod. When the breaker opens, the two arcs between the two stationary contacts and moving contact bar are quenched by the physical separation and the cooling effect of the surrounding oil.

To avoid burning and pitting the main contacts, auxiliary arcing contacts across which the arc breaks are usually included. These arcing contacts, made of special metals, do not have the low contact resistance nor the high conductivity of the main current-carrying contacts but do have high resistance to burning or melting under the high temperatures of electric arcs. The arcing contacts break contact *after* the main contacts separate.

The current-interrupting capacity of the oil circuit breaker is usually raised by increasing the physical size and enclosing the arcing contacts in an arc chamber. The chamber confines the arc within a limited space so that the gas pressure developed forces a jet of cold oil either radially across or along the arc path.

The principle of the air circuit breaker is basically the same as that of the oil type except that the arc breaks in air instead of being quenched in oil. The air circuit breaker is most often used at 600 volts or less, while the oil circuit breaker is most often used above 600 volts. Improvements, in recent years, in design and arc extinction have produced air circuit breakers for distribution use at 15,000 volts and up and with interrupting capacities comparable to the oil type. At the lower voltages the arc is extinguished in an arc chute by stretching the arc across strips of fiber forming the chute. This deionizes and breaks up the arc. At the higher voltages and interrupting ratings, various auxiliaries increase the interrupting rating and the speed with which the arc breaks and clears. Among these are improved arc chutes, a jet of compressed air across the arc path to push the arc across the fiber baffles of the chute and literally blow the arc out, a strong magnetic field to deflect and stretch the arc to breaking in the arc chute. Oil circuit breakers are built for either indoor or out-door use without separate housing. Air circuit breakers are not usually made for outdoor use except where the breaker is installed within an outdoor-type housing or switch house to protect it from the elements.

Both types of circuit breakers usually close against spring pressure, held in closed position by a latch or toggle joint. The breaker may be opened or closed either by a hand-operated lever or automatically. When manually controlled, it acts as a disconnect switch only. When protecting the circuit against overcurrent, its operation may be made automatic and completely independent of manual control by electric control devices.

A number of tripping methods are used for overcurrent protection. With low-voltage low-capacity air circuit breakers, thermal action on a bimetal strip or a magnet coil trips the holding latch. In thermal tripping the holding latch action is initiated by the bending of a bimetal strip from the heat of I^2R loss under heavy overcurrents. In magnetic action the breaker acts from overcurrent flowing through a magnet coil producing a strong magnetic field to attract the tripping clapper or plunger of the solenoid magnet. In some types, both thermal and magnetic trips are used. Trip coils or small solenoids open higher capacity and voltage circuit breakers. A solenoid consists of turns wound around an iron core surrounding a movable iron bar or plunger. Current flowing through the winding magnetizes the iron core which attracts and lifts the plunger or forces it into the center of the core and coil assembly. The moving plunger strikes the lightly balanced trigger, latching the breaker in the closed position, tripping the breaker open. These trip coils or tripping solenoids are small, since only a light blow striking the trip latch opens the breaker. They may be either current or potential coils.

In a few cases where both the voltage and current are low and accurate control of time delay in tripping is not important, series trip coils through which the line current flows may be used, Fig. 7-7*a*. In other cases the series or current trip coil may be used with a current transformer in the line, Fig. 7-7*b*. The current transformer avoids high line currents through the trip coil and full line voltage. The series or current trip may be used in conjunction with a current transformer and a circuit-opening relay, Fig. 7-7*c*. The relay contacts are in shunt with the trip coil and in series with the current-transformer secondary circuit so that with normal line current flowing the major portion of the current passes through the shunting-relay contacts and very little through the trip-coil winding. Consequently, when a heavy overcurrent flows, a proportional overcurrent in the current-transformer secondary circuit lifts the relay plunger and opens its contacts. Opening the shunting-relay contacts permits full overcurrent in the transformer secondary circuit to flow through the series trip coil, opening the circuit breaker.

The series or current trip coil has several disadvantages compared with the potential trip coil. Among these are the need for (1) larger conductors and (2) at least two trip coils in a polyphase ac circuit if series trip is used. A potential trip coil will open the breaker whenever voltage within its rated operating range is applied to the coil. This method is by far the most commonly used, but it requires a relay which connects the trip coil to a voltage supply. Where reliability is most essential, dc potential trip coils are commonly used. When an ac trip coil is used,

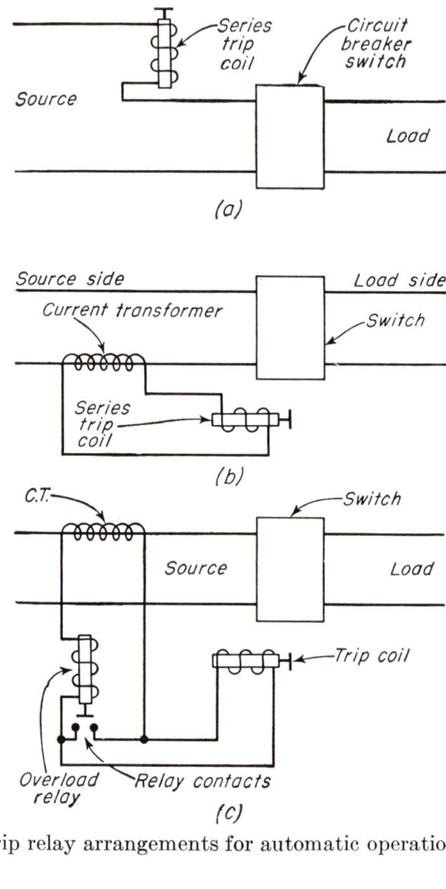

Fig. 7-7. Various trip relay arrangements for automatic operation of circuit breakers.

energized from the same source as the protected circuit, a heavy short circuit at or near the circuit breaker may reduce the ac voltage to such a low value that it will not operate the trip coil. To avoid this difficulty a separate and entirely independent potential source for operating the trip coils is necessary. The most satisfactory independent source is a storage battery. A single dc trip coil may trip out a circuit breaker on a polyphase ac circuit, but at least two overload relays will be required to assure the breaker tripping out under any fault condition as shown in

Fig. 7-8. A storage battery, however, has the disadvantage of requiring considerable additional investment and maintenance since the dc power supply must be available at all times without interruption. This is easy to justify in a large station where a single storage battery can supply dc power for remote control of many circuit breakers. In small stations an independent dc supply may be difficult to justify, and a compromise between an ac and a dc is generally used. For instance, in a small outdoor distribution substation supplying one or two distribution primary feeders, each protected by a feeder circuit breaker, some means of utilizing ac from the substation bus rather than a storage battery is usually found.

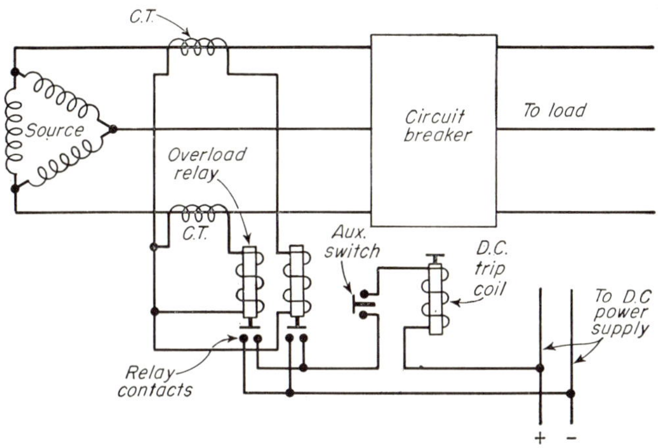

FIG. 7-8. Tripping arrangement using dc power for operation.

Among these methods are series current trip coils, Fig. 7-7. Another method provides a dc control circuit from ac bus supply voltage by a station power-supply transformer to step down the bus voltage and a dry-type rectifier. One type of dry rectifier for circuit-breaker control uses groups of copper oxide coated disks which have the peculiar property of low resistance to current flow in one direction and very high resistance to current flowing in the reverse direction. The copper oxide coated disks perform as a half-wave rectifier. They permit current to flow in only one direction. One of the most satisfactory methods of utilizing ac for circuit-breaker tripping is the capacitor-trip method, which is not affected by the momentary drop in ac voltage when a nearby heavy short circuit occurs. The capacitor-trip method uses overload relays, a dry-type rectifier, and a capacitor, Fig. 7-9. The capacitor is dc charged through the rectifier under normal operation, but no current flows through the trip coil. Under fault conditions, when overcurrent in the line causes the contacts of the overload relays to close, the charged capacitor discharges through the trip coil and trips out the circuit breaker.

In the majority of cases relays control the tripping of circuit breakers (see "System Operation," Chap. 2*). Where extreme accuracy in trip current values or in time delay is not of great importance or where tripping time does not have to be closely coordinated with other overcurrent protective devices in series, a simple plunger-type relay, Figs. 7-7 and 7-8, may be used. Where a series trip scheme is used without an overcurrent relay, the current value at which the breaker trips can be varied from about 50 to 200 percent of nominal current rating. The plunger-type relay operates at a definite minimum current value and may be set for instantaneous operation or for inverse time-delay operation. Time delay is most often used since instantaneous tripping may cause interruption under transient overloads. Time delay is adjusted by setting a needle valve controlling the discharge of oil in a dashpot.

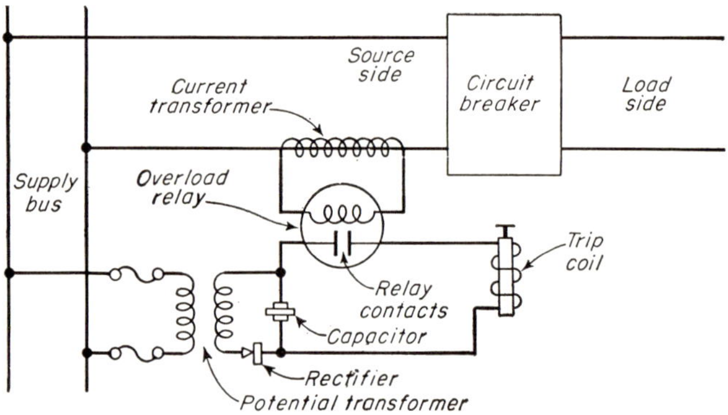

FIG. 7-9. Capacitor trip arrangement uses overload relays, dry-type rectifier.

In the larger installations where greater accuracy is required, induction-type overcurrent relays are used. The induction-type relay bears considerable similarity to the induction watthour meter. The relay has a disk carrying a contact. A turning torque is set up by current flowing in the relay current coils supplied from a current transformer in series with the line. A spirally wound spring on the disk shaft opposes disk rotation. If sufficient current flows through the relay-current coils, the torque developed turns the disk until the movable contact on the disk touches the stationary contact and energizes the trip circuit of the breaker. The minimum current at which the relay closes its contacts is varied by taps from its current coil. Time delay to close contacts is selected by changing the position of a lever controlling the distance the moving contact must travel. The line current at which the relay will pick up and close its contacts equals the current-transformer ratio multiplied by the relay-current rating. For example, in a 300/5-amp cur-

* B. G. A. Skrotzki, "Electric System Operation," McGraw-Hill Book Company, Inc.

rent transformer when using the 6-amp tap of the relay, the minimum line current at which the relay will close its contacts equals 300/5 times 6 amp, or 360 amp. Under certain conditions other types of circuit protection than overcurrent may be desirable. Among these are tripping due to low voltage, high voltage, reverse power, reverse current, etc.

After tripping, the circuit breaker may be reclosed manually, by remote control, or automatically. Where an operator is always on hand, it may be satisfactory to have manual reclosing. In isolated or unattended stations, reclosing must be completely automatic to avoid a long-time interruption from transient or nonpermanent line faults. Reclosure of the breaker, either automatically or manually by remote push-button control, may utilize a reclosing solenoid, electric motor, or compressed air.

In automatic schemes the reclosing relay is a timing device driven by a small motor energized through the closing of an auxiliary switch on the circuit-breaker opening mechanism. This switch opens when the breaker closes, and vice versa. The timer has several sets of contacts, arranged to close successively, so that the breaker can be automatically reclosed after tripping as many as three times before finally locking out.

These timing contacts are arranged so that the breaker will be reclosed within the selected time interval after the relay is energized by the opening of the breaker. The first reclosure after the first trip out can be made practically instantaneously or after a definite time interval, and succeeding reclosures after trip out from a continuing fault can be made at variable time intervals up to 2 min. The closing of the reclosing relay contact energizes the circuit to the reclosing solenoid, or reclosing motor, etc., and recloses the circuit breaker.

If the fault or line disturbance which caused the trip out has been cleared so that the breaker stays closed, the timer will reset itself at its original position. If the line disturbance has not been cleared, the breaker will be tripped out again and the reclosing relay will again cause the breaker to reclose. If, after a predetermined number of reclosures, the fault is not cleared and the circuit breaker still trips out, the reclosing relay circuit will be left deenergized after the last breaker trip out.

When the cause of the breaker lockout has been cleared, the reclosing relay must be reset to the starting position by hand. The time interval to initiate the first reclosure by this type of relay can be varied from a setting of zero seconds to as much as 15 sec by adjustment of the timer contacts. Succeeding reclosures are made only after a definite time delay which can be varied by adjustment of the timer contacts also. A frequently used cycle on distribution primary circuits provides a total of three reclosures before lockout with an immediate or instantaneous reclosure after the first trip out and two succeeding reclosures at intervals of 15, 30, or 60 sec. This type of reclosing relay can be set for one, two, or three reclosures before breaker lockout. There are also other, more simple, types such as the single-shot recloser which will reclose the circuit breaker after the first trip out but which will not immediately reset itself and reclose the breaker a second time if a second trip out occurs immediately following the first.

The circuit breaker should be mechanically and electrically trip free. That is, the closing mechanism and controls should be mechanically and electrically arranged so that it is impossible to hold the circuit breaker closed against a short circuit with either automatic or manual operation.

7-6. Automatic Circuit Reclosers. Circuit breakers are generally reserved for circuit protection where load and short-circuit currents are heaviest and where the service loss or the destruction of facilities would be most damaging. A circuit breaker, together with current transformers, relays, and other control devices, represents considerable investment even though voltage, load, and current-interrupting rating is low. The automatic circuit recloser, Fig. 7-10, performs the automatic functions of the oil or air circuit breaker at important sectionalizing points on distribution primary circuits and provides feeder protection against overcurrent at small substations, where the load and interrupting current rating required is relatively small. In the recloser both the automatic tripping and reclosing features are built in and full line current flows through the operating or trip coil. The interrupting rating depends upon its series coil size or its normal current rating. The recloser is a smaller and more simple, self-contained version of the circuit breaker. Recloser operation is initiated by solenoid action of overcurrent flowing through the series operating coil which lifts a plunger and separates two contacts in series with the operating coil and the line. The power arc is quenched in oil.

FIG. 7-10. Circuit recloser automatically recloses circuit after tripping out on short circuit or overload. (*Courtesy of General Electric Company.*)

The reclosure after trip out is controlled by a mechanical timing device which releases a latch and permits the plunger to drop under the influence of gravity and spring action to reengage its main-line contacts. The first trip out of the recloser is practically instantaneous, while succeeding trip outs may be set for either definite time delay or inverse time delay with a minimum time for tripping. The recloser can also be adjusted for lockout after either one, two, or three unsuccessful reclosures. The circuit recloser does not have nearly so wide a range of current settings at which the switch will open as the circuit breaker since the minimum tripping current is basically tied in with the rated current of the device or of its operating coil. The minimum current at which most types of reclosers will trip lies between 175 and 250 percent of the rated full-load current. Reclosers are primarily used for automatic sectionalizing and overcurrent protection on overhead distribution primary circuits since they are not adequately protected against occasional submersion or available in the higher ratings necessary for larger urban underground distribution systems.

Reclosers installed on the primary feeder at some distance from the station feeder breaker prevent possible lockout of the station breaker and interruption of service to the entire feeder due to line faults on single-phase branches or minor outlying parts of the circuit. With practically instantaneous operation on the first trip out, the recloser may open and reduce the fault current to zero before sectionalizing fuses after the recloser have time to melt. Since succeeding trip outs will be made after a much longer time delay, the fuses after the recloser will operate and clear the faulted section before the recloser can complete additional trip outs, provided the sectionalizing fuse sizes are coordinated with the characteristic of the recloser. This feature may reduce the number of sectionalizing fuse operations by as much as 75 percent since the majority of line faults causing fuse operation on lines without reclosers are transient and of short duration.

Reclosers are built in either single-pole or three-pole units. Where single-pole reclosers protect a three-phase delta-connected primary circuit, one recloser in each of any two phases, instead of three reclosers with one in each phase, is frequently used since all line faults will involve at least two of the three phases and consequently will clear the faulted section under either single- or three-phase faults. If single-phase reclosers are to protect a three-phase wye-connected primary circuit, however, a recloser should be installed in each phase since line faults to ground may involve only a single-phase wire and the common neutral return.

7-7. Coordination of Protective Devices. Circuit breakers, fuses, and circuit reclosers on distribution primary circuits protect circuits and equipment from destruction by short-circuit currents, protect consumers from service interruptions caused by transient line faults, and limit the number of consumers affected by line faults at one time to as few as is economically feasible. To provide all this, it is often necessary to place a number of overcurrent protective devices in series along the line. Such a series cascade of overcurrent protective devices may include the station-feeder circuit breaker, together with one or more circuit reclosers and several fuses in series at different points along the line.

These series protective devices must be of such a size that they will not operate under normal load current, and the time delay must be different for each device in series, decreasing progressively outward from the source, so that with the same current flowing through each, the protective device nearest the fault will operate first. Assume that one section

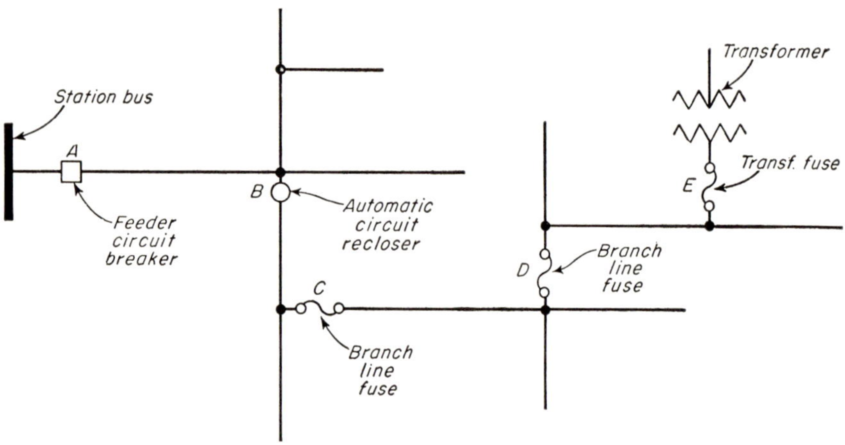

FIG. 7-11. Sequence of placing various protection devices on a feeder circuit.

of a long distribution primary feeder is arranged, as in Fig. 7-11, down to a single transformer protected by a fuse on the high-voltage side. If the transformer at E develops a short circuit, the fuse in its high-voltage leads should disconnect the transformer and isolate the fault before any of the other protective devices at D, C, B, or A can operate. Service interruption from this fault would thus be confined to consumers supplied by transformer E. Likewise, a line fault beyond point D should operate the fuse at D before any of the other protective devices at C, B, or A can lock open. A line fault between C and D should operate the fuse at C before the recloser at B can lock open, etc.

This is done by coordinating the overcurrent devices so that the time delay between the start of the high current and their opening will give the proper sequence of operation. The station-feeder circuit breaker at A, Fig. 7-11, will have the highest normal current rating, since all the feeder load current flows through it, and will be set to trip out at the highest minimum value of overcurrent. The recloser at B will require a somewhat lower minimum normal current rating since all the feeder load current will not pass through it and should trip out at a lower minimum overcurrent than the breaker at A to assure selectivity in tripping between the two.

Likewise, the fuse at C should afford selectivity between itself and the recloser at B, the fuse at D should offer the proper range compared with C, etc. The selection of the proper sizes of protective devices to be installed in series along a distribution primary feeder is based upon the time between the beginning of overcurrent in the line and operation of the circuit interrupting device.

Typical time-current characteristic curves for a station-feeder circuit breaker controlled by overcurrent relays, for an automatic circuit recloser, and for fuses on overhead-distribution primary circuits are shown in Fig. 7-12. These curves illustrate and compare (1) the time required for an overcurrent relay with inverse-time delay to close its contacts and initiate the tripping of the breaker, (2) the time required for a circuit recloser to open instantaneously for the first trip out and with inverse time delay for succeeding trip outs, (3) and the time required for various fuse sizes to operate with different currents flowing. These curves show that the contacts of the overcurrent relay will close in OC' sec when the minimum trip current OA amp flows opening the feeder circuit breaker. Current values greater than OA will cause the relay contacts to close in a shorter time than OC' sec or in inverse proportion to the increase in current to a point approaching a definite minimum time. The time-current curves of the recloser show that it will initially trip out in OE' sec when its minimum trip current of OD amp flows. If the fault remains, succeeding trip outs up to lockout will be in the relatively longer time of OI' sec. The initial trip time of the recloser decreases with increase in current and approaches instantaneous operation. With current values greater than OD, there will still be a relatively large difference between the initial and final trip times of the recloser, as can be seen by comparing the time of OB' sec for final lockout with the time required for initial trip when OA amp flows.

Protective devices in series should be coordinated so that they will operate in proper sequence. In Fig. 7-12, any device whose time-current characteristic curve is wholly to the left of the relay characteristic will open ahead of the station-feeder breaker controlled by the relay. Likewise, any fuse whose characteristic falls between the initial, almost instantaneous, trip curve of a recloser and its characteristic curve for succeeding trip outs will coordinate successfully. With a line fault, the recloser will initially open before the fuse can melt. It will reclose immediately and if the fault has not cleared will start to reopen but with time delay. The fuse will now melt, before the recloser can reopen, and remove the faulted section from the circuit. This feature prevents excessive fuse operation from transient line faults which clear quickly. Records indicate that about 80 percent of primary circuit trip outs can be reclosed immediately. In 10 to 15 percent of the cases the circuit

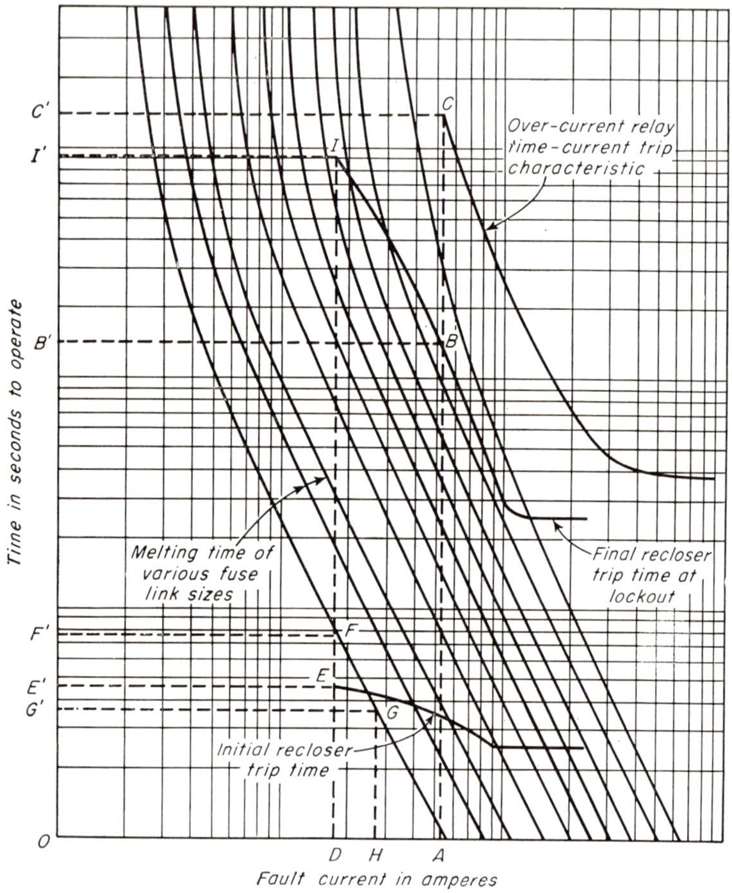

FIG. 7-12. Comparison of feeder circuit breaker, recloser, and fuse link operating times.

remains closed without further disturbance after one or two additional reclosures. In only 5 to 10 percent of all the cases will the line fault cause a reclosing breaker to lock out with a long-time service interruption. Where an automatic circuit recloser is not suitable, a reclosing circuit breaker or a reclosing fuse can be coordinated with the station-feeder circuit breaker and used instead.

In coordinating overcurrent devices sufficient allowance in operating time should be made so that the desired sequence will be realized and the remaining device will not be overheated or damaged. In a fuse, the fusible section melts in a given time, but the time to halt the current flow completely is slightly longer because of the arc between the ends of the melted link. The melting time as well as the small increment of arcing time will be much smaller for heavy currents than for low-current overloads. In selecting two fuses to operate in series, sufficient difference in size should be allowed so that under a heavy overload the smaller link (following the larger link) will completely interrupt the current without approaching the melting time of the larger link.

When coordinating fuses, if the clearing time of the smaller fuse, for any current, does not exceed 75 percent of the melting time of the larger fuse, the two will usually operate successfully in series. For fuse sizes up to 200 amp, used at distribution primary voltages, a difference of 25 amp in size will provide selectivity and satisfactory coordination in the majority of cases. The operating time of the various overcurrent devices normally decreases with the increase of overload, and if several are in series the difference in time for coordination may be measured in fractions of a second. A difference in operating time between adjacent overcurrent devices of 0.10 to 0.25 sec will usually prove satisfactory, but no arbitrary or rule-of-thumb method will satisfy all possible conditions.

7-8. Calculation of Short-circuit Currents. Before planning overcurrent protection, two things about the circuit should be known: (1) the normal maximum load current and (2) the maximum short-circuit current likely to occur. The first of these determines the nominal size of equipment, the second determines the interrupting capacity required. Of the two, the second is much more difficult to determine.

Circuit-breaker current ratings include (1) the safe rms continuous current, (2) the short-time total rms current the breaker will safely carry up to 1 sec, (3) the maximum current the breaker will safely interrupt at its rated voltage, and (4) the maximum current the breaker will safely interrupt at any voltage below its rated voltage. Ability of a breaker to interrupt current is based on kva, but as the current goes up with reduced voltage for constant kva the current interrupted *must not exceed* the maximum interruption-current rating. Neither may the current exceed the momentary current rating of the breaker.

The determination of impedance for the calculation of short-circuit currents is more difficult if both resistance and reactance are included. In many cases total reactance is much larger than total resistance, and slight error will result if resistance is completely ignored. When both resistances and reactances are present in a circuit and the total value of one is at least three times as great as the other, the smaller of the two can be neglected and the impedance taken as equal to the larger of the two. The error will not be more than about 6 percent and will be on the safe side since the rms short-circuit current calculated will be greater than the actual.

Normally a 60-cycle sine-wave current is symmetrical so that the positive half cycle is a duplicate of the negative half cycle. When a short circuit is slapped across an ac circuit, the impedance controlling the current reduces suddenly changing the current instantaneously. This momentarily upsets the balance between the several reactances within the generator so that the suddenly increased current may be shifted from symmetry about the axis. This is similar to the effect of adding a component of dc current which decays with time to a symmetrical ac current. If the dc component is large enough, the total resultant current may be shifted so that it is completely offset, either above or below, the axis. The ac current, under short-circuit conditions, is highest when the short-circuit current is first applied and then also decays for a few cycles at a rapid rate and then more slowly for a greater number of cycles. Then the resultant current settles to a steady symmetrical value represented by the normal synchronous impedance.

To represent these transient ac currents, three values of reactance are used. These are the subtransient, the transient, and the synchronous reactances of the machine. Of these three, the subtransient reactance is the one in which we are principally interested. In a generator it controls the initial output of short-circuit current, determining both the momentary current a circuit breaker must withstand before its contacts open and the power it must interrupt at the moment its contacts separate.

To take into account the greater duty which may be imposed on a breaker under short circuit by the dc component, the calculated short-circuit current is multiplied by 1.6 unless the voltage is 5,000 or below. At 5,000 volts or less this multiplying factor may be reduced to 1.4 except when the current is fed predominantly by direct-connected synchronous machines or through current-limiting reactors. In determining the interrupting-current rating required of a circuit breaker on a distribution system, there is sufficient time between the initial tripping impulse and separation of the breaker contacts so that no allowance need be made for the rapidly decaying dc component. The multiplying factor of 1.6 or 1.4, will decay to unity if the time between tripping impulse and opening exceeds about 8 cycles. Since the majority of distribution-type circuit breakers equal or exceed 8 cycles operating time, the calculated short-circuit current can be used for determining interrupting rating without any multiplying factor.

A greater proportion of resistance in a circuit accelerates the rate of dc component decay. At points remote from the generator and at the lower distribution voltages the ratio of X to R is generally much lower, and consequently a lower factor can be used to correct for the dc component since it will decay still more rapidly. If the entire system impedance, including generator reactances, is to be included in selecting fuses and circuit reclosers, the calculated symmetrical rms value of short-circuit current should be multiplied by 1.2 in determining the current to be interrupted or the operating time under heavy short circuit.

When a short circuit occurs on a line, the voltage may approach zero at the fault if the fault impedance is low. Any synchronous or induction motors near the fault will continue to rotate at nearly rated speed for a short time after the fault occurs and the supply voltage drops. The motors act as generators for a very short time and help to feed current into the fault.

This generator action decays more rapidly in an induction motor than with a synchronous motor. The additional current contributed to a fault may be important in industrial stations where the total horsepower of motors may be a large proportion of the total load. Where the motor load is far enough from the fault so that the drop in voltage will not cause the circuit voltage at the motor to approach zero, the motor contribution to short-circuit current may be neglected. The motor contribution may also be neglected where the connected motor horsepower is small in comparison to the total kva-supply capacity. In determining the required interrupting capacity, where the motor contribution is included, the transient reactance of synchronous motors should be used and the effect of induction motors neglected because of the rapid rate of decay of their current contribution. For determining the momentary-current duty the subtransient reactance of both synchronous and induction motors should be used.

Generators, large motors, induction voltage regulators, large power transformers, line reactors, etc., have resistances small in comparison to their reactance and can be neglected. The rated impedance of power transformers can be figured from name-plate data on the transformer. If required, the transformer resistance can be determined from the rated copper loss and the values of R, X, and Z in percent can be converted to their equivalent ohmic values by the equations in Art. 1-30.

The percent subtransient and transient reactance of 60-cycle rotating machines, used in the calculation of short-circuit currents, will normally fall within the following ranges:

	Subtransient reactance, %	Transient reactance, %
Turbine generators, 1,800 rpm	10–17	
Turbine generators, 3,600 rpm	7–13	
Salient-pole generators:		
Waterwheel driven	20–35	
Engine driven	15–30	
Synchronous motors, 720 rpm and up	10–20	15–35
Synchronous motors, 600 rpm and below	20–35	20–50
Induction motors	15–25	

These values are based upon the kva rating of the machine, the resistance is considered negligible.

The name-plate data on current-limiting reactors usually includes the voltage drop across the reactor with rated current flowing. The reactance of the reactor is

$$X = \frac{\text{volts drop across reactor}}{\text{rated current of reactor}} = \frac{XI}{I}$$

Induction voltage regulators with a regulating range of 10 percent raise and 10 percent lower have an average impedance under short circuit of approximately 1 percent over the full range. This means that, if the regulator impedance alone impedes the current under short circuit, the current at rated line voltage will be limited to 100 times the rated full-load line current of the regulator. The percent impedance of the regulator is based upon the kva with rated regulator line current and voltage. Since resistance in the regulator is negligible, reactance and impedance are synonymous and the value in ohms is equal to

$$X = \frac{E}{100I}$$

I is the rated regulator line current and, in a three-phase circuit, E is the phase-to-neutral voltage of the circuit. The short-circuit impedance of a step-voltage regulator is lower than that of an induction type and to all practical purposes may be ignored in short-circuit calculations. In many cases the impedance of induction voltage regulators can also be ignored with very slight error.

Power transformers usually have a higher value of impedance than distribution transformers. The percent impedance of the transformer is usually given on its name plate and usually ranges from about 5 to 10 percent, although somewhat higher values may be characteristic at the highest voltages. The resistance usually ranges from 1 percent to as low as 0.45 percent and tends to decrease as the kva transformer size increases. These percentage values are based upon the kva rating of the transformer.

The impedance of machines and transformers is usually expressed in terms of percent, while the impedance of wires and cables is given in ohms. The percent values are based on a rated kva and represent the voltage drop, from no load to full load, expressed as a percentage of the rated voltage. Per unit values of resistance, reactance, or impedance are equal to the percent values divided by 100. Thus, a machine which has an impedance of 10 percent has a per unit impedance of 0.10. Under short circuit a current equal to 100/10 or 1/0.10 times normal full-load current would flow. Impedance values may be expressed in percent, per unit, or in ohmic values, but all must be reduced to the same terms.

To combine percent or per unit values placed in series through the circuit, they must first be expressed in terms of the same kva base. For instance, if a 10,000-kva and a 5,000-kva transformer bank are in series in the same circuit and both have an impedance of 6 percent, both impedances will have to be expressed in terms of the same kva base before being combined to determine the total impedance. Thus, if a base of 10,000 kva were chosen, the impedance of the first would be unchanged but the impedance of the second would be increased to 12 percent since the current and impedance voltage drop will be twice as great with 10,000 kva through the bank as with 5,000 kva. On the other hand, if a 5,000-kva base were chosen, the impedance of the large bank would be reduced to 3 percent while the impedance of the smaller bank would remain at 6 percent. Likewise, on a 20,000-kva base the impedance of the large bank would be doubled and the impedance of the smaller bank would be four times as much since the impedance voltage drop will be, respectively, two and four times as great with 20,000 kva through the bank as with the rated kva of the banks.

Percent resistance values may be added and percent reactance values may be added, if they are in terms of the same kva base and in series relation. Since resistance and inductance in an ac circuit act at right angles electrically, impedance values containing different proportions of R and X will differ angularly and cannot be added arithmetically without some error. In many cases, however, where percentage impedance values are to be combined for determination of short-circuit currents in distribution circuits, the angular difference between impedances in series can be neglected and added together directly with an acceptable error. The short-circuit kva of a circuit can then be expressed

$$\text{Short-circuit kva} = \frac{100 \times \text{kva base}}{\text{total } \% \, Z}$$

$$\text{Short-circuit kva} = \frac{1 \times \text{kva base}}{\text{total per unit } Z}$$

The short-circuit current in amperes equals normal current at the base kva multiplied by the ratio $100/\% \, Z$ or $1/\text{per unit } Z$.

If any of the impedances are given in ohms, they may be converted to percent values on the chosen kva base as

$$\% \, R, \ \% \, X, \text{ or } \% \, Z = \frac{\text{ohms} \times \text{kva base} \times 10^5}{E^2}$$

where the base is in three-phase kva and E is the voltage between phases.

If the percentage value is known and the corresponding value in ohms is required, the above equation can be transposed as follows:

$$\text{Ohms on voltage base } E = \frac{\%\ Z \times E^2}{\text{kva base} \times 10^5}$$

Impedance elements in ohms are always expressed in terms of a voltage base. Where more than one voltage is concerned, the equivalent ohms change in proportion to the square of the voltage ratio. This is illustrated by the transformer. Consider a 100-kva 2,400-to-240-volt distribution transformer which has an impedance of 3.2 percent. Using the above equations, the impedance measured on the high-voltage side, corresponding to 3.2 percent, equals 0.61 ohm and that on the 240-volt side equals 0.0061 ohm. Thus, Z_1 ohms at a voltage E_1 is equivalent to Z_2 ohms at voltage E_2 so that the relation between the two is

$$\frac{Z_1}{Z_2} = \left(\frac{E_1}{E_2}\right)^2$$

Where more than one voltage is concerned in a circuit, all values of R, X, and Z should be reduced to the same voltage base before being combined.

In determining short-circuit currents, the maximum current is usually the value desired. On three-phase circuits the highest value of short-circuit current, although not the most likely to occur, will usually result from a three-phase fault involving all three phases simultaneously. In other cases, as in the selection of fuses and reclosers for single-phase lines tapping from a three-phase line, the current from a single-phase fault involving two phase wires of a delta circuit or one phase wire and the grounded neutral of a wye circuit may be required.

In any circuit where more than one voltage is involved, both delta and wye circuits may be included. If the fault is three-phase, equal currents will flow in each of the three phase wires. If the fault is single-phase, the short-circuit current will be unbalanced and will flow in two phase wires or in one phase wire and the neutral return, depending upon whether the circuit is delta- or wye-connected, as shown in Fig. 7-13.

Suppose that a group of transformer banks connect in series a group of lines operating at different voltages and that a three-phase fault occurs on the line running from the last bank so that equal current flows in each phase wire. Balanced current will flow in each line and transformer bank from the fault to the generator, although the current will vary inversely with the voltage. Reactances are phase to neutral, and line currents are those which would flow in an equivalent wye circuit. If the phase currents are equal, their vector sum will equal zero and no current will flow in the neutral return of a wye circuit. Any three-phase circuit, either delta or wye, is equivalent to three single-phase circuits connected so that one wire is common to all three. Consequently, a three-phase circuit under three-phase short circuit can be represented by a single-phase circuit in which the series impedances of lines and transformer banks, representing phase-to-neutral values of one phase, are placed in one leg and the return leg (representing the common return of the three single-phases) is represented as containing zero impedance, as shown in Fig. 7-14*a*. The voltage applied across this equivalent single-phase circuit equals either the real or the equivalent phase-to-neutral voltage of the three-phase circuit. In Fig. 7-14 the total impedance to neutral Z equals the sum of all the impedances Z_1, Z_2, Z_3, etc. representing the equivalent impedances of generators, transmission lines, power

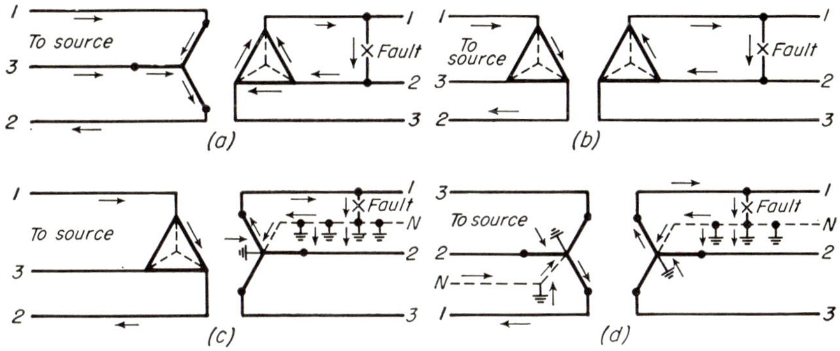

FIG. 7-13. Short circuits on any three-phase system always affect all three phases even though the short occurs on only one phase.

transformers, distribution lines, etc., from the source to the fault while E is the voltage between phases.

If the fault is single phase, as in Fig. 7-13*a* and *b*, current flows out through one phase leg and returns through another so that the impedances opposing the current are in both legs and may be represented as shown in Fig. 7-14*b*. If the single-phase fault is between phase and ground of a grounded-wye circuit, involving a return path of the neutral wire and the earth in parallel, the circuit impedances may also be represented as in Fig. 7-14*b*, but impedance of the return leg may not equal that of the phase leg. If impedance values are in ohms, the value of short-circuit current is

$$I_{3\phi} = \frac{E}{\sqrt{3}Z} = \text{amperes in each phase from three-phase short circuit}$$

$$I_{LL} = \frac{E}{2Z} = \text{amperes from single-phase short circuit between two phases}$$

$$I_G = \frac{E}{\sqrt{3}(2Z_1 + Z_2 + Z_G)} = \text{amperes from single-phase short circuit between one phase and ground}$$

where E = voltage between phases
Z = total equivalent line-to-neutral impedance of one phase to the point of fault, ohms
Z_1 = total equivalent line-to-neutral impedance of one phase, but excluding the impedance in the phase wire of a grounded wye circuit, ohms
Z_2 = total equivalent line-to-neutral impedance of one phase of a grounded-wye circuit which was excluded from Z_1, ohms
Z_G = total equivalent line-to-neutral impedance of the neutral return of a grounded wye circuit, ohms

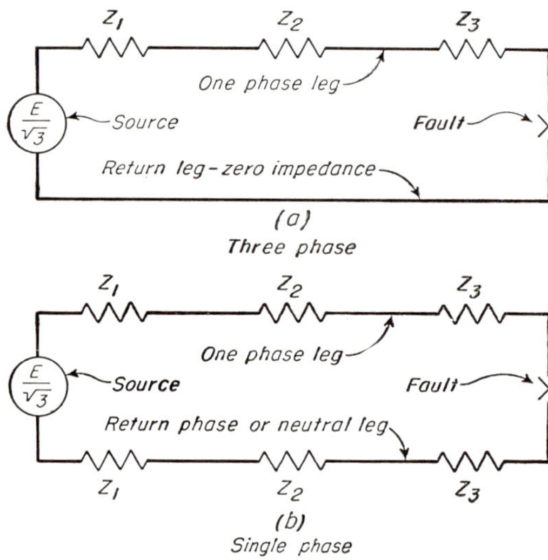

FIG. 7-14. Equivalent circuits for three-phase and single-phase short circuits.

In the case of a multigrounded neutral return, the impedance of the neutral Z_G will be somewhat less than that of a phase wire of the same size because of the parallel path through the earth. The resistance of such a parallel wire and earth path will range from about 12 to 15 percent of small phase wires up to about 35 percent of large phase wires, while the reactance will range from about 65 to 70 percent that of the phase wire. For distribution-fault calculations, the impedance of the neutral return path may be taken as equal to the reactance with slight error in the majority of cases. In the case of an earth-return neutral path due to a single ground at the station only, these values will be slightly higher because of the absence of the parallel neutral wire path.

If the circuit impedances involved are in percent or per unit values, the short-circuit current may be approximated by

$$I_{3\phi} = \frac{100 \times I}{\% Z}$$

or

$$I_{3\phi} = \frac{I}{\text{per unit } Z} = \text{amperes per phase for three-phase short circuit}$$

$$I_{LL} = \frac{\sqrt{3} \times 100 \times I}{2 \times \% Z}$$

or

$$I_{LL} = \frac{\sqrt{3} \times I}{2 \times \text{per unit } Z} = \text{amperes for single-phase short circuit between two phases}$$

$$I_G = \frac{100 \times I}{(2 \times \% Z_1) + \% Z_2 + \% Z_G}$$

or

$$I_G = \frac{I}{(2 \times \text{pu}Z_1) + \text{pu}Z_2 + \text{pu}Z_G} = \text{amperes for single-phase short between one phase and ground}$$

In the above equations the percent or per unit values of impedance have the same general meaning as in the equations for short-circuit current using ohmic values. I represents the normal current of the kva base to which the percent and per unit values are referred. If parallel impedance paths are involved, such as might be the case with parallel transmission or distribution lines, they may reduced to equivalent values of a single impedance path, as was shown in Arts. 1-28 and 1-29. Where short-circuit currents are to be determined on distribution circuits, or in industrial plants, supplied from an extensive utility high-voltage transmission system, the kva capacity of the substation stepping down from transmission to distribution voltage will usually be small in comparison to the kva capacity of the entire utility system. In such a case the impedance of the entire system between source and distribution step-down substation can usually be ignored, because the supply-system impedances down to the distribution substation will be quite small when reduced to equivalent values at the much lower distribution voltage. These values may be insignificant when compared to the impedances of the distribution lines to the point of fault or to the equivalent impedance of the substation-transformer bank. Omission of the supply-system impedances up to the distribution substation will give slightly higher values of calculated short-circuit current which will be on the safe side.

Example: Two generators are connected as in Fig. 7-15 to supply a plant at 2,400-volts delta. Part of the load consists of a group of 2,400-volt synchronous motors of 7,500 kva connected capacity. What current-interrupting capacity and momentary-current rating will be required in the feeder circuit breaker at D with a short circuit at the point indicated?

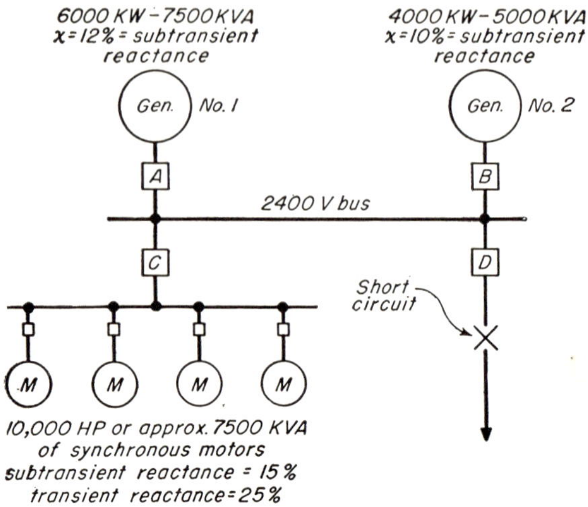

FIG. 7-15. Circuit diagram for example.

Answer:

1. Reduce percentage values of impedance to equivalent values on a 10,000-kva base:

$$X \text{ of generator 1} = \frac{10{,}000 \text{ kva}}{7{,}500 \text{ kva}} \times 12\% = 16\% \text{ subtransient reactance}$$

$$X \text{ of generator 2} = \frac{10{,}000 \text{ kva}}{5{,}000 \text{ kva}} \times 10\% = 20\% \text{ subtransient reactance}$$

$$X \text{ of synchronous motors} = \frac{10{,}000}{7{,}500} \times 15\% = 20\% \text{ subtransient reactance}$$

$$= \frac{10{,}000}{7{,}500} \times 25\% = 33.3\% \text{ transient reactance}$$

2. Equivalent impedance of the two generators in parallel:

$$\text{Equivalent } \% X = \frac{1}{1/16\% + 1/20\%} = 8.89\%$$

3. Total equivalent impedance for calculating interrupting capacity required, including parallel transient reactance of motors to provide for generator action of motors:

$$\text{Total equivalent } \% X = \frac{1}{1/8.89\% + 1/33.3\%} = 7\%$$

4. Total equivalent impedance for calculating momentary current rating required, including the paralleled subtransient reactance of the motors to provide for the generator action of the motors:

$$\text{Total equivalent } \% X = \frac{1}{1/7\% + 1/20\%} = 5.185\%$$

5. Normal line current on 10,000-kva base at 2,400 volts:

$$I = \frac{10{,}000 \text{ kva} \times 1{,}000}{\sqrt{3} \times 2{,}400 \text{ volts}} = 2{,}404 \text{ amp}$$

6. The short-circuit current the circuit breaker must interrupt:

$$I = \frac{100}{7\%} \times 2{,}404 \text{ amp} = 34{,}343 \text{ amp at } 2{,}400 \text{ volts}$$

7. The momentary current the breaker must withstand:

$$\text{Symmetrical ac component} = \frac{100}{5.185\%} \times 2{,}404 \text{ amp} = 46{,}363 \text{ amp}$$

To account for momentary dc offset

$$I_{\text{M}} = 46{,}363 \text{ amp} \times 1.6 = 74{,}200 \text{ amp}$$

since the fault is fed directly from synchronous machines.

CHAPTER REVIEW TOPICS

1. What are the elements and action in a lightning storm?
2. How does a lightning stroke cause an electric distribution line to fail?
3. What potentials are developed in an electric line by a direct stroke and by an induced surge?
4. What determines the point of breakdown in a distribution line carrying a lightning surge?
5. Why is a simple arc gap insufficient protection against lightning surges?
6. How does the valve-type arrester work?
7. How does the expulsion-type arrester work?
8. How much current may be involved in lightning strokes?
9. How are lightning arresters connected to distribution lines?
10. What are three reasons for grounding distribution systems?
11. How are ground connections made?
12. What precautions must be observed to achieve low ground resistance?
13. What are the operating principles of fuses and fused cutouts?
14. What device can be used to reduce long-time service interruptions from blown fuses?
15. What conditions govern the selection of fuse sizes?
16. What are the elements of a circuit breaker?
17. Describe the principles of oil-circuit-breaker operation.
18. Describe the principles of air-circuit-breaker operation.
19. Outline the principles used in relays to trip circuit breakers automatically.
20. Why are batteries used with some circuit breakers?
21. How does the capacitor trip work?
22. Why is time-delay preferable to instantaneous tripping in circuit breakers?
23. How do reclosing relays operate?
24. Describe the operation of automatic reclosers.
25. Do reclosers have to be installed in all phases of a three-phase circuit?
26. How should circuit breakers, reclosers, and fuses be coordinated on a circuit?
27. How are circuit breakers rated?
28. How does a short-circuit current behave from the instant the short develops?
29. How do motors increase the flow of current to a short circuit?
30. How are short-circuit currents computed for a circuit?
31. Under what conditions can either resistance or reactance be ignored in calculating short-circuit conditions?

32. When are subtransient, transient, and synchronous reactances used?
33. What factors are used in figuring short-circuit duties on circuit breakers?
34. Find the range within which subtransient and transient reactances will fall for turbine generators, salient-pole generators, synchronous motors, and induction motors.
35. How may impedance values be expressed for various types of equipment?
36. What precautions must be taken before combining impedances when making calculations?
37. What condition causes the maximum possible short-circuit current?
38. Describe how short-circuit calculations can be simplified by reducing to equivalent single-phase circuit.

PROBLEMS

Problem: Suppose that the two generators in Fig. 7-15 are connected to a 34.5-kv transmission line through a 5,000-kva step-up substation, that the transmission line

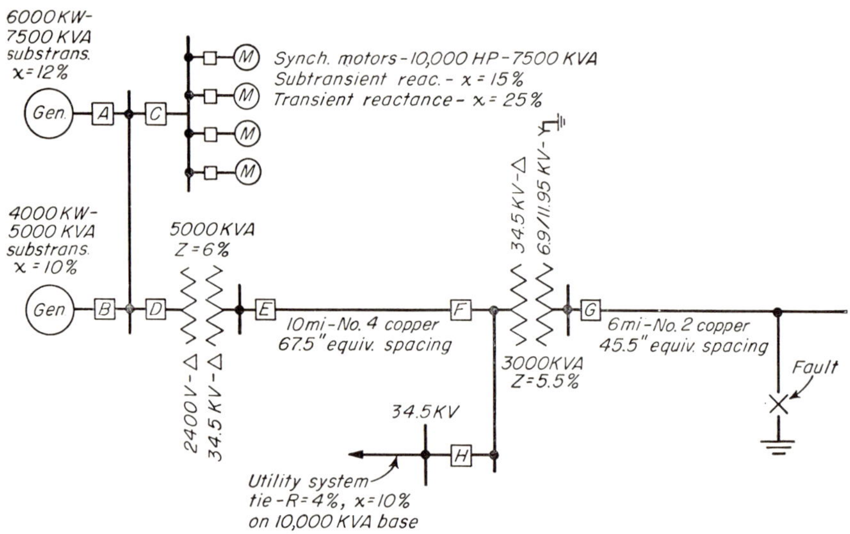

FIG. 7-16. Circuit diagram for Prob. 7-1.

feeds into a 6.9/11.95-kv distribution system through a 3,000-kva step-down substation, and that a tie is provided with another utility system, on the high-voltage bus of the 3,000-kva substation, as in Fig. 7-16. With the impedance characteristics and supply capacities shown, what will be the current interrupting capacity required if line fuses or automatic circuit reclosers are to be installed in the 6.9/12-kv line at a point 6 miles from the 3,000-kva substation?

Solution: The maximum fault current will occur with a short circuit on the load side of the fuses or reclosers at, or near, their point of installation. Since the fault will be remote from the 7,500-kva of synchronous motors installed at the generating station, their contribution to short-circuit current may be neglected.

Before percentage values can be combined they must be reduced to a common kva base, and before ohmic values can be combined they must be reduced to a common voltage base. The percentage values of impedance on a 10,000-kva base and the ohmic values on a 6.9/12-kv base, from the generating station to the tie point with the parallel utility supply system on the high-voltage bus of the 3,000-kva substation, are equal to the following:

	Resistance		Reactance	
	R ohms	% R	X ohms	% X
Paralleled generators	.....	.....	1.269	8.89
5,000-kva transformer bank	.....	.....	1.714	12.00
34.5 kv transmission line	1.742	12.20	0.982	6.88
Total	1.742	12.20	3.965	27.77
Parallel system tie	0.571	4.00	1.428	10.00
Combining these parallel impedance paths: resultant	0.431	3.01	1.051	7.36
Adding impedance of 3,000-kva substation bank	.....	.....	2.617	18.33
Total	0.431	3.01	3.668	25.99
The impedance values in the 6.9/12-kv line from the 3,000-kva substation to the point of fault is given as follows:				
Line-phase conductor	5.483	38.35	4.433	31.00
Estimated neutral return path impedance, using 0.18 and 0.67 as multipliers	0.987	6.90	2.970	20.80

Combining the above impedance values, the total impedance for a three-phase fault, comprising the impedance of one phase, is equal to

$$Z = \sqrt{(0.431 + 5.483)^2 + (3.668 + 4.433)^2} = \sqrt{5.914^2 + 8.101^2} = 10.024 \text{ ohms}$$

$$\% Z = \sqrt{(3.01 + 38.35)^2 + (25.99 + 31.00)^2} = \sqrt{41.36^2 + 56.99^2} = 70.10\%$$

In the case of a single-phase-to-ground short circuit the total impedance controlling the current will include the impedance of a return path for the flow of the unbalanced single-phase fault current, so that

$$\begin{aligned} Z &= \sqrt{(2 \times 0.431 + 5.483 + 0.987)^2 + (2 \times 3.668 + 4.433 + 2.97)^2} \\ &= \sqrt{7.332^2 + 14.739^2} = 16.46 \text{ ohms} \end{aligned}$$

or

$$\begin{aligned} \% Z &= \sqrt{(2 \times 3.01 + 38.35 + 6.90)^2 + (2 \times 25.99 + 31 + 20.80)^2} \\ &= \sqrt{51.27^2 + 103.78^2} = 115.75\% \end{aligned}$$

For a three-phase short circuit, the fault current is

$$I_3 = \frac{6{,}900 \text{ volts}}{10.024 \text{ ohms}} = 690 \text{ amp at } 6{,}900/11{,}950 \text{ volts}$$

$$I_3 = \frac{10{,}000 \text{ kva} \times 1{,}000}{3 \times 6{,}900 \text{ volts}} \times \frac{100\%}{70.1\%} = 690 \text{ amp}$$

For a single-phase-to-ground short circuit, the fault current is

$$I_G = \frac{6{,}900 \text{ volts}}{16.46 \text{ ohms}} = 419 \text{ amp at } 6{,}900 \text{ volts to ground}$$

or

$$I_G = \frac{10{,}000 \text{ kva} \times 1{,}000}{3 \times 6{,}900 \text{ volts}} \times \frac{100\%}{115.75\%} = 419 \text{ amp}$$

Since the line voltage is below 15,000 volts and since the ratio of X to R is less than 4, these calculated values should be multiplied by the factor 1.2, so that

$$I_3 = 690 \times 1.2 = 828 \text{ amps interrupting capacity required}$$

and

$$I_G = 419 \times 1.2 = 503 \text{ amps interrupting capacity required}$$

CHAPTER 8

STREET LIGHTING

8-1. Lighting Systems. Electric lighting of streets and roads was one of the first duties performed by central-station electric systems. The greater efficiency, flexibility, and controllability of electric lights gave them a great advantage over the old illuminating-gas lighting systems. The equipment for electric street lighting is similar to distribution-system apparatus. In the majority of installations, street-lighting systems are fed by a distribution system. Street-lighting equipment usually occupies the same poles and ducts, and its wires run parallel to those of the distribution system.

The energy consumed by lighting systems depends on the intensity of illumination needed. This varies from about 200 watts per 1,000 ft of street length in suburban districts to over 5,000 watts per 1,000 ft on boulevards carrying heavy traffic at night through city centers. The low lighting intensity used in suburban areas makes objects visible by silhouette rather than showing details and color. On the other hand, high-intensity lighting attempts to achieve levels comparable to daylight, giving full visibility in the area. This contributes greatly to safety and security.

Up to 1950 about 90 percent of street lights in the United States were supplied through series circuits. The remainder were on multiple circuits. The vast majority of lighting systems use the tungsten-filament lamp, also known as the incandescent lamp. Mercury-vapor lamps are occasionally used for high-intensity lighting, and in a few areas the old type arc lamps are still used.

8-2. Series System. Figure 8-1 shows the hookup of a series lighting system. The loop system *a* needs the least amount of wire, but it offers difficulty in finding open circuits when they occur and may cause telephone interference because it doesn't have an adjacent return wire to counterbalance its magnetic flux. The closed-loop system in *b* overcomes these two disadvantages but needs almost twice as much wire. Often a combination of the loop and closed-loop arrangement is used as in *c*.

As the circuit shows, the same current passes through each lamp. From the characteristics of the circuit, the source voltage must be equal to the sum of all the voltage drops across the individual lamps and the lengths of conductors between lamps. Depending on the number of lamps and the length of the circuit, the source voltage needed varies from 750 to 3,000 volts and perhaps more. Ordinarily we might think that,

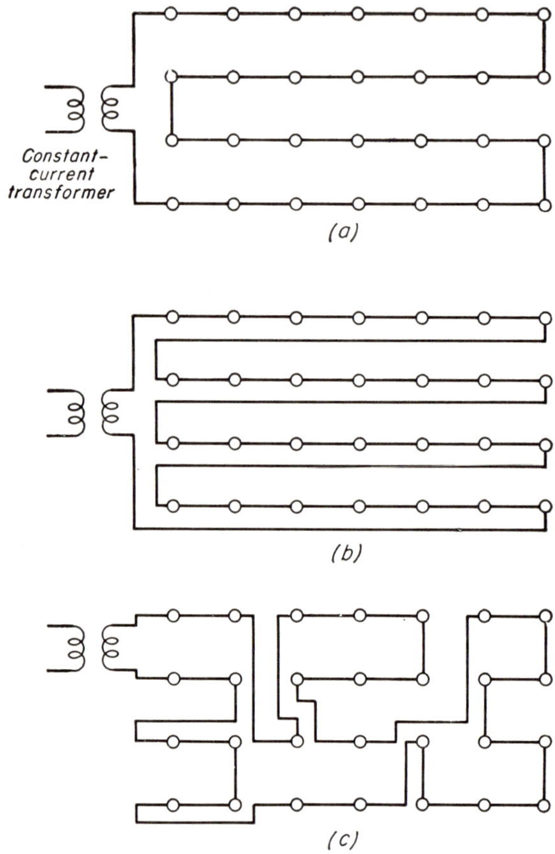

FIG. 8-1. Series lighting circuits may be arranged in a number of ways. In (*a*) the minimum length of wire is needed, but this layout is difficult to test in looking for open circuits. In (*b*) open circuits can be quickly located, but the hookup needs maximum amount of conductor. In (*c*), a compromise between (*a*) and (*b*), testing convenience is realized without using maximum length of wire.

for a given circuit supplied by ac, we could choose a transformer with the proper secondary voltage rating for a source. However, to keep lamps burning at maximum efficiency, the current flow must be maintained at the rated level, 6.6, 7.5, or 20 amp. To do this on a series circuit which might change resistance or impedance, the voltage must be varied. Too high a current burns out the bulbs; too low a current seriously reduces illumination.

8-3. Constant-current Transformer. For this duty we need a special unit called a constant-current transformer, Fig. 8-2. This transformer, available in a number of designs, consists of a magnetic core about which the primary winding remains fixed. The secondary winding or coil, though, is movable; it can slide up and down on the center core piece. A weight counterbalances the movable secondary coil. Movement of the secondary coil is automatic. To understand how the transformer works, first let's examine the flux distribution in the core, Fig. 8-2*a*.

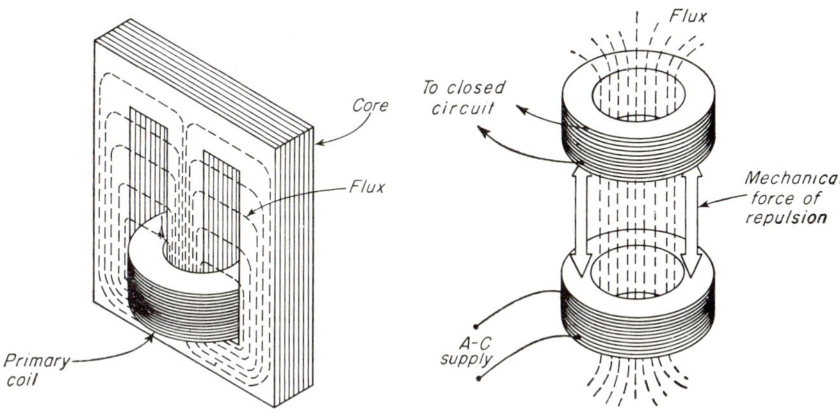

FIG. 8-2*a*. In the laminated core of a c-c transformer the primary coil sets up a magnetic field flux to link with the secondary coil (not shown). Flux leaks across the air space from the center to the outer portions of the core, so that flux progressively diminishes in the center piece.

FIG. 8-2*b*. Magnetic field set up by a primary coil, below, that links with a secondary coil, above, connected to a closed circuit, exerts a force between coils that tries to separate them physically. This is the fundamental working principle used in the c-c transformer.

The magnetic flux is a maximum at the center of the primary coil. In the upper part of the core, however, some of the flux leaks through the air space so that the flux density falls off uniformly toward the top of the core.

Now let's look at what happens to two coils when one (primary) induces a current in the other (secondary). The alternating magnetic flux from the primary sets up a current in the secondary in such a direction that the flux of the secondary current opposes the primary flux. When two magnetic fields oppose each other, the coils producing these fluxes or fields experience a mechanical force tending to push them apart Fig. 8-2*b*. This force is the operating basis of the constant-current transformer.

When the transformer secondary in Fig. 8-2*c* feeds the rated number of lamps, it operates at its maximum voltage and the coil takes a position nearest the primary. Now, if a lamp in the series circuit is shunted out, the resistance of the circuit decreases. This causes a rise in current flow since the secondary voltage momentarily remains constant. But the rise in secondary current increases the secondary-coil magnetic field. This increases the mechanical force tending to push apart the primary and secondary coils. The counterbalancing weight of the secondary coil makes the latter coil sensitive to this force, so that the coil can move up and down as the force varies.

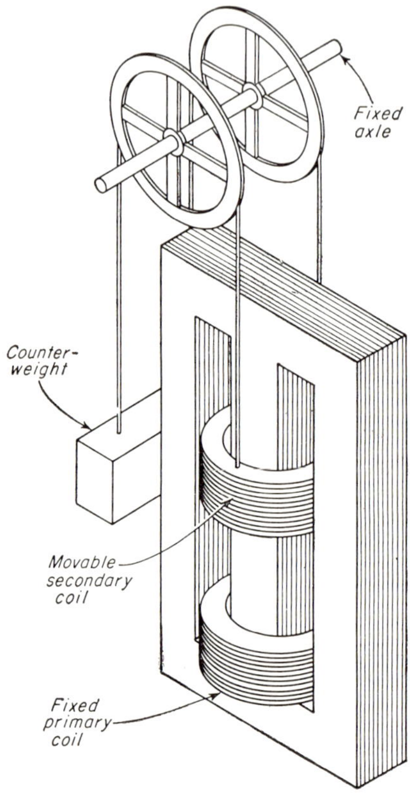

FIG. 8-2c. Schematic layout of a c-c transformer for a series lighting-circuit supply. The secondary coil is free to move up and down over the center core piece, counterbalanced by a weight. A magnetic repulsion force between coils makes the secondary move away from the primary coil when the secondary current exceeds rated value.

Increased repulsion moves the secondary coil upward, farther from the primary. But as the secondary coil moves up less flux from the primary field links with it, because some of the flux leaks across the air space. This reduces the voltage induced in the secondary. Since the resistance of the secondary circuit stays constant at its new lower value, the current flow reduces, and hence the opposing magnetic field of the secondary coil also diminishes.

This means that, as the secondary coil moves upward, the repulsion force between the coils decreases until it again equals the net weight of the secondary-coil weight minus the counterbalance. The secondary coil then comes to rest in this new position, where it's induced voltage has fallen so that it maintains the rated current in the secondary.

When resistance of a circuit is increased by adding more lamps, the reactions just described occur in a reverse sense so that the transformer coils come closer together and the secondary voltage increases so that the same current can be sent through the higher resistance.

8-4. Transformer Design. Constant-current (c-c) transformers fundamentally use the same principles as power transformers, but they do have a higher reactance. Power transformers always have a low leakage reactance. C-c transformers usually have standard primary voltage ratings of 2,400 or 4,800 volts.

Pole-mounted c-c transformers are oil immersed, ranging from 2 to 30-kw capacity. Larger types are available. The subway type is the same as the pole-mounted, except that its case can prevent standing water from seeping into the mechanism. The station-type c-c transformer uses air cooling, the entire assembly usually being protected just by a screen. Most capacities of this type range between 5 and 70 kw.

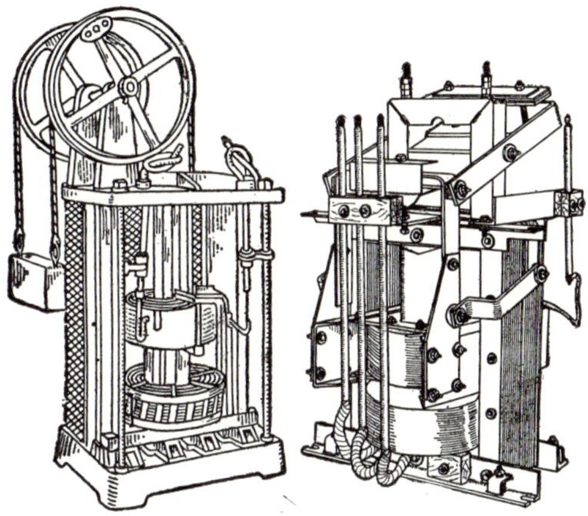

FIG. 8-2*d*. Indoor- or substation-type c-c transformer, left, is air-cooled, usually has wire cage around it. Pole-type c-c transformer, right, may be immersed in oil held in tank.

Static-type c-c transformers, with no moving parts, use a resonant network consisting of two inductive and two condensive reactors each of equal reactance. In such a network, the secondary current is directly proportional to the primary voltage, regardless of the impedance or load of the secondary circuit. An insulating transformer is tied between the network of condensers and reactors and the lighting circuit. This type of c-c transformer has a high operating power factor, but it does not compensate for variations in feeder line voltages.

The moving-coil c-c transformer regulates secondary current within 1 percent for all loads within its rating, when primary voltage is within 5 percent of normal. At full load the transformer draws its energy at about 75 percent power factor which drops to about 56 percent for three-quarter load. The input kva remains practically constant over the full operating range. In certain designs, capacitors installed integral with the transformer improve the power factor, raising it to near unity.

8-5. Advantages of the Series System. Perhaps the primary advantage of the series circuit is the small conductor needed to serve the circuit. Since line loss depends on the I^2R relationship, it is obvious that since only 6.6, 7.5, or 20 amp passes through the entire circuit it is relatively small. In a multiple circuit each lamp of the larger sizes would take this amount of current, and hence, near the source at least, the conductor would have to be much larger than in a series circuit to keep line losses to a practical minimum.

The series circuit, on the other hand, needs to have all its parts insulated for full transformer terminal voltage. This offsets somewhat the advantage of needing only a small conductor cross section. One break on a series circuit usually interrupts a large number of lamps, and therefore a large area. Because of the high voltage the lighting circuit must be placed on the same level with the high-voltage primaries in overhead distribution. This often causes pole-top congestion, requiring more expensive construction.

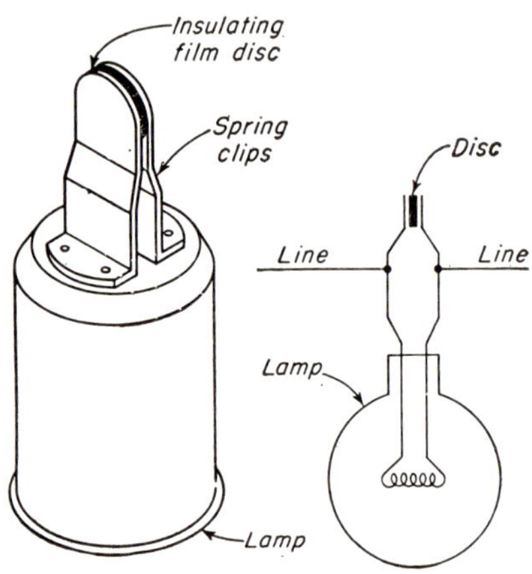

FIG. 8-3. To prevent interruptions from lamp burnouts in a series lighting circuit, an insulating film held in a disk between the spring-clip lamp terminals burns out from full voltage appearing across the lamp when it burns out. This allows the terminals or spring clips to make contact, completing the circuit through the remainder of the lamps in service. The disk must be replaced when the lamp is renewed.

8-6. Lamp Outages. Lamps for series circuits carry a higher current than equal wattage multiple lamps; hence they have a stronger, heavier filament, making the lamp more rugged. Series lamps prove more efficient, giving more light per watt consumed. They also have a better efficiency over their life span, maintaining higher efficiency throughout than the multiple lamps, the light output dropping off but little.

In series circuits, when lamps burn out, they interrupt the continuity of the circuit. Without special provisions this would put the entire circuit out of service. Since there are a large number of lamps in one circuit, and they are likely to burn out one at a time, steps must be taken to avoid numerous interruptions.

Whenever a lamp burns out in a series circuit, current flow stops and full secondary voltage appears across the break in the lamp. This automatic production of full transformer voltage at the break in the circuit solves the problem of immediate restoration of service after a momentary interruption. Figure 8-3 shows a pair of shunting clips built into the socket of every series lamp. These clips are under a tension that tends to bring them into contact and shunt out the lamp since they connect across the lamp or socket terminals. A thin film of insulating material inserted between the points of the clips prevents them from making contact. Under ordinary operation the voltage across the film is only the drop through the lamp, about 10 to 40 volts depending on lamp rating.

When the lamp burns out, however, full transformer voltage (800 to 3,000 volts) breaks down the insulating film. The arc burns out the film allowing the clips to make contact which shunts out the break in the lamp and again completes the series circuit. Thus a lamp burn out means only a slight interruption in service. When the lamp is replaced, the film must be replaced also. When a lamp goes out of service, the total resistance of the series circuit becomes smaller and a higher current flows. This causes the c-c transformer to adjust the position of its secondary coil and bring back current flow to its rated level.

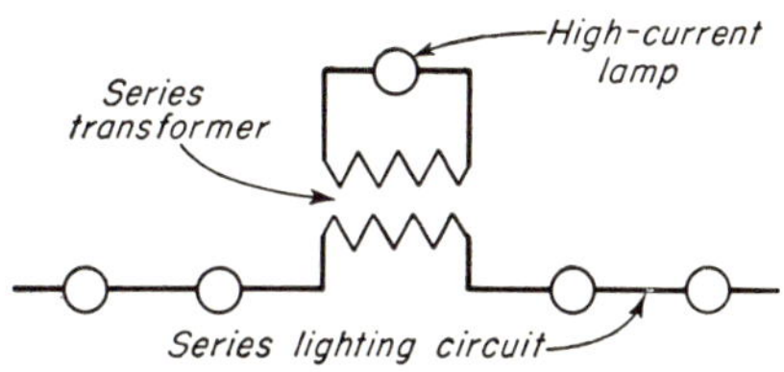

FIG. 8-4. When a higher rated lamp is needed in one point on a series circuit, it can be supplied through a series transformer hooked into the circuit. This high-current lamp does not need an insulating film disk to prevent interruption from lamp burnout.

When an occasional lamp of higher light rating is needed in a series circuit, it must have a higher current input than the circuit current. Figure 8-4 shows how such a lamp may be connected in via a series transformer. The primary connects directly into the series circuit, while the secondary produces a high enough voltage to develop the required current through the lamp. One advantage of this arrangement is that in case of a lamp burnout no interruption occurs in the rest of the circuit.

In small series circuits another scheme may be used to compensate for lamp burnouts automatically. This consists of a reactance connected in shunt across each lamp at the socket terminals. The reactance has a value that makes the voltage drop across it alone practically equal to the voltage drop across both the reactance and the lamp connected in parallel. When the lamp burns out, the circuit current stays practically constant without even a momentary interruption in service. This system can use standard constant-voltage transformers for supply since the lamp shunt reactances maintain constant current.

8-7. Multiple Circuits. Where relatively few lamps need to be supplied in a small area, it usually pays to use 115- or 120-volt multiple circuits tied into the normal distribution circuits, Fig. 8-5. These circuits are individually controlled by a variety of methods which we'll talk about later. In downtown areas of cities it is advantageous to supply street lamps directly from the regular distribution system.

Several advantages are claimed for the multiple circuit, which is being increasingly applied in new extensions of street lighting. The lamps cost less than for the series type and are easier to replace. Lower operating voltages don't call for linemen skilled in handling high voltages for installation and repairs. A third circuit is not involved which must be accommodated next to distribution circuits on poles and in ducts. Since wires must run parallel, it's much easier to locate faults.

Losses in a lighting system are heat or I^2R and wattless losses. The series system generally operates with a higher current and has higher heat loss than the multiple circuit with constant voltage and larger copper sizes. The total light output per watt of the series lamps being higher than the multiple circuit offsets the higher heat loss. In general, mul-

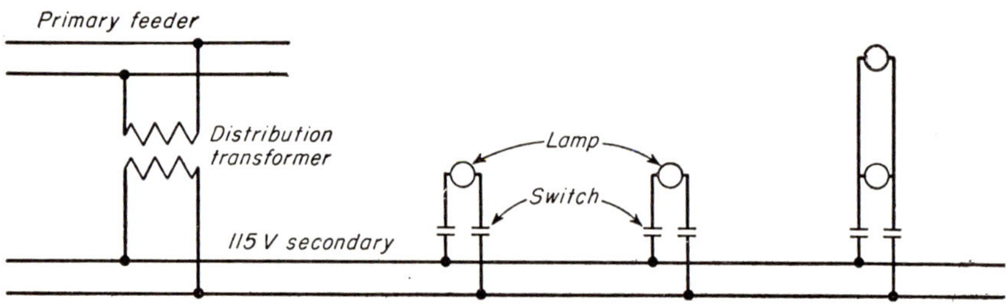

FIG. 8-5. Multiple lighting circuits are fed directly from 115-volt service secondaries supplying other loads.

tiple circuits operate at a higher power factor which contributes to lower over-all system losses.

8-8. Lighting Control. Series circuits may be turned on and off in several ways: (1) manually, (2) with time clocks, (3) by relays using the cascade principle. When the c-c transformers are located in attended substations or power stations, the series lighting circuits are often controlled by hand throwing of switches. If attendants are needed anyway for other purposes, this is the most economical method of control, since only the simplest type of switch is needed in the transformer primary circuit.

When circuits are located in areas where attendants normally are not needed, time switches may be used. These switches are automatically turned on and off at certain hours by electrically driven time clocks. The simpler types of clocks must be adjusted manually for operating times as the hours of sunset and sunrise change with the season. More advanced types incorporate astronomical dials which automatically vary the tripping time as needed with the change in season. Such time switches may control one or more c-c transformers and one or more multiple circuits. Whenever an extended outage in service occurs, time clocks must be reset for proper tripping times.

Cascade control uses the operation of one lighting circuit to control also one or more adjacent lighting circuits in the field. Thus a first series lighting circuit may be controlled manually in a substation. Operation of the first lighting circuit may be used simultaneously to control a second lighting circuit that does not enter the substation, but does run close to the first circuit at some one point. Figure 8-6 shows how the two circuits are interconnected for control purposes. The first series circuit at the left energizes the coil of the controller or oil switch to close the primary circuit of the second circuit c-c transformer. When the first circuit is deenergized by opening a switch, the solenoid coil of the con-

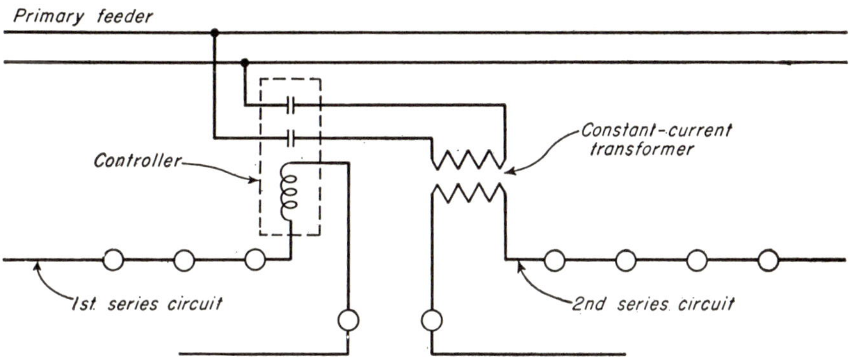

FIG. 8-6. One series circuit can be made to automatically control a second series circuit through the controller. A third can be controlled from the second and so forth. This is a cascade-control arrangement.

troller also becomes deenergized, causing the controller to open the primary circuit of the second-circuit c-c transformer. By such controllers the second circuit can be made to control a third circuit, and so on. In this scheme you can see that interruption of current in any one series circuit will turn off all succeeding circuits depending on that one for control.

Pilot wires and relays can be arranged to control large numbers of lamps, either series or multiple. The pilot wire running from the control station, either manual or automatic, connects to relays at each lighting-circuit switch. The relays in turn control the lamp-circuit-switch operation. In some schemes the pilot wire carries current during the time that the lighting circuits are active. That is, energizing them closes the lighting-circuit switches, deenergizing them opens the switches. In other schemes the pilot wire and relays work on signal impulses. One set of impulses closes the relays, and another set opens them. Different relays can be set to respond to various sets of impulses, so that some circuits can be closed and opened independently of the others from the central control station.

On some central-station systems, separate primary feeders supply pole-mounted c-c transformers. This allows manual control of all lighting circuits from the substation. While some of the other arrangements mentioned may be somewhat less costly, the separate feeder has the advantages of central control, separation from the distribution system disturbances, and no need for feeder automatic regulation since c-c transformers provide their own automatic voltage regulation.

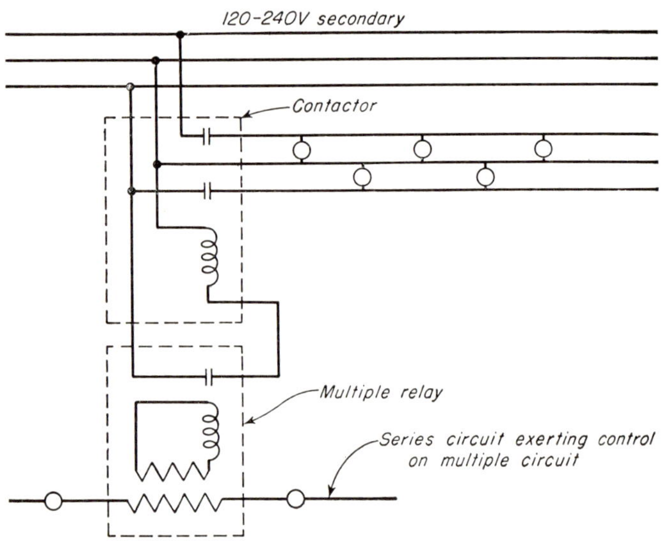

FIG. 8-7. Cascade control can also be made of a multiple lighting circuit from a series circuit, or another multiple circuit.

Carrier-current systems may be used to control 2,000 or more multiple lamps simultaneously. A receiver relay controls each lamp switch. At the substation a high-frequency generator couples to the primary feeders through capacitors or transformers. The generator sends out impulses of short duration at specified frequencies, much higher than standard. These signals operate the relays and their switches to open or close the lighting circuits.

This type of signaling system is also used to control certain loads on distribution systems from the substations. Such loads as water heating may be disconnected in this way during peaks on the system and reconnected after the peak demand has passed. This permits the water-heating load to have a preferred lower rate by giving better utilization of system generating and transmitting equipment. The same generator and carrier system to control lighting may be used for water-heating loads or other loads that may be dumped.

Photoelectric cells sensitive to actual light conditions may also be used to control lighting systems. These have the advantage of turning lights on according to actual need rather than according to a fixed schedule. The cell operates a relay that controls the switch operation.

Similar to the cascade control for series-lighting circuits, the latter can also be made to control adjacent multiple circuits, Fig. 8-7. A multiple relay inserted in the series circuit controls the operation of a contactor on the multiple circuit. This arrangement again depends on operation of the series circuit; if it fails, the multiple circuit opens up also.

8-9. Lamps for street lighting vary in light output from 500 to 20,000 lumens. Since 1 candlepower equals 10 lumens, the lamps vary from 50 to 2,000 cp. Tungsten-filament lamps find the most frequent application in street lighting. Their greater flexibility and lower maintenance needs enabled them to replace the old magnetite and arc lamps that were first used for street lighting. More than 60 percent of lamps in use are 1,000 lumens or smaller. Table 8-1 gives some of the characteristics of series lamps.

TABLE 8-1. CHARACTERISTICS OF SERIES LAMPS

Rated lumens	1,000	2,500	5,000	5,000	10,000	15,000
Rated candlepower	100	250	500	500	1,000	1,500
Current, amperes	6.6	6.6	6.6	20.0	20.0	20.0
Lamp wattage	64.5	151.8	281	279	524	746

Mercury-vapor lamps rated 16,000 lumens using 400 watts have been used in some cities. The efficiency as measured by lumen output per watt is about twice as high as incandescent lamps. The average life of the 16,000-lumen mercury lamp runs about 5,000 hr compared with 1,600 hr for the series incandescent lamp, 2,400 hr for the multiple lamp; all have the same light output.

CHAPTER REVIEW TOPICS

1. What is the approximate wattage needed per 1,000 ft for suburban streets? For heavy traffic streets?
2. Describe the two methods by which street lighting makes objects visible.
3. What type of street-lighting circuits are used mostly in the United States?
4. Describe the loop-type series circuit, the closed-loop type, combination.
5. How is the supply voltage calculated for a series-type circuit?
6. Describe the electrical characteristics of a series circuit.
7. Draw a simplified sketch of a c-c transformer; describe its operation.
8. What happens to the flux in the center core of a c-c transformer?
9. How are the secondary and primary coils of a c-c transformer moved apart?

10. Describe the types of installations of c-c transformers.
11. What are the components of a static-type c-c transformer?
12. What are the regulation characteristics of c-c transformers?
13. How can the power factor of a c-c transformer be improved?
14. Compare the conductor sizes needed by series and multiple lighting circuits.
15. Compare the insulation needs of series and multiple lighting circuits.
16. Compare the operating characteristics of series and multiple lamps.
17. How is circuit continuity reestablished in a series circuit when a lamp burns out? Describe the operation of the two methods.
18. How can one lamp needing higher current than the circuit current be fed by a series circuit?
19. What are the advantages of multiple lighting circuits?
20. How are series lighting circuits manually controlled?
21. What are the basic elements in a clock-controlled circuit?
22. How are tripping times adjusted for seasonal variation in lighting needs?
23. What is the arrangement in a cascade control?
24. How are pilot wires and relays used to control lighting circuits?
25. Do some systems use separate primary feeders for series lighting circuits?
26. What are the elements of carrier current systems used to control lighting circuits and other types of remotely controlled loads?
27. How can street lights be automatically controlled according to actual light need?
28. What types of lamps are used in street lighting?

CHAPTER 9

PLANNING AND LOAD CHARACTERISTICS

By E. M. Adkins

Jersey Central Power and Light Company

Electrical energy must be ready to meet the needs and requirements of all users, from the largest to the smallest, upon demand. To do this, the growing requirements of electricity users must be anticipated well in advance and provision made to meet these upon their earliest need. Design of the system depends upon the characteristics of individual loads and how they combine.

9-1. Electrical Load. In Art. 1-5, it was noted that electric power represents the rate at which work is done by an electric circuit. The commonly used unit of power measurement is the watt or the kilowatt, the kilowatt being equal to 1,000 watts. The term "rate at which work is done" introduces an element of time in defining electric power. Thus, a kilowatt for a definite period of time represents a definite rate at which work can be done. The kilowatthour, representing electric power of one kilowatt acting over a time interval of one hour, is a unit measurement of total work done by the electric circuit. If 60 kw are delivered by the circuit for 1 min, the same amount of work, 1 kwhr $= 60$ kw $\times \frac{1}{60}$ hr, will be done, but the rate at which the circuit is doing the work will be sixty times as great. An electrical load defines the rate at which the supply system is required to do work.

9-2. Maximum Demand. Electrical loads are generally measured in terms of amperes, kilowatts, or kilovolt-amperes. Before the whole or any part of an electric system can be efficiently designed and built, these characteristics of the electrical loads must be known. In general, electrical loads are seldom constant for any appreciable length of time, but fluctuate constantly. Figure 9-1 shows a 24-hr load curve for a distribution transformer. The load varies between a maximum at about 7:30 o'clock in the evening to a minimum at about 3:30 o'clock in the early morning. Between these times the load is constantly changing. This pattern of change is constantly repeated, although the values may differ. A similar pattern of variation between maximum and minimum is present in each part of the electric system.

The greatest value of load, Fig. 9-1, is called the peak or maximum demand on the transformer for the 24-hr period or day. If similar curves were collected for 7 consecutive days, the largest load shown would be the maximum demand or peak load on the transformer for a week. Similarly, the greatest load for a month or year is the maximum demand or peak load for a month or year. The maximum demand over a year is the value most frequently used in planning for electric-system growth and expansion. The term *demand* is frequently used in a sense meaning the maximum demand for the period of time specified. The accurate

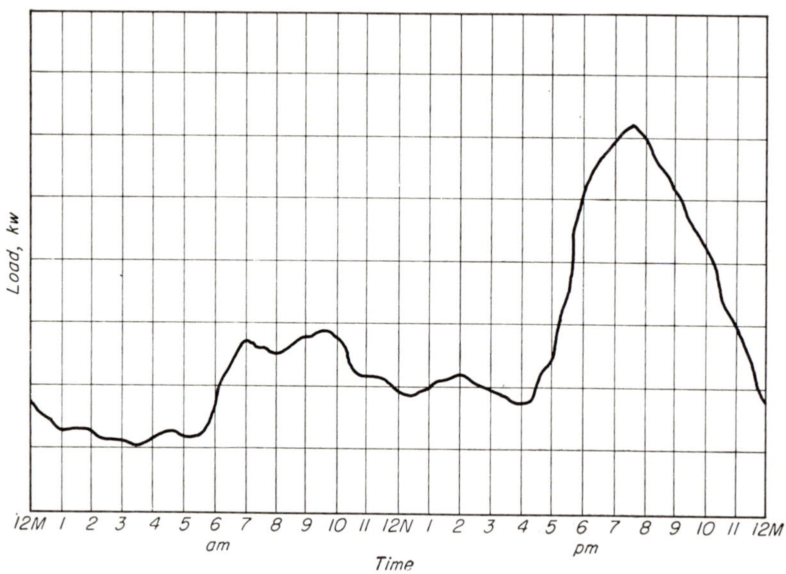

FIG. 9-1. Typical load curve for a distribution-system transformer.

determination of the maximum demand of individual loads is, of course, necessary where a demand charge is included in the billing for electric service.

Knowledge of the maximum demand of groups of loads and their combined effect throughout the electric system is also of the greatest importance, since it is the group maximum demand which determines the size throughout the electric system. Thus, a group of small consumer loads combine to form a maximum demand which determines the size of distribution transformer required; the loads supplied by a group of distribution transformers may result in a maximum demand which determines the conductor size, regulator size, circuit-breaker size, etc., making up a primary-feeder circuit; the combined maximum demand of a group of primary feeders may determine the size of substation transformers required, and so on back to the final determination of the size of generating facilities required. In any case the determination of maximum demand is of vital importance, and if accurate measurements of demand cannot be made, the best estimates possible must be used in forecasting future loads.

9-3. Time Intervals for Demand Measurement. Peak-load demands are not usually measured instantaneously but are averaged over a definite interval of time. This average value is used as the demand for the time interval specified. The time interval chosen for demand measurement will vary somewhat with how the demand figures are to be used. If fuse sizes or circuit-breaker settings are to be selected, practically instantaneous values of current demand are required. The load capacity of most pieces of electrical equipment is based upon safe temperature rise, and since the temperature does not instantaneously change with load but with load and time, such pieces of electrical equipment as wires, cables, transformers, etc., have a considerable time lag in temperature rise and appreciable heat-storage capacity. Since loads in many parts of the electric system vary between a comparatively sharp peak and an appreciably lower minimum value, and since the load capacity of conductors and other equipment is based upon continuous unvarying load, conductors and equipment have considerable overload capacity with sharp peaks. If the load of Fig. 9-1 consists principally of an induction motor, the instantaneous value of starting current will be five or six times the normal full-load running current and probably several times the rated continuous full-load current of the transformer but will last for a very short interval of time, usually less than a second. Since the rated load capacity of the transformer is based upon temperature rise with continuous load, and since the temperature rise is determined by heat energy which may be measured by watt-seconds, kilowatthours, etc., high values of current of very short duration may not produce an appreciable temperature rise, and consequently it would be uneconomical to determine the transformer size required upon such higher current values of very short duration.

Demands are usually measured over some definite interval of time such as 15 min, 30 min, or 1 hr. The 15- or 30-min interval is generally used for billing or for determining circuit-equipment capacity. In Fig. 9-2 a load curve rises to a sharp peak and rapidly drops off. If, instead of showing instantaneous values, the curve was drawn from average demands, averaged over 15-min intervals, the curve would indicate lower demands, including a lower maximum demand. Likewise, lower values would be indicated with 30-min demand intervals and still lower values with a 1-hr demand interval. The average demand over a 15-min interval, for example, represents the continuous unvarying load which would represent the same energy as the actual varying load over the 15-min interval. The same is true for a 30-min or 1-hr time interval.

The terms kilowatt and kilowatthours are sometimes confused. It should be remembered that kilowatt represents the rate at which work can be done, while a kilowatthour represents the amount of energy used or the amount of work done in an interval of time. The average demand over any period of time is equal to the number of kilowatthours used divided by the number of hours in the period of time.

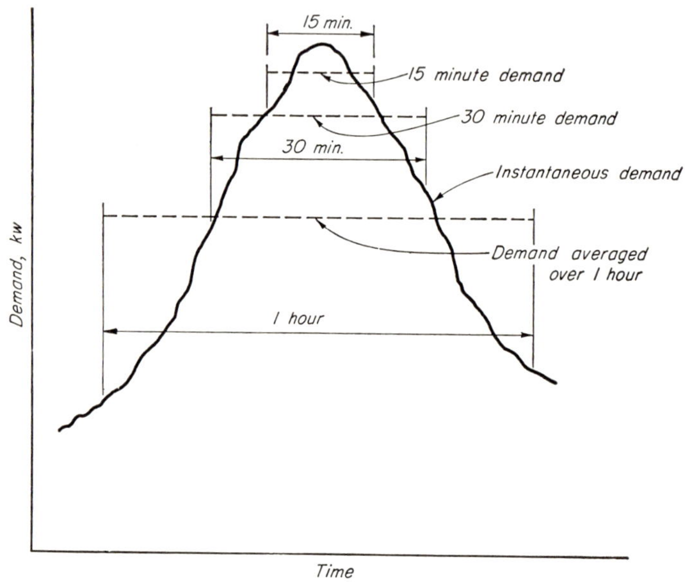

Fig. 9-2. Magnitude of maximum demand varies with the time period over which it is measured. Instantaneous peak is always highest; as time interval increases, the value decreases.

9-4. Connected Load and Demand Factor. Connected load is the sum of the rated maximum values of all loads on a consumer's premises which *might be connected* in service at the same time to produce a maximum load demand. The connected load may refer to the whole or any part of the system. The connected load may be expressed in watts, kilowatts, amperes, horsepower, kilovolt-amperes, etc., in accord with the preferred unit for expressing demand. Connected load represents the greatest load demand possible. If a customer has a connected load consisting of twenty 100-watt and thirty 250-watt incandescent lights plus a total of 80-hp of motor load, for example, his connected load would be equal to 9,500 watts of lighting load plus 80 hp of motor load. Expressed in kilowatts, his connected load is equal to 9.5 kw of lighting plus

$$80 \text{ hp} \times 0.746 = 59.7 \text{ kw}$$

or 69.2 kw total. Expressed in kva the connected load in kilowatts should be converted by dividing by the power factor of the combined connected load.

The ratio between the actual maximum demand and the connected load is called the demand factor. Demand and connected load should be in the same units. If a group of loads were all connected to the supply source at the same time and drew their name-plate demand, the demand factor would be unity, or 100 percent. In most cases, however, the demand factor is less than 100 percent. There are two main causes which usually reduce the demand factor below 1.0. First, some load devices are slightly larger than the minimum size needed in order to use standard units or to provide overload capacity. Consequently they normally draw less than rated load. Secondly, all load devices will seldom be connected to the supply system at the same time, or if connected, will seldom reach maximum demand at the same time. The second condition refers to *diversity* between individual loads of a group, see Art. 9-6, and is the major reason for demand factors of less than 100 percent. While the demand factor of a group of loads is usually less than 100 percent, instances will be found where the demand factor is more than 100 percent. In such cases the load elements are overloaded.

The larger the connected load and the greater the number of individual devices making up the connected load, the smaller the demand factor. Knowledge of average demand factors helps in estimating future maximum demands where the total connected load and the operations are known.

A great many things influence demand factor, and as a result it varies considerably for different types of load and classes of service. If a connected load is made up of a group of small motors plus one large motor, the maximum demand will probably occur when the large motor runs. The demand factor will probably be appreciably higher than with the same total horsepower of connected motor load without a large motor. The demand factor of a 4- to 6-room residence will probably be larger than that of a 10- or 12-room residence. If a residence has the usual number of lights and household appliances, its demand factor may be lowered by addition of an electric range even though the demand of the range alone is high compared with the demand without the range. Representative demand factors for lighting and appliance loads are given in the table on page 394.

The demand factors of motor loads are influenced by the number of motors and by the conditions under which they are used. If a plant has a large number of motors and a group of motors always operate together as a unit, each group should be regarded as a single motor of larger size and a higher demand factor will result. Representative demand factors

Small residences—lights and appliances without ranges	50–75%
Large residences—lights and appliances without ranges	40–65
Residences—lights, appliances, and ranges	35–60
Offices	60–80
Small stores	40–60
Department stores	70–90
Small industrial plants	35–65
Large industrial plants	50–80
Hotels	35–60

for groups of motors where each motor operates independently are shown in the accompanying table.

Total Connected Hp	*Percent*
1–10	70–80
10–20	60–70
20–50	55–65
50–100	50–60
Above 100	45–55

9-5. Load Factor. Load factor is the ratio of average load to maximum demand with a maximum of unity or 100 percent. The average and maximum loads may be expressed as kilowatts, kilovolt-amperes, amperes, etc., but both must be expressed in the same units. Load factors may be calculated for any given period of time but are usually for a day, a month, or a year. The yearly, or annual, load factor is the most useful since a year represents a full cycle of time.

The average load may be described as the continuous and unchanging load which would draw the same amount of energy over a given period of time as the actual load. Also, the load factor represents the percent-

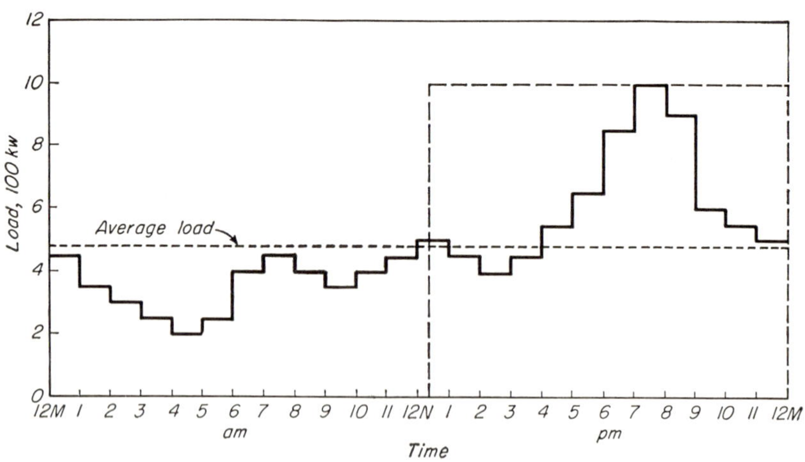

FIG. 9-3. Ratio of average load to maximum demand for the period equals the load factor for the day.

age of the total hours over which the maximum demand would have to last in order to draw the same amount of energy as the actual load. Thus, suppose that a load varies as shown in Fig. 9-3 for a 24-hr period. The hourly load is

12 Mid. – 1 A.M.	450 kw
1 A.M. – 2 A.M.	350 kw
2 A.M. – 3 A.M.	300 kw
3 A.M. – 4 A.M.	250 kw
4 A.M. – 5 A.M.	200 kw
5 A.M. – 6 A.M.	250 kw
6 A.M. – 7 A.M.	400 kw
7 A.M. – 8 A.M.	450 kw
8 A.M. – 9 A.M.	400 kw
9 A.M. –10 A.M.	350 kw
10 A.M. –11 A.M.	400 kw
11 A.M. –12 Noon	450 kw
12 Noon– 1 P.M.	500 kw
1 P.M. – 2 P.M.	450 kw
2 P.M. – 3 P.M.	400 kw
3 P.M. – 4 P.M.	450 kw
4 P.M. – 5 P.M.	550 kw
5 P.M. – 6 P.M.	650 kw
6 P.M. – 7 P.M.	850 kw
7 P.M. – 8 P.M.	1000 kw
8 P.M. – 9 P.M.	900 kw
9 P.M. –10 P.M.	600 kw
10 P.M. –11 P.M.	550 kw
11 P.M. –12 Mid	500 kw
Sum of hourly demands	11,650 kw

Total energy = 11,650 kw × 1 hr = 11,650 kwhr

The average demand per hour equals

$$11{,}650 \text{ kwhr} \div 24 \text{ hr} = 485.4 \text{ kw}$$

Dividing the total kilowatt-hours by the demand gives the number of hours out of the period which the demand would have to last so that the total energy equals the actual kwhr of energy:

$$11{,}650 \text{ kwhr} \div 1{,}000 \text{ kw} = 11.65 \text{ hr}$$

In Fig. 9-3, the dotted line indicates the average demand of 485.4 kw for the 24-hr period, the vertical and horizontal dashed line indicates the number of hours (11.65) the maximum demand of 1,000 kw would have to last to draw the same amount of energy. In Fig. 9-3, the load factor for the 24-hr period equals

$$485.4 \text{ kw} \div 1{,}000 \text{ kw} = 48.54\%$$

If the maximum demand is expressed in some other unit than kilowatts, the total energy must be expressed in similar units with time, or the maximum demand must be converted to an equivalent value in kilowatts. Thus, if the maximum demand is in amperes, the total energy should be expressed in ampere-hours or the maximum demand in amperes should be converted to its equivalent in kilowatts. Load factor may be approximated with sufficient accuracy for most purposes by any method which approximates average demand when the maximum demand is known. For instance, a reasonable approximation of annual load factor can be made from hourly log-sheet readings of demand in amperes by (1) separating the year into the four seasons; (2) selecting representative weekdays, Saturdays, Sundays, and holidays for each season; (3) averaging the hourly ampere demands for each of these representative days for each season; (4) multiplying these average daily ampere demands by the number of corresponding days in each season; (5) adding these products and dividing the resulting total by the total number of days in the year. This result approximates the average ampere demand for a year and when divided by the maximum hourly reading in amperes approximates the annual load factor. A similar approximation of load factor can be made if load curves are available showing the variation of demand for the desired period of time by either working out average demand for representative days or by estimating average demand from an approximation of the area under the curve.

Knowing the load factor gives a picture of the degree of utilization of equipment. In an electric system size of load facilities depends upon the maximum demand to be supplied. If a machine or a complete electric system continuously supplies a load equal to its rated capacity, the system, machine, or equipment will be utilized 100 percent in supplying a load of 100 percent load factor. This is highly desirable but seldom, if ever, achieved. Some types of machines, equipment, or parts of a supply system can't be utilized 100 percent because excess capacity is needed to allow taking parts out of service for repairs or maintenance. Also, loads aren't continuous but usually vary between a daily maximum and a minimum so that load factor is something less than 100 percent. If the load factor is less than 100 percent, then part of the supply system required to supply the maximum demand is idle part of the time. Load factor measures the proportion of maximum capacity which is idle part of the time.

Each part of a supply system should be supported by financial returns at least equal to its cost of operation, maintenance and investment. When a part of the capacity of a system is idle because of reduced demand, no revenue comes in but the cost of supporting this capacity continues to mount regardless of whether it is working or idle. This cost when idle includes fixed or continuous charges as interest which the money invested might be expected to earn; the cost of maintenance to keep the system in good working condition; the cost of depreciation so that sufficient capital can be accumulated to at least replace the original equipment when worn out; and taxes, insurance, costs of standby capacity, etc. If part of the capacity is idle part of the time and consequently earning a return only part of the time, the remaining part working all the time or at least a greater proportion of the time must earn a return large enough to make up for the money lost by idle plant. The net result is that the cost per kilowatthour to supply an electric-system load is lower with a high than with a low load factor.

Example: A load of 100 kw maximum demand at 10 percent annual load factor is to be supplied. This means that 100 kw of system capacity must be ready and available at all times but that only 10 kw, or 10 percent, will be in service on the average. This also means that the full maximum capacity of 100 kw could supply energy requirements for a full year by operating 10 percent of the time, or 876 out of 8,760 hr, and remaining idle during the remaining 7,884 hr. The total energy delivered would amount to 87,600 kwhr during the year. If the maximum demand remained at 100 kw but the load factor increased to 80 percent, the energy delivered to the load from the same system capacity would increase eight times, or to 700,800 kwhr. Since the proportion of idle plant is greatly reduced, the system capacity is unchanged, and the fixed charges are unchanged, the average cost per kilowatthour to supply energy will be appreciably lower at the high load factor.

This is illustrated by the off-peak electric hot-water heater. A lower rate per kilowatthour is charged for off-peak hot-water heating since the load will be using system capacity which would otherwise be idle. The addition of such off-peak load increases the system load factor without increasing the maximum demand by filling in the valley portion of the daily load curve. Care must be exercised, however, that the off-peak load added does not create a new peak so that the off-peak load becomes an "on-peak" load.

9-6. Diversity and Coincidence Factors. Electrical loads of the same type are generally similar in that the daily load curve has the same shape. For instance, residential lighting and appliance loads will have a peak demand in the early evening, then drop steadily to a minimum value around 3 to 4 A.M.; rise to another, but much lower, peak about 8 or 9 A.M.; then drop off to a lower value until mid-afternoon at which time a gradual increase occurs until 4 or 5 P.M. when a steep rise toward the evening peak takes place.

This same general variation recurs each day though load values and times may differ. This characteristic is true of individual residences as well as of the combined load of a group, even though there may be intervals during each day when all lights and appliances in some individual residence are disconnected.

Combining the demands of a group of residences, as on a transformer, gives a combined maximum demand less than the sum of the individual maximum demands because each individual doesn't turn on lights or use appliances at exactly the same time. This lack of coincidence between the loads of individual consumers, as well as between separate groups of consumers, has an important bearing upon electric-system economics. The importance of this *diversity* between demands can be appreciated

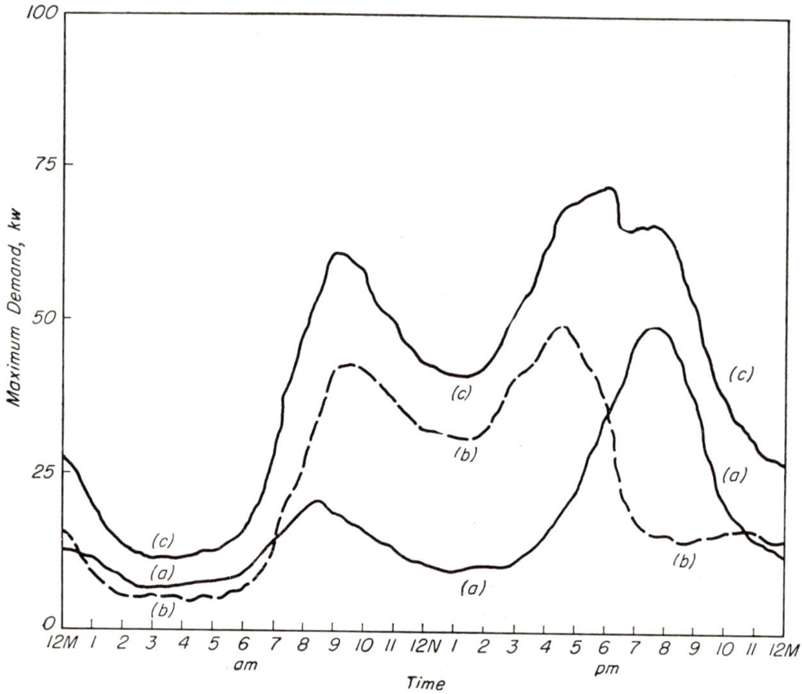

FIG. 9-4. At any instant, total load c equals the sum of component loads a and b. Because of diversity, maximum load of c is less than the sum of maximum loads of a and b.

when considering the tremendous increase in system capacity, from the generator to the individual consumer's meter, which would be necessary if the individual maximum demand of every consumer occurred simultaneously.

This diversity between maximum demands is measured by diversity factor. This may be defined as the ratio of the sum of the individual maximum demands to the maximum demand of the whole group. Diversity factor may refer to any two or more separate loads or may include all loads in any part of an electric system or the entire system as a whole. Diversity factor is always greater than unity. If the individual maximum demands of any group of loads and the diversity factor are known, the demand of the group equals the sum of the individual demands divided by the diversity factor.

Example: A load having a maximum demand of 50 kw and varying as shown by curve a of Fig. 9-4 is to be combined with another load also having a maximum demand of 50 kw as indicated by curve b. These two loads have a maximum demand of 72.5 kw as indicated by curve c. At each instant curve c equals curve a plus curve b. The diversity factor for the two loads is

$$\frac{50\text{ kw} + 50\text{ kw}}{72.5\text{ kw}} = 1.38$$

The average demand of load a is 17.75 kw, and the load factor is 35.5 percent. The average demand of load b is 23.25 kw, and the load factor is 46.5 percent. Combining loads a and b to give c results in an average demand of 41 kw and a load factor of 56.5 percent. The higher load factor of c stems from the lack of coincidence between the maximum demands of a and b, their diversity of demand. If loads a and b are to be supplied separately by individual transformers, each will require a capacity of 50 kw, or a total capacity of 100 kw. On the other hand, if a and b are to be supplied from a common source, only 72.5 kw of capacity will be required. The combination saves 27.5 kw of capacity.

Diversity factor is used to determine maximum demand resulting from combining a group of individual loads, or from combining two or more such groups. These combinations may represent a group of consumers supplied by a single transformer, a group of transformers supplied by a primary feeder, a group of primary feeders supplied by a substation, etc., clear back to the ultimate point of supply. Many individuals prefer a factor of multiplication rather than division. In this case the coincidence factor, reciprocal of diversity factor, can be used. Maximum demand then would be calculated by multiplying the sum of the individual or group demands by the coincidence factor. In the example of Fig. 9-4,

$$\text{Coincidence factor} = \frac{72.5\text{ kw}}{50\text{ kw} + 50\text{ kw}} = 0.725$$

or

$$\text{Coincidence factor} = 1 \div \text{diversity factor} = \frac{1}{1.38} = 0.725$$

While diversity factor is never less than unity, coincidence factor is never greater than unity. Coincidence factor can be regarded as the average percentage of each individual maximum demand of a group which is coincident at the time of the group maximum demand, or the contribution of each individual load, in percentage of its demand, toward the combined demand.

Diversity and coincidence factors are affected by such things as the number of individual loads, load factor, habit characteristics, etc. Diversity factor tends to increase with the number of consumers in a group, quite rapidly at first and then less rapidly as the number of consumers increase. Similarly, coincidence factor decreases rapidly at first and less rapidly as the number of consumers increases.

If the load factor of an individual load is low, the maximum demand will be of short duration and sharply peaked. If a group of such individual loads is combined, a comparatively slight difference in the time at which the individual maximum demands occur produces a high degree of diversity or lack of coincidence and the diversity factor will be high. If the average individual load factor is high, the valleys in the load curve will not be so deep, the difference between the average and the maximum demands will not be so great, the rise to a maximum will be broader across the base, the curve at maximum demand will be less peaked, and the degree of coincidence between maximum demands will be greater so that the diversity factor will be smaller or the coincidence factor greater. The rate at which diversity changes with load factor is quite rapid up to individual load factors of about 30 percent. Beyond this point, diversity changes at a much slower rate. This might be illustrated by a plant operating a single 8-hr shift per day changing to three 8-hr shifts, or 24-hr operation. The maximum demand probably will not change appreciably, and the diversity between a group of such loads will not change very much, but the load factor will increase greatly.

The diversity between individual loads or separate groups also tends to increase if the class characteristics of the loads differ. Thus, if one group made up of individual loads normally having their maximum demand in the evening (such as residential loads) is to be combined with an equal group made up of individual loads normally having their maximum demands in the morning (such as small and medium industrial loads), the diversity factor will be higher than if all loads had their maximum in the evening or all had their maximum in the morning.

Figure 9-5 shows the approximate range of coincidence factors for residential consumers based upon annual maximum demands. The upper and lower curves represent the limits of the approximate range of coincidence factors for groups composed of average residential consumers. Local habit and local characteristics of residential loads may cause such variations in diversity. Since the load factor of commercial light and power loads and of industrial light and power loads is usually somewhat higher than that of residential lighting and appliance loads, the amount of diversity between such loads generally will be appreciably lower than the diversity between residential loads. The coincidence factor for commercial or industrial loads may run to as much as twice that for residential loads. The average monthly coincidence factor usually will be higher than the corresponding factor for a year. This is because of seasonal changes in load and the fact that monthly diversity is based upon 12 different maximum demands during the year while annual diversity is based upon only the largest of these. In estimating system loading for design the annual diversity or coincidence factor is commonly used.

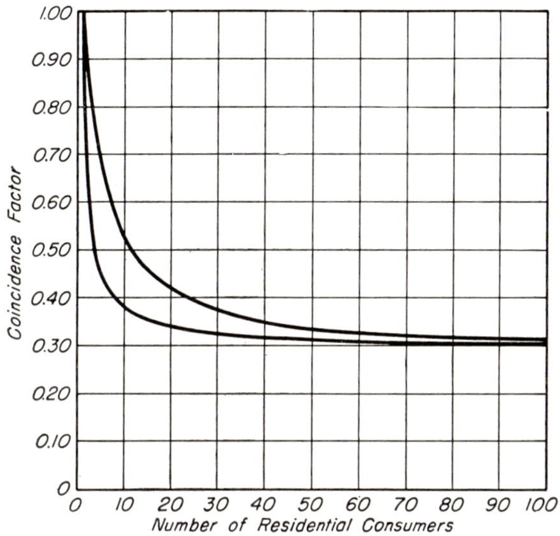

FIG. 9-5. These curves show the limits of coincidence factor variation for average groups of residential consumers.

If a large number of small transformers is combined, as on a rural-type primary feeder, the diversity factor between transformers will be higher than between a group of larger transformers on an urban-type primary feeder supplying a high-class residential load or a commercial-and-industrial-type lighting and power load in a city. In general, diversity will usually fall within the limits shown in the accompanying table for

	Diversity factor	Coincidence factor, %
Between distribution transformers	1.20–1.35	74–83.5
Between distribution primary feeders	1.08–1.20	83.3–92.5
Between distribution substations	1.05–1.25	80.0–95.5

these larger subdivisions of the supply system. The total diversity equals the product of these factors for the component parts back to the source. Thus, suppose that the diversity factor between the individual loads on a distribution transformer is 3, between transformers 1.25, between primary feeders 1.11, and between substations 1.08. The total diversity factor from the consumers to the common point of supply will be $3.0 \times 1.25 \times 1.11 \times 1.08 = 4.50$. The total coincidence factor is equal to 1/4.5, or 22.22 percent. This means that for each kilowatt of individual maximum demand by these consumers the generating or transmission supply system will need to supply only 0.2222 kw at the time of the maximum demand for the entire system.

9-7. Copper-loss Factor and Equivalent Hours. Energy losses in a supply system are composed primarily of iron loss (Art. 1-30) and copper loss (Art. 1-5.). The iron losses originate in inductive equipment, such as transformers and regulators, and are independent of load variations.

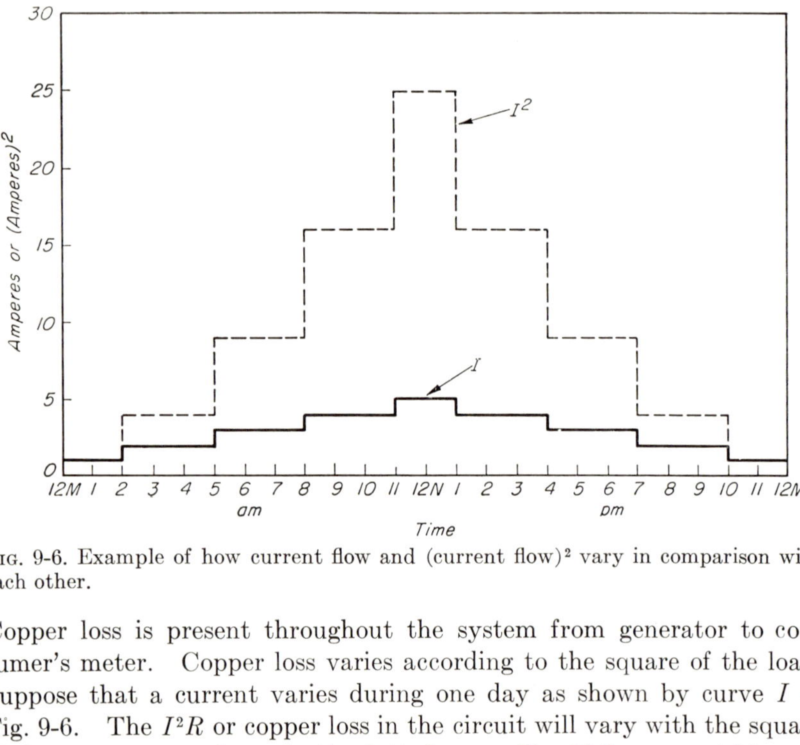

FIG. 9-6. Example of how current flow and (current flow)^2 vary in comparison with each other.

Copper loss is present throughout the system from generator to consumer's meter. Copper loss varies according to the square of the load. Suppose that a current varies during one day as shown by curve I of Fig. 9-6. The I^2R or copper loss in the circuit will vary with the square of the current or as shown by the dotted curve I^2. If the current I is read at regular intervals during a specified period of time and these readings are added and divided by the number of readings, the result approximates the average current during the specified period of time. The average value of Fig. 9-6 can be found by adding the I at each hour and dividing by 24, the number of hours in the day. Figure 9-6 shows that I equals 1 amp for 4 hr, 2 amp for 6 hr, 3 amp for 6 hr, 4 amp for 6 hr, and 5 amp for 2 hr. The sum of these 24 hourly readings equals 68, and the average I during the day equals $^{68}\!/_{24}$, or 2.833 amp. The sum of the hourly readings is expressed in ampere-hours which equal the area under the curve representing the variations of I. The ratio of the average to the maximum current equals the load factor. The load factor for the day equals $2.833 \div 5$, or 56.66 percent.

The copper loss in the circuit at any instant equals the resistance times I^2. Thus, with a resistance of 1 ohm, the I^2R loss in Fig. 9-6 varies between a minimum of 1 watt and a maximum of 25 watts while the current varies between a minimum of 1 amp and a maximum of 5 amp. The I^2R loss in the circuit equals 1 watt for 4 hr, 4 watts for 6 hr, 9 watts for 6 hr, 16 watts for 6 hr, and 25 watts for 2 hr. The sum of the hourly values of I^2, or I^2R with a resistance of 1 ohm, equals 228 and the average value equals $228 \div 24$ hr, or 9.5. The average I^2 is found in the same way that the average I was found. If the resistance of the circuit is known, the variation of copper loss in watts can be determined from the values of I^2. Since the curve representing variations in I^2R loss varies with time as well as with amperes squared, the summation of the watts I^2R loss per hour for any definite period of time equals the total energy lost as heat during the specified period of time, expressed as watthours. In the case of Fig. 9-6, with a circuit resistance of 1 ohm the total I^2R loss during the day is 228 watthours and the average I^2R energy loss is 9.5 watts per hr. Thus, the average I^2R loss in watts or kilowatts can be determined if the total I^2R loss in watthours or kilowatthours is known for any given period of time such as for a day, a month, or a year, etc. A factor relating average to maximum values can be established for I^2R losses in a manner similar to load factor. This factor is

$$\text{Copper-loss factor} = \frac{\text{average } I^2}{\text{maximum } I^2} = \frac{\text{average } I^2R \text{ loss}}{\text{maximum } I^2R \text{ loss}}$$

These average and maximum values refer to the same period of time. The interval (15 min, 30 min, 60 min, etc.) of maximum demand should also be specified.

Note that the *average value of* I^2 is quite different from the square of the average value of I. The average I in Fig. 9-6 equals 2.833 amp and its square equals 8.026, while the average I^2 equals 9.5. While the loss factor is similar to a load factor of copper losses and is affected by variations in load factor, the distinction between the two should be clearly understood. Loss factor is never greater numerically than load factor and never less than (load factor)^2. These extreme limits are demonstrated by comparing load factor and loss factor for loads, as shown in Fig. 9-7. In Fig. 9-7a, the demand is constant for part of the time and zero the rest of the time:

$$\text{Load factor} = \frac{\text{average load}}{\text{maximum load}} = \frac{Lt}{T} \div L = \frac{t}{T}$$

where L = maximum demand
t = period of time load is connected
T = total time interval concerned
Lt = energy delivered to load
$\frac{Lt}{T}$ = average demand during time T

If it is remembered that copper loss is proportional to (load)^2, then,

$$\text{Loss factor} = \frac{\text{average loss}}{\text{maximum loss}} = \frac{I^2Rt}{T} \div I^2R = \frac{t}{T} = \text{load factor}$$

where I^2R = maximum copper loss where I is current corresponding to load L
I^2Rt = total energy lost in circuit resistance during time t
$\frac{I^2Rt}{T}$ = average copper loss during time T

A series street-lighting load, constant for about 12 hr each day and disconnected the remainder of the time, is an example of this type of load in which loss factor equals load factor.

The other extreme limit where loss factor equals load factor squared is illustrated by Fig. 9-7b. The load has a sharp peak of very short duration and a constant demand of lower value during the remainder of the

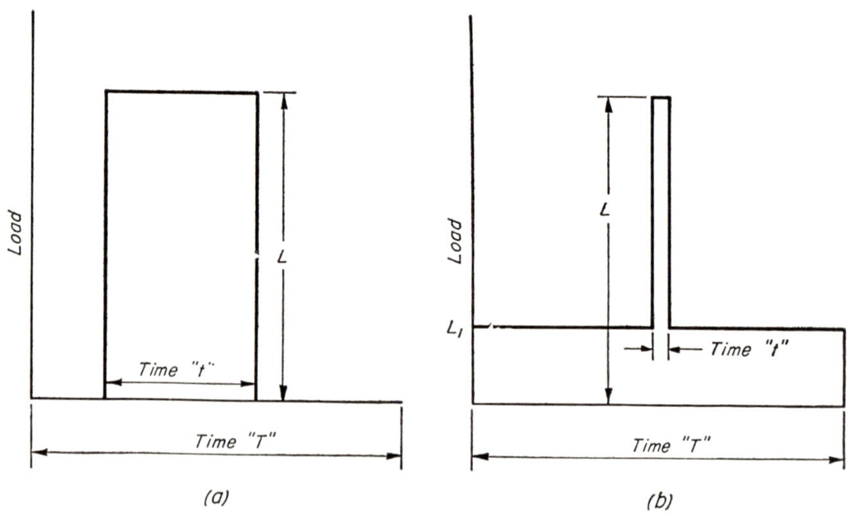

FIG. 9-7. For constant maximum load similar to (a), the copper-loss factor equals the load factor. For relatively constant minimum load with short sharp peak as in (b), the copper-loss factor equals the square of the load factor.

load period. The increase of load from L_1 to L during time t will have an insignificant effect upon the total energy taken by the load because of

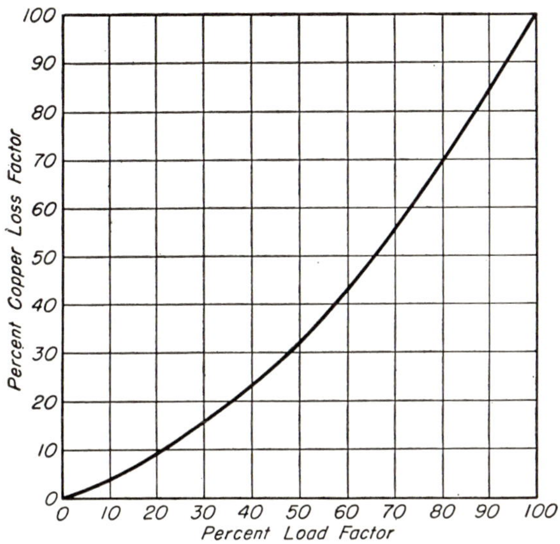

FIG. 9-8. Approximate relation between load factor and copper-loss factor for most distribution systems.

the shortness of time t compared with time T, and consequently, the average demand will nearly equal L_1 so that

$$\frac{L_1}{L} = \text{load factor}$$

In this case,

$$\text{Copper loss factor} = \frac{\text{average loss}}{\text{maximum loss}}$$
$$= \frac{I_1{}^2R}{I^2R} = \frac{I_1{}^2}{I^2} = \frac{L_1{}^2}{L^2} = (\text{load factor})^2$$

where $I_1{}^2R$ = average copper loss where I_1 is the current corresponding to load L_1

I^2R = maximum copper loss where I is the current corresponding to maximum demand L

$$\frac{I_1}{I} = \frac{L_1}{L} = \text{load factor}$$

Loads like Fig. 9-7 are seldom encountered. In the great majority of cases the loss factor equals some intermediate value between the limits. The following empirical equation greatly simplifies finding the loss factor from known values of load factor*

* From F. H. Buller and C. A. Woodrow, Load Factor—Equivalent Hour Values Compared, *Electrical World*, July 14, 1928.

$$\text{Loss factor} = 0.7(\text{load factor})^2 + 0.3(\text{load factor})$$

Note that the loss factor, the ratio of average to maximum copper loss, also represents the percentage of total hours which the maximum loss would have to last to equal the actual loss. Thus, if the loss factor over a year's time equals 20 percent, the equivalent peak-load copper loss lasts for 20 percent of the 8,760 hr in a year, or 1,752 hr. This percentage when expressed as hours of the year is called copper-loss equivalent hours. Figure 9-8 shows the curve of the above empirical equation relating loss factor to load factor. Using the annual loss factor, the annual copper energy loss in kwhr can be closely approximated from the loss at peak load and the load factor:

$$\begin{aligned}\text{Annual copper loss} &= \text{loss factor} \times 8{,}760 \text{ hr} \times \text{kw peak-load copper loss} \\ &= \text{equivalent hours} \times \text{kw peak-load copper loss}\end{aligned}$$

9-8. Planning and Load Forecasting. If load growth stopped, the electrical-utility industry would become static and planning would not be needed. New users, however, and new uses for electricity are found each year. These require continuous analyzing of changing characteristics and trends of electrical loads. The scope may involve plans to meet load needs several years in the future, or it may involve provision for loads next week. Rate of load growth in the past must be analyzed and projected into the future to give a reasonably accurate picture of what future loads will be. The formulation and execution of plans for construction of a new generating station or high-voltage transmission line may take several years. Such large elements must be completed in advance of the loads needing them and should add capacity to meet load growth for some years in the future.

Planning must provide not only for slowly developing but also for immediate system changes. This last is typical of the daily changes in a distribution system involving short primary or secondary line extensions and the addition or change of small distribution transformers. While these individual changes are small, their cumulative total in a year is large and represents a considerable part of planning.

The forecast of future loads is a continuing project, and the penalty for inaccuracy may be high. If the load estimate is too high, the interest, maintenance, and depreciation charges on idle investment may be excessive. If the estimate is too low, the penalty may include loss of revenue or loads due to lack of capacity or excessive cost of new capacity because of premiums and overtime for the rushed construction of new facilities. The forecast of future load requirements may be for an entire electric-supply system or for a comparatively small part. Thus, planning for future distribution primary-feeder changes, local secondary-circuit changes, local distribution-transformer changes, etc., will involve some degree of load forecasting for the local distribution area concerned and the distribution substation facilities supplying the area. The successful prediction of future load requirements, for either a whole system or only a part, depends upon knowledge of past trends of load growth, knowledge of the characteristics of different types of load and how they combine to affect future demands, knowledge of economic trends, and the exercise of good judgment.

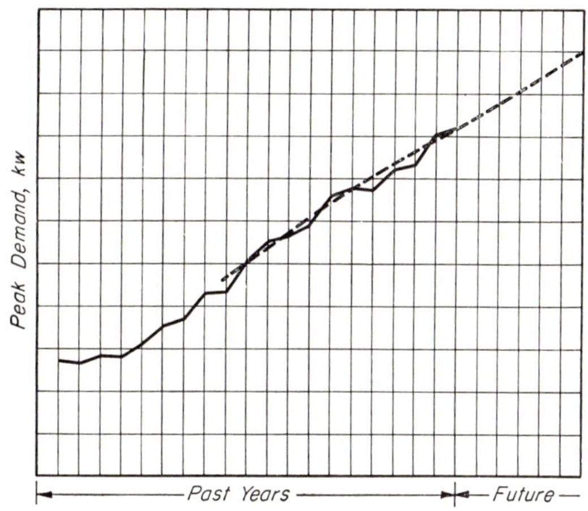

FIG. 9-9. Maximum demands in the near future can be closely estimated by projecting the trend of growth during the immediate past.

The simplest method of load forecasting projects the curve of load data from past years into the future. Thus, in Fig. 9-9 the curve represents annual peak demands for a number of years in the past. The solid line represents the actual variations in load in past years, while the dashed line shows the trend of these past changes projected into the future. This forecast may be made by drawing a smooth curve through the past yearly demands and projecting a few years into the future or by estimating future demands from the average percent change in recent past years. It is assumed that the trend of load changes in the most recent years will continue into the future. Since this method assumes no knowledge of the numerous factors which affect load growth, it is unsafe to depend upon such a forecast for more than a few years in the future. If a load forecast is projected at a flat percentage of increase each year, any error is magnified by the same percentage in each succeeding year. Such a method may be satisfactory for short-term forecasts, but the margin for error is too great to make it useful for long-term forecasts. The latter must be based upon analysis of different classes of loads making up the total load, how these various classes will combine, the probable change in population density, the probable rate at which energy usage will change, the future trend of industrial and business activity in the area, the effect which promotional activity and the development of new types of load may have upon future loads, etc. Such load predictions are usually based upon forecasts of total energy requirements translated into demand. These require a thorough knowledge of the trend of system load and load factor and the relation of individual and group diversity factors.

When determining the capacity to be installed in excess of present requirements to provide for future loads, the question is largely one of economics. The cost of carrying excess capacity as idle investment is compared with the cost of replacing smaller equipment with larger and other methods of increasing capacity when required. If the average rate of past load growth has been low and there will be a continued low rate of growth, provision for excess capacity far in the future usually is not justified. For a high rate of growth, a greater amount of excess capacity is justified since too frequent replacement of overloaded equipment by larger capacity may far exceed the carrying cost of excess capacity.

CHAPTER REVIEW TOPICS

1. What does a watt mean?
2. Define the meaning of a kilowatthour.
3. What is maximum demand?
4. How do you describe a load curve?
5. Does daily, weekly, monthly, or yearly maximum demand determine equipment sizes to be used?
6. How are peak loads measured in point of time?
7. Why isn't instantaneous maximum demand used in sizing distribution apparatus?
8. What is connected load?
9. Define the demand factor.
10. How does demand factor vary with connected load and number of devices?
11. How is load factor defined?
12. On what time periods are load factors based?
13. How can load factors be approximated from log-sheet readings?
14. What are the advantages of off-peak loads?
15. Define diversity factor.
16. Define coincidence factor. How is it related to diversity factor?
17. To what parts of a system are diversity and coincidence factors applied in computing capacities?
18. How are copper and iron losses evaluated in a distribution and transmission system?
19. What is the relationship between loss and load factors?
20. Describe some of the considerations used in load forecasting.

PROBLEMS

Problem 9-1: Suppose that a load rises from zero to 10 kw instantaneously and stays constant for 1 min, then rises to 20 kw and remains constant for 1 min, continues at this rate of rise until it reaches a maximum value of 50 kw for 1 min, then instantly drops to zero for 1 min, after which it again rises in 10-kw steps at 1-min intervals to a maximum of 50 kw and returns to zero for 1 min. If the load continues to vary in these steps, (1) what is the maximum demand averaged over the first 15 min? (2) Over the second 15 min? (3) Over a 30-min demand interval?

Solution: 1. As seen by sketching out the load curve during the first 15 min, there will be

10 kw for three 1-min intervals	=	30 kw min
20 kw for three 1-min intervals	=	60 kw min
30 kw for three 1-min intervals	=	90 kw min
40 kw for two 1-min intervals	=	80 kw min
50 kw for two 1-min intervals	=	100 kw min
0 kw for two 1-min intervals	=	0 kw min
		360 kw min

The demand averaged over 15 min = 360 ÷ 15 = 24 kw

2. In a similar manner, the sum of the demands for the second 15 one-min intervals is equal to 390, and the average demand is equal to 390 ÷ 15 = 26 kw.

3. For the full 30-min interval the sum of the demands for the 30 one-min intervals is equal to 750, and the average demand is equal to 750 ÷ 30 = 25 kw.

Problem 9-2: A small industrial consumer is to have a total connected load composed of 37 motors, totaling 161 hp, and 30 kw of incandescent lighting. The motors will be separately controlled, and it is anticipated that the power factor of the combined load at maximum demand will be 87 percent. What will be the probable range of maximum demand?

Solution:

161 hp × 0.746 = 120.10 kw of connected motor load
30.00 kw of connected lighting load
150.10 kw = total connected load

From the table in Art. 9-4, demand factor will probably range between a minimum of 35 percent and a maximum of 65 percent. The demand will probably range between

150.1 kw × 0.35 = 52.5 kw, 52.5 kw ÷ 0.87 = 60.4 kva minimum
150.1 kw × 0.65 = 97.6 kw, 97.6 kw ÷ 0.87 = 112.1 kva maximum

Problem 9-3: A store has a 15-min maximum demand of 380 kw at 90 percent power factor and a connected load of 500 kw. The connected load is to be increased to 635 kw. What will be the probable maximum demand if it is assumed that conditions of use will be unchanged?

Solution:

$$\text{Demand factor of present connected load} = \frac{380 \text{ kw}}{500 \text{ kw}} = 76\%$$

Probable demand with additional connected load = 635 kw × 0.76 = 482.6 kw

Problem 9-4: A certain customer used 361,588-kwhr in a 365-day year. His 15-min demand was 118 kw. (1) What was his average load? (2) What was his annual load factor? He expects to add sufficient load to increase his average annual use of energy to 429,240 kwhr and his load factor to 40 percent. (3) What will be his maximum demand?

Solution:

1. 361,588 kwhr ÷ 8,760 hr = 41.3 kw average load
2. 41.3 kw ÷ 118 kw = 35% load factor
3. 8,760 hr × 0.40 = 3,504 equivalent hours of demand
   429,240 kwhr ÷ 3,504 = 122.5 kw demand

Problem 9-5: A distribution transformer has a 15-min maximum demand of 24.5 kva at 90 percent power factor. Its annual load factor is 27 percent. (1) What is its annual output in kwhr? (2) How many hours would the transformer have to operate at maximum demand to equal its actual kwhr output?

Solution:

1. 24.5 kva × 0.90 × 8,760 hr × 0.27 = 52,153 kwhr
2. 8,760 hr × 0.27 = 2,365.2 hr per year.

Problem 9-6: The sum of the annual individual maximum demands of a group of residential consumers supplied from a single distribution transformer equals 78.0 kva. The average diversity factor between the group of consumers is 3.2. (1) What is the coincidence factor? (2) What is the nearest standard size transformer to supply the load?

Solution:

1. Coincidence factor $= \dfrac{1}{3.2} = 0.313$
2. Maximum demand = 78.0 ÷ 3.2 = 24.4 kva or 78.0 × 0.313 = 24.4 kva
   Nearest transformer size = 25 kva

Problem 9-7: A distribution primary feeder has a large number of distribution transformers connected. The recent annual 15-min maximum demand was 1,190 kw at 85 percent power factor. What was the probable total installed transformer capacity supplied by the feeder if the transformers averaged demands equal 95 percent of rated capacity?

Solution:

Maximum demand of feeder = 1,190 kw ÷ 0.85 = 1,400 kva

From the table of Art. 9-6, diversity factor between transformers will range between 1.20 and 1.35.

Sum of transformer maximum demands will probably range from

1,400 kva × 1.20 = 1,680 kva to 1,400 kva × 1.35 = 1,890 kva

Total installed transformer capacity will probably fall between

1,680 kva ÷ 0.95 = 1,768 kva and 1,890 kva ÷ 0.95 = 1,989 kva

Problem 9-8: A 5,000-kva bank of transformers has a rated copper loss of 30,118 watts. The maximum demand on the bank within the past year was 81 percent of rated capacity and the load factor was 44 percent. (1) What was the copper loss in watts at peak load? (2) What was the total annual copper energy loss in the bank?

Solution:

1. Since copper loss varies in proportion to the square of the load,
   Copper loss = 30,118 watts × 0.81 × 0.81 = 30,118 × 0.656 = 19,760 watts
2. From Fig. 9-8, percent copper-loss factor for load factor of 44% = 26.75%
   Loss factor equivalent hours = 8,760 × 0.2675 = 2,343 hr
   Copper energy loss = 19.76 kw × 2,343 hr = 46,298 kwhr

CHAPTER 10

AUXILIARIES

By E. L. Michelson

Commonwealth Edison Company

10-1. AC-DC Converters. At the present time all the power generated in United States public-utility systems is ac, and most of the power consumed in factories and homes is also ac. There are, however, a few special cases requiring dc power. For example, some electrically operated railway systems used dc to take advantage of the favorable characteristics of the dc series-type motor for motive power. Most of these railways run at 600 volts.

In addition to railway load, a considerable amount of dc energy is consumed by residential and commercial consumers in downtown sections of many large cities. In the early days of the electric industry, all generation and distribution was by the Edison three-wire dc system. Batteries could be used as emergency capacity and make the system more reliable. After changing generation to ac, many distribution systems continued to operate as dc for many years. In recent years expansion of dc distribution systems has been halted. They are gradually being replaced by ac systems.

To obtain power for dc distribution systems needed for railway and remaining Edison three-wire systems the following apparatus may be used: (1) motor-generator sets, (2) synchronous converters, (3) mercury-arc rectifiers.

10-2. Motor-generator Sets. For conversion service these would consist of an ac motor of the synchronous or induction type coupled directly to a dc generator. While a simple arrangement, and advantageous from an operating standpoint, it is not used to any great extent because of the greater economy of the other two methods available.

10-3. Synchronous Converters. *General.* The synchronous converter is a rotating machine similar in appearance to the synchronous motor and the dc generator. It consists of a field and armature. On one side of the armature are connections to the ac system through slip rings, on the other side is a commutator with connections to the dc system, Fig. 10-1.

The principle of the synchronous converter is based on the fact that the typical dc generator actually has ac flowing in the armature. The ac is converted to dc by the commutator. By making connections directly from the dc armature to the ac system, the motor currents necessary to drive the converter are introduced directly into the armature of the machine. The current that actually flows in the armature is the difference between the motor current and the generator current.

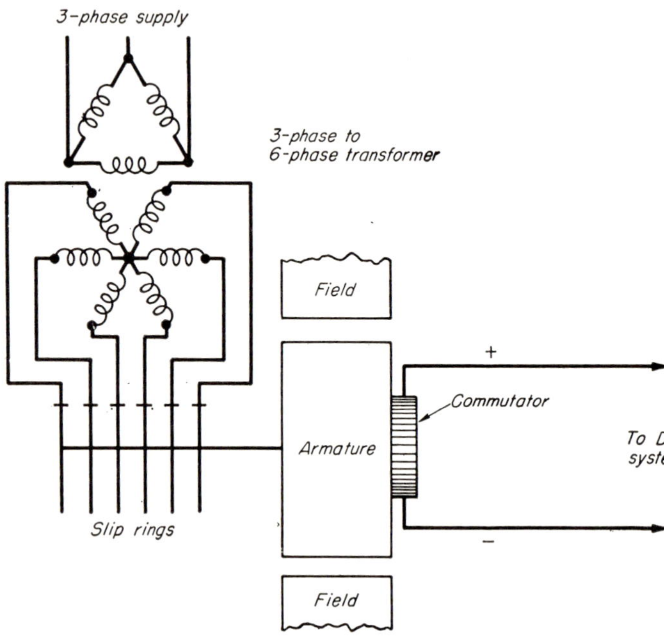

FIG. 10-1. Principal connections of a six-phase synchronous converter for changing ac to dc or dc to ac.

Another way of looking at the properties of the synchronous converter shows that the armature winding of the dc generator can also be used as the armature winding for the synchronous motor. In the synchronous converter, the same armature serves both purposes, and in that way a more economical method is obtained for conversion of ac to dc.

As stated before, the current actually flowing in the armature conductors is the difference between the motor and generator currents. The resulting current has interesting properties. The actual magnitude of the current depends upon the amount of load supplied by the converter and the power factor at which the ac end operates. The number of field poles on the synchronous converter is determined by the frequency of the ac system and the rotary speed for which the converter is designed. For example, on a 60-cycle system, a 300-rpm converter would have 24 field poles.

The design of the converter depends on the number of phases on the ac system. The higher the number of phases, the more economical will be the design and operation of the machine. For this reason most synchronous converters are designed as six-phase machines, the six-phase system being obtained from the three-phase system through special transformer connections. Figure 10-1 shows a typical connection diagram for a six-phase converter.

Voltage Ratios. One of the important characteristics of the synchronous converter is the ratio between the ac and dc voltages. The theoretical voltage ratios of synchronous converters are definitely fixed by the number of phases and the method by which the transformer terminals are connected to the slip rings. In a single-phase converter the two collector rings connect to armature points 180 electrical degrees apart. Since the commutator brushes are also located at points 180 electrical degrees apart, the dc voltage equals the maximum value of the ac voltage, or the effective value of the ac voltage equals 0.707 times the dc voltage. For the six-phase converter, the effective value of the ac voltage from each slip ring to neutral equals 0.354 times the dc voltage. For example, in a 240-volt converter, the voltage applied to the ac end should be 85 volts from each slip ring to neutral. The correct voltage for the ac end is obtained by designing the transformer ratio for the required values.

The actual voltage ratio will be somewhat different from the theoretical due to various factors, such as resistance drop in the converter, waveform of the impressed voltage, and the machine itself. However, departure from the theoretical value will be relatively small.

Current Ratios. The ratio between the ac current fed into the converter and the dc current taken out depends on the voltage ratio, the power factor at which the machine operates, and the losses in the machine. For a machine operating at unity power factor, the most advantageous condition, the six-phase converter will have an ac current equal to 0.47 times the dc current. At any other power factor, the ratio equals the above ratio divided by the power factor.

The power factor of the machine has an even greater effect on the heating and carrying capacity of the converter. Because of the manner in which the motor and generator currents combine in the armature, the minimum heating occurs at unity power factor. For power factors other than unity, the resultant current in some of the coils increases, thereby increasing the heating in these coils. For example, if a machine operates at 90 percent power factor, the maximum heating is increased by 25 percent, with a corresponding reduction in the carrying capacity of the synchronous converter.

Voltage Control. Since the dc voltage output of the synchronous converter is determined only by the voltage applied to the ac end of the machine, some means must be provided to vary the dc voltage as needed. While it may appear that the dc output could be varied by controlling the field current of the machine, as in a dc generator, actually the change in field current will not directly affect the dc voltage. Its only effect is to change the power factor at which the machine operates.

The most common method for controlling the ac voltage is the use of an ac synchronous booster with the synchronous converter. This booster consists of a mechanically connected ac generator, with its armature winding connected in series with the armature winding of the rotary converter so that the voltage generated in it either adds to or subtracts from the voltage supplied to the converter. This has a direct effect on the dc voltage. The booster must have the same number of poles as the converter, and the manner in which it affects the voltage of the converter is determined by the magnitude and direction of the current through its field winding.

Another method for controlling the dc voltage makes use of a voltage regulator on the ac end of the machine. This regulator could consist of a tap-changing arrangement, as described in another section of this chapter, or an induction regulator of the three-phase type.

As stated above, when the dc field current in the converter is varied, the power factor at which the machine operates changes. This change in power factor will cause the ac voltage supplied to the converter to change because of the difference in voltage drop through the transformer. While this method cannot be used to effect large changes in voltage, it is useful where a relatively small change in the dc output voltage will satisfactorily regulate the system.

Starting of Synchronous Converters. The synchronous converter can be started from either the ac or the dc end. In some respects the dc method is the simpler, but it requires that a source of dc be available. In this case, the converter is started as a shunt motor, with the starting current limited by a starting resistor. When the converter is brought up to speed, it is synchronized on the ac end in the same way that a generator would be synchronized, using the starting resistor to vary the speed.

The converter may also be started from the ac end similar to the way in which a synchronous motor is brought up to speed. To reduce the starting currents, general practice arranges for starting at reduced voltage by a special connection of the transformer. During the starting period, rather high voltages will be induced in the field winding of the machine, and it is necessary to provide protection against this effect. After the current converter has been brought up to speed and connected to the ac lines, the polarity on the dc end may be wrong, this polarity being dependent upon a random effect. The polarity is determined by the instant at which the machine pulls into step on the ac end. If the dc polarity is wrong, it is necessary to make the converter slip a pole, which can be done by opening and closing the main switch in the ac lines. Another method is to reverse the current through the shunt-field circuit. Figure 10-2 shows a typical example of a synchronous converter.

In recent years very few converters have been installed on power systems, although a large number are still in operation. The supplying of customers with dc lighting power is not being expanded; in fact, the existing dc systems are being curtailed. In new installations of dc power supply for railway systems, the mercury-arc rectifier is being used

FIG. 10-2. Typical small-capacity converter. (*Courtesy of General Electric Company.*)

generally, because of its higher efficiency for the voltage at which these systems operate.

10-4. Mercury-arc Rectifiers. This more recent type of equipment has largely supplanted synchronous converters for supplying power to dc railway load.

The mercury-arc rectifier, a static device, has various advantages over the synchronous converter. As a result, railways have been turning to mercury-arc–rectifier equipment for converting ac to dc. Following are four good reasons for this trend:

1. The reliability of this type of apparatus shows an availability of 99.9 percent. That is, it is forced out of service only 0.1 percent of the time over long periods.

2. The efficiency of rectifier equipment at 600 volts is remarkably high. This is particularly true at light loads, since the main loss consists of the transformer excitation losses. At higher voltages, the rectifier efficiency is even better, because equipment losses vary almost in inverse ratio with the operating voltage.

3. Rectifier equipment has very low maintenance expense and utilizes extremely simple switching and control ideally suited to automatic operation. Installation costs are minimized, since no heavy foundations are required.

4. This latter factor, together with the absence of noise, makes possible the installation of rectifiers in locations which otherwise would not be considered. The ability to use such locations is often an important factor in operating economy, saving copper and reducing transmission losses.

The mercury-arc rectifier works on the same principles as ordinary radio tubes. Two electrodes, one of which can emit electrons while the other cannot, will serve as a rectifier since current can flow in only one direction. As applied to the mercury-arc rectifier, the operation can be explained as follows:

All power rectifiers operate on the basic principle of an electron-emitting cathode and a "normal" or non-electron-emitting anode. These two parts are physically separated and enclosed in a vacuum chamber filled with mercury vapor. The vacuum chamber must be practically free of other gases, such as oxygen, nitrogen, etc.

A cathode spot first must be established at the surface of the mercury pool to provide an initial source of electrons. The cathode spot also generates some of the required mercury vapor. The electrons emitted into the space above the cathode are negatively charged and ionize the mercury vapor by collision. These electrons are attracted by the positively charged anode and flow toward it in sufficient numbers to carry the required current. At the same time, the positively charged mercury-vapor ions flow toward the negatively charged cathode, bombarding the mercury pool and driving off more electrons. The flow of electrons in one direction and positive ions in the opposite direction constitutes a flow of current. This action continues until the end of the conducting period. As current conduction by an anode ceases, it becomes negatively charged, repels electrons, and collects positive ions. However, as a normal anode does not emit electrons, no current flows during the interval in which the cathode is positive with respect to the anode.

Under these conditions, current passes between a normal anode and the cathode during the positive half of the cycle—that is, when the anode is more positive than the cathode. The anode current flows during a time interval determined by the number of anodes, the main rectifier transformer connections, and the counter electromotive force of the load. This time interval is known as the conduction period of the anode.

During the negative half cycle (that is, when the cathode is positive instead of negative with respect to the anode), current does not flow because the anode is still normal (that is, incapable of providing a source of ionization). Therefore, the rectifier output is a pulsating unidirec-

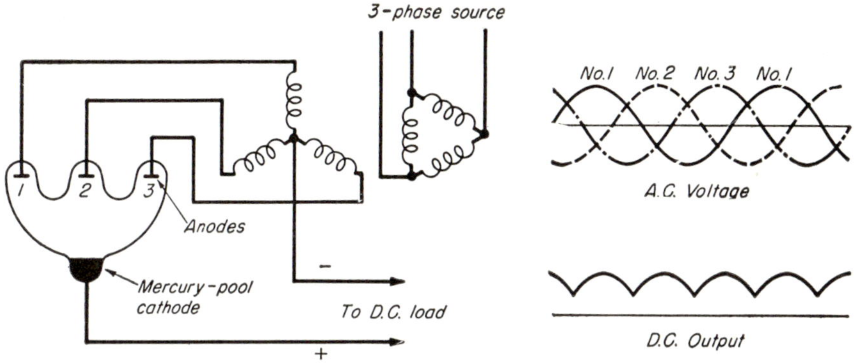

FIG. 10-3. Connections for a three-phase mercury-vapor rectifier for changing ac to dc.

FIG. 10-4. A 3,000-kw 600-volt pumped ignitron rectifier. (*Courtesy of General Electric Company.*)

tional current, and the action above is commonly referred to as the *check valve* action of the mercury-vapor–filled rectifier.

The above explanation is based on the operation of one anode with its corresponding cathode. In actual practice, all rectifiers are supplied with polyphase power, six phase or twelve phase, which results in a more uniform dc output voltage. Figure 10-3 shows the electrical connections for a three-phase unit. As shown in the figure the polyphase type of rectifier requires a multiplicity of anodes. In the older types of rectifiers, these anodes are all contained in the one rectifier bowl.

There is only one pool of mercury in such a rectifier, and the arc travels from one anode to another as required. In this type of unit, special means are needed to start the arc when the rectifier is being connected to the system. In a later development rectifiers with each anode in a separate bowl are used. One example of this type is the General Electric Company Ignitron rectifier, Fig. 10-4. In this rectifier it is necessary to establish the cathode spot, or source of electrons, each positive half cycle since in each tank of the ignitron there is only one anode. Therefore, during each cycle of operation there is a period when current flows in a particular tank and there are periods when current does not flow. This requires that the cathode spot be energized once every cycle, which is accomplished by the ignitor, a rod in intimate contact with the mercury pool. When current passes through the ignitor, a high-temperature spot created in the mercury pool becomes the source of electrons to carry the load current. Special control circuits start the flow of current through the ignitor at the proper point in the cycle of ac voltage.

10-5. Operating Characteristics of Rectifiers. The output voltage of an unregulated mercury-arc rectifier is similar to that of a shunt-connected converter or generator. Depending on the transformer connections used and the reactance of the supply line and transformer, the rectifier will have 4 to 6 percent voltage regulation, and the regulation curve can be represented by a straight line throughout the working range of the rectifier.

When some other output voltage characteristic is required, such as constant voltage from no load to full load, or even overcompounding (a rise in voltage with load), the desired results can be accomplished by controlling the firing of the rectifier anodes (phase control). This phase control can be continuously varied by a voltage regulator, to obtain the desired output-voltage characteristic. Any continuous characteristic, such as obtained from dc generators, is available, since the phase control of anode firing is analogous to field control of a dc generator. Phase control adversely affects rectifier wave shape and power factor. However, where extreme variations in the output voltage of a rectifier are required, tap changing on the rectifier transformer may be used to give the major part of the desired voltage change, thus limiting the amount of phase control necessary.

The power factor of a rectifier is defined as the ratio of the power input to the volt-ampere input. Reduction from unity power factor is due partly to the reactance of the transformer and partly to the harmonic components caused by rectifier operation. Rectifier power factor is not susceptible to regulation; it is affected by the reactance of the power transformer and supply circuits, the number of rectifier phases, and the amount of dc voltage reduction by phase control. The power factor of mercury-arc–rectifier equipment varies from approximately 0.97 to 0.92 lagging, from 15 to 200 percent load. An increase in the number of rectifier phases improves the power factor; that is, the power factor of a 12-phase rectifier is better than that of a 6-phase rectifier. Phase control used to reduce the output voltage may make reductions in power factor that will considerably exceed the limits given above, which are representative of unregulated rectifiers. Power factor, as defined above for mercury-arc–rectifier equipment, will not be indicated properly by the common dynamometer type of power-factor meter, because this meter will not respond to the harmonic components caused by rectifier operation.

An arcback in a rectifier causes loss of the rectifier valve action and a short circuit on the rectifier transformer and the dc circuit supplied by the rectifier. Sufficient operating experience has been gained with modern rectifiers to demonstrate that arcbacks are a relatively unimportant factor in their operation. Occasional arcbacks of a random nature may be expected. When an arcback occurs, the rectifier is automatically disconnected from the ac power supply and the dc bus. After a few seconds, the rectifier will be restored automatically to service.

Rectifiers are available for any power-system frequency. The modifications required in the equipment on the various frequencies are chiefly in the main transformer and the motor-operated auxiliary equipment. A number of rectifiers have been supplied for selective operation on either 25 or 60 cycles.

10-6. Rectifier Substations. A rectifier substation consists essentially of six main components:

1. The incoming ac supply circuits, with their high-voltage circuit breakers, lightning arresters, and protective relaying. The ac equipment, up to 15 kv, can be installed in metal-clad factory-assembled units. These units reduce the time required for installation and the cost of installation and maintenance. Such equipment also provides safety for the operator.

2. The step-down transformer, which converts from the incoming-line ac voltage, three-phase, to the proper voltage and phases to be commutated by the rectifier and provides the required dc output. Rectifier transformers differ considerably from conventional power transformers in that their secondary windings produce an increased number of phases to achieve the desired smoothness in the dc output. The common connections of power-rectifier transformers provide for 6- or 12-phase operation of the associated rectifier. Where a greater number of phases is required, it is possible to use combinations of different transformer con-

nections to produce an over-all operation of several rectifier units which will provide the desired number of phases. Phase-shifting transformers also can be used with certain of the rectifier units in a station to produce phase displacement between the various units, thus providing over-all station operation corresponding to the desired number of phases. To illustrate, a station containing two 6-phase units can be operated as a 12-phase equipment when both units are in service. This would be accomplished by using wye-connected and delta-connected transformers. Or, a station containing three 12-phase units could be made to operate 36-phase with all units in operation by using phase-shifting transformers in series with two of the 12-phase transformers.

3. The rectifier itself is an important part of the station equipment, but performs a relatively simple function: it merely acts as a commutator for the transformer. The output characteristics of the equipment, therefore, depend largely on the design of the transformer; the only modification of characteristics which can be produced by the rectifier is to delay the firing of the anodes (phase control) in order to reduce the dc output voltage of the equipment. Continuous regulation of this nature can be provided to give a regulated dc voltage output.

4. The rectifier control. Various rectifier operations require control of one form or another, for example, temperature regulation. Such control is better accomplished by automatic relays than by an operator.

5. The dc switchgear and control of the equipment consists of a circuit breaker which connects the rectifier to the dc bus and the dc outgoing feeder circuits. It is customary to arrange the rectifier cathode circuit breaker to trip only on reverse current and to depend on the feeder circuit breaker for protection against short circuits and severe overloads. Satisfactory operation results from the use of automatic dc reclosing feeders in rectifier stations to provide adequate protection for the equipment and the overhead system.

6. Automatic control is generally employed in rectifier substations. Supervision of the automatic equipment, by centralized supervisory control and telemetering, provides means for coordinating the entire system operation.

10-7. Frequency Changers. A frequency changer, as the name implies, is a machine or group of machines for converting electric energy from one frequency to another. Machines of this type interconnect two systems of different frequencies. There exists in this country a considerable amount of generating capacity other than 60 cycles, most of which is 25 cycles.

For example, on public-utility power systems where there are both 25- and 60-cycle generators, it may be economical to install a frequency changer to obtain a better balance of load and capacity and to reduce the amount of reserve generating capacity required by making available to both systems the reserve on each system.

Another general application for frequency changers is to supply loads from a 60-cycle system which uses equipment of some other frequency. An example of this situation is railroad electrification where 25-cycle motors are more desirable than 60-cycle motors. It becomes necessary to install a frequency changer to feed the railway load from a 60-cycle power system. Another example is in steel-mill power systems which are generally 25-cycle because of the more favorable characteristics of the 25-cycle motor. Many steel mills have their own 25-cycle power systems, interconnected with the 60-cycle utility power system. Here again, a frequency changer makes the interconnection.

10-8. Types of Frequency Changers. Various types of frequency changers are in use, depending on system requirements. In general, these can be classified as synchronous and nonsynchronous ties. The synchronous-synchronous tie consists of a synchronous motor and a synchronous generator on a common shaft, each with its own exciter. For a 60/25-cycle installation, the speed is 300 rpm, which requires 24 poles on the 60-cycle end and 10 poles on the 25-cycle end. The synchronous-synchronous tie represents the simplest type of frequency-changer installation and the most economical. Figure 10-5*a* shows the schematic arrangement of this type of set. However, it has the disadvantage that it represents a rigid tie between the two systems and makes the frequency changer equivalent to a transmission line in terms of the electrical conditions. It requires both systems to run at exactly the same frequency at all times, which in some cases is a difficult requirement to meet. However, in many systems it is possible to use the synchronous-synchronous type of frequency changer, and because of its economy it is the most widely used of the various types. Other important characteristics of this type of frequency changer are:

1. Voltage regulators and excitation systems are usually provided to obtain sufficient kvar to meet the normal and emergency system requirements for maintenance of system voltage.
2. Sufficient kvar capacity can be provided to meet the load and system requirements at a relatively low cost per kvar.
3. A frame shifter is usually provided on the 60-cycle end for parallel load division between frequency-changer sets. Frame shifters can be provided which allow for a shift of 16 to 24 mechanical degrees for a 300-rpm set. This range is sufficient for synchronizing and paralleling.

Frequency-changer sets of this type have been manufactured in sizes as high as 40,000 kw.

10-9. Nonsynchronous Ties. One of the disadvantages of the synchronous-synchronous tie is that it requires the two interconnected systems to run at a fixed ratio of frequencies. In the case of some steel mills, for example, where the principal load is rolling mill load, the variations in power requirements can be very great and can change very quickly, with the result that it is difficult to maintain constant frequency on the system. If such a system were interconnected with a large 60-cycle system running at a practically constant frequency, it would cause large swings of power through the frequency changer.

In such cases, it becomes advantageous to use a nonsynchronous type of machine which permits the two interconnected systems to operate at

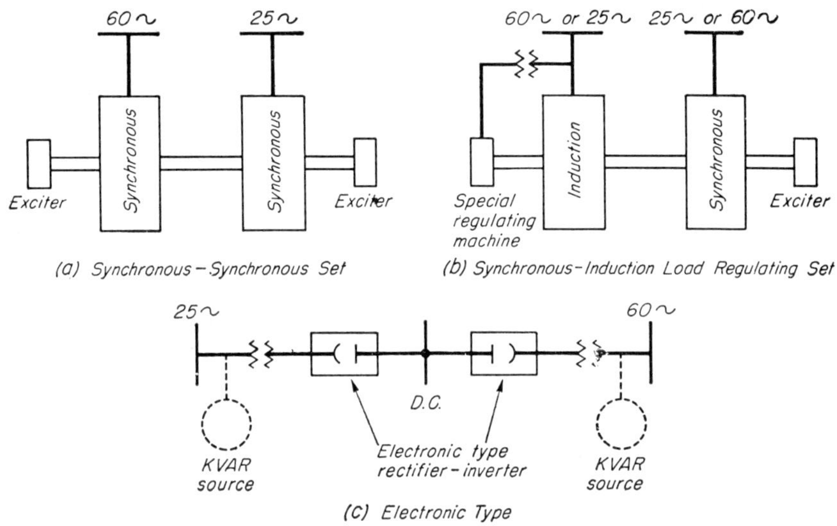

FIG. 10-5. Schematic arrangements of frequency changers connecting 25- and 60-cycle systems. (*a*) Synchronous-synchronous set. (*b*) Synchronous-induction load-regulating set. (*c*) Electronic type.

a somewhat variable ratio of frequencies. Two general types are in use, one based on rotating machines and the other on electronic equipment. The machine type of nonsynchronous tie is shown in Fig. 10-5*b*, an induction-synchronous load regulating system. This type of machine has the following characteristics:

1. The set consists of a synchronous machine connected to one system with an induction machine connected to the other. The synchronous machine, of normal design, incorporates a usual excitation system for controlling the reactive output of the machine.

The induction machine is of the wound-rotor type. Regulating characteristics of the set depend on a relatively complicated excitation system which introduces currents of slip frequency into the rotor of the induction machine.

The distinguishing feature of this set is the variable frequency ratio, that is, frequency of one system may vary in relation to that of the other. Voltage and load disturbances reflect through the set to a lesser degree than through other types. Energy flow is reversible, and load is controlled at the set itself, rather than depending upon the system, as does the synchronous-synchronous set.

2. Control of voltage and kvar supply can be obtained at either end. It is somewhat easier to obtain high power factor at the synchronous end.

3. Load control can be made sufficiently fast to meet the most severe requirements usually met in practice.

4. It can be used to supply single-phase load from the synchronous end.

5. Controlled frequency range is usually about ± 2 percent.

The main advantage of this type of set lies in its stability and the ease with which it may be started and the load controlled. Automatic control of load is readily obtained.

The synchronous-induction frequency changer can also function without the special regulating equipment. In this case an ordinary squirrel-cage type of induction motor can be used. The frequency ratio varies according to the load on the set and is not controllable at the set. The set is reversible when sufficient synchronous capacity is available to supply the exciting kva of the induction machine. Power-factor control is available on the synchronous end of the set but not on the induction end. Because of the poor control, this type of set is not widely used.

10-10. Electronic Frequency Changers. This type is shown in Fig. 10-5*c*, and its principal characteristics are:

1. It consists of rectifier and inverter elements with transformers, switches, and control.

2. Load through the frequency changer can be controlled to hold constant with rapid changes in system frequency. This equipment can isolate load changes of one system from another.

3. System kvar requirements as well as the excitation kvar requirements of the frequency-changer set must be supplied by synchronous condensers or from system sources.

4. Because this set is excited from a system, it requires system synchronous capacity to maintain the voltage and frequency within limits for proper rectifier and inverter excitation.

5. In case of power failure on one or both systems, the service may be quickly reestablished by reclosing the line circuit breakers, provided the system voltages have returned to the range required for commutation. No resynchronization is needed as with many types of rotating equipment.

6. This type of static frequency changer is well suited for operation in an unattended station.

10-11. Voltage Regulators. In distribution systems, it is necessary to maintain the voltage on the customer's service at a reasonable level.

Since voltage drop in transmission and distribution systems is generally high, it is necessary to provide some method for changing system voltage in accordance with the load to obtain a fairly constant voltage at the customer's end of the system. In some cases this can be done by changing the generator voltage, either manually or automatically, through adjustment of the field currents of the generators. However, in most cases the various circuits on the system have dissimilar load characteristics that make it necessary to provide individual means of regulation for each circuit or group of circuits. Two types of regulators are in common use at the present time—the induction voltage regulator and the step regulator.

10-12. Induction Voltage Regulators. The induction voltage regulator is essentially a two-winding transformer in which the voltage induced in

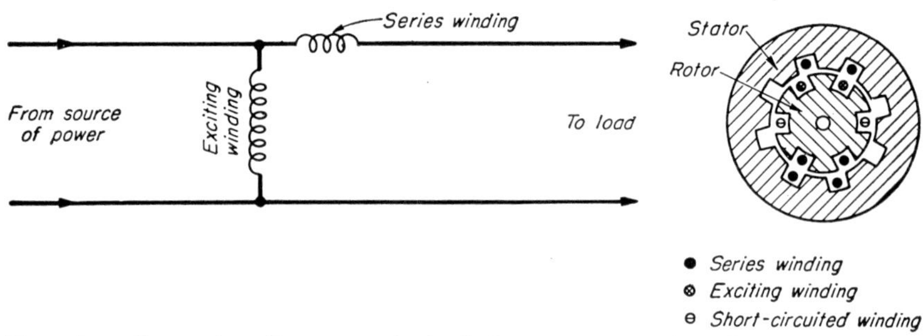

FIG. 10-6. Connection diagram and physical arrangement of an induction voltage regulator.

the secondary winding can be varied by moving one winding with respect to the other. Figure 10-6 shows the basic arrangement.

The exciting winding is on the rotor and a series winding is on the stator. Varying the rotor position with respect to the stator causes a variation in voltage of the series winding which raises or lowers the output voltage gradually as required by changes in load. In the station-type regulator, this regulation is controlled automatically.

A closed-circuit winding is permanently fixed on the rotor with a phase displacement of 90° from the shunt windings. It has the effect of equalizing the losses and the reactance of the regulator when the rotor is in various positions of raise and lower so that the load current can flow through the regulator with minimum voltage drop.

The rotor of the regulator is driven by a simple mechanism consisting of a vertically mounted motor and a double worm gear reduction. The driving motor, of the reversible capacitor type, has high starting torque and rapid acceleration. These give quick response to the controls and produce smooth, rapid regulation. On a typical modern unit, the average rate of voltage correction is two-thirds percent per second, so that a voltage change of 10 percent can be made in 15 sec.

A contact-making voltmeter controls the regulator position. In this instrument, the iron core of a solenoid is suspended from a compensating spring and is flexibly spring-coupled to one end of a contact beam. This beam pivots at its center, and with a given voltage applied to the solenoid

FIG. 10-7. Interior part of 4,800-volt induction voltage regulator. (*Courtesy of General Electric Company.*)

winding, the system is held in mechanical equilibrium by the magnetic pull of the solenoid on one end and the compensating spring on the other. This beam is fitted with contacts. As the solenoid core moves upward or downward with changing line voltage, either one or the other of the contacts will close the circuit to the motor-control relay, causing the motor to operate and return the line voltage to its normal value.

A line-drop compensator helps the regulator to maintain the feeder voltage within narrow limits at some predetermined load center at a distance from the station. The compensator does this by introducing in the voltmeter circuit resistance and reactance drops proportional to those in the feeder up to the load center. This type of compensation gives accurate results for all conditions of voltage, current, and power factor on the feeder.

The regulator described is a single-phase device. For regulation of a three-phase circuit, three units would be required, one in each phase. Although it is possible to construct a three-phase induction regulator,

FIG. 10-8. Exterior of completely assembled 4,800-volt induction voltage regulator. (*Courtesy of General Electric Company.*)

these units are not used extensively because they introduce a shift in the phase angle. Also, where regulators of this type are used, it is generally necessary to regulate each phase separately because of variations in load on the three phases.

Figures 10-7 and 10-8 show interior and exterior views of typical induction regulators.

10-13. Step Voltage Regulators. While the induction regulator has many advantages, such as smooth regulation in very fine steps and rapid response, it has the disadvantages of relatively high cost in both investment and maintenance and also requires relatively high exciting current

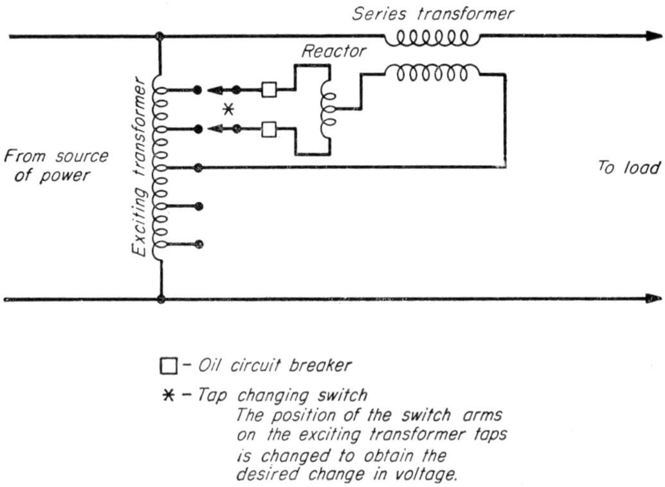

FIG. 10-9. Wiring diagram of tap-changing transformer showing series and exciting components with reactor and oil switches in tap circuit.

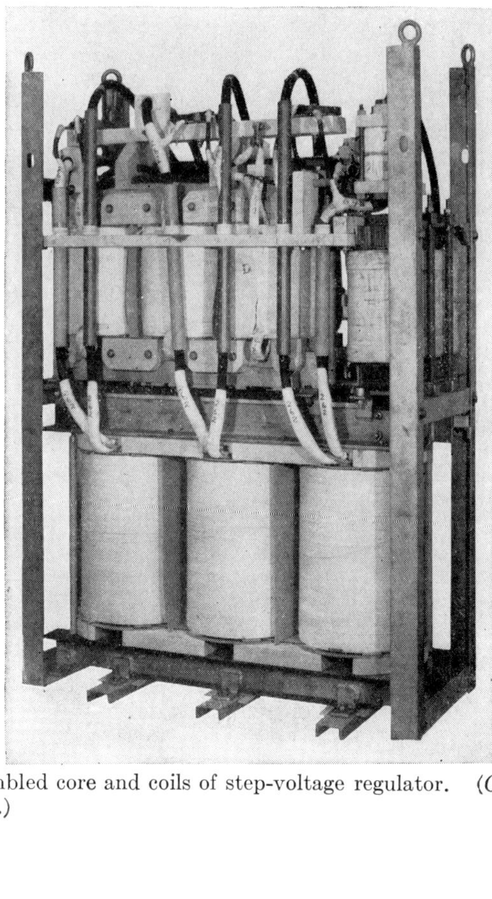

FIG. 10-10. Assembled core and coils of step-voltage regulator. (*Courtesy of General Electric Company.*)

because of the air gap needed in its construction. In recent years there has been considerable development of the step-type regulator as an alternative to the induction regulator. At the present time these units are widely applied for regulation of distribution circuits. In principle, a step-type regulator is simply a series transformer with a number of taps that can be changed under load. Only the tap-changing mechanism has

FIG. 10-11. Step-voltage regulator with control compartment door open, at right. Radiator for cooling, at left. (*Courtesy of General Electric Company.*)

moving parts. As a result, this type of regulator has a relatively low cost, is smaller in size than the induction regulator, and requires less maintenance and lower exciting current.

The method of operation is explained by Fig. 10-9. This shows the exciting winding and the series winding, with its several taps. The control equipment, similar to that for the induction regulator, is based principally on a contact-making voltmeter for determining the need to raise or lower the voltage. When the voltage must be changed, the motor-operated tap-changing switch is energized, and it moves from one tap position to another. To prevent the circuit from being opened while the taps are being changed, a by-pass reactor carries the load current during this interval. Figures 10-10 and 10-11 show typical examples of this type of regulator.

10-14. Shunt Capacitors on Distribution Circuits. The flow of power on a distribution system is accompanied by a flow of reactive power required for various types of load, such as motors and fluorescent lighting. As a result, the power factor of distribution circuits may be relatively low, 70 to 80 percent, or even lower.

Flow of reactive power on the distribution system loads up the cables, transformers, and overhead wires and causes the following disadvantages:

1. Reactive power reduces the ability of the distribution system to carry larger amounts of active power. For a power factor of 70 percent, 37 percent of the capacity of the distribution system handles the reactive current.

2. For a given load being supplied, the copper loss in the system is increased by the flow of reactive power. For example, for a power factor of 70 percent, the system copper loss is twice the loss for a power factor of 100 percent.

3. Reactive power causes a large voltage drop in the distribution system, especially on overhead lines. Therefore, in those cases where the capacity of the circuit is limited by voltage drop, the flow of reactive power reduces the amount of load which can be carried on the circuit.

Instead of supplying reactive power from the generators and transmitting it to the load through the distribution system, it is possible to obtain the necessary reactive power by using static capacitors. They connect directly to the distribution circuits close to the load. Since capacitors operate at a leading power factor, they neutralize the lagging reactive current supplied to inductive loads on the system. Capacitors can be considered as generators of wattless current to supply the reactive power required by the system load.

The use of capacitors on distribution systems in large numbers is a relatively new development. The cost of capacitors has been reduced materially in recent years, while the cost of other types of power-system equipment has been increasing, with the result that their use has become economical. Studies show it is economical to use capacitors in amounts of 20 to 30 percent of the total load on the distribution system.

Figure 10-12 shows a typical installation of shunt capacitors on an overhead distribution system. The capacitors are generally installed in three-phase banks. The size of the individual units is 25 kva, and banks of any capacity can be installed by using a number of individual capacitors.

Figure 10-13 shows a typical case of a shunt capacitor in a distribution circuit. The currents in the system before and after installing the capacitor are shown. Addition of the capacitor greatly reduces the amount of current carried in the circuit and in this manner makes possible the supply of a larger kilowatt load by the same circuit.

Figure 10-14 shows the amount of capacitor kva required to improve the power factor of a given circuit. For example, if the power factor is

FIG. 10-12. A 300-kvar 4,160-volt capacitor installation on pole top (in rectangular cases). (*Courtesy of General Electric Company.*)

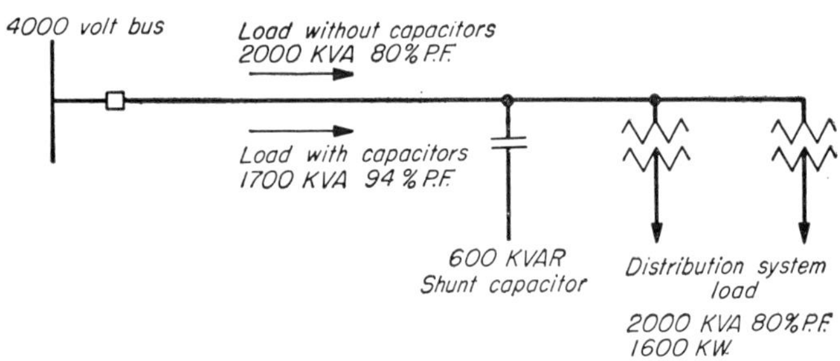

FIG. 10-13. Typical improvement in distribution conditions with and without capacitor installation near transformers.

raised from 75 percent to 95 percent by the capacitor, the gain in kilowatt capacity of the system will be 26.5 percent, or 0.265 kw gain per kilowatt. The capacitor kva required would be 2.63 kva for each kilowatt gain in circuit capacity. These answers are found on the ordinates where the 75 and 95 percent curves intersect.

The improvement in power factor will greatly reduce the voltage drop on the distribution system. In many cases where the circuit is relatively long, but consists of overhead wire, the voltage drop determines the amount of load that can be supplied by the circuit. Therefore, improvement in power factor will greatly increase the amount of load which can be supplied for a given amount of voltage drop.

As shown in Fig. 10-13, the shunt capacitors are usually installed on the high-voltage side of the distribution circuit, generally 4,000 volts

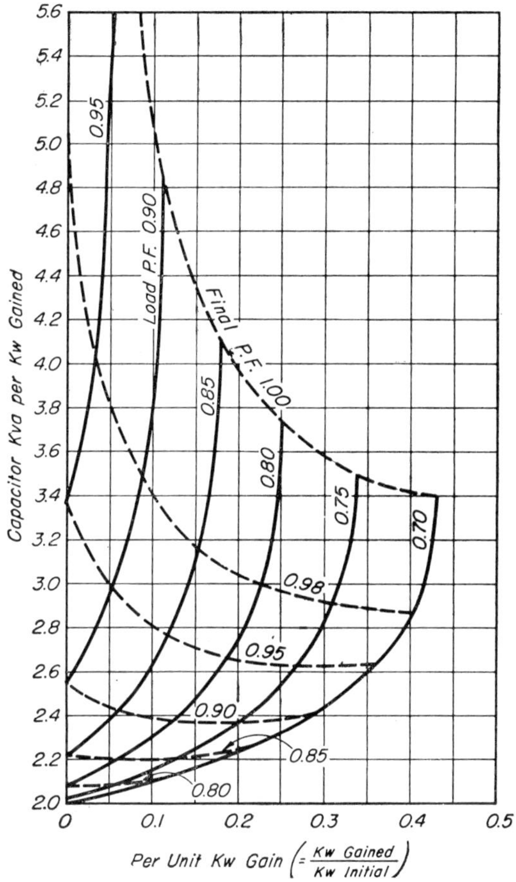

FIG. 10-14. How distribution circuit capacitors increase load limits where current-carrying capacity is the controlling factor.

phase-to-phase. Capacitors can be installed on the low-voltage side of the distribution transformer and in that way increase their usefulness by reducing the load on the transformer and the leads into the customer. However, the cost of low-voltage capacitors is considerably higher per kva than for high-voltage capacitors; therefore, it has been the practice to install the capacitors on the high-voltage side of the circuit.

Where the capacitors are installed to improve voltage conditions, it is frequently found that during times of light load the voltage on the distribution system will be excessively high. This difficulty can be eliminated by automatic switches installed with the capacitors which disconnect them from the system during light-load periods. In this manner the benefits made possible by capacitors can be obtained without creating objectionable conditions during the light-load period. The switching can be accomplished either through time switches or by voltage relays

FIG. 10-15. Automatic-switching capacitor unit for 4,800 volts, three-phase, delta. (*Courtesy of General Electric Company.*)

which open the switches when the voltage becomes too high and close the switches when the voltage is low. Figure 10-15 shows an example of an automatically switched capacitor bank.

10-15. Series Capacitors. Capacitors can also be used in series with the distribution circuit. In this type of installation the capacitor introduces negative reactance which neutralizes the inductive reactance of the distribution circuit. It is possible to choose the size of the capacitor so that its negative reactance is exactly equal to the inductive reactance of the line, and the effect is then approximately the same as if there were only resistance in the line.

The most important application of series capacitors is for reduction of voltage flicker due to variable loads, such as welders and furnaces. The characteristic of the capacitor results in instantaneous compensation for the voltage drop which would otherwise occur in the circuit due to sudden increase of current flow. Other methods of voltage regulation require a change in voltage to actuate the device so that some time delay is introduced. With the series capacitor, this delay is eliminated.

10-16. Synchronous Condensers. In the transmission of power over long distances, some method of power-factor correction is required. It is not feasible, in general, to transmit the reactive power required by the load over long distances, because of the large voltage drop which accompanies such transmission. Another factor is the large loss of reactive power inherent in overhead transmission systems owing to the high reactance of high-voltage lines and transformers.

As a result, some method of generating reactive power close to the load is required by power systems which receive their supply over long-distance lines. Where the magnitude of the reactive power requirement is relatively small (10,000 kva or under), it is practical and economical to use static capacitors, as described before. For larger installations, the synchronous condenser is generally used. This type of machine is essentially a synchronous motor without any mechanical load imposed upon it. The dc field current is varied to obtain the amount of reactive power required to satisfy voltage and power-factor conditions. This method of voltage and power-factor control has inherent advantages over static capacitors, because of finer control and increased system stability.

A typical machine is shown in Fig. 10-16. Machines as large as 75,000 kva have been constructed for use as synchronous condensers. The electrical characteristics of the synchronous condenser are similar to synchronous generators and motors. Generally, the field current is regulated automatically to obtain a desired voltage level, using the same type of equipment as for generators.

A typical example of the application of a synchronous condenser is shown in Fig. 10-17. In this case a load of 150,000 kw is to be transmitted over a distance of 150 miles. If the power is to be received at a lagging power factor of 85 percent, it would be necessary for the sending-end voltage to be 50 percent higher than the receiving-end voltage, and the power factor at the sending end would be 68 percent lagging. This voltage drop is larger than can be tolerated for a practical system. By installing a 75,000-kva synchronous condenser at the receiving end, the voltage drop in the line is reduced to approximately one-half, bringing it within a practical operating value. The power factor at the sending end is increased to 88 percent, a satisfactory level.

In this case the synchronous condenser also serves a secondary function. This transmission line has a charging current of 40,000 kva, and its sudden connection to the system, or disconnection, would result in an appreciable voltage surge. The condenser is equipped with a starting motor which can be used to bring the machine up to synchronous speed. In this way it is possible to energize the line from the condenser without any voltage surge on the system.

FIG. 10-16. Two air-cooled synchronous condensers, 25,000-kva (leading), 12,500-kva (lagging), 12,000-volt, three-phase, 60-cycle, installed outdoors. (*Courtesy of General Electric Company.*)

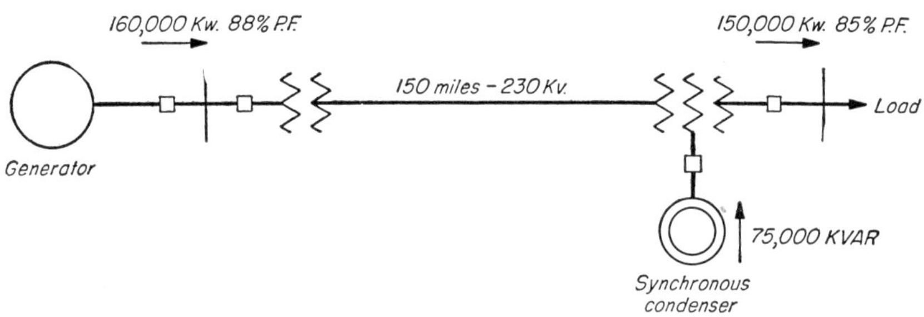

FIG. 10-17. Single-line diagram of synchronous condenser application on high-voltage transmission line.

Synchronous condensers increase the stability of the transmission system by maintaining a higher voltage on the system during and after a short circuit on the system. High-speed voltage-regulating systems, similar to those used on generators, make it possible to increase rapidly the field current of the synchronous condenser and thus maintain a higher average voltage on the system.

10-17. Reactors. Current-limiting reactors are used in power systems principally to reduce the current which flows under short-circuit conditions. Through the reduction of short-circuit current, reactors decrease stresses on conductors and supports, circuit breakers and other apparatus carrying the short-circuit current. Also, because of the lowered short-circuit current, voltage in various parts of the power system can be maintained at a higher level during the period of the short circuit, making operation of the system more stable. One of the principal benefits of reactors is that they allow installing circuit breakers of lower interrupting capacity.

Reactors can be classified according to the part of the system on which they are installed. Thus, we have feeder reactors, generator reactors, bus sectionalizing reactors, and neutral reactors. Let us discuss these various applications and the benefits gained by the use of reactors.

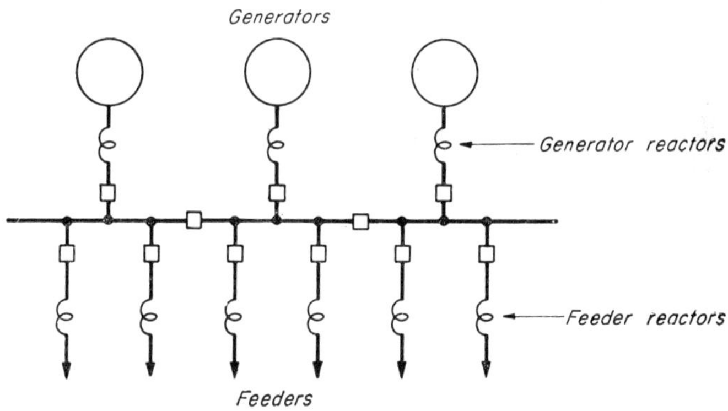

Fig. 10-18. Typical generating-station distribution system with generator and feeder reactors.

10-18. Feeder Reactors. In large generating and substations, common practice installs reactors on each outgoing feeder. These reactors permit the installation of lower capacity circuit breakers, both at the generating station and at the stations supplied by the feeders. In this manner considerable savings can be achieved in station investment. Figure 10-18 shows a typical arrangement of generating-station buses with outgoing feeders supplying substations. Note that the reactor is connected on the line side of the circuit breaker. With this arrangement, if a fault occurs between the circuit breaker and the reactor, no benefit will be gained from the reactor. This could be avoided by installing the reactor on the bus side of the circuit breaker. However, this would be disadvantageous since reactor faults would be equivalent to a bus fault. At the same time, the probability of a fault between the circuit breaker and the reactor is extremely small so that the connections in Fig. 10-18 are acceptable even with the circuit breaker designed only to interrupt currents on the line side of the reactor.

Feeder reactors simplify relay protection since they provide a large difference between the short-circuit magnitude for faults on the two sides of the reactor. This makes possible relay schemes which are highly selective.

Feeder reactors improve the stability of the power system and are very valuable from this viewpoint. When a fault occurs on a feeder, the reactor maintains a relatively high voltage on the generating-station bus. This, in turn, provides for stable operation of the generators and the system interconnecting with the generating station. Another improvement effected by reactors is that the higher bus voltage reduces the effect of a fault in causing motors to drop out while the short is on the system.

In some cases, feeder reactors improve the balance of loading between two lines connected in parallel between stations. The size of the reactor can be designed to provide for the required division of load between the two lines.

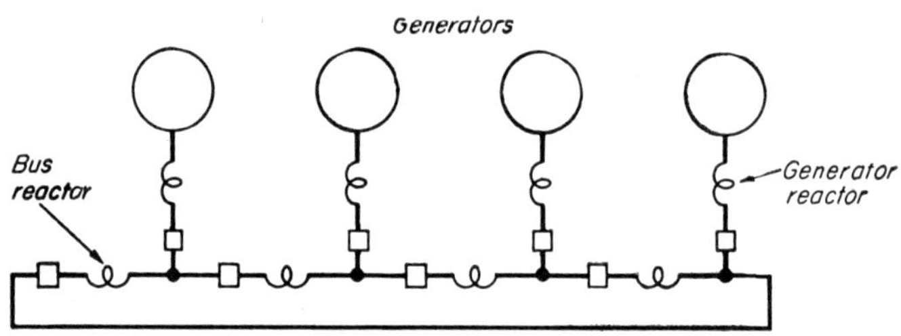

FIG. 10-19. Typical ring bus with generator and bus reactors.

10-19. Bus-sectionalizing Reactors. In large generating stations where several units are installed, it is common practice to sectionalize the bus so that faults at one point on the bus will not remove the entire station from service. Also, in a large station, the magnitude of fault currents may have to be controlled to allow using circuit breakers of reasonable ratings. These desirable features can be achieved by a bus separated into sections by circuit breakers and with reactors connected between sections. Many different arrangements have been used, such as the ring bus, straight bus, and star bus. An example of a ring bus with reactors between sections is shown in Fig. 10-19. A ring-type bus has many advantages, and its popularity has become widespread. These bus-sectionalizing reactors will reduce the short-circuit current for a fault on any of the bus sections and thereby improve the stability of the generating system. It is also a way of obtaining simplified relaying schemes because of the large difference in current for faults at different points in the bus.

10-20. Generator Reactors. In most cases where generators are connected directly to a distribution bus, it is common practice to install a generator reactor. This reactor provides for reduction of the fault current being fed into the bus and, more important, will protect the bus in the event of a short circuit within the generator.

10-21. Neutral Reactors. In many power systems, impedance is used in the neutral of the system to reduce the phase-to-ground fault current without influencing normal operation of the system. Either a resistor or a reactor can be used for this purpose, and both are being generally used. The use of reactors for grounding the neutral must be watched carefully, since it is possible to create oscillatory conditions on the system under which a phase-to-ground fault would create dangerously high voltages on the system. Generally, reactors can be used when their ohmic value is low in relation to the total impedance of the system.

FIG. 10-20. Current-limiting reactor in enclosure for 2,300-volt circuit. (*Courtesy of General Electric Company.*)

10-22. Description of Reactors. In general, there are two types of reactors, air-cooled and oil-filled. Figure 10-20 shows a typical three-phase reactor, designed for indoor installation. Reactors used for the purpose described in this chapter always have air cores. Iron-core reactors are not used because of their tendency to saturate under heavy fault conditions.

Reactors are generally rated in terms of percent reactance. The percent reactance is the ratio of the voltage drop across the reactor (when the rated current of the circuit at rated frequency is flowing through the reactor) to the voltage between line and neutral on three-phase circuits or the voltage between the lines on single-phase circuits.

10-23. Shunt Reactors. In some high-voltage transmission systems, the charging current is very high, because of the long distances that power is transmitted. During heavy-load periods, this high charging current has beneficial effects, but during light-load conditions it creates difficulties because of the voltage rise which occurs when charging current flows through the transmission system.

One method to control this situation is to install shunt reactors, Fig. 10-21, which are connected to the system only during light-load conditions. These reactors neutralize the effect of the transmission-line charging current and thereby aid in stabilizing system voltage.

FIG. 10-21. Shunt reactor rated at 5,000 kva, 66,400 volts to operate in 115-kv circuit. (*Courtesy of General Electric Company.*)

CHAPTER REVIEW TOPICS

1. What frequencies are mostly used in the U.S.?
2. How may dc distribution systems be served from ac generation?
3. Why aren't motor-generator sets generally used as ac-dc converters?
4. What basic principle is used in the synchronous converter?
5. How is the armature current in a synchronous converter related to the ac and dc in the external circuits?
6. How are six phases obtained to drive a synchronous converter?
7. What is the relation between ac and dc voltages in a synchronous converter?
8. How are ac and dc values related in a synchronous converter?
9. Describe two methods of controlling dc voltage in a synchronous converter.
10. How may synchronous converters be started?
11. What advantages do mercury-arc rectifiers offer for ac-dc conversion?
12. Describe the operating principle of the mercury-arc rectifier.
13. How does voltage vary with load in a mercury-arc rectifier?
14. How many phases are used in a mercury-arc rectifier?
15. What is meant by an arc back?
16. What elements are needed to make up a rectifier substation? Describe them

17. What types of loads use 25-cycle ac?
18. How many types of frequency changers are used in modern systems?
19. Describe the principal features of a synchronous-synchronous tie.
20. What elements are used in a machine-type nonsynchronous tie?
21. Describe the elements and operation of an electronic frequency changer.
22. What types of voltage regulators are used on distribution systems?
23. Describe the operation of an induction voltage regulator.
24. How do step voltage regulators work?
25. What function do shunt capacitors perform on a distribution system?
26. Why are high-voltage capacitors used in preference to low-voltage types?
27. How may capacitors be disconnected during light load periods?
28. Why are series capacitors used on a distribution circuit?
29. What is the duty of synchronous condensers?
30. How may synchronous condensers be started without causing a voltage surge on the system?
31. What is the purpose of applying reactors to a distribution system?
32. How many applications may be found for reactors? Name them.
33. Describe the duty of the reactor or its advantage in each of the above applications.

APPENDIX

TRIGONOMETRIC FUNCTIONS

Angle, °	Sine	Cosine	Tangent	Cotangent	Angle, °	Sine	Cosine	Tangent	Cotangent
0	0.0000	1.0000	0.0000	Infinity	45	0.7071	0.7071	1.0000	1.0000
1	0.0175	0.9998	0.0175	57.29	46	0.7193	0.6947	1.036	0.9657
2	0.0349	0.9994	0.0349	28.636	47	0.7314	0.6820	1.072	0.9326
3	0.0523	0.9986	0.0524	19.081	48	0.7431	0.6691	1.111	0.9004
4	0.0698	0.9976	0.0699	14.301	49	0.7547	0.6561	1.150	0.8693
5	0.0872	0.9962	0.0875	11.430	50	0.7660	0.6428	1.192	0.8391
6	0.1045	0.9945	0.1051	9.514	51	0.7772	0.6293	1.235	0.8098
7	0.1219	0.9926	0.1228	8.144	52	0.7880	0.6157	1.280	0.7813
8	0.1392	0.9903	0.1405	7.115	53	0.7986	0.6018	1.327	0.7536
9	0.1564	0.9877	0.1584	6.314	54	0.8090	0.5878	1.376	0.7265
10	0.1737	0.9848	0.1763	5.671	55	0.8192	0.5736	1.428	0.7002
11	0.1908	0.9816	0.1944	5.145	56	0.8290	0.5592	1.483	0.6745
12	0.2079	0.9782	0.2126	4.705	57	0.8387	0.5446	1.540	0.6494
13	0.2250	0.9744	0.2309	4.332	58	0.8481	0.5299	1.600	0.6249
14	0.2419	0.9703	0.2493	4.011	59	0.8572	0.5150	1.664	0.6009
15	0.2588	0.9659	0.2680	3.732	60	0.8660	0.5000	1.732	0.5774
16	0.2756	0.9613	0.2868	3.487	61	0.8746	0.4848	1.804	0.5543
17	0.2924	0.9563	0.3057	3.271	62	0.8830	0.4695	1.881	0.5317
18	0.3090	0.9511	0.3249	3.078	63	0.8910	0.4540	1.963	0.5095
19	0.3256	0.9455	0.3443	2.904	64	0.8988	0.4384	2.050	0.4877
20	0.3420	0.9397	0.3640	2.748	65	0.9063	0.4226	2.145	0.4663
21	0.3584	0.9336	0.3839	2.605	66	0.9136	0.4067	2.246	0.4452
22	0.3746	0.9272	0.4040	2.475	67	0.9205	0.3907	2.356	0.4245
23	0.3907	0.9205	0.4245	2.356	68	0.9272	0.3746	2.475	0.4040
24	0.4067	0.9136	0.4452	2.246	69	0.9336	0.3584	2.605	0.3839
25	0.4226	0.9063	0.4663	2.145	70	0.9397	0.3420	2.748	0.3640
26	0.4384	0.8988	0.4877	2.050	71	0.9455	0.3256	2.904	0.3443
27	0.4540	0.8910	0.5095	1.963	72	0.9511	0.3090	3.078	0.3249
28	0.4695	0.8830	0.5317	1.881	73	0.9563	0.2924	3.271	0.3057
29	0.4848	0.8746	0.5543	1.804	74	0.9613	0.2756	3.487	0.2868
30	0.5000	0.8660	0.5774	1.732	75	0.9659	0.2588	3.732	0.2680
31	0.5150	0.8572	0.6009	1.664	76	0.9703	0.2419	4.011	0.2493
32	0.5299	0.8481	0.6249	1.600	77	0.9744	0.2250	4.332	0.2309
33	0.5446	0.8387	0.6494	1.540	78	0.9782	0.2079	4.705	0.2126
34	0.5592	0.8290	0.6745	1.483	79	0.9816	0.1908	5.145	0.1944
35	0.5736	0.8192	0.7002	1.428	80	0.9848	0.1737	5.671	0.1763
36	0.5878	0.8090	0.7265	1.376	81	0.9877	0.1564	6.314	0.1584
37	0.6018	0.7986	0.7536	1.327	82	0.9903	0.1392	7.115	0.1405
38	0.6157	0.7880	0.7813	1.280	83	0.9926	0.1219	8.144	0.1228
39	0.6293	0.7772	0.8098	1.235	84	0.9945	0.1045	9.514	0.1051
40	0.6428	0.7660	0.8391	1.192	85	0.9962	0.0872	11.430	0.0875
41	0.6561	0.7547	0.8693	1.150	86	0.9976	0.0698	14.301	0.0699
42	0.6691	0.7431	0.9004	1.111	87	0.9986	0.0523	19.081	0.0524
43	0.6820	0.7314	0.9325	1.072	88	0.9994	0.0349	28.636	0.0349
44	0.6947	0.7193	0.9657	1.036	89	0.9999	0.0175	57.290	0.0175
					90	1.0000	0.0000	Infinity	0.0000

INDEX